T0177056

Quantitative Analysis of Geopressure for Geoscientists and Engineers

Geopressure, or pore pressure in subsurface rock formations impacts hydrocarbon resource estimation, drilling, and drilling safety in operations. This book provides a comprehensive overview of geopressure analysis, bringing together rock physics, seismic technology, quantitative basin modeling, and geomechanics. It provides a fundamental physical and geological basis for understanding geopressure by explaining the coupled mechanical and thermal processes. It also brings together state-of-the-art tools and technologies for analysis and detection of geopressure, along with the associated uncertainty. Prediction and detection of shallow geohazards and gas hydrates are also discussed, and field examples are used to illustrate how models can be practically applied. With supplementary Matlab codes and exercises available online, this is an ideal resource for students, researchers, and industry professionals in geoscience and petroleum engineering looking to understand and analyze subsurface formation pressure.

Nader C. Dutta is a recognized industry expert on geopressure and was the senior science advisor of Schlumberger prior to retiring in 2015. He is currently a visiting scholar at the Geological Sciences Department, Stanford University. He is the editor of SEG's book *Geopressure* (1987) and has been a member of United Nations Environmental Program (UNEP) and USA-DOE Gas Hydrate Assessment Committee.

Ran Bachrach is a scientific advisor for Schlumberger, supporting both research and operations and specializing in various geoscience topics, including high-resolution geophysics, rock physics and geomechanics, 3D/4D imaging of subsurface processes, seismic inversion, and seismic reservoir analysis.

Tapan Mukerji is a professor in the Department of Energy Resources Engineering and by courtesy in the Departments of Geophysics and Geological Sciences at Stanford University. He received the Society of Exploration Geophysicists' Karcher Award in 2000 and shared the 2014 ENI award for pioneering innovations in theoretical and practical rock physics for seismic reservoir characterization. He is also a coauthor of *The Rock Physics Handbook* (2020), *Value of Information in Earth Sciences* (2015), and *Quantitative Seismic Interpretation* (2005), Cambridge University Press.

Quantitative Analysis of Geopressure for Geoscientists and Engineers

NADER C. DUTTA
Stanford University

RAN BACHRACH
Schlumberger

TAPAN MUKERJI
Stanford University

CAMBRIDGE
UNIVERSITY PRESS

CAMBRIDGE
UNIVERSITY PRESS

University Printing House, Cambridge CB2 8BS, United Kingdom

One Liberty Plaza, 20th Floor, New York, NY 10006, USA

477 Williamstown Road, Port Melbourne, VIC 3207, Australia

314–321, 3rd Floor, Plot 3, Splendor Forum, Jasola District Centre,
New Delhi – 110025, India

79 Anson Road, #06–04/06, Singapore 079906

Cambridge University Press is part of the University of Cambridge.

It furthers the University's mission by disseminating knowledge in the pursuit of
education, learning, and research at the highest international levels of excellence.

www.cambridge.org
Information on this title: www.cambridge.org/9781107194113
DOI: 10.1017/9781108151726

First published 2021

Printed in the United Kingdom by TJ Books Limited, Padstow Cornwall

A catalogue record for this publication is available from the British Library.

Library of Congress Cataloging-in-Publication Data
Names: Dutta, Nader C., author. | Bachrach, Ran, 1967– author. | Mukerji, Tapan, 1965– author.
Title: Quantitative analysis of geopressure for geoscientists and engineers / Nader C. Dutta, formerly
of Schlumberger, currently at Staford University, California, Ran Bachrach, Schlumberger, Tapan
Mukerji, Stanford University, California.
Description: Cambridge, United Kingdom ; New York, NY, USA : Cambridge University Press, 2021.
| Includes index.
Identifiers: LCCN 2020026301 | ISBN 9781107194113 (hardback) | ISBN 9781108151726 (ebook)
Subjects: LCSH: Reservoir oil pressure – Mathematical models. | Petroleum – Migration. | Fluid
dynamics.
Classification: LCC TN871.18 .D88 2021 | DDC 553.2/82–dc23
LC record available at https://lccn.loc.gov/2020026301

ISBN 978-1-107-19411-3 Hardback

Additional resources for this publication at www.cambridge.org/geopressure.

Contents

Color plates can be found between pages 276 and 277.

Preface

How did we come to write this book? Our research suggested that the discipline of geopressure started based on fundamentals of geology (such as the pioneering works of Ruby and Hubbert and Dickinson during the middle of the twentieth century), with an excellent promise of delivery of applications to the hydrocarbon industry. As the quest for hydrocarbon exploration and exploitation required more and more integrated approaches, contributions from many diverse fields of sciences, such as geology, geophysics, petrophysics, applied physics, engineering, and applied mathematics, became the norm. However, quick-fix engineering approaches to tackle challenging problems at hand resulted in fragmented knowledge building and lack of emphasis on fundamentals. This was noted in an earlier publication (Dutta, 1987a, vii): "understanding of the geopressuring phenomenon is worth vigorous pursuit because that understanding calls for an integrated approach in unraveling its mysteries." We felt that the field of geopressure required another look – one that would culminate in a comprehensive discussion of the subject, including the industrial applications and an assessment of the road ahead. This is the goal of this book. Whether this goal is met awaits the judgment of our readers and peers.

During our professional careers, we have been fortunate enough to have witnessed some remarkable achievements in the field of geophysics, in particular, in the seismic subdiscipline. It has been propelled by high-speed computing with concomitant development of complex algorithms, such as tomography and full waveform inversion (FWI), by some brilliant geoscientists. This resulted in a step change in the subsurface seismic image quality. Therefore, some timely questions needed to be asked: Have we taken advantage of these opportunities in geopressure analysis that requires earth model building rather than velocity modeling? Are subsurface images at the right depths? Well, partly yes, but not consistently. Building *an earth model* requires a thorough understanding of the underlying basic physics to describe important subsurface phenomena, such as geopressure, among others. Just what would the effect on imaging be if we were to get this description on the right footing? This requires analysis of the geopressure phenomenon quantitatively, reliably, and making it accessible to all geoscientists and engineers so that integration with other viable model-building processes can take place. We hope the readers will appreciate the attempt undertaken in the book to address this issue.

The book has fourteen chapters that describe the geopressure phenomenon – fundamentals, models and mechanisms, and tools to predict and detect it – from borehole centric to seismic, taking care to explain the basic physics behind these tools, including limitations of their operating envelopes. We have come to understand better the physics of rocks through careful measurements, both in the laboratory and in the field, and through theoretical analysis. This has enabled us to develop and test subsurface models with more confidence and has helped us to extract rock properties from seismic attributes using sophisticated inversion technologies. The knowledge captured from this directly impacts our understanding of geopressure. Therefore, in this book, an attempt is made to put some of the known subsurface pore pressure models on firmer ground by providing a rock physics basis for some of these models. This allowed us to extend the traditional scale of geopressure prediction envelope from exploration – say, several hundred feet – to drilling around a borehole, at a few feet. To bridge this *scale* is to lay the foundation for a best-practice approach in geopressure and to enable us to extrapolate into what is yet to come. Nonetheless, it is a snapshot at the present and obviously colored by our own biases. We hope the future generation will build on it.

A unique feature of this book deals with applications to illustrate how the geopressure models can be used not only for energy resources assessment but also for environmental issues. In this context, our experiences in dealing with prediction of subsurface geohazards, such as possible existence of shallow aquifer pressured sands in deepwater (aka shallow-water-flow sands), gas charged sands and gas hydrates, and various seabed hazardous features, will be beneficial not only to the energy resource developers and operators but also to regulatory agencies. The approach discussed in the book enables us to go beyond *color coding* a geohazard map – the current practice – to adding *qualifiers*, such as just how red is *red*, what is the extent of the *yellow*, and what is the comfort zone of the *green*? We address these geohazard issues quantitatively so that our sister community of drilling can benefit from closer interaction with geoscientists.

Several books are devoted to subsurface pressure; however, while they were classics during their times, their contents are now mostly outdated. Some other compilations consist of conference proceedings and reviews of papers dealing with special aspects of geopressure and do not include many recent and important developments. These may not be appropriate for students and researchers beginning their careers. This book aims to bridge the gap. To help the readers self-assess their understanding of various subjects addressed in the book, we have a companion website (see the Cambridge University Press site) with suggested exercises and Matlab codes.

By the time we finished the manuscript of the book, the world that we knew had changed. We are witnessing a pandemic incurred by COVID-19, resulting in many deaths and lockdowns in our homes. So the environment has changed drastically between the time we started the project some three years ago and the time when we finished it. However, the project provided some solace to us!

Now comes the most pleasant part of this preface – acknowledgments. There are so many that to mention all the names is practically impossible. Therefore, we sincerely apologize to those who contributed over the years but whose names are not mentioned. We benefited greatly from the scientific training received in the academy and the industry to the tune of more than 75 years of cumulative experience – through knowledge sharing with students, staff, and industry partners, often with hands-on experience and project management. This has broadened our curiosity, given us strength to march on, and empowered us with tools that resulted in this book. We are very grateful to those who gave us this opportunity. Special thanks are due to Jianchun Dai, Yangjun (Kevin) Liu, and Sherman Yang – all were dear colleagues of the senior author while he was employed at Schlumberger. Thanks, guys! On a personal note, Nader Dutta presented a good portion of this book in a training course at Stanford University in 2016. Feedback from students greatly impacted the presentation of the subject matter in the book. In particular, Anshuman Pradhan – soon to be Dr. Pradhan – deserves special thanks. He was the teaching assistant when the course in geopressure was taught at Stanford. Some of Anshuman's work is included in this book in Chapters 10 and 13. Thanks, Anshuman. Thanks are also due to Dr. Huy Le, who graduated recently with a PhD from Stanford and addressed a good part of his dissertation to link seismic imaging with pore pressure constraints using FWI. The methodology is partly based on some of the material that we discuss in Chapter 6. Thanks, Huy. The encouragement of his thesis advisor at Stanford, Professor Biondo Biondi, to share knowledge is greatly appreciated. Dr. Allegra Hosford-Scheirer at Stanford provided a very constructive environment to carry on integrating basin modeling to imaging through pore pressure. Her enthusiasm and energy are legendary and inspirational to all. Thanks, Allegra! Gary Mavko provided great encouragement and practical advice – finish the book first! Thanks, Gary! Here it is! We are grateful to the members of the following affiliate groups at Stanford University for sponsoring our work over the years and for funding Nader Dutta's stay at Stanford: Stanford Rock and Borehole Geophysics (SRB), Stanford Exploration Project (SEP), Basin and Petroleum System Modeling (BPSM), and the Stanford Center for Earth Resources Forecasting (SCERF). We acknowledge additional funding from Prof. Steve Graham, Dean of the Stanford School of Earth, Energy, and Environmental Sciences. We acknowledge Schlumberger for donations of software and data used in the work of Anshuman Pradhan and Huy Le, described in this book. A special thanks to Susan Francis and Sarah Lambert of Cambridge University Press for guiding us through this project – a long and arduous journey that finished with exhilaration. Thanks! Last, but not the least, Nader Dutta is grateful to his loving spouse, Chizuko, for providing gentle and timely criticism of the manuscript and sharing his joys as well as his frustrations – there were many!

Good reading, folks! Have fun!

1   Basic Pressure Concepts and Definitions

1.1   Introduction

Geopressure is the pressure beneath the surface of the earth. It is also known as the formation pressure. This could be lower than, equal to, or higher than the normal or hydrostatic pressure for a given depth. Hydrostatic or normal pressure is the force exerted per unit area by a column of freshwater from the earth's surface (e.g., sea level) to a given depth. Geopressures lower than the hydrostatic pressures are known as underpressures or subpressures, and they occur in areas where fluids have been drained, such as a depleted hydrocarbon reservoir. Geopressures higher than hydrostatic pressures are known as overpressured, and they occur worldwide in formations where fluids are trapped within sediments due to many geologic conditions and support the overlying load. Overpressured formations are also known as formations with abnormally high pore pressure. The lithostatic (or overburden) pressure at a given depth is due to the *combined* weight of the overlying rock and fluids. The fracture pressure is the pressure that causes the formation rock to crack. Figure 1.1 shows these concepts in graphical terms.

If the overlying fluid is composed of hydrocarbon as well as water (brine), the pressure versus depth plot will look like that shown in Figure 1.2.

The slope changes in the plot are due to density differences between brine, oil, and gas. The overpressure phenomenon is well known throughout the world. Among other things, the magnitude and distribution of overpressure in sedimentary basins have been known to critically impact the evolution of hydrocarbon provinces, control the migration of fluids within a basin, and affect the processes that are used to mine the subsurface resources, such as oil and gas. The most discussed and well-known cause of overpressure is the rapid burial of low-permeability water-filled sediments (e.g., clay) at a rate that does not allow the fluid to escape fast enough to maintain hydrostatic equilibrium upon further burial. Thus, further burial causes geopressure to rise even more. This is known as compaction disequilibrium. This is the leading cause of overpressure in most of the Tertiary clastic basins of the world, such as the Gulf of Mexico. This and many other mechanisms of overpressure are discussed in detail in Chapter 3.

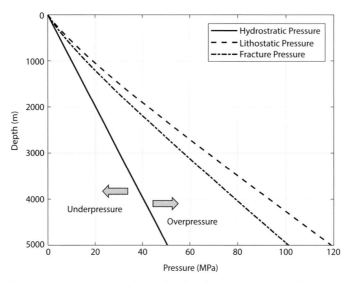

Figure 1.1 Pressure versus depth plot showing geopressure regimes.

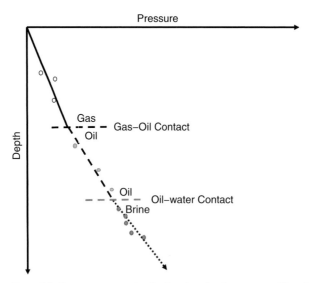

Figure 1.2 Pressure versus depth plot showing buoyancy effect due to hydrocarbons.

1.2   Basic Concepts

1.2.1   Units and Dimensions

Before we proceed, a word on units and dimensions is in order. All quantities in physics must either be dimensionless or have dimensions. All units can be expressed in terms of mass [M], length [L], and time [T]. In equations, the units must be consistent; there is no need for conversion factors. However, care is needed for quantities, such as pressure. It is force per unit area. The dimension of force is $[\text{ML}^{-1}\text{T}^{-2}]$; but if the unit of length is feet and the unit of pressure is pounds per square inch, or psi (as is commonly used in the US drilling community), a conversion factor is required, since these are inconsistent mixed units. In SI units, such conversion factors are not needed because all the units are consistent. The lack of inconsistency of units using the American system is known to have created a massive headache between the two drilling communities – those who use the SI units (some communities in Europe, for example) and those who do not use SI units. We shall discuss this further in this chapter in the context of pore pressure measurements.

As mentioned, pore pressure has the dimension of force per unit area. In the SI system, the unit of pressure is pascal (Pa), and in the British system, the unit is pounds per square inch (psi). We note that $1\,\text{Pa} = 1.4504 \times 10^{-4}$ psi. This is a rather small unit, and for most practical applications, it is customary to use either kilopascal (KPa) or megapascal (MPa). Drillers, engineers, and well loggers still use the British system, while the academicians prefer the SI system. Therefore, a fluency in both type of units is a must. We will be using the mixed system throughout the book. However, whenever possible, we will provide the SI or British system equivalents.

1.2.2   Hydrostatic Pressure

Sedimentary rocks in formations are composed of solid material and fluids in the porous network. Hydrostatic or normal pressure, P_h, is the pressure caused by the weight of a column of fluid and is given by

$$P_h = \int_0^z \rho_f(z)gdz + Pair \approx \rho_f gz + P_{air} \tag{1.1}$$

where z is the column height of the fluid, ρ_f is the density of the fluid, g is the acceleration due to gravity, and P_{air} denotes the pressure due to the atmosphere. The size and shape of the fluid column have no effect on hydrostatic pressure. The approximation on the right-hand side of equation (1.1) assumes that ρ_f is constant and z is the depth below sea level or the land surface. Hydrostatic pore pressure increases with depth; the gradient at a given depth is dictated by the fluid density at that depth. This is because the water or brine density is not constant. Water tends to expand with rising temperature but contracts with rising pressure. As we shall see later, between the two processes in the subsurface (i.e., increase in temperature and pressure), thermal expansion with increasing depth is greater than the mechanical compression. There are other factors that affect the water density, such as dissolved salt – the solubility of salt also increases with depth. Subsurface brines are more saline than the ocean water. This increase in total dissolved salt increases the density of water. The net effect is that the water (brine) density is a complex function of temperature, pressure, and total dissolved solids. If a subsurface formation is in the hydrostatic condition, it implies that there is an interconnected and open pore system from the earth's surface to the depth of measurement. To summarize, the fluid density depends on various factors, such as fluid type (oil, water, or gas), concentration of dissolved solids (i.e., salts and other minerals), and temperature and pressure of the fluid and gases dissolved in the fluid column. In Appendix A we provide some practical empirical relationships for physical properties of brine, gas, and oil needed for quantitative analysis of geopressure.

We strongly recommend that those who wish to pursue quantitative evaluation of geopressure use those equations for density and velocity of brine, gas, and oil ("dead" and "live") in a computer code. This will enable them to evaluate the true hydrostatic pressure as well as the pressure due to gas and oil columns of various heights. Here "dead" oil designates oil without any dissolved gas, whereas "live" oil means it contains dissolved gas.

What would a pressure versus depth plot such as the one in Figure 1.1 look like for a reservoir rock containing gas, oil, and water? An example is given in Figure 1.2 for the case of a reservoir filled with gas, oil, and water (brine). The slope changes are due to the density contrast between different kinds of fluids (gas, oil, brine), as discussed earlier. This kind of plot is very useful for determining the height of a hydrocarbon column in a reservoir. The discrete data points show actual measurements of pore pressure in a reservoir. Typically, not many measurements are carried out, as measurements are expensive in a real petroleum well; petroleum engineers make these discrete measurements and then look for slope changes to determine gas–oil and oil–water contacts, which yield the hydrocarbon column height. Eventually, seismic data are used along with geologic structure maps of a prospect to map these contacts in 3D. Volumetric calculations resulting from these measurements, along with uncertainty estimates, are used to determine the ultimate value of the asset.

1.2.3   Head

We introduce some terminology commonly used in fluid dynamics and relevant to geopressure. In the subsurface, fluids always move, although the speed at which they move is small in the human timescale. It is not so in the geologic timescale. The definitions that we gave in Section 1.2.2 are for fluids in the static condition. In fluid mechanics literature, the word *head* is commonly used. *Head* refers to a vertical dimension and has the dimension of length [L]. There are various types of heads.

A *pressure head* (also termed as *static pressure head* or *static head*) is the vertical elevation of the free surface of water above the point of interest. It is given by

$$\psi = P/\gamma = P/\rho g \qquad (1.2)$$

where

ψ is the pressure head (length, typically in units of m)
P is the fluid pressure (force per unit area, typically in units of Pa)
γ is the specific weight (force per unit volume, typically in units of Newton/m^3)
ρ is the density of the fluid (mass per unit volume, typically in kg/m^3)

The term *hydraulic head* or *piezometric head* is used to specify a specific measurement of liquid pressure above a *datum*. It is composed of three terms: velocity head (h_v), elevation head ($z_{\text{elevation}}$), and pressure head (ψ). The following is referred to as the *head equation*:

$$C = h_v + z_{\text{elevation}} + \psi \qquad (1.3)$$

Here C is a constant for the system (referred to as the *total head*) that appears in the context of the Bernoulli equation for incompressible fluids in hydrodynamics, which states that an increase in fluid speed occurs simultaneously with a decrease in pressure (Streeter, 1966). The *velocity head* (also referred to as *kinetic head*) is the head due to the energy of movement of the water. (In subsurface flow through porous rocks, this is negligible.) The *elevation head* is the elevation of the point of interest above a datum, usually sea level or the land surface.

1.3   Pore Pressure Gradient

A gradient is the first derivative of a physical quantity. The pressure gradient, dp/dz, is the *true* gradient of pore pressure, p, versus depth at a given point z. It shows change of pore pressure in a small scale. It is the rate at which pressure varies along a uniform column of fluid due to the fluid's own weight. Thus, a change in gradient implies a change in fluid density. Local gradients are most useful when working with the absolute pressure. However, the drilling community uses a term called *pore pressure gradient* to denote the density of fluid. It is the ratio of the pore pressure (p at a depth z) to the depth z. This is usually expressed in pounds per square inch per foot (abbreviated by psi/ft) in the British system of units and MPa/m in the SI system. It is clear that this gradient is *datum* dependent. Furthermore, pore pressure gradient is *not the true gradient* of p as a geoscientist or an engineer would define. It is simply pressure/depth. The conversion between fluid density and fluid pressure gradient is

$$1 \text{ psi/ft} = 2.31 \text{ g/cm}^3$$
$$\text{Thus } 1 \text{ psi/ft} = 2.31 \text{ g/cm}^3 = 0.0225 \text{ MPa/m} = 22.5 \text{ kPa/m} \qquad (1.4)$$

Thus, the fluid density, can be defined as

$$\text{fluid density (g/cm}^3) = 0.433 \text{ (psi/ft)} \qquad (1.5)$$

The drilling community uses a term called *equivalent mud weight* (EMW) to denote the density of fluid (mud) required to drill a well. It is expressed in pounds per gallon, abbreviated as ppg. A conversion factor for equivalent mud weights is

$$1 \text{ lb/gal or } 1 \text{ ppg} = 0.0519 \text{ psi/ft} \qquad (1.6)$$

Weight in itself is not a gradient. If we relate weight to a volume, however, we have density, and density does convert to a gradient. When we refer to mud weights as 10 pounds, we mean the mud *density* is 10.0 lb/gal or ppg. This is a density. (In this nomenclature, pure water density would be 8.344 ppg.) Fertl (1976) suggested the following relation for hydrostatic pressure in psi, as is commonly used in drilling operations:

$$P_h = CM_w z \tag{1.7}$$

where z is the vertical height of fluid column in feet, M_w is fluid density for mud weight expressed in lb/gal (or ppg) or pounds per cubic feet (lb/ft^3), and C is a conversion constant equal to 0.0519 if M_w is expressed in pounds per US gallon and 0.00695 if M_w is expressed in lb/ft^3. The conversion factor 0.0519 (inverse of 19.250) is derived from dimensional analysis as follows:

$$\frac{1\,\mathrm{psi}}{\mathrm{ft}} = \frac{1\,\mathrm{ft}}{12\,\mathrm{in}} \times \frac{1\,\mathrm{lb/in}^2}{1\,\mathrm{psi}} \times \frac{231\,\mathrm{in}^3}{1\,\mathrm{USgal}} = 19.250\,\mathrm{lb/gal} \tag{1.8}$$

It would be more accurate to divide a value in lb/gal by 19.25 than to multiply that value by 0.052. The magnitude of the error caused by multiplying by 0.052 is approximately 0.1 percent. Let us take an example: for a column of freshwater of 8.33 pounds per gallon (lb/US gal or ppg) standing still hydrostatically in a 21,000 ft vertically cased wellbore from top to bottom (vertical hole), the pressure gradient would be

$$\text{Pressure gradient} = 8.33/19.25 = 0.43273 \text{ psi/ft} \tag{1.9}$$

and the hydrostatic *bottom hole pressure* (BHP) is then BHP = true vertical depth × pressure gradient = 21,000 (ft) × 0.43273 (psi/ft) = 9087 psi. However, the formation fluid pressure (pore pressure) is usually much greater than the pressure due to a column of freshwater, and it can be as much as 19 or 20 ppg. For an onshore vertical wellbore with an exposed open hole interval at 21,000 ft with a pore pressure gradient of 19 ppg (or 19 × 0.0519 (psi/ft)), the BHP would be BHP = pressure gradient × true vertical depth = 19.0 × 0.0519 (psi/ft) × 21,000 (ft) = 20,708 psi. (It would be 20,727 psi if we replace 0.0519 by 1/19.25.) The calculation of a bottom hole pressure and the pressure induced by a static column of fluid (drilling mud) are the most important and basic calculations in the petroleum industry. In summary, pore pressure gradient is a *dimensional* term used by drilling engineers and mud engineers during the design of drilling programs for drilling (constructing) of oil and gas wells into the earth. In Table 1.1 we give some useful conversion factors.

In the Gulf Coast of the United States, a fluid pressure gradient of 0.465 psi/ft is considered to be normal or hydrostatic; it corresponds to a salt concentration of 80,000 ppm and a temperature of 77°F. However, the hydrostatic pressure gradient is variable depending upon the temperature, pressure, and salinity, as noted earlier. An increase in salt concentration at a given temperature and pressure would increase the hydrostatic gradient. Dissolved gas in water, for example, methane, also affects the density of water – it lowers the density – and hence the hydrostatic gradient will be lower. We

Table 1.1 Units and conversions

Psi	$= 0.070307$ kg. force/cm^2
Atm	$= 1.033$ kg force/ cm^2
Atm	$= 14.6959$ psi
Psi	$= 0.006895$ MPa
Psi/ft	$= 2.31$g/cm3
	$= 1\,4\,4\ lb\,/\,ft^3$
	$= 19.25$ lb/gallons or ppg

$1\ Pa = 1 N/m^2 = 1.4504 \times 10^{-4}$ psi
$1\ Mpa = 10^6\ Pa = 145.0378$ psi
$1\ Mpa = 10$ bars
$1\ N = 1$ kg. m/s^2
$1\ kbar = 100$ MPa
1 psi / ft. $= 2.31$ g/cm^3
$\approx 0.0225 MPa/m = 22.5 kPa/m$

note that the solubility of methane in water is a function of salt concentration at a given temperature and pressure – it increases with increasing salt concentration. Thus, dissolved gases would cause the hydrostatic gradient to be lower. In the vicinity of salt domes, salt concentration could be markedly higher, leading to a higher hydrostatic gradient. In Table 1.2 we show typical values for density and pressure gradients for oil, brine, and some drilling fluids. In Table 1.3 we show typical hydrostatic pressure gradients for several areas of active drilling.

It is clear from these discussions that hydrostatic (or normal) pressure for a static water column of height z is equivalent to a water-saturated porous medium such as clean sandstone of the same height with the assumption that the sandstone consists of interconnected pores. Formation pressure (or geopressure) that *differs* from hydrostatic pressure is defined as *abnormal pressure*. Formation pressure (or geopressure) *exceeding* hydrostatic pressure is defined as *overpressure*, whereas formation pressure *lower* than hydrostatic is defined as *subpressure*. Therefore, we emphasize that before embarking on any computation involving determination of subsurface pore pressure, we must establish a proper baseline – deciding on the "accurate" hydrostatic pressure gradient with as much accuracy as possible.

1.4   Overburden Stress

The *overburden* or *lithostatic stress*, S, at any depth, z, is the stress that results from the *combined* vertical weight of the rock matrix and the fluids in the pore space overlying the formation of interest as well as the weight of the static water column, if in an offshore environment, and the atmospheric air pressure. This can be expressed as

$$S = P_{\text{air}} + g \int_0^{z_w} \rho_{sw}\,(z)dz + g \int_{z_w}^{z} \rho_b\,(z)dz \qquad (1.10)$$

Table 1.2 Fluid densities and corresponding pressure gradients

Fluid	Total solids (ppm)	Density (g/ml)	Fluid pressure gradient (psi/ft)	(kPa/m)
Freshwater	0	1.0	0.433	9.8
Brine	28,000	1.02	0.441	10.0
	55,000	1.04	0.450	10.2
	84,000	1.06	0.459	10.4
	113,000	1.08	0.467	10.6
	144,000	1.10	0.476	10.8
	176,000	1.12	0.485	11.0
	210,000	1.14	0.493	11.2
Oil	API° (60°F)			
	70.6	0.70	0.303	6.90
	45.40	0.80	0.346	7.80
	25.70	0.90	0.390	8.80
	10.00	1.00	0.433	9.80
Drilling mud	lb/gal or ppg			
	8.35	1.00	0.433	9.80
	10.02	1.20	0.520	11.8
	11.69	1.40	0.607	13.70
	13.36	1.60	0.693	15.70
	15.03	1.80	0.780	17.70
	16.70	2.00	0.867	19.60
	18.37	2.20	0.953	21.60
	20.04	2.40	1.040	23.50
	21.71	2.60	1.126	25.50
	23.38	2.80	1.213	27.50
	25.05	3.00	1.300	29.40

Note: Pressure gradients are related to the specific gravity (γ) rather than the density (ρ), where $\gamma = \rho g$, g = 9.81 m/s^2.
Source: Modified after Gretener (1981)

Table 1.3 Normal pore pressure gradients for several areas

Area	Pressure gradient (psi/ft)	Pressure gradient (g/cc)
West Texas	0.433	1.000
Gulf of Mexico (coastline)	0.465	1.074
North Sea	0.452	1.044
Malaysia	0.442	1.021
Mackenzie Delta	0.442	1.021
West Africa	0.442	1.021
Anadarko Basin	0.433	1.000
Rocky Mountains	0.436	1.007
California	0.436	1.014

where P_{air} is the pressure due to the atmospheric air column (typically 14.5 psi or 1 bar), ρ_b is the bulk density, ρ_{sw} is the sea water density (both depend on depth), z_w is depth to the ocean bottom, and g is the acceleration due to gravity. The bulk density of a fluid-saturated rock is given by

$$\rho_b = (1 - \phi)\rho_r + \phi \rho_f \qquad (1.11)$$

where ϕ is the fractional porosity (the void space in the rock), ρ_f is the pore fluid density, and ρ_r is the density of the matrix (grain density). It should be noted that the overburden stress computation in the context of drilling wells should always account for the air gap or the atmospheric pressure. Although this is small, it could be significant while dealing with overburden stress in shallow formations, such as pressured aquifer sands or methane hydrates, as discussed later. Overburden stress is *depth dependent* and increases with depth in a nonlinear fashion. In some of the older literature on geopressure, a default value of 1.0 psi/ft for overburden stress gradient (overburden stress divided by depth) has been recommended for the "average" Tertiary deposits off the Texas–Louisiana coast. This corresponds to a force exerted by a formation with an *average* bulk density of 2.31 g/cm^3. However, this is not true in reality, where we always deal with rocks of variable bulk densities. At shallower depths, the overburden gradient would be less than 1.0 psi/ft, while at deeper depths, it could be larger than 1.0 psi/ft. In Figure 1.3 we show typical overburden gradients from selected basins.

1.5   Effective Vertical Stress and Terzaghi's Law

When a rock is subjected to an external stress, it is opposed by the fluid pressure of pores in the rock. This is due to Newton's law of classical mechanics. More explicitly, if P is the formation or pore fluid pressure at a depth where the vertical component of the total stress (namely, the vertical overburden stress) on it is S, then the vertical effective stress σ is defined as (see Figure 1.4)

$$\sigma = S - P \qquad (1.12)$$

This is known as the Terzaghi's principle or law (Terzaghi, 1923). This principle was invoked to describe the consolidation of soil in the context of geotechnical engineering (soil consolidation) (see Chapter 2). Compaction is due to the vertical effective stress – it is the stress that is transmitted through the solid framework. This (vertical effective stress) is a very important parameter to describe geopressure phenomenon quantitatively, especially when geophysical methods such as seismic or sonic logs are used to quantify geopressure. A relationship between velocity and overpressure is intuitively expected, since acoustic velocity and vertical effective stress are related closely. This will be discussed in Chapter 3. We note that

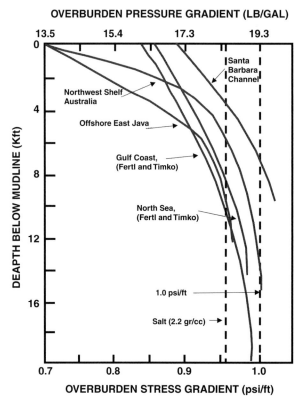

Figure 1.3 Typical overburden stress gradients versus depth in psi/ft and ppg. Modified after Dutta (1987a).

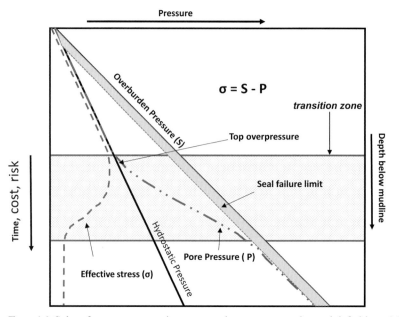

Figure 1.4 Subsurface pressure environment and some commonly used definitions. Modified after Dutta (1987a).

Terzaghi's Law is only an approximation because the vertical component of the total stress S does not remain constant in a sedimentary basin that is actively developing and accumulating sediment, nor does it remain strictly constant during compaction of sedimentary rocks (see Chapter 2). It has several other assumptions as well: The soil is isotropic and homogenous; the solid particles and fluids are incompressible; the fluid flow occurs in one dimension only (vertical direction); the strains in the soil are very small (linearized strain model); Darcy's law for fluid flow is valid for all pressures (linearized fluid flow model); the permeability remains constant throughout the process; and lastly, compaction process is independent of time. While some of the assumptions are reasonable and borne out by experiments, the remainder of the assumptions such as the linearity of strain, one-directional flow model as well as the time-independence of the compaction process (relationship between porosity and vertical effective stress) may be questioned.

Subsequent literature such as Hornby et al. (1994, and references therein) as well as Sarker and Batzle (2008, and references therein) suggested using the Biot model for the effective stress as given below, instead of Terzaghi's Law (see Chapter 2):

$$\sigma = S - \alpha P \qquad (1.13)$$

where α is termed as Biot's consolidation coefficient and its value lies between 0 and 1. Further, it was mentioned that this coefficient could depend on the lithology also. Sarker and Batzle (2008) and Hornby et al. (1994) suggested that α is close or equal to unity for near-surface sediments such as soil and clay – the subject of measurements by Terzaghi (1923) but for consolidated rocks, it could be less than unity and its value is ~ 0.90 to 0.93 as determined by laboratory measurements using ultrasonic waves (Hornby et al., 1994). Suffice it to say at this point that the uncertainty regarding α is related to three major factors: (1) what are the "assumed" physical processes by which unconsolidated sediments morph into rocks under gravitational loading and many chemical effects? (2) How are the measurements conducted, namely, under static or dynamic conditions? and (3) what kind of rock types are considered in the measurements? It must be noted that there is no direct way to measure α; it is derived indirectly from measurements of physical quantities which are related to the effective stress such as pore pressure and velocity. We note that Biot's consolidation coefficient plays an important role in building geomechanical models (see Chapter 12). *In our discussions in this book we shall continue to use $\alpha = 1$ and the original version of the Terzaghi's Law as given in* equation (1.12) *unless otherwise stated.* In Chapter 2 we will show a simple derivation of Terzaghi's Law based on Archimedes' principle for solid particles suspended in a fluid.

There are many field evidences of Terzaghi's Law. It is the reason why the surface over oil fields subsides after hydrocarbon or other fluids are produced (Kugler, 1933; Gabrysch, 1967; Mayuga, 1970). In these examples (and there are many examples, such as the subsidence of the Ekofisk oil field in the Norwegian Sector of the North Sea and various offshore fields in the Gulf of Mexico and Louisiana, USA), the extraction of liquids led to a reduction in pore pressure, an increase in the effective stress (σ), and hence further compaction and subsequent subsidence of the reservoirs. In these cases, the reservoir pressure is *less* than the normal or hydrostatic pressure and gas and water are injected to counter the subsidence. In the case of Ekofisk Field, the drop in pore pressure was shown to be balanced by an increase of the effective stress in accordance with the Terzaghi's Law (Chapman, 1994a). Such effects can be monitored by 4D-seismic techniques (see Chapter 12).

1.6   Formation Pressure

As defined earlier, formation pressure is the pressure acting on the fluid contained in the pore spaces of sediments or rocks. Figure 1.1 illustrates a typical geopressure versus depth profile. The geopressure regime could be loosely classified in three zones: Subnormal or subpressure where pore pressure is below hydrostatic pressure; hydrostatic or normal pressure; and overpressure where the pore fluid pressure is higher than the hydrostatic pressure. The transition from hydrostatic to overpressure may be fairly well-defined as in the Miocene sections of the Texas–Louisiana Gulf Coast, or gradual as in the case of deepwater Pliocene and Pleistocene sections of the many Tertiary Clastic provinces in the world. The depths at which these transitions occur and the shape of the transition zone depend mostly on the permeability (it controls the fluid flow out of the sediments) and the rate of deposition of the sediments (it provides the source of the fluid in the sediments). Typically when the *outflow* of fluids (brine) is less than the *inflow*, the system is overpressured as the fluid begins to support the load. This mostly dictates the magnitude and the distribution of pore pressure. The extent of the transition zone can vary from a few hundred feet to many thousands of feet. Although there is no universally accepted scale expressing the degree of geopressuring, the nomenclature introduced by Dutta (1987a, Table 1, p. 5) is commonly accepted. This is reproduced in Table 1.4.

An important concept seen in Figure 1.4 is that pore pressure typically *does not* reach overburden stress. As pore pressure approaches overburden stress (actually, the least principal confining stress which is usually less than the overburden stress as discussed in Chapter 2), fractures in the rock open and release fluids and pressures. The pressure at which this happens is termed as the "fracture pressure." Thus, every rock has a characteristic limit defined as the *seal limit* in Figure 1.4. Fracture pressure at a given depth divided by the depth is known as the fracture gradient at that depth. There

Table 1.4 Geopressure characterization

Fluid pressure gradient (psi/ft)	Overpressure characterization
Hydrostatic $< FPG \leq 0.65$	Soft or mild
$0.65 < FPG \leq 0.85$	Moderate
$FPG > 0.85$	Hard

are three *important* quantities that the drilling community is mostly concerned with prior to drilling a well. These are*: pore pressure gradient, fracture gradient and overburden gradient.* All gradients are typically converted to EMW and the drilling plan deals with obtaining these quantities for the entire well.

Variation in the effective stress is also shown in Figure 1.4. It is the difference between overburden or lithostatic stress and pore pressure, i.e., essentially the amount of overburden stress that is supported by the porous network of rock grains as is allowed by the Terzaghi (1923) principle. If we assume that the pore fluid consists of brine with the density of 1.07 gm/cc (equivalent to a pressure gradient of 0.465 psi/ft), the normal or hydrostatic fluid pressure (in psi) will be given by

$$p_h = 0.465(\text{psi/ft})z(\text{ft}) \tag{1.14}$$

In this case, the effective stress (σ) will be given by

$$\sigma = 0.535(\text{psi/}ft)z(\text{ft}) \tag{1.15}$$

if we assume an overburden stress gradient of 1.0 psi/ft (i.e., equivalent to an assumed average and constant bulk density of 2.31 gm/cc). Hubbert and Rubey (1959) defined a quantity, λ, as the ratio between the formation pressure and the vertical overburden stress given by

$$P = \lambda S \tag{1.16}$$

Then,

$$\sigma = (1 - \lambda) S \tag{1.17}$$

Some authors (e.g., Fertl, 1976) used $\lambda = 0.9$ as an estimate of the upper limit of pore pressure in regressive sequences such as in the Tertiary rocks in the Gulf of Mexico, USA. They suggested that beyond this limit (seal limit) hydraulic fracture of the sedimentary rocks was a distinct possibility. Some practitioners use this as a measure of fracture pressure. However, the subject of subsurface fracturing due to overpressure (a natural cause) is more complex and it depends on many other factors besides pore pressure, such as lithology and tectonic stress and its orientation (Zoback, 2007). We shall discuss this in more detail in Chapter 2. Hubbert and Rubey (1959) also introduced a useful concept – *equilibrium depth* Z_E. This is defined as the depth where the effective stress is equal to what it would be at a shallower depth, had the rocks compacted normally. Swarbrick et al. (2002) introduced the term *fluid retention depth* (FRD) – it is the depth at which pore pressure begins to get higher than the normal pressure; it is also the depth where the effective stress (σ) reaches its maximum value (see Figure 1.4). We note that as pore pressure increases, so does the drilling time, cost and risk.

Overpressure implies low effective stress – lower than the effective stress for hydrostatic pressure conditions. Thus, it is maximum at the depth where the departure from hydrostatic to overpressure occurs. Drilling experiences have shown that the lowering of the effective stress in the overpressured zone is not "sharp" – it is typically preceded by a zone of almost constant effective stress. This has to do with various competing pressure mechanisms for generating geopressure as we shall see later in Chapter 3. The variables required for predicting and assigning prospect risks for prospectivity of hydrocarbons are (see Figure 1.4)

– the depth of the top of the overpressured zone,
– the depth of the top of the "hard overpressured" zone,
– the shape of the transition zone, and
– seal failure limit (the pressure needed to induce hydraulic fracturing of the "seal or cap rocks").

Let us consider the geometry of a brine-filled reservoir as shown in Figure 1.5. The reservoir is truncated at a fault and we assume that the fault is a sealing fault, namely, it does not allow any fluid to flow across the fault. In the absence of fluid flow, the difference in pore pressure between points A and B is simply the weight of the fluid in the vertical reservoir column (Figure 1.5a). If this fluid is water, pore pressure at any elevation in the reservoir will follow a hydrostatic slope as shown in Figure 1.5b. If the reservoir is overpressured and filled with brine, pore pressure will track a line parallel to the normal hydrostatic pressure curve for brine, which means that overpressure at each depth is the same as shown in Figure 1.2. This is important because it means that overpressure in a continuous reservoir unit must be constant throughout the water-bearing portion of the reservoir. This situation occurs because the permeability of the reservoir sand is much higher than that of the encasing impermeable rock (shale). In Figure 1.5, the pore pressure at the updip location (B) is related to the pore pressure in the downdip direction (A) by

$$P_A = P_B + (Z_A - Z_B)g\rho_f \tag{1.18}$$

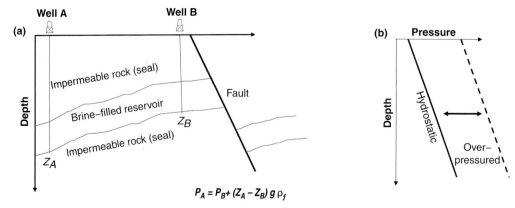

Figure 1.5 (a) Pore pressure profile of a reservoir sand with structure embedded in an overpressured shale. (b) For an overpressured sand filled with brine, the pore pressure tracks a line parallel to the normal hydrostatic pressure curve for brine.

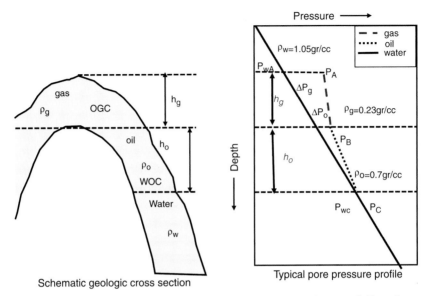

Figure 1.6 Schematic of a reservoir saturated with gas, oil, and water (left) and pore pressure elevated by oil and gas columns and density contrast between water, oil, and gas in a reservoir. Typical density contrasts are given in the figure on the right. (left) A hypothetical geologic cross section. (right) Pore pressure elevated by hydrocarbon columns in a hydraulically connected formation as depicted in cross section on the left. Modified after Zhang (2011).

In Figure 1.6 we show on the left a schematic geologic cross section of a hydraulically connected reservoir filled with gas of column height h_g, oil of column height h_o and brine. The right figure shows pore pressure elevated by hydrocarbon columns. This pressure profile is due to the buoyancy effect of the fluids. Typical values of fluid densities are 1.05 g/cm^3 for water (brine), 0.7 g/cm^3 for oil, and 0.23 g/cm^3 for gas as indicated by the slope changes of the pressure profiles on the right. It is clear from equation (1.18) and the buoyancy effect of hydrocarbons that at the crest of the structure, the pore pressure gradient would be larger than the case when the reservoir was brine saturated. This increases the possibility of the caprock failure. It is for this reason, drillers usually do not spud a well directly at the crest of a structure.

1.6.1 Subpressure

Subpressure formations are those in which the pore pressure is below the hydrostatic pressure. Such pressure conditions are known to exist in many depleted oil reservoirs, in areas with withdrawal of ground water and attendant local subsidence and in lenticular reservoirs closely associated with shales in areas that have undergone erosion. Since pore pressure gradient is datum dependent, the local topography plays a significant role. Pressure gradients can be either lower or higher than the hydrostatic pressure gradient.

In Figures 1.7a–c, we show three possibilities where the outcrop elevation and the well site elevation determine the pore pressure gradient as measured in that well bore. Figure 1.7a shows the normal pressure situation where the well elevation is the *same* as the outcrop elevation. The pressure gradient at the wellbore is 0.465 psi/ft (0.0105 MPa/m). Figure 1.7b shows the abnormally high pore pressure situation where the well elevation is *lower* than the outcrop elevation. The pore pressure at the wellbore is 0.465 psi/ft × 13,000 ft = 6045 psi (41.6787 MPa), leading to a pressure gradient of 6045 psi/10,000 ft = 0.6045 psi/ft (0.01360 MPa/m). Figure 1.7c shows a situation where abnormally low or subnormal pressure occurs where the well elevation is *higher* than the outcrop elevation. In this case, the pore pressure at the well site is 0.465 psi/ft × 7000 ft = 3255 psi (22.4483 MPa) leading to a pressure gradient of 3255/10,000 ft = 0.3255 psi/ft (0.007323 MPa/m) which is lower

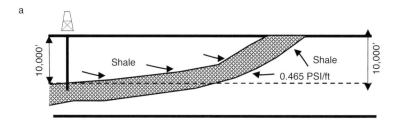

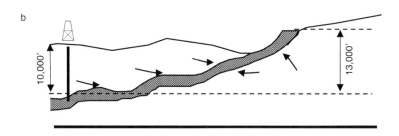

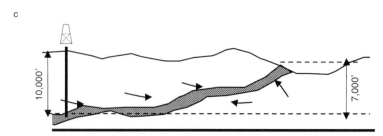

Figure 1.7 Effect of datum on pore pressure gradient. (a) Well elevation is the *same* as that of the outcrop elevation. (b) Well elevation is *lower* than the outcrop elevation leading to an abnormally high-pressure gradient. (c) Well elevation is *higher* than the outcrop elevation, leading to an abnormally low-pressure gradient.

than the normal pressure gradient. We note that subpressures are relatively uncommon except for depleted reservoirs. Proximity to a mountain range is also a feature that relate to subpressures – it provides a sink through which water is abstracted from the basins.

1.6.2   Fracture Pressure

In oil field terminology, fracture pressure is the pressure that causes a formation to fracture and the circulating fluids to be lost. Normally it is expressed as the fracture gradient or the fracture pressure divided by the depth. In this way, it determines the maximum mud weight that can be used to drill a well bore at a given depth. Hence it is an important parameter for mud weight design in both the drilling planning stage and in the drilling stage. If the mud weight is higher than the fracture gradient of the formation the well bore will undergo tensile failure, causing losses of drilling mud or even lost circulation. In practice, fracture pressure is measured from various types of leak-off tests (LOT). A leak-off test is performed to estimate the maximum amount of pressure or fluid density that the test depth can hold before leakage and formation fracture may occur. This measurement is usually made at the casing points. There are several approaches to calculate fracture gradient. We shall discuss this in details later (Chapters 2 and 4).

1.6.3   Equivalent Circulation Density (ECD)

This is an important concept in drilling. Hydrostatic weight of the mud is intended to balance the formation pressure under "static" condition. During drilling, mud pumps are turned on and the situation is no longer static. In this case, the borehole geometry introduces *pressure loss* based on drag on the fluid as it passes through the various components of the fluid flow path during drilling such as standpipe, drillstring components, open hole, and casing. The mud pumps supply the pressure that forces drilling mud down the drill string to the bottom of the hole and up again to the surface. As the drilling mud exits the bit nozzles, the mud has to flow through the annular space between the drill string and the borehole wall. Contact is made between the drilling mud and the borehole wall as the drilling mud flows upward to the surface. This contact creates "drag" as the result of friction and the drilling mud loses some of the pressure supplied by the pump in order to overcome this frictional drag. This pressure loss is absorbed by the formation. Equivalent circulating density (ECD) is the effective density that combines current mud density and annular pressure drop. Thus, ECD in ppg = (annular pressure loss in psi) ÷ 0.052 ÷ true vertical depth (TVD) in ft + (current mud weight in ppg). This is why the ECD is always *greater* than the mud density under static condition. The greater the vertical distance through which the drilling mud has to travel until it reaches the surface, the higher are the pressure drop and ECD. Note that ECD is a function of the *true vertical depth* and *not the measured depth*. Measured depth includes the horizontal section in deviated wells but ECD only depends on the true "vertical" depth. Acquired solids (cuttings), the type of mud and its temperature also affect ECD.

1.7   Casing Design

Both pore pressure and fracture pressure dictate the casing design in drilling. A casing is a large heavy steel pipe which is lowered into the well. Generally, a casing is subjected to various physical and chemically related loads during its lifetime. Its purpose is to prevent collapse of the borehole while drilling, hydraulically isolate the wellbore fluids from formations and formation fluid, minimize damage of both the subsurface environment from the drilling process and from extreme subsurface environment, provide a high strength flow conduit for the drilling fluid, and provide safe control of formation pressure. If the casing is properly cemented it can also help to isolate communications between different perforated formation levels. Selection of the number of casing string and their respective setting depths generally is based on a consideration of the pore pressure gradients and fracture gradients of the drilling area. In Figure 1.8, we show a typical casing design for a hydrocarbon well.

A casing string consists of (1) surface casing, (2) intermediate casing, (3) liners, and (4) production casing. The main purpose of the surface casing is to prevent the shallow water region from contamination, the structural support for weak soil areas near the subsurface, and also protect the casing strings inside. Again, this can prevent blowout and can close the surface casing in the event of a kick or explosion. When drilling

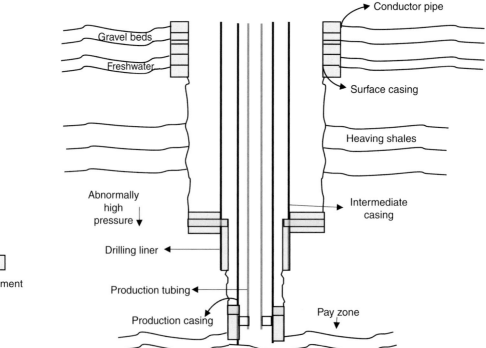

Figure 1.8 Typical casing strings used in the hydrocarbon industry. Modified after https://petrowiki.org/o_and_tubing.

deeper through weak zones like salt sections and abnormally pressurized formations, these unstable sections need more pipe sections, in the form of intermediate casings between the surface casing and the final casing. When abnormal pore pressures are present below the surface casing, intermediate casings are needed to protect the formation. The liner is a casing string that does not extend to the surface. It is suspended from the bottom of the next large casing string. The principal advantage of a liner is its lower cost. It serves as a low cost intermediate casing. This casing string provides protection for the environment in the event of a failure during production. Generally, there are two types of wells. The first are exploration wells that are drilled and abandoned within a few months. The second are production wells that are used continuously through their life. Production casings are connected to wellhead using a tie-back when the well is completed. This casing is used for the entire interval of the drilling.

Drilling environments often require several casing strings to reach the total desired depth. Some of the strings are: drive, or conductor, surface, intermediate (also known as protection pipe), liners, and production (also known as an oil string). Figure 1.8 shows the relationship of some of these strings. All wells will not use each casing type as shown. The conditions encountered in each well must be analyzed to determine types and amount of pipe necessary to drill it. Pore pressure and fracture pressure are the two most important parameters that dictate the casing design. For example, selecting casing seats for pressure control starts with formation pressures (expressed as EMW) and fracture – mud weight. This information is generally needed prior to designing a casing program. Quantitative evaluation of geopressure is essential to determine the exact locations for each casing seat. This procedure is implemented from the bottom to the top as shown in Figure 1.9. Setting-depth selection is made for the deepest strings to be run in the well, and successively designed from the bottom to the surface. Although this procedure may appear at first to be reversed, it avoids several time-consuming iterative procedures. Errors in estimates of pore pressure and fracture pressure affect the casing design significantly. Surface casing design procedures are based on other criteria also, such as shallow hazards. We shall discuss this in Chapter 11.

As noted earlier the first criterion for selecting deep casing depths is for mud weight to control formation pressures *without* fracturing shallow formations. It is a common practice to establish a "safe mud window" as shown in Figure 1.9. This window is typically 0.2–0.3 ppg *lower* than the fracture pressure and 0.2–0.3 ppg *higher* than the "anticipated" formation pressure. It is clear that as the well is drilled to deeper depths, the width of the "safe mud widow" will become narrower; this can cause a severe problem if the program is not managed properly. In Kankanamge (2013), readers will find an interesting case study in how to design a casing for a deep gas well.

1.8   Importance of Geopressure

Geopressure is a worldwide phenomenon as shown in Figure 1.10 where many of the formations are in overpressured conditions. There is a good description of these in Fertl

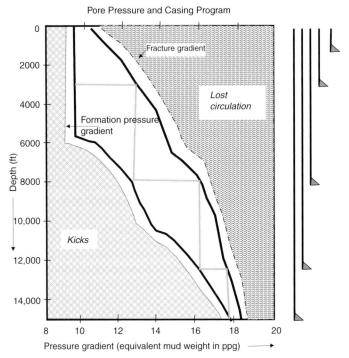

Figure 1.9 Pore pressure gradient versus depth and setting a casing program for a hypothetical well. Note that drillers use a safety margin prior to designing a casing program. It is typically 0.2–0.3 ppg higher than the formation pressure and 0.2–0.3 ppg lower than the fracture pressure gradient.

(1976), Chapman (1994b), and Chilingar et al. (2002). A quantitative study of geo-pressure is essential for the following reasons:

- guide safe drilling activity (proper mud and casing program and blowout prevention),
- provide exploration support for hydrocarbon (trap/risk identification and hydrocarbon migration path assessment; seismic imaging improvements; economic basement assessment), and
- assess environmental risks (shallow hazards identification and mitigation including overpressured aquifer sands and gas hydrates).

Most sedimentary basins exhibit characteristics of overpressured formations to varying degrees. Although overpressure is more pronounced in young basins, they are known to occur in formations with highly varied lithologies such as sandstone, shale, limestone, and dolomite anywhere between Pleistocene and Cambrian (Fertl, 1976; Law and Spencer, 1998). They are also known to occur in igneous environments such as in the Gulf of Bohai in China (Chilingar et al., 2002).

The United States Geological Survey sponsored a Conference on the Mechanical Effects of Fluids in Faulting under the auspices of the National Earthquake Hazards

Abnormally High Pore Pressure

- It is a worldwide occurrence

- Cause of large accidents and a lot of nonproductive time (NPT)

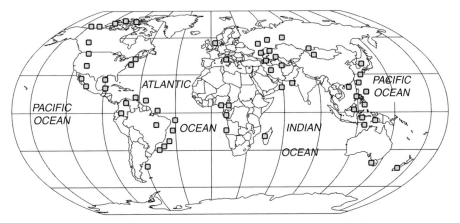

Figure 1.10 A world map showing occurrences of geopressure. It is a worldwide phenomenon. It causes big accidents and significant nonproductive time (NPT) during drilling.

Reduction Program at Fish Camp, California, from June 6 to 10, 1993. At that conference a growing body of evidence suggested that fluids were intimately linked to a variety of faulting processes (Hickman et al., 1995). The authors noted that these included the long term structural and compositional evolution of fault zones; fault creep; and the nucleation, propagation, arrest, and recurrence of earthquake ruptures. This is generally believed now. Occurrence of overpressure is not necessarily contemporaneous with the surrounding sediment. For instance, the presence of high pressured fluids in Paleozoic formation may have been developed in the Tertiary period. Fluid containment in a closed or semiclosed environment is the source of abnormally high pore pressure. It is for this reason that there is evidence of high pore pressure in thick Paleozoic shale formations in Wyoming (Hubbert and Rubey, 1959) and subsequent detachment and movement of some of the thick blocks of overpressured rocks from underlying formations. Thus, overpressures are intimately related to structural geology and it is found in various ages of formations.

1.8.1   Guide for Safe Drilling Practices

Even though the petroleum industry has a good safety record, drilling through high pressured formations is known to pose serious drilling challenges. Blowouts are also known to occur occasionally. Some of the reasons are listed above. While catastrophic events are rare, what is not rare is the nonproductive drilling time spent during a drilling operation as shown in Figure 1.11. About 95 percent of the incidents involves problems related to drilling performance. In addition, ~5–25 percent of the well cost is

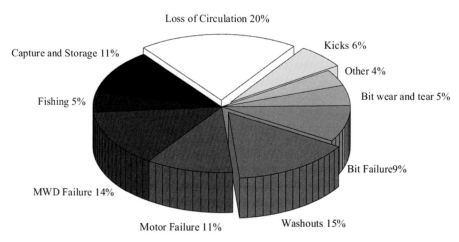

~ 30% - 40% of the downtime is related to pore pressure related issues !

Figure 1.11 Time lost during drilling (nonproductive time or NPT) as known in the petroleum industry. About 30–40 percent of NPT are due to issues related to overpressure. *Fishing* is a term used by the driller to retrieve any tool lost in a borehole.

a result of inadequate drilling performance. As the data shows, ~30–40 percent of the operations cost during drilling are related to overpressure. Some estimates put the total loss to the industry as high as $3.0 billion annually. For serious problems such as those due to loss of circulation of drilling mud into the surrounding formation or influx of the formation fluid into the borehole (Figure 1.11), the nonproductive downtime could be as long as seven days or higher – for deepwater drilling activities where daily rig rates could be as high as $200 million, the losses could be very high. In the extreme case, if the formation pressure encountered during drilling is much higher than planned, it would be impossible to drill through to the target at deeper depths and to set proper casing as there may not be enough casing string left. This would result in abandoning the well without reaching its target and thus causing huge losses.

We mentioned earlier that quantitative analysis of geopressure is important as it impacts the *mud program* while drilling a well. Mud program refers to designing a formal plan for drilling fluid requirements in general and specific maintenance need, namely, choosing drilling fluid for a specific well with predictions and requirements at various intervals of the wellbore depth. It consists of details on the mud type; composition, density, rheology, filtration and other properties. The density is especially important because they must fit with the casing design program and to ensure wellbore pressures are properly controlled as the well is drilled deeper. Good drilling fluid or mud (as is known in the industry) with proper density is necessary to ensure that no formation fluid influx occurs into the wellbore. This is particularly important while drilling in high pressured wells that require increasing the fluid density (or mud weight) with depth. However, a best practice dictates that the drilling fluid density or the mud weight must not be too high as compared to the true formation pressure. If not, it can cause a serious drilling problem known as lost circulation with unwanted consequences such as formation of thick mud filtration between the well bore and the formation, poor cementing job, and so on. (There are other uses of a good drilling fluid: remove cuttings expeditiously from beneath the drill bit; transport cuttings to the surface without degradation; economically deliver a wellbore suitable for formation evaluation and completion and control torque on drill string, particularly in deviated wells.) Thus, variation of geopressure with depth is perhaps the key parameter that dictates designing a good mud program.

There is another importance of geopressure that impacts safety. As we will learn in this book that high pressures usually occur in thick shale formations with considerably higher water (brine) content compared with the normally pressured case. The types of drilling fluid used depend on the pressure regime. In the low-pressure regime, cost-effective water-base drilling fluid (with bentonite mud) is commonly used. For higher pressured wells, the industry uses oil-base drilling fluid. It is expensive and affects the rate of penetration of the drill bit. Thus, a transition from water-base mud to oil-base mud at a specific depth is highly influenced by the details of the quantitative profile of geopressure with depth.

1.8.2   Exploration Use

In the context of exploration for hydrocarbon, a proper quantification of pore fluid pressure in 3D is highly desirable as it can suggest a path for hydrocarbon migration through porous formations and subsequent trapping, either stratigraphically or structurally favored environments (faults, folds, etc.). Thus, the hydrodynamics of a sedimentary basin is greatly impacted by the distribution of pressured fluids, and an analysis may reveal the proper location to drill (or to avoid) and the expected height of hydrocarbon column in the prospect area. The economics of drilling for hydrocarbon is impacted in two major ways by the presence of overpressure in sedimentary basins. First, one would like to know the depth of an economic basement, if any. This is defined as the depth where the pore pressure is so high (and the vertical effective stress is so low) that the likelihood of fracturing the rock by natural causes such as hydraulic fracturing or rock movement due to earthquake can cause the hydrocarbon trap to mitigate and the oil and gas accumulation to escape from the reservoir. Figure 1.12, developed for a part of the Gulf of Mexico (Dutta, 1997b), shows such a map in which color codes indicate potential areas of possible *seal failure* due to high pore pressure. Each square block of that figure represents a federal lease area – a block that is 9 sq mi. The red colors indicate areas with possibility of low effective stresses (~500 ± 250 psi). Maps such as the one in Figure 1.12 could be used to high-grade the prospect inventory (possibility of seal mitigation) and assign a potential hazard index to those areas with extremely *low* effective stresses for drilling. Second, the explorer of hydrocarbons would like to know the depth of the *top* of the onset of overpressure (Figure 1.13) and the distribution of overpressure in 3D (Figure 1.14). As we shall see

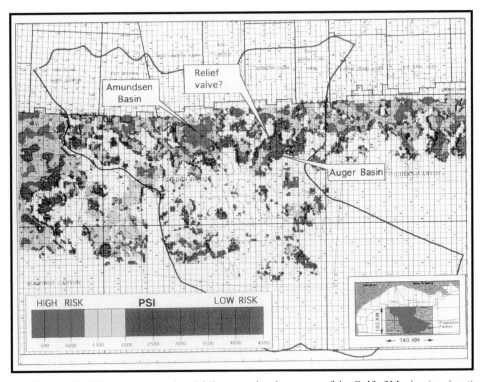

Figure 1.12 Effective stress and seal failure map in a large area of the Gulf of Mexico (see inset) as predicted from a combination of surface seismic and basin modeling. Red indicates high probability of seal failure. Taken from Dutta (1997b). The range of effective stress is from 0 to 5000 psi. (A black and white version of this figure will appear in some formats. For the color version, please refer to the plate section.)

later, the hydrocarbon column height in a reservoir is greatly impacted by the magnitude of overpressure in the reservoir – *higher* the overpressure, the *lower* will be the total hydrocarbon column height. This is because the buoyancy force due to hydrocarbon will contribute further to the existing high pore pressure of the fluids, thus raising the likelihood of *seal leakage* and eventually *seal failure.*

Recent studies have indicated another use for quantification of geopressure for exploration. This is related to defining a better seismic velocity model for seismic imaging. Traditionally, seismic velocities are obtained by various inversion processes (we will study this in Chapter 5) that produce velocity models that are inherently nonunique and potentially ambiguous. Recent studies (Dutta et al., 2014, 2015a, 2015b; Le et al., 2018) have shown that imposing a pore pressure *constraint* on the derived seismic velocity model and demanding that the velocity model yield a *physically plausible* pore pressure model, namely, the predicted pore pressure be equal or higher than the hydrostatic pore pressure and be less than the fracture pressure, for example, yields a better seismic image of the subsurface. An example is provided in Figure 1.15. Figure 1.15a shows the legacy image using conventional velocity analysis (tomography based anisotropic velocity analysis) while the one in Figure 1.15b (that

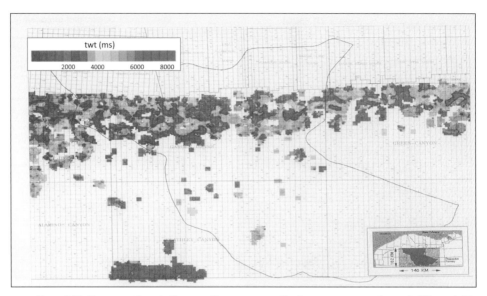

Figure 1.13 Two-way time to the top of overpressure for the same area as shown in Figure 1.12. Blue indicates shallow overpressure. Taken from Dutta (1997b). The scale is two-way time (twt); the range is 0-8000 ms. (A black and white version of this figure will appear in some formats. For the color version, please refer to the plate section.)

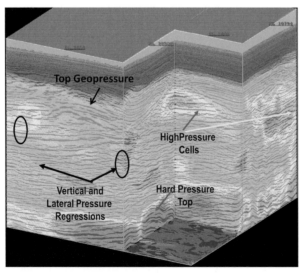

Figure 1.14 Seismic-based map of distribution of pore pressure in 3D. Annotation shows "sweet" spots for exploration and identifies areas for "drilling" through risky zones. The size of the cube is ~600 sq. mi. in area and 5 s deep (in time). From Dutta and Khazanehdari (2006). (A black and white version of this figure will appear in some formats. For the color version, please refer to the plate section.)

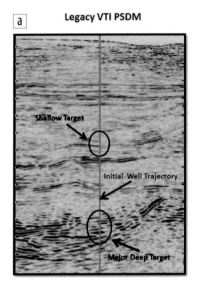
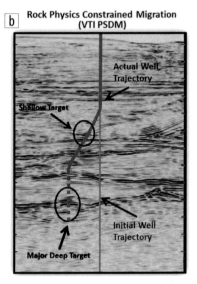

Figure 1.15 Use of seismic data to improve the image of subsurface formations in a velocity model. (a) An image based on a legacy anisotropic velocity model. (b) An improved image based on an anisotropic velocity model that uses pore pressure constraint on the velocity model. The image is better focused. This led to a better well path trajectory to reach the deeper target. After Dutta et al. (2014).

used a *constrained* anisotropic velocity model (tomography) – constrained by plausible range of pore pressure) shows a marked improvement. Not only is the velocity lowered (a consequence of high pore pressure) but the seismic energy is more focused resulting in a better illumination of the image of the potential deep targets for exploration. We shall discuss this approach for building a common velocity modeling for pore pressure and imaging in detail in Chapter 6.

Over the last several decades, our quest for hydrocarbon exploration and drilling has taken us to the frontiers of the deepwater where sizable accumulations have been found. These activities have pointed to a link between high pore pressures and potential risk to the environment. We have seen the risk of blowouts of pressured aquifer sands in the shallow part of the stratigraphy below seabed (Ostermeier et al., 2002). These are known in the industry as *shallow-water-flow* (SWF) *sands* – it is a deepwater phenomenon and deals with aquifer pressured sands held by a thin veneer of clay. Drilling through these sands can cause blowouts and serious damages to the environment. This is a near-surface geologic phenomenon and the quantification of pore pressure is difficult but necessary so that adequate precautions (avoidance, for example) can be undertaken. In Figure 1.16, we show a compilation of many other geological causes of geohazards – not all of which are related to high pore pressure. An example is the gas hydrates in deepwater that exist in the environment similar to the place where SWF sands occur. In Chapter 11 we shall discuss how to quantify some of these geohazards and the role of geopressure.

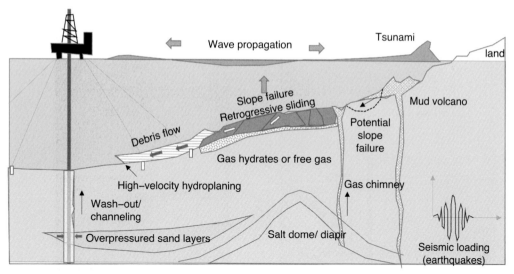

Figure 1.16 A schematic compilation of geological causes of geohazards.

1.8.3 Seals, Seal Capacity, and Pore Pressure

Hydrocarbon exploration is fundamentally about recognizing three things in a sedimentary basin: existence of source rocks (source), a reservoir or container for hydrocarbon to migrate into (reservoir), and a seal or a lid to contain the hydrocarbon (seal). Geopressure impacts all of these and in addition, it facilitates the primary and secondary migration of hydrocarbon (Tissot and Welte, 1978). In a scholarly article, Watts (1987) describes three types of seals: hydrodynamic, caprock, and fault. Hydrodynamic seals are controlled by excess hydrodynamic head above hydrocarbon accumulation. In this case hydrocarbon column reaches equilibrium when hydrocarbon buoyancy pressure is balanced by a downward hydrodynamic flow force, as shown in a schematic diagram in Figure 1.17 in an anticlinal reservoir under the condition of increasing action of flowing water (brine). Under hydrostatic conditions (no flow), oil and gas would simply rise (top figure), according to the principle of buoyancy, to the highest available part of the trap. Under hydrodynamic conditions, however, it is not necessary that hydrocarbon (oil or gas) rises in the crest position of the structure. Fault-related seals are those that prohibit fluid migration through the faulted regions and these are controlled by the entry pressure of largest interconnected pore throats across the fault planes. Sealing faults are caused mainly by clay smear (very low permeability) or a variety of diagenetic processes (Tissot and Welte, 1978).

In the petroleum industry, *caprock* is defined as any nonpermeable formation that may trap oil, gas or water, thus preventing it from migrating to the surface. Caprock is essential to create a reservoir of oil, gas or water beneath it and is a primary target for the petroleum industry. It can be overpressured as is the case for almost all Tertiary Clastic basins. According to Tissot and Welte (1978), caprock seals can be of two types: *membrane and hydraulic*. The membrane seal integrity is mostly controlled by

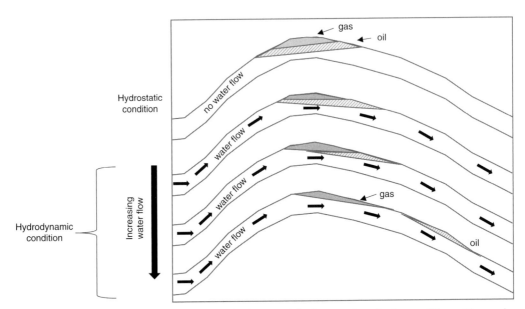

Figure 1.17 Hydrocarbon distribution in an anticlinal reservoir under the conditions of increasing action of flowing water (brine). Modified after Tissot and Welte (1978).

the entry pressure of largest interconnected pore throats, while this is not the case for hydraulic seals where seal breach occurs mainly due to hydraulic fractures (typically due to high overpressure).

Seal capacity refers to the hydrocarbon column height that the caprock can retain before capillary forces allow migration of the hydrocarbon into, and possibly through, the pore system of the caprock. Seal capacity is affected by both the physical properties of seal including its pore pressure state and the properties of the hydrocarbon. When hydrocarbon begins to fill in a reservoir, the pore space of that reservoir is usually filled with water (formation water). As hydrocarbon has a lower density than the formation water occupying the pore space, the hydrocarbon rises upward through the reservoir due to buoyancy (the density difference between hydrocarbon and water). Greater the density difference between the two phases is, so is the buoyant force that pushes the less dense, more buoyant hydrocarbon-phase upward.

The upward movement of the hydrocarbon through the pore system is resisted by capillary pressure. Capillary pressure is defined as the pressure required to displace the formation water from the pores and pore throats of the seal. The capillary pressure P_c is known as the displacement pressure in the petroleum industry, and it is given by (Berg, 1975)

$$P_C = \frac{2\gamma \cos\theta}{r} \qquad (1.19)$$

where γ is the interfacial tension (dynes/cm) between hydrocarbon (oil or gas) and brine, θ is the contact angle (degrees), and r is the pore throat radius (cm). Clayton and

Hay (1994) showed the following relation for the thickness of the hydrocarbon column T_H:

$$T_H = \frac{2\gamma \cos\theta}{r(\rho_w - \rho_h)g} - \frac{\Delta P}{(\rho_w - \rho_h)g} \tag{1.20}$$

where ρ_w is the density of the formation water (brine), ρ_h is the density of the trapped hydrocarbon, g is the acceleration due to gravity, and ΔP is the overpressure in reservoir *relative* to the seal (caprock). The first part of the right-hand side of equation (1.20) gives the column height under normal or hydrostatic pore pressure conditions (seal capacity), and the second part gives a correction for excess reservoir overpressure. In Figure 1.18 we show the range of seal capacities of different rock types as documented in AAPG Wiki. This figure was compiled from published displacement pressures based upon the mercury capillary curves. Column heights were calculated using a 35°API oil at near-surface conditions with a density of 0.85 g/cm³, an interfacial tension of 21 dynes/cm, and a brine density of 1.05 g/cm³. Data were compiled from Smith (1966), Thomas et al. (1968), Schowalter (1979), Wells and Amaefule (1985), Melas and Friedman (1992), Vavra et al. (1992), Boult (1993), and Shea et al. (1993). The figure suggests that (1) shales can trap thousands of feet of hydrocarbon (normally pressured case), (2) most clean sands can trap up to 50 ft or less column of oil, and (3) poor quality sands and siltstones can trap 50–400 ft of oil. Carbonates have a wide range of displacement pressures. Some carbonates can seal as much as 1500–6000 ft of oil.

We note that higher the overpressure, lower is the hydrocarbon column height. This is clear from the following:

$$H_{hc,\text{max}} = \frac{FP - RP}{(\rho_w - \rho_{hc})g} \tag{1.21}$$

where

$H_{hc,\text{max}}$ = maximum hydrocarbon column height
FP = fracture pressure or the pore pressure at which hydraulic fracturing occurs
RP = reservoir pressure or the pore pressure of the brine-filled reservoir
ρ_w = density of brine
ρ_{hc} = density of hydrocarbon

The fracture and reservoir pressure should be estimated or measured at the crest of the structural closure. Equation (1.21) shows why a study of geopressure is so important – as reservoir pressure increases (due to some overpressure mechanisms), the hydrocarbon column height decreases. It impacts both hydrocarbon accumulation and migration from source to reservoir. It also defines the seal integrity of caprock, namely, whether hydraulic fracture would occur or not. Compaction and diagenesis during burial cause a progressive reduction in pore throats in most seal lithologies. This affects seal capacity. In addition, the interfacial tension of the hydrocarbons changes

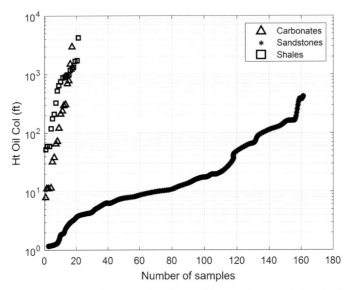

Figure 1.18 Range of measured seal capacities of oil accumulation for different rock types. Modified after a figure from the AAPG Wiki at http://wiki.aapg.org /Seal_capacity_of_different_rock_types.

with depth and affects seal capacity. Most importantly, the interfacial tension of oil and gas changes at different rates and impacts the seal capacity.

1.8.4   Permeability and Fluid Flow in Porous Rocks

Rocks are porous material composed of solid grains, void spaces and fluids in the void spaces. Darcy (1856) showed that the fluid flow rate in a porous rock is linearly related to the pressure gradient. The flow equation can be expressed generally as

$$Q = -\frac{\kappa A}{\mu} gradP \tag{1.22}$$

where Q is the vector fluid volumetric velocity, A is the cross-sectional area normal to the pressure gradient, μ is the fluid viscosity, and κ is the permeability (it is a tensor) with units of area. The unit of permeability commonly used is Darcy. We shall discuss how permeability affects fluid flow and contributes to geopressure in Chapter 3 in detail. The ability of fluids to flow is controlled by the permeability κ that is of great importance in the petroleum industry. Formations that transmit fluids readily, such as sandstones, are described as permeable and tend to have many large, well-connected pores. Impermeable formations, such as silts and shales, tend to be finer grained or of mixed grain sizes, with smaller, fewer, or less interconnected pores. *Absolute permeability* is defined as that permeability measured when a single fluid is present in the rock. Its dimension is of an area. *Relative permeability* is the ratio of permeability of a particular fluid at a particular saturation to the absolute permeability of that fluid at full saturation. It is a dimensionless quantity. Calculation of relative permeability enables us to compare different abilities of each fluid to flow in the presence of the other. The presence of more than one fluid generally hinders flow. *Effective permeability* is a measure of permeability when two immiscible fluids occupy the same pore space of a rock, the movement of each is influenced by the other, and by their saturations. The following are the typical ranges of the individual permeability of common rock types:

Conventional oil and gas reservoirs ~ milli-Darcy (10^{-3} Darcy)
Tight sands, coal and some shales ~ micro-Darcy (10^{-6} Darcy)
Some coals and most shales ~ nano-Darcy (10^{-9} Darcy)

Permeability of sediments is related to porosity ϕ and therefore, it will decrease with compaction as rocks are buried. The Kozeny–Carman equation (Kozeny, 1927; Carman, 1937) describes the relationship between permeability κ, porosity, and grain diameter d, as follows:

$$\kappa = B \frac{\phi^3}{(1-\phi)^2} d^2 \tag{1.23}$$

where B is a constant that includes tortuosity of the rock. The porosity exponent of 3 in equation (1.23) is valid for very clean sandstones. Higher exponents are more appropriate of clays and shales (Bourbie et al., 1987).

In this chapter we introduced a host of definitions related to pore pressure and drilling. In Appendix B the readers will find a glossary of some of the commonly used terms introduced in this chapter as well as those used in the industry.

2 Basic Continuum Mechanics and Its Relevance to Geopressure

2.1 Introduction

Geopressure treatment relies on the joint analysis of stress and pore pressure. As our concern here is the treatment of the macroscopic behavior of earth materials, the continuum approach is appropriate to the analysis of both fields within the earth. Therefore, we begin by reviewing basic concepts and defining the general mathematical and physical entities used in the analysis and study of stresses and strains in a continuum. We then review some concepts of poromechanics – continuum mechanics applied to porous materials – that are relevant for understanding stresses and pore pressure in fluid-filled rocks and sediments in the earth. We provide here some results that can be reviewed in more detail in specialized books on the subject, such as Malvern (1969). We end the chapter with a discussion on the relevant rock physics basis for detection and estimation of geopressure.

2.2 Stresses and Forces in a Continuum

2.2.1 The Continuum Concept

When we are not concerned about the molecular structure of matter, we need a macroscopic explanation for the behavior of the material. Thus, we assign properties to the material as if it is continuously distributed throughout the volume of interest and completely fills the space it occupies. This *continuum concept* is a fundamental postulate in continuum mechanics.

> **Continuum Concept Example: Mass Density**
>
> The average density of a material is defined as
>
> $$\rho_{av} = \frac{\Delta M}{\Delta V} \qquad (2.1)$$
>
> where ΔM is the mass contained within the volume ΔV. The density of the point in the continuum inside the volume is given mathematically (in accordance with the continuum concept) as
>
> $$\rho = \lim_{\Delta V \to 0} \frac{\Delta M}{\Delta V} = \frac{dM}{dV} \qquad (2.2)$$

2.2.2   Homogeneity, Isotropy

A *homogenous* material is one having identical properties at all points. With respect to one property, a material is *isotropic* if that property is the same in all directions at a point. A material is called *anisotropic* if those properties are directionally dependent at a point.

Tensors

We have seen that a physical law must be coordinate independent. A vector is a quantity that has a direction and magnitude in an $[x_1, x_2, x_3]$ system. To transform the vector **V** into a new coordinate system, we use the transformation matrix **Q**. The vector **V′** in the new coordinate system $[x'_1, x'_2, x'_3]$ is given by

$$\underline{V}' = \underline{\underline{Q}}\,\underline{V} \qquad (2.3)$$

A vector can be defined as a first-rank tensor based on the transformation law of equation (2.3). A second-rank tensor will follow a transformation law of the form

$$\underline{\underline{\sigma}}' = \underline{\underline{Q}}\,\underline{\underline{\sigma}}\,\underline{\underline{Q}}^T \qquad (2.4)$$

To better understand tensors of second and higher rank, we analyze stress as an example for a physical law that relates force at a point (traction) to the state of stress in the solid. It will be shown that the stress is a second-rank tensor. Higher-rank tensors will be introduced through the text as needed.

Cauchy's Stress Tensor

Two types of forces can act on an object: body forces, which act everywhere within the object, resulting in a force proportional to the volume of the object (e.g., gravity, inertia), and surface forces. The body forces are given by their relation to density or mass as

$$\underline{p} = \rho\underline{b} \ \text{ or } \ p_i = \rho b_i \qquad (2.5)$$

The vector **p** is force per unit volume and **b** is force per unit mass.

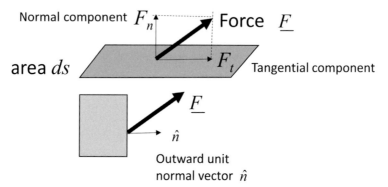

Normal component F_n Force $\underline{F}$

area ds F_t Tangential component

$\underline{F}$

$\hat{n}$

Outward unit
normal vector $\hat{n}$

Figure 2.1 Forces acting on a surface.

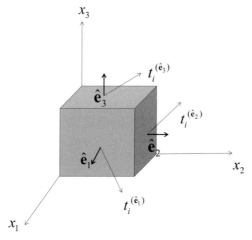

Figure 2.2 Traction forces.

Surface forces act on the surface of the volume, yielding a net force proportional to the surface area of the object.

We will define the surface forces as traction [force/unit area] as shown in Figure 2.1:

$$\underline{t}\,(\hat{n}) = \lim_{ds \to 0} \frac{\underline{F}}{ds}$$

$$t_i^{(\hat{n})} = \lim_{ds \to 0} \frac{F_i}{ds} \tag{2.6}$$

The traction has the same orientation as the force $\mathbf{F}$ and is a function of the unit normal vector $\hat{n}$. If we denote the vector $\hat{n} = \hat{e}_1, \hat{e}_2, \hat{e}_3$, then the traction vector $\mathbf{t}$ can be written as follows (Figure 2.2):

$$\underline{t}^{(\hat{e}_1)} = t_1^{\hat{e}_1}\hat{e}_1 + t_2^{\hat{e}_1}\hat{e}_2 + t_3^{\hat{e}_1}\hat{e}_3$$
$$\underline{t}^{(\hat{e}_2)} = t_1^{\hat{e}_2}\hat{e}_1 + t_2^{\hat{e}_2}\hat{e}_2 + t_3^{\hat{e}_2}\hat{e}_3$$

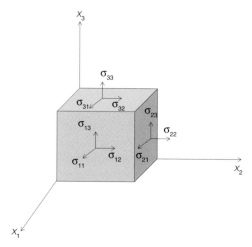

Figure 2.3 Stress components.

$$\underline{t}^{(\hat{e}_3)} = t_1^{\hat{e}_3}\hat{e}_1 + t_2^{\hat{e}_3}\hat{e}_2 + t_3^{\hat{e}_3}\hat{e}_3 \tag{2.7}$$

The stress tensor completely describes the surface forces acting on a body. We can describe the force on a surface oriented in a direction $\hat{n}$ as (Figure 2.1)

$$\underline{t} = \underline{\underline{\sigma}} \cdot \hat{n}, \ or \ t_i = \sum_{j=1}^{3} \sigma_{ij}n_j = \sigma_{ij}n_j, \ [\sigma_{ij}] = \begin{bmatrix} \sigma_{11} & \sigma_{12} & \sigma_{13} \\ \sigma_{21} & \sigma_{22} & \sigma_{23} \\ \sigma_{31} & \sigma_{32} & \sigma_{33} \end{bmatrix} \tag{2.8}$$

We note that repeated indices imply summation. Note that stress is a tensor of second order. A tensor of second order dotted into a vector yields a vector (tensor of order 1). The dot product of tensors is equivalent to the reduction in the order of the tensor. In this case, the dot product of stress (second-rank tensor) with unit normal $\hat{n}$ (vector or first-rank tensor) will be a traction vector (first-rank tensor).

The diagonal elements of the stress matrices are defined as the normal stresses, and the off-diagonal element are the shear stresses.

Equilibrium

Equilibrium at an arbitrary volume V of the continuum, subjected to a set of surface tractions t_i and body forces b_i (including inertia forces, if present), is given by the integral relation

$$\int_S t_i^n \, dS + \int_V \rho b_i \, dV = 0 \tag{2.9}$$

The divergence theorem of Gauss relates the volume integral to a surface integral so that $\int_S \underline{v} \cdot \hat{n} dS = \int_V \nabla \cdot \underline{v} dV$. This relation can be expanded to a tensor field T_{ijkl} of any rank such that $\int_S \underline{T} \cdot \hat{n} dS = \int_V \nabla \cdot \underline{T} dV$. Thus the equilibrium relation can be expressed as

$$\int_V (\sigma_{ji,j} + \rho b_i) dV = 0 \tag{2.10}$$

As this must be true for every volume V, the equilibrium relation can be stated as

$$\sigma_{ji,j} + \rho b_i = 0 \ or \ \nabla \cdot \underline{\underline{\sigma}} + \rho \underline{b} = 0 \tag{2.11}$$

Stress Tensor Symmetry

The stress tensor must be symmetric for a body to be in equilibrium and have no net rotation in it, i.e., $\sigma_{ij} = \sigma_{ji}$. Thus the number of independent stress components is six.

Transformation of Vectors and Tensors

Transformation of vectors from one coordinate system to another are given by equation (2.3). Transformation of stress from $[x_1, x_2, x_3]$ into a new coordinate system $[x_1', x_2', x_3']$ is given by $\underline{\underline{\sigma}}' = \underline{\underline{Q}} \, \underline{\underline{\sigma}} \, \underline{\underline{Q}}^T$ in equation (2.3) and proved in equation (2.12):

$$\underline{T} = \underline{\underline{\sigma}} \hat{n}$$
$$\underline{T}' = Q\underline{T} = Q(\underline{\underline{\sigma}} \hat{n}) = Q\underline{\underline{\sigma}} Q^T Q\hat{n}$$
$$Q\hat{n} \equiv \hat{n}'; \ Q\underline{\underline{\sigma}} Q^T \equiv \underline{\underline{\sigma}}' \ \Rightarrow \ \underline{T}' = \underline{\underline{\sigma}}' \hat{n}'. \tag{2.12}$$

Any symmetric tensor can be transformed into a coordinate system where the off-diagonal elements are zero. Thus, there is a coordinate system where the stress is given by a diagonal 3×3 matrix:

$$\underline{\underline{\sigma}}' = Q\underline{\underline{\sigma}} Q^T, \ [\underline{\underline{\sigma}}'_{ij}] = \begin{bmatrix} \sigma'_{11} & 0 & 0 \\ 0 & \sigma'_{22} & 0 \\ 0 & 0 & \sigma'_{33} \end{bmatrix} \tag{2.13}$$

This coordinate system is called the principal one, and $[x_1', x_2', x_3']$ are called the principal axes. The diagonal elements of the stresses in the principal axes are defined as principal stresses.

Principal Stresses

In a principal stress space (i.e., a space whose axes are in the principal stress directions) the traction is given by

$$\begin{bmatrix} \sigma_1 & & \\ & \sigma_2 & \\ & & \sigma_3 \end{bmatrix} \begin{bmatrix} n_1 \\ n_2 \\ n_3 \end{bmatrix} = \begin{bmatrix} t_1 \\ t_2 \\ t_3 \end{bmatrix} \tag{2.14}$$

or $\underline{t}^{(\hat{n})} = \underline{\underline{\sigma}} \hat{n}$

Maximum and Minimum Shear Stress

Consider the traction vector t resolved into the orthogonal normal and tangential components. The magnitude of the shearing or tangential component is given by

$$\sigma_S^2 = t_i^{(\hat{n})} t_i^{(\hat{n})} - \sigma_N^2 \tag{2.15}$$

The normal stress is given by

$$\sigma_N = (\boldsymbol{\sigma} \cdot \hat{n}) \cdot \hat{n} = \sigma_1 n_1^2 + \sigma_2 n_2^2 + \sigma_3 n_3^2 \tag{2.16}$$

Substituting equation (2.16) into (2.15), we get that the shear stress as a function of the direction cosines of the unit normal is given by

$$\sigma_S^2 = \sigma_1^2 n_1^2 + \sigma_2^2 n_2^2 + \sigma_3^2 n_3^2 - (\sigma_1 n_1^2 + \sigma_2 n_2^2 + \sigma_3 n_3^2)^2 \tag{2.17}$$

The maximum and minimum shear stress values may be obtained from equation (2.17) (Malvern, 1969). It can be shown that the minimum shear stress is zero in the principal coordinate system (i.e., where $\sigma_1, \sigma_2, \sigma_3$ are in the normal direction). The maximum shear stress occurs when the direction cosines are at $\pm 45°$.

Mohr's Circles for Stress

From equations (2.16) and (2.17) we get that

$$\sigma_N = \sigma_1 n_1^2 + \sigma_2 n_2^2 + \sigma_3 n_3^2$$
$$\sigma_N^2 + \sigma_S^2 = \sigma_1^2 n_1^2 + \sigma_2^2 n_2^2 + \sigma_3^2 n_3^2 \tag{2.18}$$

Combining equation (2.18) with the fact that $(n_1)^2 + (n_2)^2 + (n_3)^2 = 1$, we can solve for the direction cosines and get the following equations, which are the basis of Mohr's stress circles:

$$n_1^2 = \frac{(\sigma_N - \sigma_2)(\sigma_N - \sigma_3) + \sigma_S^2}{(\sigma_1 - \sigma_2)(\sigma_1 - \sigma_3)}$$
$$n_2^2 = \frac{(\sigma_N - \sigma_1)(\sigma_N - \sigma_3) + \sigma_S^2}{(\sigma_2 - \sigma_3)(\sigma_2 - \sigma_1)} \tag{2.19}$$
$$n_3^2 = \frac{(\sigma_N - \sigma_2)(\sigma_N - \sigma_1) + \sigma_S^2}{(\sigma_3 - \sigma_1)(\sigma_3 - \sigma_2)}$$

Mohr's stress circles are a convenient 2D graphical representation of the 3D stress tensor as demonstrated in Figure 2.4.

All stress points must be in the shaded region in Figure 2.4. This can be shown using geometrical reconstruction of directional cosines on a unit sphere (Malvern, 1969).

Mohr–Coulomb Failure Criteria

The Mohr–Coulomb failure criterion represents the linear envelope that is obtained from a plot of the shear strength of the material versus applied normal stress. The relation is written as

$$\sigma_c = c_0 + \sigma_N \tan\varphi \qquad (2.20)$$

where c_0 is the cohesion and φ is known as the angle of internal friction.

In its basis, Mohr–Coulomb failure is nothing but a static internal friction law where shear failure occurs when the shear force on a plane reaches the failure line σ_c. Here c_0 is the cohesion, and it defines the strength of the rock (in terms of resistance to shear failure) when the normal stress is zero. For unconsolidated sediments and soils, $c_0 = 0$, which means that under no stress, there is no resistance to shear.

Figure 2.5 demonstrates the shear failure associated with equation (2.20). When the normal stress is negative, the failure mode is considered as tensile failure, which may not follow the same line defined in equation (2.20). Also, it is important to note that failure and deformation are complicated processes that will be discussed in more detail in the next sections.

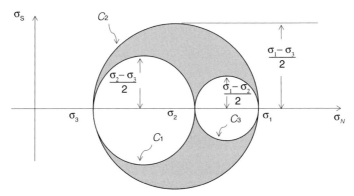

Figure 2.4 Mohr's stress circle.

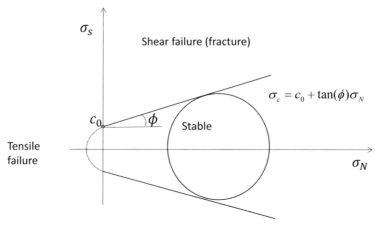

Figure 2.5 Mohr–Columb failure criteria are for stresses whose normal and shear components lie above the shear failure line.

Plane Stress

The state of stress when one principal stress is zero is known as plane stress. At this point the Mohr's stress circle will have one of the principal stresses aligned with the origin, and therefore they will plot as in Figure 2.6.

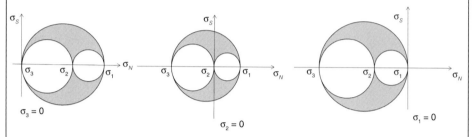

Figure 2.6 Mohr's circle and plane stress.

2.3   Deformation and Strain

In general we distinguish finite strain from infinitesimal strain. Large strains which are associated with large deformation processes in the earth such as failure and compaction often require us to use either large strain and, or discontinuity assumptions such as fractures and cracks. Small deformations are associated with the elastic region and specifically are appropriate when analyzing elastic wave propagation. When dealing with finite strain tensors, one needs to consider a Lagrangian or Eulerian description point of view and define the finite strain tensors between the undeformed and deformed configuration. In linear elasticity and especially elastodynamics it can be shown that the Eulerian and Lagrangian descriptions are equivalent (see Ben Menahem and Singh, 1980, for discussion). Further discussion on finite strain is not within the scope of this book; we refer the reader to a standard textbook on continuum mechanics for more information.

2.3.1   Infinitesimal Strains

When an elastic body is subjected to stresses, deformation results. The strain tensor describes the deformation resulting from differential motion within the body. Figure 2.7 shows the deformation possible to a body, with the initial (undeformed) and deformed Cartesian coordinate axes $[X_1, X_2, X_3]$ and $[x_1, x_2, x_3]$, respectively, sharing the same origin. The particles Q_0 and P_0 in the undeformed configuration move to Q and P in the deformed configuration. We can describe the first-order deformation by expanding the distortion of the body into a Taylor series:

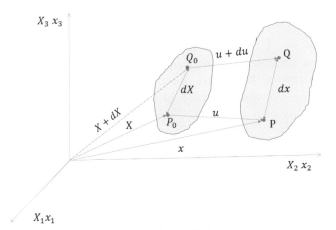

Figure 2.7 Deformation of a continuous body.

$$u_i(\underline{x} + \delta\underline{x}) \approx u_i(\underline{x}) + \frac{\partial u_i(\underline{x})}{\partial x_j}\delta\underline{x} = u_i(\underline{x}) + \delta u_i$$

$$\delta u_i = \frac{\partial u_i(\underline{x})}{\partial x_j}\delta\underline{x} \tag{2.21}$$

The relative displacement near $\underline{x}$ to the first order is δu_i.

Equation (2.21) involves the 3×3 matrix $\dfrac{\partial u_i(\underline{x})}{\partial x_j}$ that is dotted into the $\delta\underline{x}$ vector. The expression can be always decomposed into a symmetric and antisymmetric parts as follows:

$$\delta u_i = \frac{1}{2}\left[\frac{\partial u_i}{\partial x_j} + \frac{\partial u_j}{\partial x_i}\right]\delta x_j + \frac{1}{2}\left[\frac{\partial u_i}{\partial x_j} - \frac{\partial u_j}{\partial x_i}\right]\delta x_j = (e_{ij} + \omega_{ij})\delta x_j \tag{2.22}$$

where the ω_{ij} term corresponds to a rigid body rotation without deformation. The e_{ij} term is the strain tensor describing the internal deformation of the material. Its components in the $[x, y, z]$ system are:

$$[e_{ij}] = \begin{bmatrix} \dfrac{\partial u_x}{\partial x} & \dfrac{1}{2}\left[\dfrac{\partial u_x}{\partial y} + \dfrac{\partial u_y}{\partial x}\right] & \dfrac{1}{2}\left[\dfrac{\partial u_x}{\partial z} + \dfrac{\partial u_z}{\partial x}\right] \\[3mm] \dfrac{1}{2}\left[\dfrac{\partial u_y}{\partial x} + \dfrac{\partial u_x}{\partial y}\right] & \dfrac{\partial u_y}{\partial y} & \dfrac{1}{2}\left[\dfrac{\partial u_y}{\partial z} + \dfrac{\partial u_z}{\partial y}\right] \\[3mm] \dfrac{1}{2}\left[\dfrac{\partial u_z}{\partial x} + \dfrac{\partial u_x}{\partial z}\right] & \dfrac{1}{2}\left[\dfrac{\partial u_z}{\partial y} + \dfrac{\partial u_y}{\partial z}\right] & \dfrac{\partial u_z}{\partial z} \end{bmatrix} \tag{2.23}$$

The strain tensor can be expressed in a principal coordinate system as well (symmetric tensor). In this coordinate system the trace or sum of the eigenvalues (diagonal elements), is known as the *dilatation*, and it is equal to the divergence of the displacement field $u(x)$:

$$\theta = e_{ii} = \frac{\partial u_1}{\partial x_1} + \frac{\partial u_2}{\partial x_2} + \frac{\partial u_3}{\partial x_3} = \nabla \cdot \underline{u} \tag{2.24}$$

Volumetric strain: Dilatation

Dilatation represents to the first order the change in volume $V = dx_1 dx_2 dx_3$ associated with the deformation. This is shown in the derivation below:

$$V + \Delta V = \left[1 + \frac{\partial u_1}{\partial x_1}\right] dx_1 \left[1 + \frac{\partial u_2}{\partial x_2}\right] dx_2 \left[1 + \frac{\partial u_3}{\partial x_3}\right] dx_3$$

$$\approx \left[1 + \frac{\partial u_1}{\partial x_1} + \frac{\partial u_2}{\partial x_2} + \frac{\partial u_3}{\partial x_3}\right] dx_1 dx_2 dx_3$$

$$= [1 + \theta]V \Rightarrow V + \Delta V \approx [1 + \theta]V, \ \theta = \frac{\Delta V}{V} \tag{2.25}$$

Additional strain components of interest are the extensional strain, defined as the relative change in length of a material undergoing one dimensional extension.

$$\Delta L = L - L_0 = u(x + dx) - u(x) \approx \frac{\partial u}{\partial x} L_0 \Rightarrow$$

$$\frac{\Delta L}{L_0} \approx \frac{\partial u}{\partial x} \equiv e_{xx} \tag{2.26}$$

Therefore the elements on the diagonal of the strain tensors are related to the deformation of the body along the principal directions. Figure 2.8 shows dilatation and extensional strains.

Shear Strain

The off-diagonal components of the strain tensor are interpreted as pure shear. This can be demonstrated as follows: Consider the deformation described in Figure 2.9 where a square body element is deformed by angle α Then $\tan \alpha = \frac{\partial u_x}{\partial y} = \frac{\partial u_y}{\partial x}$. For small angles we get $\tan \alpha \approx \alpha \approx \frac{1}{2}\left[\frac{\partial u_x}{\partial y} + \frac{\partial u_y}{\partial x}\right]$.

2.4 Fundamental Laws of Continuum Mechanics

We state here without proofs the fundamental laws of continuum mechanics. We refer the reader to continuum mechanics textbook (e.g., Malvern, 1969) for a complete discussion.

2.4.1 Conservation of Mass

Total mass in a volume will not change in time, in the absence of sources or sinks. This can be posed mathematically as

Dilatation   Extensional strain

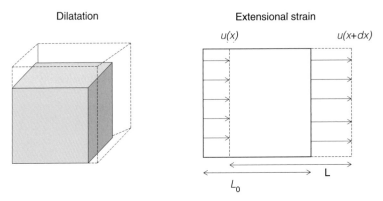

Figure 2.8 Dilation (left) is a volumetric deformation of a solid element. Extensional strain is the deformation of the element along one direction such that $du/dx \sim (L_0 - L)/L_0$.

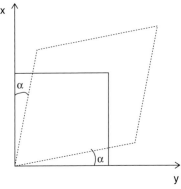

Figure 2.9 Shear strain: A square body element is deformed by angle α. Then $\tan \alpha = \dfrac{\partial u_x}{\partial y} = \dfrac{\partial u_y}{\partial x}$. For small angles we get $\tan \alpha \approx \alpha \approx \dfrac{1}{2}\left[\dfrac{\partial u_x}{\partial y} + \dfrac{\partial u_y}{\partial x}\right]$, which is the shear strain.

$$m = \int_V \rho(x,t)dV$$

$$\frac{dm}{dt} = \frac{d}{dt}\int_V \rho(x,t)dV = \int_V \left(\frac{d}{dt}\rho(x,t) + \rho\nabla \cdot v\right)dV = 0 \qquad (2.27)$$

where v is the velocity.

2.4.2   Equation of Motion

Change in momentum P is equal to the sum of body forces and surface tractions acting on the material

$$P = \int_V \rho(x, t)v dV$$

$$\int_S t^{(\hat{n})} dS + \int_V \rho(x, t)b dV = \frac{dP}{dt} \longleftrightarrow \int_V \left(\nabla \cdot \sigma + \rho(x, t)b \right) dV = \frac{dP}{dt} \tag{2.28}$$

In equilibrium we get

$$\int_V \left(\nabla \cdot \sigma + \rho(x, t)b \right) dV = 0$$

$$\nabla \cdot \sigma + \rho(x, t)b = 0 \tag{2.29}$$

or

$$\sigma_{ij,j} + \rho b_i = 0$$

where repeated indices imply summation.

2.4.3 Conservation of Angular Momentum

Change in angular momentum N is equal to the sum of surface integral over traction moments and volume integral over body force moment:

$$N = \int_V (x \times \rho v) dV$$

$$\int_S \left(x \times t^{(\hat{n})} \right) dS + \int_V \left(x \times \rho(x, t)b \right) dV = \frac{dN}{dt} \tag{2.30}$$

2.4.4 Conservation of Energy: First Law of Thermodynamics

The first law of thermodynamics relates the work done on the system and the heat transfer into the system to the change in energy of the system.

Considering K as the elastic energy in the material, U the potential energy, and total work done W, the first law of thermodynamics can be written as

$$\frac{dK}{dt} + \frac{dU}{dt} = \frac{d^*W}{dt}$$

The conservation of energy implies the following equilibrium:

$$\overbrace{\sigma : \varepsilon}^{\text{strain energy}} + \overbrace{Q}^{\text{internal heat generation}} \overbrace{- \nabla \cdot q}^{\text{heat flux}} = \overbrace{\rho \frac{du}{dt}}^{\text{kinetic energy}} \tag{2.31}$$

The term d*W/dt denotes that it is not an exact differential.

2.4.5 Equation of State, Second Law of Thermodynamics

Entropy grows with irreversible process and the change in entropy is equal to zero in reversal processes. In loose terms, entropy defines a measure of disorder in a system and is often related to the fact that conversion between mechanical energy and heat increase is imperfect.

Define L as volumetric entropy and s as entropy density. Then

$$L = \int_V \rho s \, dV$$

(2.32)

$$ds = ds^{(e)} + ds^{(i)}$$

where $ds^{(e)}, ds^{(i)}$ are increments in specific entropy due to interaction with exterior (e) and internal (i) increments. Then the second law of thermodynamics requires that

$$ds^{(i)} > 0 \ \text{(irreversible process)}$$
$$ds^{(i)} = 0 \ \text{(reversible process)}$$

(2.33)

i.e., if we assign external entropy to a system, its total entropy will increase if the process is irreversible or, if the process is reversible, will be exactly equal to the external increment.

2.4.6 Clausius–Duhem Inequality, Dissipation Function

According to the second law of thermodynamics, the rate of entropy increase is greater than or equal to the entropy input rate. This is known as the Clausius–Duhem inequality. From statistical mechanics, there is an energy dissipation function associated with irreversible processes. In a continuum undergoing a reversible process such as a perfect spring undergoing deformation and expansion there will be no energy dissipation. For irreversible thermodynamics an energy dissipation function proportional to the entropy production rate can be defined. This function is positive definite. An example of using this concept will be presented in a later section when poroelasticity is discussed.

2.5 Hooke's Law and Constitutive Equations

The relation between the stresses applied to the material and the strains resulting from it is called a *constitutive equation*. Each material behaves differently, and there are many models to describe different materials.

One of the simplest types of materials is called *linearly elastic* material. For this material the relation between stress and strain can be expressed by a linear relation called *Hooke's law*:

$$\sigma_{ij} = C_{ijkl} e_{kl} \leftrightarrow \underline{\underline{\sigma}} = \underline{\underline{C}} \ \underline{\underline{e}}$$

(2.34)

The elastic constants C_{ijkl} are called the elastic moduli of the material. C_{ijkl} is a fourth-order tensor (relates a second-order tensor to a second-order tensor in the same way that a second-order tensor relates a first-order tensor (vector) to a first-order tensor).

A fourth-order tensor in a 3D system can have 3^4 independent variables. But the elastic tensor is symmetric ($C_{ijkl} = C_{jikl}$; $C_{ijkl} = C_{jilk}$), which makes the number of independent constants to be 36 (6 independent stress components and 6 independent strain components). Further reduction in the number of independent coefficients to 21 can be shown when considering the strain energy symmetry. The strain energy is defined as

$$u^* = \frac{1}{2}\sigma_{ij}e_{ij} = \frac{1}{2}C_{ijkl}e_{kl}e_{ij} \tag{2.35}$$

While in its most general form the elastic stiffness matrix consist of 21 constants, the simplest type of elastic material is the one that is *isotropic*. For isotropic media the elastic tensor C_{ijkl} can be expressed as

$$C_{ijkl} = \lambda\delta_{ij}\delta_{kl} + \mu(\delta_{ik}\delta_{jl} + \delta_{il}\delta_{jk})$$

where λ and μ are Lamé constants. The relation between stress and strain can be expressed as

$$\sigma_{ij} = \lambda e_{kk}\delta_{ij} + 2\mu e_{ij} = \lambda\theta\delta_{ij} + 2\mu e_{ij} \tag{2.36}$$

or

$$\begin{aligned}\sigma_{ij} &= \lambda\theta + 2\mu e_{ii}, \quad i = (x,y,z) \\ \sigma_{ij} &= 2\mu e_{ij}, \qquad\quad i = (x,y,z), i \neq j\end{aligned} \tag{2.37}$$

When material properties are directional, the material is said to be anisotropic. Lame's coefficient μ, is also known as the shear modulus. In engineering notation the shear modulus is often referred to as G. To facilitate communication between the geophysics and engineering terminology, we will refer in the following chapters to the shear modulus as G while Lame's parameter μ, will be used interchangeably, but will be referred to as Lame's parameter.

2.5.1   Elastic Constants for Isotropic Media

Although Lamé coefficients are very convenient we often use other elastic moduli or constants. The most common are *Young's modulus E*, *bulk modulus K*, and *Poisson's ratio v*. These represent specific tests.

Consider a medium in which all stresses are zero but σ_{xx}. Then for positive σ_{xx}, e_{xx} will increase and e_{yy} and e_{zz} will typically decrease. *Young's modulus* is defined as

$$E = \frac{\sigma_{xx}}{e_{xx}}. \tag{2.38}$$

Poisson's ratio is defined as

$$v = -\frac{e_{yy}}{e_{xx}} = -\frac{e_{zz}}{e_{xx}} \tag{2.39}$$

To define bulk modulus, we consider a material acted upon only by a hydrostatic stress P:

$$\sigma_{xx} = \sigma_{yy} = \sigma_{zz} = -P, \quad \sigma_{xy} = \sigma_{yz} = \sigma_{zx} = 0. \tag{2.40}$$

This stress causes a dilatation θ.

Then the *bulk modulus K* is defined as

$$K = -P/\theta \tag{2.41}$$

It can be shown that for isotropic materials

$$E = \frac{\mu(3\lambda + 2\mu)}{\lambda + \mu}$$

$$K = \frac{1}{3}(3\lambda + 2\mu) \tag{2.42}$$

$$v = \frac{\lambda}{2(\lambda + \mu)}$$

2.5.2   Elastodynamics Review: Equation of Motion

The equilibrium equation of motion is based on Newton's second law: $F = ma$.

We consider a cube of volume $dV = dxdydz$ and calculate the forces acting in the x direction only (Figure 2.10).

Stresses on the front face are opposing the ones on the back face. In the x direction, for example, we have σ_{xx}, σ_{xy}, σ_{xz} on the back face and $\sigma_{xx} + \dfrac{\partial \sigma_{xx}}{\partial x}dx$, $\sigma_{xy} + \dfrac{\partial \sigma_{xy}}{\partial y}dy$, $\sigma_{xz} + \dfrac{\partial \sigma_{xz}}{\partial z}dz$ on the front face, using a Taylor series approximation. The balance of force is done by multiplying the stress by the area normal to the stress. Then the net balance is equal to the acceleration in the x direction:

$$\left[\frac{\partial \sigma_{xx}}{\partial x} + \frac{\partial \sigma_{xy}}{\partial y} + \frac{\partial \sigma_{xz}}{\partial z}\right]dxdydz = \rho\frac{\partial^2 u_x}{\partial t^2}dxdydz \tag{2.43}$$

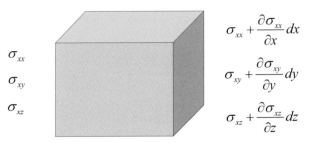

Figure 2.10 Stress balance along the x direction.

Doing the same for the y and z directions, we get a set of three scalar equations:

$$\left[\frac{\partial \sigma_{xx}}{\partial x} + \frac{\partial \sigma_{xy}}{\partial y} + \frac{\partial \sigma_{xz}}{\partial z}\right] = \rho \frac{\partial^2 u_x}{\partial t^2}$$

$$\left[\frac{\partial \sigma_{yx}}{\partial x} + \frac{\partial \sigma_{yy}}{\partial y} + \frac{\partial \sigma_{yz}}{\partial z}\right] = \rho \frac{\partial^2 u_y}{\partial t^2} \qquad (2.44)$$

$$\left[\frac{\partial \sigma_{zx}}{\partial x} + \frac{\partial \sigma_{zy}}{\partial y} + \frac{\partial \sigma_{zz}}{\partial z}\right] = \rho \frac{\partial^2 u_z}{\partial t^2}$$

Now we apply Hooke's law,

$$\begin{aligned} \sigma_{ij} &= \lambda\theta + 2\mu e_{ii}, \quad i = (x, y, z) \\ \sigma_{ij} &= 2\mu e_{ij}, \qquad i = (x, y, z), i \neq j \\ e_{ij} &= \frac{1}{2}\left(\frac{\partial u_i}{\partial x_j} + \frac{\partial u_j}{\partial x_i}\right) \end{aligned} \qquad (2.45)$$

and rewrite equation (2.44) as

$$\left[(\lambda + \mu)\frac{\partial \theta}{\partial x} + \mu \nabla^2 u_x\right] = \rho \frac{\partial^2 u_x}{\partial t^2} \qquad (2.46a)$$

$$\left[(\lambda + \mu)\frac{\partial \theta}{\partial y} + \mu \nabla^2 u_y\right] = \rho \frac{\partial^2 u_y}{\partial t^2} \qquad (2.46b)$$

$$\left[(\lambda + \mu)\frac{\partial \theta}{\partial z} + \mu \nabla^2 u_z\right] = \rho \frac{\partial^2 u_z}{\partial t^2} \qquad (2.46c)$$

The displacement field **u** satisfies equation (2.46). Taking the divergence of equation (2.46) and summing the three equations, we can derive the classical wave equation for the dilatation θ:

$$\rho \frac{\partial^2 \theta}{\partial t^2} = (\lambda + 2\mu)\nabla^2 \theta \qquad (2.47)$$

We can also derive another wave equation from equation (2.46) by taking the derivative of equation (2.46b) with respect to z, $\left(\frac{\partial}{\partial z}\right)$ and subtracting from $\frac{\partial}{\partial y}$ (equation (2.46c)). Then we get

$$\rho \frac{\partial^2}{\partial t^2}\left(\frac{\partial u_z}{\partial y} - \frac{\partial u_y}{\partial z}\right) = \mu \nabla^2\left(\frac{\partial u_z}{\partial y} - \frac{\partial u_y}{\partial z}\right) \qquad (2.48)$$

Recalling that $\underline{w} \equiv \nabla \times \underline{u}$, we see that equation (2.48) is a wave equation for the rotation:

$$\rho \frac{\partial^2}{\partial t^2} w_x = \mu \nabla^2 w_x \qquad (2.49)$$

The two wave equations (2.47) and (2.49) have the form of the classic wave equation,

$$\frac{\partial^2}{\partial t^2}\psi = c^2 \nabla^2 \psi \tag{2.50}$$

and the velocity of the dilatational wave is given by V_p and the rotational wave by V_s:

$$\begin{aligned}
\alpha^2 &= \frac{(\lambda + 2\mu)}{\rho} = V_P^2 \\
\beta^2 &= \frac{\mu}{\rho} = V_s^2
\end{aligned} \tag{2.51}$$

Note that $V_p > V_s$ always!

P-waves are longitudinal waves with particle displacement in the direction of propagation. The shear waves (S-waves) are polarized like electromagnetic (EM) waves, perpendicular to the direction of propagation. P- and S-wave trains from a three-component experiment are shown in Figure 2.11. S-waves are the tangential displacement field, and P-waves are visible on the vertical displacement field (after Bachrach et al., 2000). Both P- and S-waves are useful for geopressure prediction. We will discuss these in subsequent chapters.

2.5.3 Anisotropic Elasticity

Hooke's law states that for sufficiently small stresses, strain is proportional to stress. For an anisotropic medium, Hooke's law may be written as

$$\varepsilon_{ij} = S_{ijkl}\sigma_{kl} \tag{2.52a}$$

$$\sigma_{ij} = C_{ijkl}\varepsilon_{kl} \tag{2.52b}$$

The Matrix Notation

The tensors $\varepsilon, S, \sigma, C$ possess the following symmetries:

$$\begin{aligned}
\varepsilon_{ij} &= \varepsilon_{ji}, \ \sigma_{ij} = \sigma_{ji} \\
S_{ijkl} &= S_{jikl}, \ S_{ijlk} = S_{jikl}, \\
C_{ijkl} &= C_{jikl}, \ C_{ijlk} = C_{jikl}
\end{aligned} \tag{2.53}$$

This reduces the number of independent elements of **C** and S from 81 to 36 and makes it possible to introduce the conventional matrix notation (Voigt, 1910; Nye, 1985) in which pairs of subscripts ij and kl are abbreviated by single subscripts according to the convention

$$11 \rightarrow 1, 22 \rightarrow 2, 33 \rightarrow 3, 23,32 \rightarrow 4, 31,13 \rightarrow 5, 12,21 \rightarrow 6 \tag{2.54}$$

Each of the tensor elements ε_{ij}, σ_{ij} are associated with the following matrix elements:

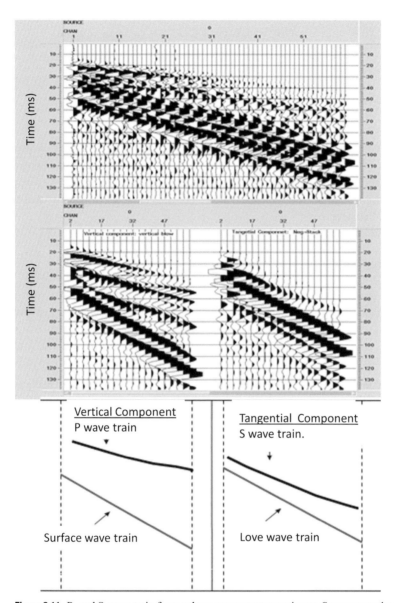

Figure 2.11 P- and S-wave train from a three-component experiment. S-waves are the tangential displacement field, and P-waves are visible on the vertical displacement field. After Bachrach et al. (2000).

$$
\begin{bmatrix} \sigma_{11} & \sigma_{12} & \sigma_{13} \\ \sigma_{21} & \sigma_{22} & \sigma_{23} \\ \sigma_{31} & \sigma_{32} & \sigma_{33} \end{bmatrix} \rightarrow \begin{bmatrix} \sigma_1 & \sigma_6 & \sigma_5 \\ \sigma_6 & \sigma_2 & \sigma_4 \\ \sigma_5 & \sigma_4 & \sigma_3 \end{bmatrix}
$$

$$
\begin{bmatrix} \varepsilon_{11} & \varepsilon_{12} & \varepsilon_{13} \\ \varepsilon_{21} & \varepsilon_{22} & \varepsilon_{23} \\ \varepsilon_{31} & \varepsilon_{32} & \varepsilon_{33} \end{bmatrix} \rightarrow \begin{bmatrix} \varepsilon_1 & \frac{1}{2}\varepsilon_6 & \frac{1}{2}\varepsilon_5 \\ \frac{1}{2}\varepsilon_6 & \varepsilon_2 & \frac{1}{2}\varepsilon_4 \\ \frac{1}{2}\varepsilon_5 & \frac{1}{2}\varepsilon_4 & \varepsilon_3 \end{bmatrix} \tag{2.55}
$$

$$C_{ijkl}, \ (i,j,k,l=1,2,3)=C_{mn} \ (m,n=1,..,6) \tag{2.56}$$

$$
\begin{aligned}
&S_{ijkl}, \ (i,j,k,l=1,2,3) \to S_{mn} \quad (m \ \& \ n=1,2,3)\\
&S_{ijkl}, \ (i,j,k,l=1,2,3) \to \frac{1}{2}S_{mn} \ (\text{either } m \text{ or } n=4,5,6)\\
&S_{ijkl}, \ (i,j,k,l=1,2,3) \to \frac{1}{4}S_{mn} \ (m \ \& \ n=4,5,6)
\end{aligned} \tag{2.57}
$$

The factors of 2 and 4 in equations (2.55) and (2.57) are probably responsible for much of the confusion regarding the relations between the elastic stiffness components and the elastic compliance components. Alternative choices (such as the Kelvin notation) are possible, but the Voigt scheme outlined above is the most commonly used convention in the literature.

With this convention, equations (2.52a) and (2.52b) take the form

$$
\begin{aligned}
\varepsilon_i &= S_{ij}\sigma_j\\
\sigma_i &= C_{ij}\varepsilon_j
\end{aligned} \tag{2.58}
$$

Finally, if there exists an elastic potential, the following additional symmetries hold,

$$
\begin{aligned}
S_{ij} &= S_{ji}\\
C_{ij} &= C_{ji}.
\end{aligned} \tag{2.59}
$$

and the number of independent elements of S_{ijkl} and C_{ijkl} is further reduced from 36 to 21 for the most general anisotropic linear elastic medium. This number is reduced further if the medium displays symmetry. The most common types of symmetry used to describe geological materials are *isotropy*, *transverse isotropy*, and *orthotropy* (*orthorhombicity*) and are described further below.

Isotropy in Voigt's Notation

An isotropic material has two independent elastic constants, with the elastic stiffness matrix taking the form

$$
C_{ij} = \begin{bmatrix}
C_{11} & C_{12} & C_{12} & 0 & 0 & 0\\
C_{12} & C_{11} & C_{12} & 0 & 0 & 0\\
C_{12} & C_{12} & C_{11} & 0 & 0 & 0\\
0 & 0 & 0 & C_{44} & 0 & 0\\
0 & 0 & 0 & 0 & C_{44} & 0\\
0 & 0 & 0 & 0 & 0 & C_{44}
\end{bmatrix}, \ C_{11}-C_{12}-2C_{44}=0 \tag{2.60}
$$

Hooke's law, using the Lamé elastic constants λ and μ for the components of the elastic stiffness tensor, may be represented in the form

$$
\begin{bmatrix}
\lambda+2\mu & \lambda & \lambda & 0 & 0 & 0\\
\lambda & \lambda+2\mu & \lambda & 0 & 0 & 0\\
\lambda & \lambda & \lambda+2\mu & 0 & 0 & 0\\
0 & 0 & 0 & \mu & 0 & 0\\
0 & 0 & 0 & 0 & \mu & 0\\
0 & 0 & 0 & 0 & 0 & \mu
\end{bmatrix}
\begin{bmatrix}
\varepsilon_{11}\\
\varepsilon_{22}\\
\varepsilon_{33}\\
2\varepsilon_{23}\\
2\varepsilon_{13}\\
2\varepsilon_{12}
\end{bmatrix}
=
\begin{bmatrix}
\sigma_{11}\\
\sigma_{22}\\
\sigma_{33}\\
\sigma_{23}\\
\sigma_{13}\\
\sigma_{12}
\end{bmatrix} \tag{2.61}
$$

In geomechanics, Young's modulus E and Poisson's ratio v are more commonly used and arise naturally from the compliance form of the elastic tensor, as follows:

$$S_{11} = 1/E, \quad S_{12} = -v/E$$

The elastic compliance matrix for an isotropic medium takes the form

$$S_{ij} = \begin{bmatrix} S_{11} & S_{12} & S_{12} & 0 & 0 & 0 \\ S_{12} & S_{11} & S_{12} & 0 & 0 & 0 \\ S_{12} & S_{12} & S_{11} & 0 & 0 & 0 \\ 0 & 0 & 0 & S_{44} & 0 & 0 \\ 0 & 0 & 0 & 0 & S_{44} & 0 \\ 0 & 0 & 0 & 0 & 0 & S_{44} \end{bmatrix}, \quad S_{11} - S_{12} - S_{44}/2 = 0 \quad (2.62)$$

In terms of the E and v, the elastic compliance tensor may be represented in the form

$$S_{ij} = \begin{bmatrix} 1/E & -v/E & -v/E & 0 & 0 & 0 \\ -v/E & 1/E & -v/E & 0 & 0 & 0 \\ -v/E & -v/E & 1/E & 0 & 0 & 0 \\ 0 & 0 & 0 & 1/G & 0 & 0 \\ 0 & 0 & 0 & 0 & 1/G & 0 \\ 0 & 0 & 0 & 0 & 0 & 1/G \end{bmatrix}, \quad G = \frac{E}{2(1+v)} \quad (2.63)$$

The compliance formulation is widely used in geomechnics and engineering, so here we use the shear modulus notation G which is equal to the Lame's parameter μ.

Transverse Isotropy
A transversely isotropic (TI) medium as shown in Figure 2.12 has an axis of rotational symmetry that will be taken to define the x_3 axis. A transversely isotropic medium has five independent elastic constants, the nonvanishing components of the elastic stiffness matrix being $C_{11} = C_{22}$, $C_{12}, C_{13} = C_{23}$, $C_{44} = C_{55}$, $C_{66} = (C_{11} - C_{12})/2$. The elastic stiffness matrix is then

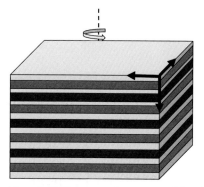

Figure 2.12 Transverse isotropy resulting from layering. When the symmetry axis is vertical, this is known as vertical transverse isotropy, or VTI. VTI is a common model for layered media.

$$
C_{ij} = \begin{bmatrix}
C_{11} & C_{12} & C_{13} & 0 & 0 & 0 \\
C_{12} & C_{11} & C_{13} & 0 & 0 & 0 \\
C_{13} & C_{13} & C_{33} & 0 & 0 & 0 \\
0 & 0 & 0 & C_{55} & 0 & 0 \\
0 & 0 & 0 & 0 & C_{55} & 0 \\
0 & 0 & 0 & 0 & 0 & C_{66}
\end{bmatrix}, \quad C_{11} - C_{12} - 2C_{66} = 0 \qquad (2.64)
$$

It is convenient to denote the velocity of a wave traveling along axis x_i and having polarization along axis x_j by V_{ij}. Thus V_{33} is the velocity of a compressional wave propagating along axis x_3 and having polarization along axis x_3, while V_{12} is the velocity of a shear wave propagating along axis x_1 and having polarization along axis x_2. The relations between the V_{ij} and the C_{ij} are as follows:

$$
V_{11} = V_{22} = \sqrt{\frac{C_{11}}{\rho}}, \quad V_{33} = \sqrt{\frac{C_{33}}{\rho}}
$$
$$
\qquad (2.65)
$$
$$
V_{12} = V_{21} = \sqrt{\frac{C_{66}}{\rho}}, \quad V_{31} = V_{13} = V_{23} = V_{32} = \sqrt{\frac{C_{55}}{\rho}}
$$

For a transversely isotropic medium, the elastic compliance matrix is

$$
S_{ij} = \begin{bmatrix}
S_{11} & S_{12} & S_{13} & 0 & 0 & 0 \\
S_{12} & S_{11} & S_{13} & 0 & 0 & 0 \\
S_{13} & S_{13} & S_{33} & 0 & 0 & 0 \\
0 & 0 & 0 & S_{55} & 0 & 0 \\
0 & 0 & 0 & 0 & S_{55} & 0 \\
0 & 0 & 0 & 0 & 0 & S_{66}
\end{bmatrix}, \quad S_{11} - S_{12} - S_{66}/2 = 0 \qquad (2.66)
$$

The S_{ij} follow from inversion of equation (2.64).

If E_i denotes the Young's modulus corresponding to axis x_i, v_{ij} denotes the Poisson's ratio that relates the expansion along axis x_j to a compression applied along axis x_i, and G_{ij} denotes the shear modulus in the plane x_i, x_j, the elastic compliance matrix for a transversely isotropic medium can be written as

$$
S_{ij} = \begin{bmatrix}
1/E_1 & -v_{21}/E_2 & -v_{31}/E_2 & 0 & 0 & 0 \\
-v_{21}/E_2 & 1/E_1 & -v_{32}/E_2 & 0 & 0 & 0 \\
-v_{13}/E_2 & -v_{23}/E_2 & 1/E_3 & 0 & 0 & 0 \\
0 & 0 & 0 & 1/G_{23} & 0 & 0 \\
0 & 0 & 0 & 0 & 1/G_{31} & 0 \\
0 & 0 & 0 & 0 & 0 & 1/G_{12}
\end{bmatrix} \qquad (2.67)
$$

Because of the symmetry of equations (2.66) and (2.67) can be simplified using the notation

$$S_{ij} = \begin{bmatrix} 1/E & -v/E_2 & -v'/E_2 & 0 & 0 & 0 \\ -v/E_2 & 1/E & -v'/E_2 & 0 & 0 & 0 \\ -v'/E_2 & -v'/E_2 & 1/E' & 0 & 0 & 0 \\ 0 & 0 & 0 & 1/G' & 0 & 0 \\ 0 & 0 & 0 & 0 & 1/G' & 0 \\ 0 & 0 & 0 & 0 & 0 & 1/G \end{bmatrix} \tag{2.68}$$

where

$$\begin{aligned} E_1 &= E_2 = E \\ E_3 &= E' \\ v_{12} &= v_{21} = v \\ v_{31} &= v_{32} = v' \\ G_{23} &= G_{13} = G' \\ G_{12} &= G = \frac{E}{2(1+v)} \end{aligned} \tag{2.69}$$

Since the x_1, x_2 plane is a plane of isotropy, the last of equations (2.69) has the same form as equation (2.63). In view of this equation, the five independent elastic stiffnesses appearing in equations (2.69) can be considered to be the two Young's moduli E and E', the two shear moduli G and G', and Poisson's ratio v'.

Thomsen's Notation

The following are useful parameters (Thomsen, 1986) that map anisotropy in TI media (see Figure 2.13):

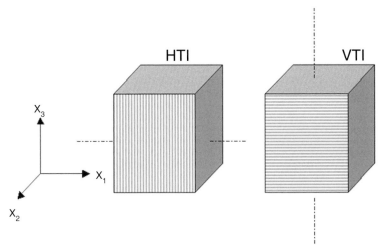

Figure 2.13 HTI and VTI media. In HTI media the axis of symmetry is the x_1 direction, whereas in VTI media the axis of symmetry is the x_3 direction. Note that vertically fractured media are often conceptualized as having HTI properties, while layered shales and shaley sands are often assumed to have VTI properties.

$$\alpha = \sqrt{\frac{c_{33}}{\rho}} \qquad \varepsilon = \frac{c_{11} - c_{33}}{2c_{33}}$$

$$\beta = \sqrt{\frac{c_{44}}{\rho}} \qquad \gamma = \frac{c_{66} - c_{44}}{2c_{44}} \tag{2.70}$$

$$\delta = \frac{(c_{13} + c_{44})^2 - (c_{33} - c_{44})^2}{2c_{33}(c_{33} - c_{44})}$$

Orthorhombic (Orthotropic) Media

Characterized by nine independent constants in its principal coordinate system, orthorhombic symmetry is a reasonable geological model for stressed or fractured layered media. When the layering and fractures (or stress direction) are orthogonal the system will have orthorhombic symmetry (see Figure 2.14).

The elastic stiffness matrix for orthorhombic symmetry is given by

$$\begin{bmatrix} c_{11} & c_{12} & c_{13} & & & \\ c_{12} & c_{22} & c_{23} & & & \\ c_{13} & c_{23} & c_{33} & & & \\ & & & c_{44} & & \\ & & & & c_{55} & \\ & & & & & c_{66} \end{bmatrix} \tag{2.71}$$

2.6   Elasticity, Stress Path, and Rock Mechanics

Linear elastic media is an idealized model for rocks. When real rocks are deformed stress–strain relations are rarely simple and linear. If we consider for example a typical uniaxial deformation test on a real rock (as demonstrated in Figure 2.15, after Gueguen and Palciauskas, 1994) where a plot of the uniaxial strain versus confining stress is presented we observe the following: initially, the rock will have a *nonlinear* response until it reaches

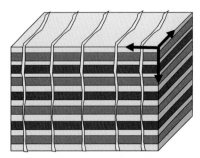

Figure 2.14 Example of orthorhombic symmetry: A single set of parallel vertical fractures embedded in a VTI medium generates orthorhombic symmetry. This special case of orthorhombic media is known as vertically fractured TI media (Schoenberg and Sayers, 1995).

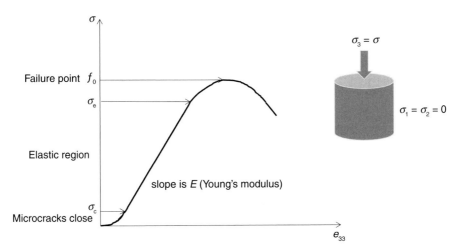

Figure 2.15 Typical stress–strain curve for a rock under uniaxial stress. A cylindrical plug subjected to vertical confining stress σ versus vertical strain. See text for details.

a certain stress σ_c, which is the stress at which small cracks will close. Above this stress the rock will have a fairly linear response until it starts deviating from the linear response. The stress at this point is σ_e which defines the transition from approximate linearity of stress to a clearly nonlinear curve. As stress increases further, the rock will not behave in a linear fashion although in general the material supports the applied stress and further loading of the sample will cause additional strain. Additional loading will bring the rock sample into its failure point f_0. Beyond this point the rock cannot support further loading. At and after failure there will be strain increase without additional loading.

We define a medium as being elastic if the stress–strain relations are reversible (i.e., if the path in the stress–strain plane is the same regardless of the directionality and/or history of the material). A medium is defined as being inelastic if the stress and the strain are path dependent as illustrated in Figure 2.16. When the material is inelastic, there will be a plastic strain (e_p), which is defined as the difference between the two loading paths. In rock mechanics it is common to perform loading cycles for the rock sample where the media will be loaded up to a certain stress level, then unloaded and again loaded. Typically when rocks and soils are being stressed via several cycles of loading and unloading plastic strain will accumulate. Constitutive laws for plasticity and elastoplasticity will be defined later after discussing failure models for rocks and soils.

2.6.1   Static and Dynamic Elastic Moduli

Measurements of elastic constants for the rock can be done using measurements of elastic wave velocities or by deforming a rock in the laboratory. There are fundamental differences between the elastic properties measured using elastic wave propagation and using deformation. The strain amplitude associated with

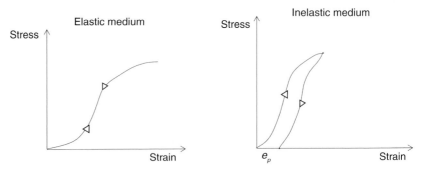

Figure 2.16 Stress path in elastic and inelastic media.

elastic wave propagation is of the order of 10^{-6} to 10^{-10} while for laboratory testing using rock deformation methods the strain amplitudes are of the order of 10^{-2} to 10^{-4}. It is important to note that in general the static and dynamic measurements are not the same for the same rock as when strain amplitudes are large rocks will become nonlinear which often will cause the elastic stiffness to be lower than the one measured by dynamic methods with small strain amplitudes. It is also important to bear in mind that different elastic constants may be used for different applications. For example, finite element modeling in geomechanics is done typically using static elastic moduli as the deformation modeled is large. Seismic velocity constants used for seismic imaging are always dynamic properties. Mapping dynamic to static properties is typically done using empirical methods which are based on many measurements in the laboratory. In many cases first-order linear regression is used to map static to dynamic moduli. The static modulus Π_{static} is related to the dynamic modulus Π_{dynamic} via a linear equation of the type $\Pi_{\text{static}} = A\,\Pi_{\text{dynamic}} + B$. Π is typically Young's modulus (E), bulk modulus (K), shear modulus (G), or Poisson's ratio v. A and B are derived from statistical regression of laboratory data. For example: Eissa and Kazi (1988) reported that for many rock types $E_{\text{static}} = 0.74\,E_{\text{dynamic}} - 0.82$. McClan and Entwisle (1992) reported that for granites and Jurassic sediments $E_{\text{static}} = 0.69E_{\text{dynamic}} + 6.4$. Mese and Dvorkin (2000) reported $E_{\text{static}} = 0.29\,E_{\text{dynamic}} - 1.1$ and $E_{\text{static}} = -0.0.00743\,E_{\text{dynamic}} + 0.34$ for shales and shaley sands. In all of the relations above all units are in GPa. For more details we refer the reader to Mavko et al. (2009). Figure 2.17 shows an example of quasi-static loading and dynamic measurements on Gulf of Mexico sands (Zimmer et al., 2002). Note that the slope associated with the dynamic measurements is steeper than the slope associated with quasi-static loading.

2.7   Poroelasticity

Poroelasticity is a joint formulation for the behavior of solid – fluid coupled porous system. Poroelasticity describes the behavior of porous continuum and pore fluid with respect to applied displacements and stresses. Poroelasticity jointly treats the solid frame and pore fluid to calculate changes in both pore pressure, solid displacements and fluid displacements due to stresses and displacements associated with external or internal processes.

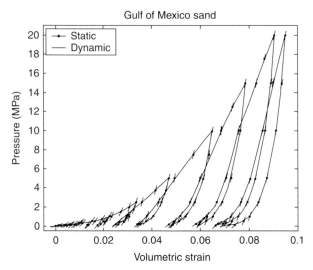

Figure 2.17 Static and dynamic Young's modulus for GOM sands at different loading stages. Data from Zimmer et al. (2002). Loading occurs when pressure increases while unloading occurs when the pressure is decreased. Dynamic modulus is the slope of the short line during the different loading stages. Note that the difference between the dynamic and static slopes (which represent the elastic modulus) is higher during loading. The unloading dynamic modulus is closer to the static one.

Thus, the theory of poroelasticity provides a theoretical base for geopressure analysis where both stress and pore pressure are coupled.

The theory of poroelasticity is based on the work of Terzaghi (1943), Biot (1941, 1956, 1962), and Gassmann (1951). Poroelasticity is often called Biot theory as the work of M. A. Biot is considered to have established the foundation for the theory. Modern treatments of poroelasticity can be found in books such as Cheng (2016), Coussy (2004), and Wang (2000). We will consider here some of the main results from poroelasticity. Their use in the study and analysis of geopressure will be demonstrated in later chapters.

2.7.1   Effective Stress

We define two effective stress formulas:

1. Terzaghi's effective stress (also known as differential pressure):

$$\sigma_{ij}^{eff} = \sigma_{ij} - P\delta_{ij} \qquad (2.72)$$

2. Biot's effective stress defined as

$$\sigma_{ij}^{eff} = \sigma_{ij} - \alpha P\delta_{ij}$$
$$\alpha = 1 - \frac{K_d}{K_0} \qquad (2.73)$$

where K_d is defined as the frame "dry" modulus and will be further discussed below. When $\alpha = 1$ (for soft materials $K_d \ll K_0$) the effective stress is equal to the differential stress (often called Terzaghi's effective stress).

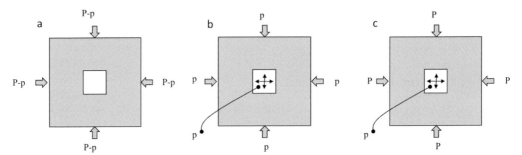

Figure 2.18 Biot effective stress derived from basic superposition principles (after Nur and Byerlee, 1971). Considering three states of solid subject to confining stress and pore pressure, one can use the superposition principle to derive Biot's effective stress law. See text for details.

A physical interpretation of Biot's effective stress law as shown in Figure 2.18b can be done as follows (Nur and Byerlee, 1971): Figure 2.18 shows three states of a material under confining and internal pore pressure. Consider a porous solid subjected to a pore pressure p and confining external hydrostatic pressure P. State (c) is obtained by superposing state (a) where pressure $P - p$ is applied on the external surface and state (b) where pressure p is applied both on the external and internal surfaces. Then from equation (2.41) the dilatation of state (a) is given by $\left(\dfrac{\delta V}{V}\right)^a = \theta^a = -\dfrac{P - p}{K}$. The dilatation of state (b) is given by $\left(\dfrac{\delta V}{V}\right)^b = \theta^b = -\dfrac{p}{K_0}$. Therefore the dilatation of state (c) is given by the principle of superposition as $\left(\dfrac{\delta V}{V}\right)^c = \theta^c = \theta^a + \theta^b = -\dfrac{1}{K}(P - \alpha p) = -\dfrac{P*}{K}$. The entity $P*$ is the effective stress that causes a given volumetric deformation of the porous system in Figure 2.18 whose effective bulk modulus is K. We will later demonstrate that the equation of linear poroelasticity can be formulated in a similar manner to the equations of linear elasticity when Biot's effective stress is considered.

2.7.2   Drained and Undrained Deformation

For a fully saturated porous material we consider two fundamental states (or thought experiments): *drained* and *undrained* (sometimes these are called *jacketed* and *unjacketed*) conditions.

Drained conditions are achieved when pore pressure in a representative elementary volume (REV) is in equilibrium with the surrounding throughout the deformation of the saturated porous medium. A special case (described as a lab experiment) is when pore pressure and fluid can freely diffuse in and out of the sample during the deformation process. Undrained (or jacketed) conditions refer to the condition in which the boundary of the porous solid (REV) is impermeable and pore fluid cannot exit the rock during the deformation process.

Physical measurements on both states (laboratory measurements) will determine the poroelastic coefficients of the rock. Often "dry conditions" are associated with the drained experiment. In poroelasticity, the term "dry" elastic modulus refers to the drained experiment and not the elastic modulus when the material is dry. This is an important distinction to remember.

2.8 Linear Stress–Strain Formulation for Poroelastic Media (Static Poroelasticity)

Following Biot (1962), we consider the displacement vector in the solid as $\mathbf{u} = [u_x, u_y, u_z]$ and the average fluid displacement vector in the pore space as $\mathbf{U} = [U_x, U_y, U_z]$. The volume of fluid displaced through a unit area normal to the x, y, z, direction is $\phi \mathbf{U}$, where ϕ is the porosity. The poroelastic stress–strain relations, which relates stress σ and pore pressure P to displacement and strain e can be written in an abbreviated notation as

$$\sigma_{ij} = 2\mu e_{ij} + \delta_{ij}(\lambda_c \theta - \alpha M \zeta)$$
$$P = -\alpha M \theta + M \zeta \tag{2.74}$$
$$\theta = \nabla \cdot \mathbf{u}, \quad \zeta = \phi \nabla \cdot (\mathbf{u} - \mathbf{U})$$

where λ_c is the Lamé coefficient for the closed (undrained) system and μ is the shear modulus. Note that the shear strain and stress are not affected by pore pressure and fluid presence for the static (low frequency) case. In the above equations M and α are the two poroelastic constants associated with the poroelastic stress–strain formulation. M is known as Biot modulus, or P-wave modulus, and relates dry response to saturated response as follows:

$$\lambda_C = \lambda_{dry} + \alpha^2 M,$$
$$\frac{1}{M} = \frac{\alpha - \phi}{K_0} + \frac{\phi}{K_f} \tag{2.75}$$

where K_f is the pore fluid bulk modulus and ϕ is the porosity; α is the Biot parameter and is defined as

$$\alpha = 1 - K_{dry}/K_0 \tag{2.76}$$

where K_{dry} is the bulk modulus of the rock measured under drained conditions and K_0 is the bulk modulus of the frame mineral.

Note that when the system is jacketed, i.e., $\zeta = 0$, and there is no relative displacement between the fluid and matrix, the pore pressure and the total stress are related, and the bulk modulus is defined in terms of "closed" elastic properties. When the matrix is open, the assumption is that there is a constant pore pressure in the system (pore pressure is in equilibrium with the surroundings) and thus fluid can flow freely in and out of the system. One special case of such conditions is when the pore pressure is equal to zero. The term "dry" frame conditions often refers to the states of zero pore pressure.

The poroelastic linear stress–strain relation can be written in terms of pore pressure as the coupling variable rather than the relative fluid displacement (or divergence). A popular form of the linear isotropic poroelastic stress–strain relations is given by

$$\sigma_{ij} + \alpha P \delta_{ij} = 2\mu e_{ij} + \lambda_{\mathrm{dry}} \theta \delta_{ij}$$
$$\zeta = (1/M)P + \alpha\theta \tag{2.77}$$

Note that in Biot's (1962) notation, the pressure direction is assumed to be negative, i.e., the pore pressure resists a positive confining stress which push the grains together.

The above equations can be written as an equivalent linear elasticity relation by substituting the total stress with Biot's effective stress $\sigma_{ij}^{\mathrm{eff}} = \sigma_{ij} + \alpha P \delta_{ij}$ as

$$\sigma_{ij}^{\mathrm{eff}} = 2\mu e_{ij} + \lambda_{\mathrm{dry}} \theta \delta_{ij}$$
$$\zeta = (1/M)P + \alpha\theta \tag{2.78}$$

In terms of the differential pressure (also known as Terzaghi's effective stress or just the "effective" stress) defined as $\sigma_{ij}' = \sigma_{ij} + P\delta_{ij}$ the linear stress–strain relations are given by

$$\sigma_{ij}' - (1 - \alpha)P\delta_{ij} = 2\mu e_{ij} + \lambda_{\mathrm{dry}} \theta \delta_{ij}$$
$$\zeta = (1/M)P + \alpha\theta \tag{2.79}$$

When $\alpha = 1$ (soft materials where $K_{\mathrm{dry}} << K_0$), the effective stress is equal to the differential stress (often called Terzaghi's effective stress).

Another equivalent form of the linear isotropic poroelastic stress–strain relations is given as

$$\sigma_{xx} = 2G\left(e_{xx} + \frac{v}{1-2v}\theta\right) - \alpha P$$

$$\sigma_{yy} = 2G\left(e_{yy} + \frac{v}{1-2v}\theta\right) - \alpha P$$

$$\sigma_{zz} = 2G\left(e_{zz} + \frac{v}{1-2v}\theta\right) - \alpha P \tag{2.80}$$

$$\sigma_{xy} = 2Ge_{xy}, \sigma_{xz} = 2Ge_{xz}, \sigma_{yz} = 2Ge_{yz}$$

Here G is the shear modulus. Note that equation (2.80) is exactly equal to the regular (linear elastic) stress–strain relations if we substitute the stress with Biot's effective stress. Equation (2.80) is heavily used in geotechnical engineering, soil mechanics and geomechanics.

2.8.1   Gassmann's Equation and Fluid Substitution

Gassmann's equation (Gassmann, 1951) relates the elastic constants of saturated material in closed or undrained state to the drained or open conditions. In its isotropic form Gassmann's equation can be written as

$$\frac{K_{sat}}{K_0 - K_{sat}} = \frac{K_{dry}}{K_0 - K_{dry}} + \frac{K_{fl}}{\phi(K_0 - K_{fl})} \tag{2.81}$$

$$G_{sat} = G_{dry}$$

where K_{sat} refers to the saturated bulk modulus of the closed system, K_{fl} is the pore fluid bulk modulus, and ϕ is the porosity. The shear modulus for the open system (dry) and the closed system are the same $G_{sat} = G_{dry}$ as fluid does not contribute to the shear stiffness of the system under static (low frequency) conditions. Gassmann's equation can be derived directly from Biot's equations or from basic elasticity considerations by accounting for mass conservation of pore fluid in a closed system. Gassmann's equation assumes that the pore fluid is in pressure equilibrium throughout the sample and that the pore fluid does not interact with the mineral frame. For more information, see Mavko et al. (2009).

The anisotropic extension to Gassmann's equation was given by Gassmann himself and later derived in a more general form by Brown and Korringa (1975) in terms of the elastic compliance tensor, as given below:

$$S_{ijkl}^{dry} - S_{ijkl}^{sat} = \frac{(S_{ij\alpha\alpha}^{dry} - S_{ij\alpha\alpha}^{0})(S_{kl\alpha\alpha}^{dry} - S_{kl\alpha\alpha}^{0})}{(S_{\alpha\alpha\beta\beta}^{dry} - S_{\alpha\alpha\beta\beta}^{0}) + \phi(\beta_{fl} - \beta_0)} \tag{2.82}$$

where S_{ijkl}^{dry}, and S_{ijkl}^{sat} are the dry and saturated fourth-order compliance tensors, β_{fl} is the fluid compressibility (reciprocal of bulk modulus), and β_0 is the compressibility of the mineral frame.

2.8.2   Fluid Substitution

One of the many practical applications of Gassmann's equation is the ability to predict the elastic modulus of a rock saturated with fluid 2 given the elastic moduli of the rock saturated with a different fluid, say, fluid 1. One example is the prediction of seismic velocity of a gas-saturated sediment given the seismic velocity of the same sediment saturated with brine. This fluid substitution scheme is readily derived from Gassmann's equation:

$$\frac{K_{sat1}}{K_0 - K_{sat1}} - \frac{K_{fl1}}{\phi(K_0 - K_{fl1})} = \frac{K_{sat2}}{K_0 - K_{sat2}} - \frac{K_{fl2}}{\phi(K_0 - K_{fl2})} \tag{2.83}$$

$$\mu_{sat2} = \mu_{sat1}$$

The subscripts 1 and 2 in the equation 2.83 refer to the two different fluids.

2.8.3   Skempton's Coefficient

If the poroelastic system is closed (jacketed) the pore pressure and the external stress are directly related as

$$\theta = \frac{(\bar{\sigma} - \alpha P)}{K_{dry}} \rightarrow P = \frac{1}{\alpha}(\bar{\sigma} - K_{dry}\theta), \ \bar{\sigma} = \sigma_{kk}/3 \tag{2.84}$$

$$\theta = e_{kk} = \frac{\bar{\sigma}}{K_{sat}} \rightarrow P = \frac{1}{\alpha}(K_{sat} - K_{dry})\theta \rightarrow P = B\bar{\sigma}, \ B = \frac{1}{K_{sat}}\left(\frac{K_{sat} - K_{dry}}{\alpha}\right) \tag{2.85}$$

Skempton's coefficient B relates directly the pore pressure to the confining stress, and is a quantity commonly measured in the laboratory. Therefore, knowledge of Skempton's coefficient can be used to estimate pore pressure in porous formation under the jacketed assumption. Often this assumption is not met in the field. However, a jacketed system will have a higher pore pressure response when compared to the un-jacketed system and therefore, Skempton-derived pore pressure estimates can serve as an upper bound for pore pressure estimation.

2.8.4 Darcy's Law, Fluid Flow, and Poroelasticity

The rate of fluid flow is defined as the time derivative of the vector

$$\boldsymbol{w} = \phi(\boldsymbol{u} - \boldsymbol{U}) \tag{2.86}$$

$$\dot{\boldsymbol{w}} = [\partial w_x/\partial t, \partial w_y/\partial t, \partial w_z/\partial t] \tag{2.87}$$

We define a dissipation function D as a quadratic form with the volume of flow rate as follows:

$$2D = \eta[\dot{\boldsymbol{w}}]^T[\boldsymbol{r}][\dot{\boldsymbol{w}}] \tag{2.88}$$

where η is the fluid viscosity. The force acting on the fluids in the pores causes the water to flow (i.e., when there is no force, $\boldsymbol{w}$=0). In the equation 2.88 $\boldsymbol{r}$ is the resistance to flow which is inversely proportional to the permeability (i.e., $[r] = [\kappa]^{-1}$).

The components of the force given by the pressure gradient and the body forces operating on the fluids are expressed as

$$\begin{aligned} X_i &= -\partial P/\partial x_i - \rho_f g_i \\ \boldsymbol{X} &= -\nabla P - \rho_f \nabla G \end{aligned} \tag{2.89}$$

where G is the gravitational potential (i.e., $\boldsymbol{g} = \nabla G$).

In hydrogeology this quantity is often expressed using the combined potential

$$\Phi = (P/\rho_f) + G \rightarrow \boldsymbol{X} = -\rho_f \nabla \Phi \tag{2.90}$$

Following the physics of irreversible thermodynamics, the relationship between forces and fluxes is given by the derivative of the dissipation function D,

$$\boldsymbol{X} = \partial D/\partial \dot{\boldsymbol{w}} = \eta \left(\begin{bmatrix} \kappa_{11} & \kappa_{12} & \kappa_{13} \\ \kappa_{12} & \kappa_{22} & \kappa_{23} \\ \kappa_{13} & \kappa_{23} & \kappa_{33} \end{bmatrix}^{-1} \begin{bmatrix} \dot{w}_x \\ \dot{w}_y \\ \dot{w}_z \end{bmatrix} \right) = -\rho_f \begin{bmatrix} \partial \Phi/\partial x \\ \partial \Phi/\partial y \\ \partial \Phi/\partial z \end{bmatrix} \tag{2.91}$$

which in the absence of gravitational potential is given by the well-known Darcy's law:

$$\eta([\kappa]^{-1}[w]) = -\nabla P \tag{2.92}$$

It can be shown that Darcy's law is a statement of linear relationship between fluxes and forces. The positive definite matrix κ is the permeability matrix. For an isotropic media the permeability field is given by a constant or $[\kappa] = \kappa\delta_{ij}$.

2.8.5 Field Equation for Deformation and Stress Distribution

Recall that the forces in equilibrium are given by $\nabla \cdot \boldsymbol{\sigma} + \rho\mathbf{b} = 0$ as shown in equation (2.29). From Darcy's law (equation 2.91) we get

$$\partial\mathbf{w}/\partial t = \frac{-\boldsymbol{\kappa}}{\eta}\nabla P + \rho_f\nabla G \tag{2.93}$$

where $\boldsymbol{\kappa}$ is the permeability tensor and η is the fluid viscosity. Neglecting body forces and using equation (2.79) we get the following coupled system of equations:

$$2\sum_{i,j}\frac{\partial}{\partial x_j}(\mu e_{ij}) + \frac{\partial}{\partial x_i}(\lambda_C\theta - \alpha M\zeta) = 0$$
$$\frac{\partial\mathbf{w}}{\partial t} = \frac{\boldsymbol{\kappa}}{\eta}\nabla(\alpha M\theta - M\zeta) \tag{2.94}$$

These are six equations with six unknowns, a coupled set of PDEs, which can be solved given appropriate boundary and initial conditions. We note that these equations couple the deformation with the fluid diffusion in the poroelastic medium.

2.8.6 Three-Dimensional Consolidation

Poroelasticity provides the analytical framework for three-dimensional consolidation of soils and sediments as follows.

As discussed in section 2.8.5, Darcy's law relates fluid motion to pore pressure gradient and gradient of the gravitational potential G, as written in equation (2.93) ($\partial\mathbf{w}/\partial t = (-\kappa/\eta)\nabla P + \rho_f\nabla G$). Combining it with the equilibrium equation for the stress field (equation 2.29, $\nabla \cdot \boldsymbol{\sigma} + \rho\mathbf{b} = 0$) we obtain the general consolidation equations which couple fluid and solid displacements of the vectors $\mathbf{u}$ and $\mathbf{w}$ which is written as equation (2.94). For the special case where the poroelastic coefficients are constant the equation of motion is written (after application of the divergence operator and some algebra) as

$$(2\mu + \lambda_C)\nabla^2\theta - \alpha M\nabla^2\zeta = 0$$
$$\frac{\partial\zeta}{\partial t} = \frac{\boldsymbol{\kappa}}{\eta}M_C\nabla^2\zeta \tag{2.95}$$
$$M_C = M(2\mu + \lambda)/(2\mu + \lambda_C)$$

The latter equation can be written in terms of pore pressure as the familiar diffusion equation with an added volumetric solid deformation term:

$$\frac{\partial P}{\partial t} - \frac{\kappa}{\eta} M \nabla^2 P = -\alpha M \frac{\partial \theta}{\partial t} \tag{2.96}$$

This equation is used extensively in the study of soil consolidation, compacting reservoirs and stress–fluid coupled processes in the earth.

2.8.7   Pore Space Compressibility of Porous Media

A nonporous media has a single compressibility that relates change in hydrostatic stress σ to change in volumetric stress V:

$$\beta = \frac{1}{V} \frac{\partial V}{\partial \sigma} \tag{2.97}$$

In contrast, porous material compressibilities are more complicated. We have to account for at least two pressures (confining pressure and pore pressure) and two volumes (bulk volume and pore volume). Thus, we can define at least four (4!) compressibilities for the rock. Following Zimmerman's notation (Zimmerman, 1991), we indicate by the first subscript the volume (bulk or pore) and the second subscript the change in pressure (confining or pore). The compressibilities are

$$\beta_{bc} = \frac{1}{V_b} \left(\frac{\partial V_b}{\partial \sigma_c} \right)_{\sigma_P} \tag{2.98a}$$

$$\beta_{bp} = -\frac{1}{V_b} \left(\frac{\partial V_b}{\partial \sigma_P} \right)_{\sigma_C} \tag{2.98b}$$

$$\beta_{pc} = \frac{1}{v_p} \left(\frac{\partial v_p}{\partial \sigma_c} \right)_{\sigma_P} \tag{2.98c}$$

$$\beta_{pp} = -\frac{1}{v_p} \left(\frac{\partial v_p}{\partial \sigma_P} \right)_{\sigma_C} \tag{2.98d}$$

Note that the sign convention is taken such that all compressibilities are positive under tension. From here we can immediately recognize that the dry bulk modulus K_{dry} and the pore space bulk modulus K_ϕ are given by

$$\begin{aligned} K_{\mathrm{dry}} &= 1/\beta_{bc} \\ K_\phi &= 1/\beta_{pc} \end{aligned} \tag{2.99}$$

The relation between the different pore space compressibilities in terms of porosity and mineral bulk modulus is given by Zimmerman (1991)

$$\begin{aligned} \beta_{bp} &= \beta_{bc} - 1/K_0 \\ \beta_{pc} &= \beta_{bp}/\phi \\ \beta_{pp} &= [\beta_{bc} - (1+\phi)/K_0]/\phi \end{aligned} \tag{2.100}$$

Note that Gassmann's equation defines another compressibility (undrained) where the mass of the pore fluid is kept constant as the confining pressure changes and thus

$$\beta_u = \frac{1}{V_b}\left(\frac{\partial V_b}{\partial \sigma}\right)_{m_{\text{fluid}}} = 1/K_{\text{sat}} \tag{2.101}$$

Re-writing Gassmann's equation in terms of compressibility, we can show that (Mavko and Mukerji, 1995)

$$\frac{1}{K_{sat}} = \frac{1}{K_0} + \frac{\phi}{\widetilde{K}_\phi}, \quad \widetilde{K}_\phi = K_\phi + \frac{K_0 K_{fl}}{K_0 - K_{fl}} \approx K_\phi + K_{fl} \tag{2.102}$$

2.9 Mechanical Compaction from Plastic–Poroelastic Deformation Principles

Mechanical compaction or loss of porosity due to increase in effective stress is a fundamental geological process that governs many of the rock elastic and transport parameters, all of great importance in exploring and developing subsurface reservoirs. The ability to model the compaction process enables us to improve our understanding of the seismic signature of the basin, and better relate the geology of deposition to current porosity, velocity, pore pressure, and other mechanical parameters which depend on the state of compaction of the sediment (Bachrach, 2016).

2.9.1 Differential Compaction and Natural Strain Increment

Porosity loss is a deformation process where the loss can be viewed as total volumetric strain which is accumulated in a representative elementary volume of sediment (Goulty, 2004). Plastic strain (ε^P) is defined as the irreversible part of the strain accumulated in the material while elastic strain (ε^e) refers to the reversible portion of the total strain such that the total strain $\varepsilon = \varepsilon^P + \varepsilon^e$. Figure 2.19 shows an illustration of stress–strain relation and natural strain increment concept. The differential relation models the local path in the stress–strain plane. The plastic strain is the irreversible cumulative strain and elastic strain is the reversible part of the strain. The differential volumetric total strain in sediment ε can be defined in terms of differential porosity loss as

$$d\varepsilon = d\varepsilon^P + d\varepsilon^e = d\phi/\phi \tag{2.103}$$

A constitutive relation in large deformation theory can be defined using a natural strain increment (Malvern, 1969) as

$$\frac{d\varepsilon}{dt} = f(\sigma, \varepsilon, t)\frac{d\sigma}{dt} \tag{2.104}$$

Note that natural strain increments are differential forms of constitutive equations and can be used to describe many of the familiar constitutive equations such as linear elasticity or

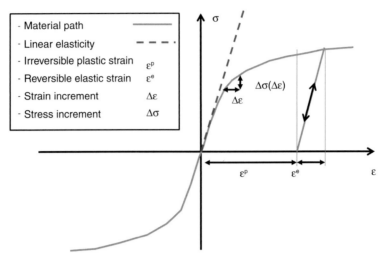

Figure 2.19 Plastic and elastic strain and natural strain increment concept in the stress–strain plane. The solid line represent a typical loading and unloading pattern where plastic strain and elastic strain accumulate as a function of the stress path.

viscoelasticity. Note also that natural strain increment formulation is written here in terms of the total strain. The benefit of using natural strain increments is that arbitrary time-dependent path in the stress–strain plane can be modeled using strain–stress increments, while conventional constitutive stress–strain relations force a specific functional form that, in many cases, is an idealized behavior. In addition, the formulation here is in terms of the total strain and is not limited to plastic only or elastic only deformation. This is illustrated in Figure 2.19. Therefore, any curve in the stress–strain path can be modeled using equation (2.104) and with it the ability to model complex deformation processes.

2.9.2   Empirical Compaction Laws

Compaction or loss of porosity with respect to effective stress is a major process that governs porosity and pore pressure evolution. We will consider here three popular equations that are used to empirically model the compaction process:

$$\phi = \phi_0 \exp(-K\sigma_{eff}) \tag{2.105a}$$

$$\sigma_{eff} = \sigma_0 \exp(-\beta \frac{\phi}{1-\phi}) \tag{2.105b}$$

$$\sigma_{eff} = \sigma_0 (1 - \phi/\phi_0)^C \tag{2.105c}$$

Equation (2.105a) is often known as Athy's compaction law, and is derived by assuming the increment of porosity is proportional to the porosity and effective stress increment (i.e., $d\phi \propto \phi d\sigma_{eff}$). Equation (2.105b) was introduced by Dutta (1986, 1987a, 1987b) in the context of compaction related to diagenesis (described by the continuously varying parameter β with temperature and geologic time) and assumes a proportionality between the stress, stress increment and void ratio increment (void ratio is defined as $\varepsilon = \dfrac{\phi}{1 - \phi}$), i.e., $d\sigma_{eff} \propto \sigma_{eff}d\varepsilon$. The use of equation (2.105b) to relate stress to the void ratio can be attributed to Skempton if β is constant. Equation (2.105c) was used by Sayers (2006) to provide a functional form that fit observed compaction data. Note that for Athy's relation, the compaction parameter that governs the rate of compaction is K. For Skempton's relation, the compaction parameter relating stress to the void ratio is β and for Sayers, the exponent is C. The critical porosity (porosity at deposition with no effective stress) is ϕ_0 and σ_0 is the stress needed to get the sediment to zero porosity. In general, these model parameters are lithology dependent. Additionally, it is important to note that the relations above are not general and are stress path dependent, as will be discussed later. However, in many cases, these relations are used under "primary loading" or "normal trend" conditions, often related to the loading experiment where either uniaxial loading or hydrostatic loading conditions are assumed.

2.9.3 Natural Strain Increment

An example of applying a natural strain increment to porosity loss can be demonstrated using simple compaction laws that are often used to model the relationship between porosity and effective stress.

Note also that, while equations (2.105a), (2.105b), and (2.105c) all have different functional dependencies for the different compaction models, they all behave similarly and can be calibrated to produce almost identical porosity-stress curves within the range of porosities between 0.4 and 0.1. This is demonstrated in Figure 2.20.

It should be noted that the range of porosities between 0.4 and 0.1 covers most of the range where mechanical compaction takes place and is observed in situ (i.e., in well log data) for both sands and shales. For shales, although initial depositional porosity (or critical porosity) can be as high as 70 to 80 percent, pore space collapses very fast, and shallow boreholes are difficult to log, so in situ measurements of porosity (such as log data) will often not measure porosities larger than 0.4. Critical porosity for sands is usually slightly lower than 40 percent. Figure 2.20 shows that the empirical phenomenological equations (2.105a), (2.105b), and (2.105c) capture the same behavior of the sediment.

From equations (2.105a), (2.105b), and (2.105c) it is possible to derive a differential relation between the increment in effective stress and the increment of porosity loss. The differential forms of the equations are

$$d\phi/\phi = -K d\sigma_{eff} \tag{2.106a}$$

$$d\phi/\phi = -d\sigma_{eff} / \left(\sigma_{eff} \beta (\phi/(1 - \phi)^2) \right) \tag{2.106b}$$

$$d\phi/\phi = d\sigma_{eff} / \left(C\phi_0 \sigma_0 (1 - \phi/\phi_0)^{C-1} \phi \right) \tag{2.106c}$$

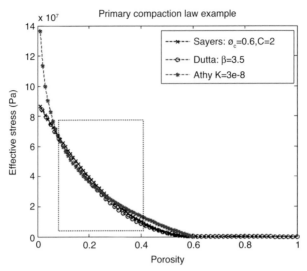

Figure 2.20 Comparison of three compaction models in equation (2.105a-c). Note that they all produce very similar curves for porosities in the range of 0.4–0.1.

Note that equations (2.106a), (2.106b), and (2.106c) are in the form of natural strain increment (equation (2.104)) when considering a steady-state solution for the volumetric plastic strain in compacting sediments. A major difference between the differential form (equations (2.106) and equations (2.105a-c)) is that the model parameters discussed above are related to a given state of stress or lithology. In general it can be shown that for varying lithology the different compaction equations can be written using lithology dependent parameters. For example, Athy's law can be written as

$$d\varepsilon = d\phi/\phi = -K(vcl)d\sigma_{\text{eff}} \qquad (2.107)$$

where the model parameter $K(vcl)$ is now dependent on the clay content vcl. As natural strain increments are defined as a differential relation between stress and strain they can be used to follow the stress path, which will translate in this case into a path-dependent compaction parameter, i.e., $K = K$ (stress path) (Bachrach, 2015).

2.9.4   The Fundamental Equation for Porosity Loss under Vertical Loading

Porosity reduction in sedimentary basins can be viewed as following primary loading curves where the vertical loading of sediments increases the effective stress and the plastic deformation of the sediment associated with the volumetric porosity loss. To model porosity evolution in sedimentary basins, we must add spatial (i.e., depth and space) aspects to the deformation. This can be done using the following poroelastic equations. Derivation of 1D compaction will be presented in detail following which an extension that accounts for the tensorial state of stress will be discussed.

The effective stress in relation to overburden stress and effective pressure using Terzaghi's relations is

$$\sigma_{\text{eff}} \equiv \sigma_{ob} - P \tag{2.108}$$

Here σ_{ob} is the vertical overburden stress. The choice of Terzaghi's effective stress is mainly because we are interested in modeling compaction in unconsolidated materials where Biot's parameter $\alpha = 1 - K_d/K_0$ is very close to unity. Extension to Biot's effective stress is straight forward here. Also, note that the overburden stress here is assumed to be the vertical stress which dominates primary compaction as will be discussed below.

Density is directly related to porosity and the density-porosity relation is given by

$$\rho(z) = \phi(z)\rho_w + (1 - \phi(z))\rho_g \tag{2.109}$$

Here, ρ_w is the brine density as we assume fully saturated sediments; ρ_g is the grain density, which, in general, can vary with depth. In the equation below, we neglect this variation.

The vertical overburden stress under the primary loading path is given by

$$\sigma_{ob}(z) = \rho_w g z_0 + g \int_{z_0}^{z} \rho(z')dz' \tag{2.110}$$

Note that equation (2.110) is a simplified 1D equation. Equation (2.110) assumes the atmospheric pressure is small as compared to water pressure and therefore is neglected. The water depth is z_0. Extension to tensorial notation will be discussed shortly.

Combining equations (2.108), (2.109), and (2.110), we can write the generalized relation between vertical effective stress, porosity, and pore pressure as

$$\sigma_{\text{eff}}(\phi) = \rho_g g(z - z_0) + (\rho_w - \rho_g)g \int_{z_0}^{z} \phi(z')dz' - P(z) \tag{2.111}$$

The differential form of equation (2.111) is given by

$$\frac{\partial \sigma_{\text{eff}}}{\partial \phi}\frac{\partial \phi}{\partial z} = \rho_g g + (\rho_w - \rho_g)g\phi - \frac{\partial P}{\partial z} \tag{2.112}$$

If we write equation (2.104) in terms of the volumetric strain for steady state, one gets

$$\frac{1}{\phi}\frac{d\phi}{d\sigma} = f(\sigma, \varepsilon, t) \tag{2.113}$$

Close inspection of equation (2.113) shows that the term $\phi(\partial \sigma_{\text{eff}}/\partial \phi) = 1/f(\sigma_{\text{eff}}, \varepsilon, t)$ is the constitutive equation in terms of natural strain increment for effective stress and volumetric porosity loss. Thus, equation (2.112), which is the differential form of equation (2.111), is the one that is best suited for a constitutive law that can handle large deformations as it is formulated in terms of natural strain increment. Using a constitutive equation such as equation (2.106), the differential equation (2.112) is written in terms of the differential constitutive compaction law. Moreover, as equation (2.112) is posed as an ordinary differential equation (ODE), one can integrate the ODE using various nonlinear ODE solvers. For a depositional sequence with a variable shale volume, one can use laboratory measurements to calibrate the constitutive laws or field measurements.

It is also important to note that the pore pressure must be explicitly given to properly model compaction. When pore pressure is not known, one can predict compaction under various pore pressure regions such as hydrostatic or a typical basin profile. It is shown here that pore pressure must be assumed or known when modeling compaction.

2.9.5   Tensorial Extension

As mentioned above, the equations presented here are formulated for the vertical component of the effective stress tensor. Below, we provide the basic equation for extension of the theory for the full stress tensor.

Because stress is a symmetric second-rank tensor, it can be displayed in the principal coordinate system using three principal stresses. Due to the strong influence of gravity, in many cases the vertical stress is one of the principal stresses. In this case, the total stress tensor $\boldsymbol{\sigma}_t$ can be written in terms of vertical stress σ_v, maximum horizontal stress σ_H, and minimum horizontal stress σ_h as

$$\boldsymbol{\sigma}_t = \begin{bmatrix} \sigma_H & 0 & 0 \\ 0 & \sigma_h & 0 \\ 0 & 0 & \sigma_v \end{bmatrix} \tag{2.114}$$

Here, σ_v is equivalent to the overburden stress defined in equation (2.110).

The effective stress is now given by

$$\boldsymbol{\sigma}_{\text{eff}} = \boldsymbol{\sigma}_t - P\mathbf{I} = \begin{bmatrix} \sigma_H - P & 0 & 0 \\ 0 & \sigma_h - P & 0 \\ 0 & 0 & \sigma_v - P \end{bmatrix} \tag{2.115}$$

The tensorial equivalent of equation (2.111) is now written as

$$\boldsymbol{\sigma}_{\text{eff}}(\phi) = \begin{bmatrix} \sigma_H(\phi) - P(z) & 0 & 0 \\ 0 & \sigma_h(\phi) - P(z) & 0 \\ 0 & 0 & \rho_g g(z - z_0) + (\rho_w - \rho_g)g \int_{z_0}^{z} \phi(z')dz' - P(z) \end{bmatrix} \tag{2.116}$$

Therefore, the differential form of equation (2.116) is given by the following three differential equations:

$$\frac{\partial \sigma_{eff\_H}}{\partial \phi} \frac{\partial \phi}{\partial z} = \frac{\partial \sigma_H}{\partial z} - \frac{\partial P}{\partial z}$$

$$\frac{\partial \sigma_{eff\_h}}{\partial \phi} \frac{\partial \phi}{\partial z} = \frac{\partial \sigma_h}{\partial z} - \frac{\partial P}{\partial z} \qquad (2.117)$$

$$\frac{\partial \sigma_{eff\_v}}{\partial \phi} \frac{\partial \phi}{\partial z} = \rho_g g + (\rho_w - \rho_g)g\phi - \frac{\partial P}{\partial z}$$

Note that the porosity now must satisfy the three equations simultaneously. Also, the differential constitutive law is written in terms of all three principal effective stresses.

2.10 Fracture Mechanics and Hydraulic Fracturing

Above a certain stress rocks fail and fractures are created. The exact stress at which a rock will fail depends on the specifics of the state of stress (e.g., extension or compression) and the state of the rock itself (i.e., existence of micro-cracks and weak contacts inside the rock). The ability of the rock to be fractured is defined loosely as its "brittleness" and is affected by other environmental factor such as temperature and overburden stress. Fractures are related to pore pressure and are used to estimate the minimum horizontal stress through leak-off tests (LOT). Hydraulic fracturing is used to stimulate reservoirs and fracture gradient is a valuable tool when assessing pore pressure within the context of borehole stability. Below we discuss the fundamentals associated with fracture mechanics and fracture propagation in rocks.

2.10.1 Fracture Modes

There are three ways (or modes) in which cracks can propagate after applying a force to the material (Figure 2.21):

Mode I fracture: Opening mode where tensile stress normal to the plane of the crack is applied

Mode II fracture: In-plane shear or sliding mode where a shear stress acts parallel to the plane of the crack and perpendicular to the crack front

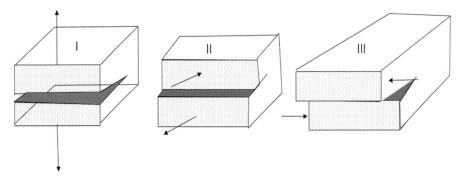

Figure 2.21 The three modes of cracks: Mode I, tensile or opening mode; Mode II, in-plane shear or sliding mode; and Mode III, antiplane shear or tearing mode.

Mode III fracture: Antiplane shear or tearing mode where a shear stress acts parallel
to the plane of the crack and parallel to the crack front.

2.10.1   Crack Stability, Griffith's Solutions, and Fracture Modes

Griffith (1920) employed an energy-balance approach to analyze the stability of an
isolated crack in a solid subjected to an applied stress field σ. Griffith's solution has
become one of the most famous developments in materials sciences.

Griffith realized that the stability criteria for a crack can be obtained by minimizing
the total free energy of the loaded system of cracks and rock. Assuming plane strain, the
strain energy U_e associated with an ellipsoidal crack of length $2a$ (Figure 2.22) can be
derived based on the Kirsh-like solution of Inglis (1913) for stress concentration around
elliptical cracks as $U_e = \dfrac{\sigma^2 \pi a^2}{2E}$, where E is the Young's modulus of the rock. The work
done by the applied loads is $W = 2U_e$ (Lawn and Wilshaw, 1975) and the mechanical
work is $U = -W + U_e$. Note that the term W is negative as it represents the decrease in
the potential energy due to the work done by the forces when the crack opens. The surface
energy with the crack is given by $S = 4a\gamma$ where γ is the thermodynamic surface energy
(in Joules/m^2). The total free energy of the system is therefore given by $U_{\text{total}} = U + S$.
In Figure 2.23 a plot of the total energy versus crack length for glass with $E = 62$ GPa,
and $\gamma = 1.75$ (J/m^2) is shown. Note that at very small a, the surface energy term
dominates while as a becomes larger the mechanical work begins to dominate. Beyond
a critical crack length a_c the system can lower its total energy by increasing the crack
length. Therefore, up to length a_c the crack will grow only if the stress applied will
increase but after the critical length the crack will grow spontaneously.

To find the critical crack length a_c one needs to set the derivative of the total
energy with respect to a to zero (i.e., $\dfrac{\partial U_{\text{total}}}{\partial a} = \dfrac{\partial(U + S)}{\partial a} = 0$) to get $a_c = \dfrac{2\gamma E}{\pi \sigma^2}$.

The maximum stress that can be applied to a rock containing a crack of length $2a$ is thus

$$\sigma_f = \sqrt{\frac{2\gamma E}{\pi a}} \tag{2.118}$$

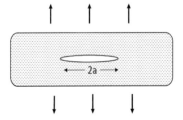

Figure 2.22 Crack subjected to extensional stress.

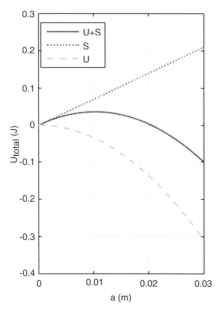

Figure 2.23 Total energy for a crack as a function of its length a.

Beyond this stress the crack will grow, and the rock will fail, and therefore σ_f can be characterized as the rock strength. Note that rocks with larger cracks are weaker.

Griffith's original work was performed on very brittle materials. When the material exhibits more ductility some of the strain energy released is absorbed not only by the new surface area created by the crack but also by energy dissipation due to plastic flow near the crack tip.

Irwin (1958) modified Griffith's theory to account for energy dissipation by plasticity. The modified Griffith's equation can be written as

$$\sigma_f = \sqrt{\frac{2\gamma G_c}{\pi a}} \qquad (2.119)$$

where G_c is a constant related to the material. Griffith's analysis can be generalized to all three modes of fractures described in Figure 2.18 where we can define the equilibrium condition for the three fracture modes to be $\dfrac{\partial(U+S)}{\partial a} = 0$, which will give $G_c = 2\gamma$ for brittle material while for ductile material the plastic dissipation term dominates.

Hydraulic Fracturing

Hydraulic fracturing (HF) occurs when pore pressure is high enough to induce sufficiently high stress field that will cause the rock to fracture (Aadnoy and Looyeh, 2010; Anderson, 1991). The geometry of the stress field depends on the geometry of the borehole, the pressure applied, and the geometry and material properties of the surrounding rocks.

Hydraulic fracturing is a technique used to measure in situ components of the stress field when a small volume of fluid is pumped into the formation in a test known as a leak-off test (LOT). Hydraulic fracturing is also a technique used to stimulate tight reservoirs achieved by using a large amount of fluid volume pumped into the formation for longer periods of time. LOT can be viewed as the early part of the hydraulic fracturing process. The fracturing process follows our general discussion on crack stability.

When the pressure at the interface with the rock reaches the extensional strength of the rock σ_1 a fracture will propagate in the plane perpendicular to the least principal stress. We define the critical pressure needed for fracturing as p_c. Then the pressure needed to keep the fracture open is p_0 which is typically smaller than p_c and is equal to the stress component normal to the fracture plane, i.e., if the minimum stress is S_h then $S_h = p_0$. For a vertical borehole where the three principal stresses are vertical and planar and the vertical stress $S_v = \int_0^z g\rho dz'$ is not the minimum stress, the fracture will be vertical as the minimum horizontal stress S_h is normal to the fracture as shown in Figure 2.24.

As we have seen before, the propagation of the hydraulic fracture results from the fact that the preexisting crack of length a has reached its stability limit.

In typical well testing and hydraulic fracturing, fluids are pumped at a constant rate into the borehole such that the pressure increases with time and volume of fluids. Figure 2.25 shows the typical pressure versus time curve. Initially the pressure in the borehole increases at a constant rate until a deflection point occurs where the rate of pore pressure build up with time decreases. This point is known as the leak off point (LOP) and represents the actual opening of the crack and in principle should be equal to p_0, the minimum principal stress. At this point the rate of fracture propagation is still slower than the rate at which fluids are being pumped into the formation. The

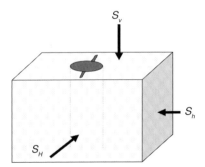

Figure 2.24 Typical fracture geometry for the case where the minimum principal stress is S_h. The fracture will strike at the maximum horizontal stress S_H direction.

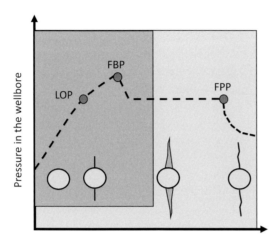

Figure 2.25 Typical LOT and hydraulic fracturing evolution as a function of pressure in the borehole and time and/or volume of fluids pumped into the formation. LOT = leak-off test; LOP = leak-off pressure; FBP = formation breakdown pressure; FPP = fracture propagation pressure. Modified after Rabia (1985). See text for details.

formation breakdown pressure represents the point at which the hydraulic fracture grows such that the pressure in the borehole decreases and the hydraulic fracture starts to propagate. The borehole pressure should drop to the fracture propagation pressure which again should be equal to p_0 after some time. The LOT is typically done within a short interval of time. Variants of LOT such as extended leak-off tests (XLOT) take longer and propagates fracture for a longer time into the formation to better identify p_0. In HF fluids are pumped such that the fracture will propagate far into the formation and will enable better access to the reservoir.

2.11   Rock Physics Basis for Detection and Estimation of Geopressure

In the previous sections we discussed the principles of continuum mechanics as related to *in situ* stress distribution, poroelasticity and fracture of rocks. Rocks are, of course, porous, heterogeneous and anisotropic in nature. We pointed out the relevance of porosity, consolidation and pore pressure in dictating the mechanical behavior of rocks by using the theory of poroelasticity. In this context Terzaghi's law (Terzaghi, 1943) describes the distribution of load on fluid-filled porous rocks and introduces the concept of effective stress defined by equation (2.115) in tensorial form. If we look only at the vertical component

$$\sigma_v^{\text{eff}} \equiv \sigma_v - P \qquad (2.120)$$

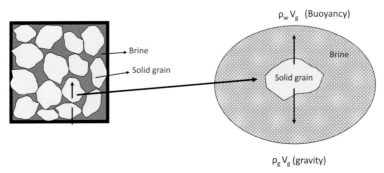

Figure 2.26 Illustration of Terzaghi's principle on a physical basis.

where σ_v^{eff} is the vertical effective stress ($\sigma_v^{\text{eff}} = \sigma_{33}^{\text{eff}}$), σ_v is the total vertical stress and P is the pore pressure. This relationship can be understood on a physical basis (after Aadnoy and Looyeh, 2010). Let us consider the case shown in Figure 2.26 for a bucket filled with water and sand grains under hydrostatic condition.

Our goal is to compute the effective stress between the sand grains at the bottom of the bucket. The total force acting at the bottom of the bucket is that due to the combined weight of the sand grains (denoted by suffix g) and water (denoted by suffix w). It is given by (ignoring the constant multiplicative factor of g, the acceleration due to gravity)

$$m_g + m_w = \rho_g V_g + \rho_w V_w = \rho_g V(1 - \phi) + \rho_w V\phi \qquad (2.121)$$

where V is the total volume, ρ_g and ρ_w are, respectively, the density of solid grains and water, ϕ is the volume fraction occupied by water (porosity), and $(1 - \phi)$ is the volume fraction occupied by the solid grains. Obviously, the total volume V is

$$V = V_g + V_w = (1 - \phi)V + \phi V \qquad (2.122)$$

Archimedes' principle states that since the sand grains are submerged in the water, each grain is lighter than its weight in the air by an amount that is equal to the displaced weight of the water. As shown in Figure 2.26 the net weight of the sand is the total weight minus the buoyancy. This means

$$\rho_g V_g - \rho_w V_g = (\rho_g - \rho_w)V_g = (\rho_g - \rho_w)(1 - \phi)V$$

$$= [\rho_g + (\rho_w - \rho_g)\phi - \rho_w]V \qquad (2.123)$$

If we denote the average density of the mixed sand and water by

$$\rho_{ave} = \frac{m_g + m_w}{V} = \rho_g + (\rho_w - \rho_g)\phi \qquad (2.124)$$

the net weight of the bucket content becomes

$$g(\rho_g V_g - \rho_w V_g) = g(\rho_{ave} - \rho_w)V \qquad (2.125)$$

Equation (2.128) states that the effective sand weight is equal to the total weight minus the weight of the water. In terms of equation (2.120), the effective stress is equal to the total stress minus the pore pressure. Thus, Terzaghi's law has a physical basis as demonstrated here. If the system (water filled bucket of sand) is not hydrostatic (as would be the case when the system is overpressured due to the inability of water to escape) the excess pressure would be a part of the pore pressure. Such will be the case if a heavy piston pushes the system shown in Figure 2.26 with no provision for the fluid to escape. (This is a common way to explain the Terzaghi principle.) Since a fluid at rest cannot transmit shear stresses, the effective stress is valid for normal stresses, and therefore the shear stress remains unchanged (Terzaghi, 1943). This is a fundamental property of the geopressured system.

Effective stress σ_v^{eff} is the most important parameter in our discussions on geopressure as it relates to both porosity reduction and velocity of sound in porous rocks. As we shall see in Chapter 3 earlier literature dealt with mostly compaction trend analysis with tools such as velocity that relied on departure from the normal compaction trend as an indicator of overpressure. This has several unwanted but significant consequences. We shall discuss this further in Chapter 4.

An alternate procedure to quantify σ_v^{eff} that uses a predictive rock model for pressure prediction has become more popular recently. This is mainly because such techniques use an integration of geology with geophysics and basics of rock physics of porous medium and hence, the procedure is more applicable in frontier areas where well control is limited or nonexistent. In this section we shall discuss some pertinent rock properties and models that are relevant to this approach of geopressure quantification (Bowers, 1994; Dutta, 2002b; Dutta et al., 2002c, 2002d). Overpressure implies lower effective stresses and increased porosity (compared to compaction under hydrostatic conditions), which in turn have a pronounced effect on geophysical properties such as seismic velocity, density, resistivity and strength, especially in soft or unconsolidated sediments. However, the relations among velocity, porosity, and pore pressure are not always simple. Effects of pore pressure may sometimes be masked by variations in other parameters. Furthermore, these parameters are often not independent of each other. Typically, compressional and shear velocities in crustal rocks increase with confining pressure as confining pressure closes the soft, crack-like porosity and increases the contact stiffnesses between grains. The effect of pore pressure is to counteract that of confining pressure by propping open the pore space. High pore pressure tends to lower the velocity. The relative change depends substantially on the type of the rock, the pore microstructure, the loading-unloading history, and the fluid saturation. Pore pressure effects are enhanced in rocks with soft pore space compressibility. Velocities also tend to decrease with increasing clay content and porosity. So, for example, an observed anomalously low velocity might be interpreted as due to high pore pressure or due to spatial variations in clay content or porosity. The uncertainty in the interpretations depends on the variance or spread in the range of clay and porosity. Interpretations in terms of pore pressure tend to be more accurate in formations that are more uniform and homogeneous with small variances in porosity and clay. Thus blind use of velocity as an indicator of geopressure without paying attention to other rock properties such as lithology and fluid content will surely result in erroneous predictions. Below we discuss a few critical issues that need to be taken into account to condition the velocity from either well log or seismic data prior to geopressure estimation. Even if the velocity is not conditioned due to unavailability of supporting data every effort must be made to estimate the associated error and uncertainty in the prediction.

In spite of experimental (Laughton, 1957; Schreiber, 1968; McCann and McCann, 1969; Hamilton, 1971a, 1971b; Elliot and Wiley, 1975; Domenico, 1977; Blangy, 1992) and theoretical work (Brandt, 1955; Walton, 1987; Digby, 1981) velocity, porosity, and pore pressures in young, unconsolidated sediments common to the offshore Gulf of Mexico is not very well understood. In such environments, seismic velocities are usually low, and the sensitivity of seismic velocity and impedance to changes in pore pressure, saturation, and stress is exaggerated. For example, the widely used Wyllie time-average equation (Wyllie et al., 1956) adequate for clean consolidated sandstones at depth, becomes inappropriate in partially gas-saturated unconsolidated sands (Elliot and Wiley, 1975). We present here some of the considerable progress that have been made during the last few decades in understanding velocity-porosity-pressure relations of unconsolidated sand-clay systems, as found in the offshore Gulf of Mexico.

In Figure 2.27, Marion (1990) and Nur et al. (1995) show velocity-porosity relation for suspended grains (porosity ≥ 0.38) and for sandstones (porosity < 0.38). The data

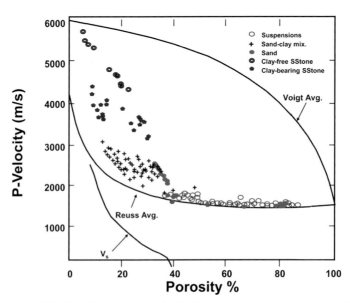

Figure 2.27 Velocity versus porosity in clastics: critical porosity (Marion, 1990; Nur et al., 1995). After Dutta et al. (2002c).

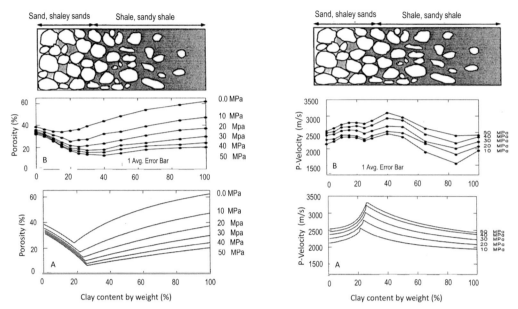

Figure 2.28 Porosity-clay and velocity-clay behavior in soft sediments. (top) Schematic cartoon of sand/shale mixture. (middle: B) Laboratory data. (bottom: A) Theoretical predictions from binary mixture model (Yin, 1992; Marion, 1990). After Dutta et al. (2002c).

fall close to the Reuss curve for porosities above 0.38. Here the Reuss (1929) average quite accurately represents unconsolidated systems under iso-stress conditions in which solid grains are totally surrounded by fluid so that the effective stress is zero (pore pressure = confining or overburden pressure). At smaller porosities, the velocity, and equivalently the effective moduli, depart abruptly from the Reuss curve as the grains come into contact to form a mechanically load-bearing skeleton. The "critical porosity" at which the suspended sediments transform from iso-stress to solid load-bearing frame is a fundamental property of unconsolidated soft sediments. Each class of unconsolidated material – defined on the basis of its common diagenetic process and mineralogy – has a typical critical porosity value. Increasing pore pressure moves seismic properties toward the Reuss bound at critical porosity. At the same time, the sensitivity of seismic velocity to changes in pore fluids and pore pressure increases. Hence, the Reuss bound may act as a reference point for understanding overpressures in these materials. Since the critical porosity is a purely geometric property, it is a robust and powerful constraint on the applicability of theoretically derived velocity-porosity relations.

To explore the effects of lithology and pressure, Yin et al. (1988) and Marion (1990) measured acoustic velocities in unconsolidated sand-clay mixtures as a function of clay content, both at dry and brine-saturated conditions. The results in Figure 2.28 reveal systematic *nonlinear* variations in porosity and velocity with *clay content* and with *pressure*. Between 0 and 25 percent clay content the compressional velocity increases as the clay content increases. Beyond 25 percent clay content, the velocity begins to decrease. The measured porosity has a minimum at 25 percent clay fraction. The minimum in porosity shifts to lower clay fraction with increased pore pressure. With increasing effective stress the mixture compacts and the porosity decreases. A simple binary mixture model (Cumberland and Crawford, 1987; Marion, 1990) allows for straightforward predictions of the velocity-porosity-clay relations in these types of sediments. The velocity and porosity variations are clearly due to the fraction of clay in the sample and the geometrical arrangement of sand and clay particles. An important conclusion from this work is that the properties of shaly sands do not necessarily lie between those of the sand and clay end members. Therefore, the *bounding surfaces* between strata may have unexpected properties, such as anomalous velocities, reflectivities, porosities, and permeabilities, not necessarily falling between the properties of the separate units. For example, the transition zone between two moderately permeable layers may in fact be a barrier or *seal* to flow.

Prasad (2002) has shown that for unconsolidated sediments, very low effective pressures can be detected by changes in V_p/V_s ratios and by changes in P- and S-wave amplitudes and frequency content. At very low pressures, both P- and S-wave amplitudes change drastically. However, for a pressure change of about 1 MPa, the frequency content of the P-waves changes only by about 8 percent; the corresponding change for S-waves is over 150 percent. Figure 2.29 shows V_p and V_s variations with pressure for two water-saturated sands: coarse grained (open circles) and fine grained (closed stars) from Prasad (2002) and Prasad and Meissner (1992). The plots show a difference in the pressure dependence between P- and S-waves; *pressure appears to have a larger effect on shear waves than on compressional waves.* This is a remarkable observation and shows the value of using shear waves for geopressure quantification.

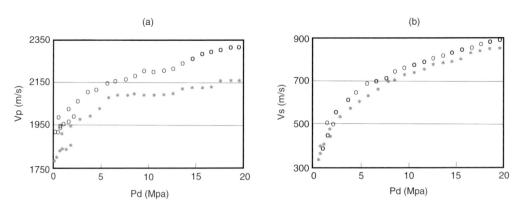

Figure 2.29 Main results from Prasad (2002) and Prasad and Meissner (1992). P- and S-wave velocities (V_p, V_s, respectively) as a function of differential pressures for two water-saturated sands with different grain sizes are plotted with (a) V_p variations and (b) V_s variations. Open circles = coarse grained; solid stars = fine grained. Modified after Prasad (2002).

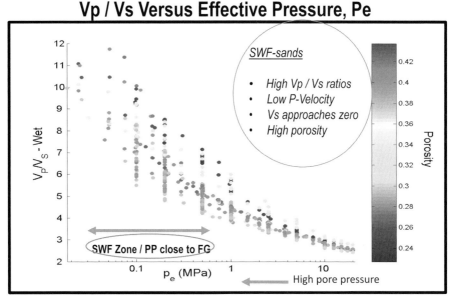

Figure 2.30 Porosity, effective stress, and V_p/V_s ratio (Zimmer et al., 2002). Figure modified after Dutta et al. (2002c). (A black and white version of this figure will appear in some formats. For the color version, please refer to the plate section.)

Zimmer et al. (2002) have shown that in addition to pressure, porosity also plays a role in the V_p/V_s ratio at low effective pressures. They have analyzed velocity data in dry glass beads and sands. The V_p/V_s ratios for the Gassmann fluid-substituted data are shown in Figure 2.30, plotted against pressure and color-coded by porosity. The water-saturated V_p/V_s ratio shows a dramatic rise as the effective pressure decreases. There is a considerable scatter in the data. This scatter is directly related to porosity variations. Very high values of V_p/V_s ratio at low effective pressures (very high pore pressure) are consistent with the occurrence of shallow water flow sands and it is a major source of geohazard (Dutta, 2002a, 2002e). Ayers and Theilen (1999) also discuss this in their work. We shall discuss this in more detail in Chapter 11.

Effective stress for a given rock and fluid combination controls the velocity of sound through it. This is shown in Figure 2.31 for Berea sandstone under water-wet conditions. The key point is that the velocity increases with increasing effective stress. The increase is at first rapid but decreases with increasing effective stress until an approximate constant, terminal velocity is attained. The velocity dependency on effective stress shown in Figure 2.31 is the key reason why velocity data are used for geopressure quantification in clastics.

Gas saturation can dramatically influence velocities and Poisson's ratios, and hence impacts the detection and quantification of overpressure in soft sediments. However, what is not so well known is that it is not only the amount of saturation, but also the *spatial distribution and scale* of saturation – homogeneous versus patchy – that can

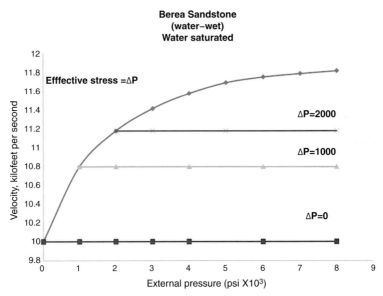

Figure 2.31 Velocity-effective stress relations for water-wet Berea sandstone.

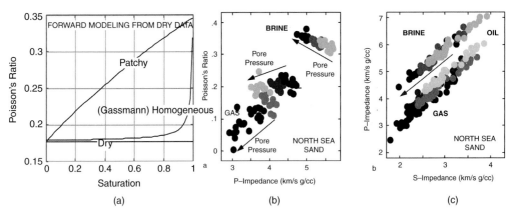

Figure 2.32 Effect of fluid type and its distribution on the rock properties (Poisson's ratio, P-impedance and S-impedance). Impedance is the product of velocity and density. (a) Poisson's ratio of patchy and homogeneous fluid (brine) saturation in porous rocks. Modified after Dutta et al. (2002c). (b) PR versus P-wave impedance for gas, oil, and brine and at varying differential pressure (5, 15, and 30 MPa). (c) Same data in a P-wave impedance versus S-wave impedance cross plot. Arrows show the direction of pore pressure increase (differential pressure decrease). Both modified after Dutta et al. (2002c, 2002d).

greatly affect the seismic response. In a homogeneous saturation distribution, gas is finely mixed with oil or brine, and is uniformly distributed within the pore space. For a patchy or heterogeneous saturation distribution, fully saturated patches of oil and brine are surrounded by dry, gas-filled regions. The difference in the Poisson's ratio between these two saturation scenarios can be dramatic (Figure 2.32a). As we shall see in Chapter 5, this difference can give rise to distinct seismic signatures that can be used for overpressure detection. Velocity and attenuation of P-waves in porous rocks with patchy saturation was discussed first by White (1975) and further examined by Dutta and Ode (1979) and Dvorkin et al. (1999b).

P-wave impedance (product of density and velocity), and Poisson's ratio can be obtained from seismic data using acoustic and elastic impedance inversion (Connolly, 1999; Zong et al., 2012, and references therein). It is, therefore, important to create diagnostic rock physics based charts that will help interpret the inversion results in terms of pore pressure and pore fluid. An example of diagnostic charts is given in Figures 2.32b and 2.32c. The data are based on laboratory measurements of the elastic wave velocity in unconsolidated North Sea sands (Blangy, 1992). Different regions in the Poisson's ratio versus P-wave impedance plane and P-impedance versus S-impedance plane correspond to different pore pressure and pore fluid. One can identify both pore pressure and pore fluid from seismic signatures (and separate the pore pressure effect from the pore fluid effect) by superimposing seismic elastic rock properties on a diagnostic chart. Such plots, calibrated to specific locations and lithologies of interest, can guide us in looking for anomalies in seismic attributes that indicate overpressures.

3 Mechanisms of Geopressure

3.1   Introduction

Overpressured formations exist worldwide. These formations have pore fluid pressure that is above the pressure that would be expected under hydrostatic conditions at a given depth. For overpressure to exist over geologic time, there must be vertical and lateral seals so that the trapped fluids cannot escape. In the US Gulf Coast, Dickinson (1953) first noted that pore fluid pressure higher than the hydrostatic condition occurred below sandy sections, below the top of massive shale units, and attributed the reason to gravitational compaction of shales due to overburden load. He further noted (Dickinson, 1953, p. 425): "the pressures in fluids within the sediments are dominated by two factors, the compression due to compaction on the one hand and the resistance to the expulsion on the other; but as compaction becomes more difficult other factors may become important."

Hubbert and Ruby (1959) noted that the *seal* to trap fluids in geopressured formations need not be totally impervious to fluid flow. In fact, the system is dynamic. He noted: "Anomalous water pressures invariably imply a hydrodynamic state. Unless the rocks are ideally impermeable, it follows that away from any pressure greater than normal pressure water must be flowing and continues to do so until excess pressure is dissipated." Therefore, in overpressured formations, such as those in the Tertiary Clastic Basins worldwide, fluid flow is still occurring, albeit too slowly to permit equilibrium conditions to prevail, and the environment is hydrodynamic. Subsequently, the terms undercompaction, mechanical compaction, and compaction disequilibrium were used to describe this phenomenon. In the field of soil mechanics, compaction means a density increase caused by mechanical or hydraulic means such as loading, vibrating, or wetting, whereas in the geological sciences, compaction means the lessening of sedimentary volume due to overburden loading, grain rearrangement, cementation, etc. While soil engineering studies contributed much to our understanding of how sediments compact under loading and subsequent expulsion of water, these studies were carried out for shallower formations and did not address issues associated with structural competency of rocks that is so relevant for deeper formations. The underlying physics and chemistry of consolidation in the latter is more complex than in soil engineering that deals mainly with the lessening of volume of porous material (such as clay) by expulsion of pore water under a load. However, the well-known Terzaghi principle (Terzaghi, 1925; Terzaghi and Peck, 1968) came out of studies in soil engineering and contributed much to our understanding of geopressure.

Since the pioneering studies by Dickinson (1951, 1953) and Hubbert and Rubey (1959) we have come to realize that there are other causes of geopressure besides undercompaction or compaction disequilibrium. Furthermore, geopressured formations have also been detected in carbonates. In the literature (Fertl, 1976; Chapman, 1994a, 1994b; Swarbrick et al., 1998), many of the causes of geopressure are mentioned. Below we have classified the geopressure mechanisms in the following three groups:

- stress related
  - vertical (compaction disequilibrium or undercompaction)
  - lateral (tectonic associated with compaction disequilibrium)
- chemical diagenesis related (thermal history dependent and activation energy controlled)
  - burial metamorphism (e.g., smectite to illite; kaolinite to chlorite, illite to mica; gypsum to anhydrite)
  - kerogen conversion and hydrocarbon generation (oil and gas)
- noncompaction related
  - charging through subsurface structures (lateral transfer of fluids)
  - hydrocarbon buoyancy
  - hydraulic head (e.g., uplift/erosion; elevation related to datum)
  - aquathermal pressuring
  - osmosis

Below we will discuss each mechanism in more detail. However, we emphasize that the listed mechanisms do not act independently of one another. In most cases, multiple mechanisms may operate. For example, water release due to compaction at depth is certainly associated with aquathermal pressuring and various chemical diagenesis processes, as listed above.

3.2 Stress Related: Vertical (Compaction Disequilibrium)

The physics behind this process is simple. As the vertical stress on a sediment column increases during active sedimentation, porosity and permeability of the sediment decrease due to compaction. This causes the pore water to escape. As long as the rate of dewatering keeps in tandem with the sediment load increase, no overpressuring results. Thus, the pore fluid pressure is hydrostatic or normal. However, for fine-grained sediments such as clay, the loss of porosity causes rapid decrease in permeability – a parameter that controls the rate of fluid flow, along with the fluid viscosity. If the rate of accumulation of sediments supports a part of the increase in overburden or vertical load, the system becomes overpressured. This is known as compaction disequilibrium – the sediments are now undercompacted as the porosity of the sediments at given depth would be larger compared to the case, if the sediments were compacted normally at that depth. The phenomenon is, however, dynamic, since no seal is perfect over geologic time. A key parameter that has a significant effect in the process is the permeability of the sediments.

   In Chapter 2 we presented Darcy's law from a thermodynamic point of view (equations (2.90), (2.92), and (2.93)) within the context of poroelasticity. We note that in most engineering applications Darcy's law is expressed as a statement that the instantaneous discharge rate through a porous medium (for a one-dimensional, homogeneous rock formation with a single fluid phase and a constant fluid viscosity) can be given by

$$Q_x = -A \frac{\kappa}{\mu} \frac{dp}{dx} \qquad (3.1a)$$

where κ is the permeability (assumed isotropic here); it has the unit of Darcy (1 Darcy = 0.986923×10^{-12} m^2); μ is the fluid viscosity (pascal or poise), Q_x denotes volumetric fluid velocity in the x direction (m/s); A is the cross-sectional area normal to the pressure gradient and $\frac{dp}{dx}$ is fluid pressure gradient in the x direction (N/m^2). The negative sign is needed because fluid flows from high pressure to low pressure. In petroleum engineering, rock permeability is usually expressed in millidarcys and viscosity in centipoise. If the linear velocity field $V = Q/A$ is considered, equation (3.1a) becomes

$$V = -\frac{\kappa}{\mu} \nabla P \qquad (3.1b)$$

Here the fluid velocity field is a vector and the permeability κ is a tensor. The petroleum industry uses a generalized Darcy equation for multiphase flow that was developed by Muskat (1937). The multiphase equation is noted as Darcy's law for multiphase flow or generalized Darcy equation. It is given by

$$V_i = -\kappa_{ri} \frac{\kappa}{\mu_i} \nabla P_i, \qquad (3.1c)$$

where the subscript i refers to each phase of the fluid and κ_{ri} is the relative permeability of phase i.

Dependence of permeability on porosity is intuitively understandable. However, it depends more strongly on the connectivity of pores in the rock. This in turn depends upon many factors, such as the size and shape of grains, the grain and pore size distributions, and other factors such as the operation of capillary forces that in turn depend upon the wetting properties of the rock. We can make some generalizations if all other factors are held constant – higher is the porosity, higher the permeability is; smaller the grains are, smaller the pores and pore throats are, and smaller is the grain size, larger the exposed surface area to the flowing fluid is, which leads to larger friction between the fluid and the rock, and hence lower permeability. The permeability of rocks varies enormously – from 1 nanodarcy, nD (1×10^{-9} D) to 1 microdarcy, μD (1×10^{-6} D) for granites, shales, and clays that form cap-rocks, to several Darcies for extremely good reservoir rocks. The low values of permeability of shales is the single most important cause of overpressure buildup during shale accretion and compaction. In addition, as discussed in Chapter 2, the permeability is in general anisotropic as the rock textures dictate often preferred orientations for fluid flow. As the permeability is a symmetric second-rank tensor it is often diagonalized to distinguish vertical from horizontal permeability and thus different values are needed to describe flow at different directions for the same elementary volume.

3.2.1   Compaction of Mudrock and Shale

Mudrock and shale are prime candidates for overpressure. (This does not rule out other sedimentary rocks like carbonates.) Gravity plays a significant role along with various geochemical phenomena. In general, compaction causes changes in the physical properties of sediments such as porosity, bulk density, compressibility, permeability, and elastic wave velocity. In clays and shales, fluid flow through porous network is greatly impacted by the chemistry of interaction between clayey sediments and that of the fluid under elevated temperature and pressure conditions. Geologically speaking, gravitational loading and geochemical phenomena are linked as the sedimentary column evolves over time. Mechanical compaction of fine-grain sediments such as clay is widely studied, the literature is abundant. Most authors (Athy, 1930a, 1930b; Baldwin and Butler, 1985; Magara, 1968; Chuhan et al., 2002, 2003; Sclater and Christie, 1980) believe that the compaction of clays is different from that of sands, shaly sands and carbonates. As mentioned earlier, higher porosity of clays, muds, and mudstones at the time of burial is significantly affected by mechanical compaction due to gravity. This is quite different for other sediments such as sandstones and carbonates where various secondary porosity generations mechanisms, such as thermal effects during burial, dominate. The literature suggests that mechanical compaction is dominant at a shallower depths to where temperature reaches $\sim 70°-80°C$ and the effective stress is the primary controlling factor of porosity. If the temperature is $> 70°-80°C$, other mechanisms such as the chemical diagenesis begin to dominate, and various other parameters such as geologic time, temperature, details of chemical reactions, and mineralogy begin to play critical roles (Bjørkum et al., 1998; Bjørlykke, 1999; Dutta, 1986; 1987b; Lander and Walderhaug, 1999; Storvoll et al., 2005). Most of our knowledge about mechanical compaction is derived from drilled and logged wells, although some controlled laboratory experiments added much to our understanding as noted below. The interplay between mechanical compaction and chemical diagenesis has a strong imprint on the models that deal with geopressuring.

Compaction affects stratigraphic thickness, water movement, and migration of hydrocarbon. Sediment begins to compact soon after it is deposited due to the weight of the overburden. Our understanding of mechanical compaction of fine-grained sediments have greatly benefited from studies on pure clays such as smectite or montmorillonite and kaolinite. In Figure 3.1, we show the compaction curves of some *pure clays* – montmorillonite, illite, and kaolinite modified after Fertl (1976). In Figure 3.2, we show more recent mechanical compaction measurements on mixture of dry and brine-saturated smectite and kaolinite aggregates under uniaxial compression (Mondol et al., 2007, 2008). Shown are the porosity reduction and bulk density variations as a function of the effective stress. Continued measurements of this type are

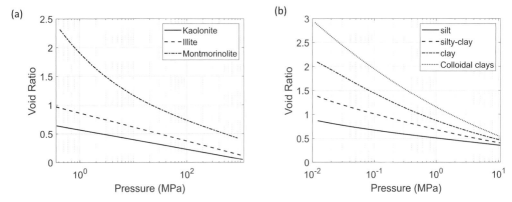

Figure 3.1 Compaction curves for pure clays. Modified after Fertl (1976).

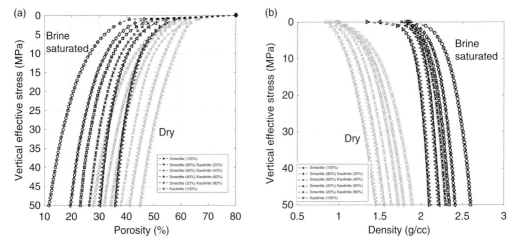

Figure 3.2 Experimental mechanical compaction of dry (in gray) and brine-saturated (in black) smectite and kaolinite aggregates and their mixtures under uniaxial compression. Porosity reduction (a) and density variations (b) as a function of vertical effective stress up to 50 MPa. Modified after Mondol et al. (2007).

highly recommended as these represent the limits of compaction in more complex geological material such as clay, shales, and shaly sands.

Porosity decrease for sands and shales is more rapid at shallow depths and slows at greater depths of burial and is consistent with many such observations in the past. Mondol et al. (2007) has compiled a good list of references (see, e.g., Bjørlykke and Hoeg, 1997; Hedberg, 1936; Athy, 1930b; Magara, 1968; Rubey and Hubbert, 1959; Dickinson, 1953; Jones, 1944; Skempton, 1944; Ham, 1966; Skempton, 1970; Baldwin, 1971; Durmishyan, 1974; Magara, 1980; Larsen and Chilingar, 1983; Dzevanshir et al., 1986; Bayer and Wetzel, 1989; Issler, 1992; Karig and Hou, 1992; Aplin et al., 1995; Vasseur et al., 1995; Hansen, 1996; Velde, 1996; Pouya et al., 1998; Yang and Aplin, 2004; Nygard et al., 2004; Aplin and Larter, 2005; Aplin et al., 2006).

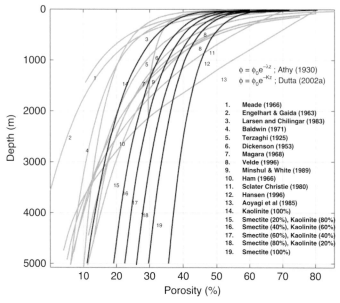

Figure 3.3 Compaction curves for shales and argillaceous sediments from various parts of the world (wireline log data). The Athy and Dutta models are discussed in the text. Modified after Mondol et al. (2007).

These studies indicate that over the last 70 years or so, a variety of empirical relationships has been proposed to express the compaction behavior of mudstones and shales as a function of increasing depth of burial or equivalently, effective stress (Terzaghi, 1925; Athy, 1930a; Hedberg, 1936; Dickinson, 1953; Rubey and Hubbert, 1959; Weller, 1959; Chilingar and Knight, 1960; Hamilton, 1976b; Baldwin and Butler, 1985; Burland, 1990; Goulty, 1998; Yang and Aplin, 2004). Since the literature is full on this subject, we decided not to reproduce these relationships here and instead encourage the readers to browse through the literature just referred and get familiar with various empirical relationships.

In Figure 3.3, modified after Mondol et al. (2007), we show a compilation of compaction curves for shales and mudstones from different parts of the world – some are data from laboratory measurements but most are trend analysis of data from wireline logs. The curves are presented as a function of depth and show a wide variability in porosity reduction, especially at shallow depths. Mechanical compaction as demonstrated in chapter 2 define porosity loss vs effective stress. However, depth is often used as a useful approximation which is easy to measure and interpret. Porosity reduction at shallow depths is, to a great extent, related to the silt content of the sediments and the ages of the sequences. These relations (porosity reduction with depth) can be described by an exponential function with depth as follows:

$$\phi = \phi_0 e^{-\lambda z} \tag{3.2}$$

where ϕ is porosity at depth z, ϕ_0 is the depositional porosity at the surface ($z = 0$), and λ is a constant with dimension of inverse length. Thus the product λz is dimensionless. Athy (1930a) is credited with proposing a model of this type. When the effective stress profile is directly proportional to depth, equation (3.2) is a useful approximation along portion of the sedimentary column as will be demonstrated below. This mathematical form is mostly valid for argillaceous clastics (shales and mudstones) and less so for sands, although the coefficient λ is a more complicated quantity for mechanical compaction of sands (Ramm and Bjørlykke, 1994). Since the introduction of Athy's model, many authors had suggested several variations, all with depth as a variable. Magara (1980) has also presented a compilation of these models.

The compaction process described by the Athy-type relations as given above cannot be transported from one basin to another or even in the same basin but at different locations since the description of the process is tied to depth. Depth embraces time (t), temperature (T), effective stress (σ), cementation, etc., all of which typically increase with depth. While temperature and overburden stress may be approximated by a linear function of depth, it is not at all certain that time or other variables are well approximated by a linear function of depth. Athy was aware of this limitation. In fact, he noted (Athy, 1930a) that the amount of compaction is not directly proportional to either reduction of pore volume or to increase in bulk density because of geologic processes such as cementation, pressure solution, and recrystallization. Therefore, it is advisable to look at the genesis of compaction and use fundamental geologic variables such as stress, temperature, and time to describe the process. This requires a brief discussion of the stress state of argillaceous sediments.

We can visualize sediments under stress (loading) as a two-phase elastic continuum – a porous space occupied by the fluid in which the pressure (P) is uniform in all directions and an elastic rock phase in which the grains are in contact and supporting loads which is usually different in each direction; σ_v is the vertical effective stress and σ_h is the horizontal effective stress. Overburden stress S is equal to the sum of vertical effective stress and pore pressure P as in Terzaghi's law:

$$S = \sigma_v + P \tag{3.3}$$

Using these fundamental variables, we can recast Athy's relation, equation (3.2), in terms of vertical effective stress σ_v. From Terzaghi's law, for normal pore pressure condition, we have

$$\sigma_n = S - P_n \tag{3.4}$$

where P_n is the hydrostatic or normal pressure at a given depth z and S is the overburden stress at that depth. For the sake of simplicity, we assume that the sediments up to the depth z can be characterized by an average bulk density $\langle \rho_b \rangle$. Then, approximately,

$$S = \langle \rho_b \rangle \, gz \tag{3.5}$$

where g is the acceleration due to gravity. Then, the expression for the normal effective stress in equation (3.4) reduces to

$$\sigma_n = (\langle \rho_b \rangle - \rho_w)gz \qquad (3.6)$$

where ρ_w is the water density. Athy's relation, equation (3.2), for the compaction can then be expressed by the following expression:

$$\phi = \phi_0 e^{-K\sigma_n} \qquad (3.7)$$

which is equivalent to equation (2.109a), where

$$K = \frac{1}{(\langle \rho_b \rangle - \rho_w)g} \qquad (3.8)$$

Taking the natural logarithm of both sides of equation (3.7), we find

$$\ln \phi = \ln \phi_0 - K\sigma_n \qquad (3.9)$$

This is a straight line when σ_n is plotted versus $\ln \phi$. A similar approach can be followed (without assuming that the shales have a constant bulk density) for any depth-dependent compaction model, and a relation will be found between porosity and effective stress under normal compaction condition. Compaction models of the type given in equation (3.7) should be preferred over those models that have explicit depth dependency such as Athy's relation (equation (3.2)). As noted earlier, depth-dependent models are difficult to transport from one basin to another.

The simple compaction model described above deals with only mechanical compaction. It ignores variables such as mineralogical variations within shales, and other effects related to temperature and burial history. Some of these are known as chemical diagenesis (see below). These models can be described as follows (Dutta, 1986, 1987a, 1987b):

$$\sigma = \sigma_0 e^{-\beta \varepsilon} \qquad (3.10)$$

which is equation (2.105b). The parameter β may depend on shale composition (mineralogy), temperature and time, σ_0 is a shale composition dependent parameter at zero porosity and ε is the void ratio – the ratio of the pore space volume to that occupied by the solid material

$$\varepsilon = \frac{\phi}{1 - \phi} \qquad (3.11)$$

Another form of compaction model is also used and may form the basis for Eaton-type models (Eaton, 1975) for geopressure analysis:

$$\sigma = A\phi^B \qquad (3.12)$$

This is a simple power law dependency of effective stress on porosity. Coefficients A and B are determined by local calibration. It ignores chemical diagenesis and shale mineralogical variations.

The compaction models discussed above involve coefficients that are not universal and their local values take local conditions into account. In all practical applications, compaction curves require measurement of in situ porosity ϕ. However, this is only possible through *indirect* measurements by using borehole logs. Also, as presented in Figure 2.20, many of the compaction models can produce similar porosity-effective stress curves when calibrated.

In the industry, both density and acoustic (sonic) logs are used for this purpose. In the absence of tectonic stress, the confining stress is the vertical overburden stress and the vertical effective stress is the difference between the overburden stress and pore pressure as discussed earlier in the context of Terzaghi's law. As the vertical effective stress increases with depth as in the case of normal compaction – a situation where the rate of expulsion of the fluid from the sediments keeps up with the rate of overburden increase, porosity decreases until it reaches a maximum value just as the pore pressure begins to increase over the hydrostatic pressure at that depth. This could happen, for example when the lithology changes from sand prone rocks to shale pone rocks. Sediment compaction depends on the loading history – it can compact inelastically (irreversibly) or elastically (reversibly) (Miller et al., 2002). Porosity is reduced irreversibly during virgin compaction but rebounds and recompacts reversibly during unloading and subsequent reloading. The porosity change follows the virgin compaction curve as long as the effective stress on the sediment is increasing and is at its maximum value. Porosity change follows the unloading/reloading curve whenever the effective stress is less than the previous maximum value. Figure 3.4, modified after Bowers (2002), shows the experimental data on Cotton Valley Shale by Tosaya (1982) to illustrate this point. Shown are the data points and Bower's (2002) model expressing velocity and density as functions of effective stress. This type of response can be caused by uplift and erosion of rocks or by active fluid source in the sedimentary column such as various thermal mechanisms within shale (e.g., shale metamorphism)

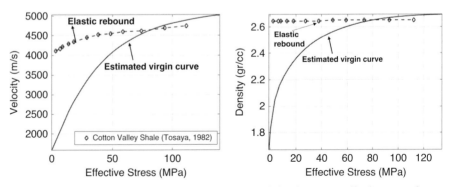

Figure 3.4 A plot of the experimental data on velocity and density versus effective stress from Tosaya (1982) and comparison with the model predictions by Bowers (2002).

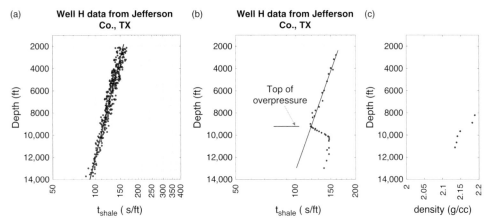

Figure 3.5 Typical response of sonic travel times versus depth of burial of shales from Gulf of Mexico for (a) hydrostatic and (b) overpressured conditions and in (c) the bulk density in the undercompacted or overpressured zone. Modified after Hottmann and Johnson (1965).

that provide additional fluid sources and thus increase pore pressure. This is termed as chemical diagenesis, and we shall discuss this in detail below.

It is worthwhile to note at this point the wireline log responses to compaction of shale. Compaction causes porosity to decrease. If the fluids are allowed to leave the system at the same rate as the rate of deposition, the pore fluid pressure is hydrostatic. If the fluid flow is retarded, the system is overpressured. These pressure signatures are often clearly visible in acoustic logs that measure the speed of sound waves in boreholes and are indicative of the velocity of rocks. In Figure 3.5, we show the typical response of sonic travel times (slowness or inverse of velocity) versus depth of burial of some shales for (a) hydrostatic and (b) overpressured conditions and in (c) the bulk density in the undercompacted zone. In Figure 3.5a all data points fall along the normal compaction trend line that is characteristic of hydrostatic pressure condition for shales. Overpressure can be detected by an *increase* in transit time (decrease in velocity) above the normal trend line values as in Figure 3.5b or a decrease in the bulk density, as shown in Figure 3.5c. This will be further discussed in detail in Chapter 5, when we discuss how various well logging methods can be used for quantitative evaluation of geopressure.

Compaction of Nonsiliceous Sediments

As described by Bjørlykke (2015), for carbonates and other nonsiliceous rocks, mechanical compaction does not play a major role on porosity loss with the exception of near-surface sediments ($\sim$ 1–1.5 km). At these shallower depths (less than $\sim$ 1–1.5 km) mechanical compaction processes control the rate of sediment compaction. At deeper depths, various other chemical processes involving fluid chemistry and rock–fluid interaction begin to play important role. Therefore, unlike siliceous sediments, carbonate compaction cannot be modeled only by mechanical compaction at depths corresponding to potential hydrocarbon reservoirs. However, the state reached by sediments after initial mechanical compaction determines the extent of grain-to-grain contacts and their sizes, which affects further porosity loss by subsequent chemical effects. In the literature, this is often referred to as chemical compaction. (We prefer the term chemical diagenesis.) Natural and experimental observations show that pressure solution creep and fluid–rock interactions are the main processes of porosity loss in carbonates (Aatshbari, 2016). Both theory and experiments show that pressure solution depends on effective stress, porosity, grain size, and pore fluid chemistry.

Compaction of chalk has been studied extensively in North Sea (Jaspen et al., 2011). These studies showed a pronounced porosity drop, from ~20 percent to less than 10 percent within a depth of less than 300 m. An interesting conclusion from these studies is that compaction behavior is closely tied to an effective stress (overburden stress minus fluid pressure) of ~17 MPa and is independent of stratigraphic position, temperature, and actual depth. High overpressures are known to occur in all types of carbonates, including chalks. A survey of literature showed that although numerous measurements on carbonates are available (rock strength, porosity, permeability, etc.), these are mostly local in nature, and there is a lack of predictive models based on the data that can be transported from a basin to another basin.

3.2.2   Fluid Flow in Low-Permeability Rocks: A Simple Hydrological Model

Permeability and porosity of compacting sediments are two of the most important parameters that affect the rate of fluid flow in rocks and hence, geopressure. How far will the fluid flow and how long will it take to achieve equilibrium? An approximate answer may be provided by a simple hydrodynamic formulation, and it is worthwhile to discuss it here. The basic fluid flow is given in equation (2.100) in terms of poroelastic constant and is further simplified for flow in 1D in Bear (1972):

$$\phi \beta_f \frac{\partial P}{\partial t} = \frac{\partial}{\partial z} \left(\frac{\kappa}{\mu} \frac{\partial P}{\partial z} \right) - \frac{1}{1-\phi} \frac{\partial \phi}{\partial t} \tag{3.13}$$

where $P(z, t)$ is the fluid pressure at a depth z and time t in an isotropic and porous medium of porosity, ϕ, and vertical permeability, κ. The viscosity of the fluid is μ, and the compressibility of the fluid is β_f. Noting that

$$\frac{\partial \phi}{\partial t} = \frac{\partial \phi}{\partial \sigma} \frac{\partial \sigma}{\partial t} \tag{3.14}$$

and using Terzaghi's law, namely,

$$\sigma = S - P \tag{3.15}$$

we get (assuming that S, the overburden stress, is constant)

$$\frac{\partial \phi}{\partial t} = \frac{\partial \phi}{\partial P} \frac{\partial P}{\partial t} \tag{3.16}$$

Thus, we have

$$\frac{1}{1-\phi}\frac{\partial\phi}{\partial t} = \frac{1}{(1-\phi)}\frac{\partial\phi}{\partial P}\frac{\partial P}{\partial t} \qquad (3.17)$$

Let us define the compressibility of the porous rock framework β_r by

$$\beta_r = \frac{1}{\phi(1-\phi)}\frac{\partial\phi}{\partial P} \qquad (3.18)$$

Then equation (3.13) can be rewritten as

$$\frac{\partial}{\partial z}\left(\frac{\kappa}{\mu}\frac{\partial P}{\partial z}\right) = \phi(\beta_f + \beta_r)\frac{\partial P}{\partial t} \qquad (3.19)$$

If we further assume that both κ and μ are constants, then we have

$$\frac{\kappa}{\mu}\frac{\partial^2 P}{\partial z^2} = \phi(\beta_f + \beta_r)\frac{\partial P}{\partial t} \qquad (3.20)$$

This can be rewritten in the form of a diffusion equation.

$$\frac{\partial^2 P}{\partial z^2} = \frac{1}{\chi}\frac{\partial P}{\partial t} \qquad (3.21a)$$

where the diffusivity χ is defined by

$$\chi = \frac{\kappa}{\mu\phi(\beta_f + \beta_r)} \qquad (3.21b)$$

Solutions of equation (3.21a) for various sources and media are discussed in Bear (1972). It is easily shown that for the one-dimensional diffusion equation with constant diffusivity, the solution is

$$P(z,t) = \frac{A}{\sqrt{t}}e^{-x^2/4Dt} = \frac{A}{\sqrt{t}}e^{-\tau/t} \qquad (3.22)$$

where A is a constant. The characteristic time τ depends on the scale length z and the diffusivity and is given by

$$\tau = \frac{z^2}{4\chi} \qquad (3.23)$$

Zoback (2007) showed that for relatively compliant sedimentary rocks in low-permeability sands with the permeability of about 10^{-15} m^2, the characteristic time for fluid transport over the length of 0.1 km is of the order of years – a relatively short time in geological terms. However, in low-permeability shales where κ is $\sim 10^{-20}$ m^2 (~ 10 nD), the diffusion time for a distance of 0.1 km is $\sim 10^5$ years, which is clearly sufficient time for increase in compressive stresses due to sediment accretion, or tectonic compression, and to enable compaction-driven pressure to build up faster than it can dissipate.

Bredehoeft and Hanshaw (1968) and Smith (1970) discuss solutions for pressure diffusion through porous rocks under simple conditions of porosity and permeability of fluid saturated medium. In Chapter 10 we discuss a solution of pore pressure buildup for a continuously accreting and dissipating medium with *variable* diffusivity and overburden under *gravitational* loading (compaction disequilibrium) and undergoing *chemical diagenesis*.

3.3   Stress Related: Lateral (Associated with Compaction Disequilibrium)

Tectonic compression is a well-known mechanism for causing overpressure. It involves large-scale stress changes that cause compaction disequilibrium. Tectonic compression can occur both locally and regionally. The movement of the Earth's plates, faulting and folding, lateral sliding and slipping, down dropping of fault blocks, and salt and shale diapirism are all causes of tectonic compression which leads to shale compaction disequilibrium and overpressure. Within active compression, concentration of abnormally high pressure occurs in the bending region. This phenomenon causes the fluid to be compressed within the pores of the rocks, inhibiting its ability to escape because of the existence of sealed rocks in the overburden. These areas, which are generally folded, contain highly overpressured rocks. The magnitude of overpressure is related to the magnitude of stress and deformation of rocks. It is important to note that horizontal stresses can cause failure and yield of the material. For mechanical compaction, equations (2.120) and (2.121) present the relation between the three components of the effective stress tensor and the pore pressure for mechanically compacting reservoirs.

Many authors (Dickinson, 1953; Dickey et al., 1968; Carver, 1968; Harkins and Baugher, 1969; Fowler, 1970) correlated various types of faults with development of abnormally high pore pressure. These authors related the pattern of abnormally high pressure zones to faulting contemporaneous with sedimentation and fluid migration due to compaction. The process creating these faults is termed as growth faults. These faults often prevent expulsion of pore water from the accumulated sediments. (It is not clear from the literature whether one can predict whether a fault is sealed or not without actual pressure measurements.) This suggests that tectonic compression in the sense of horizontal deforming stresses cannot be a general cause of overpressure, since many basins such as the US Gulf Coast and Niger Delta are tectonically passive. In these areas the evidence suggests that there has been prolonged extension and subsidence during the development of growth faults in stress fields where the maximum principal compressive stress σ_1 is vertical, σ_2 is horizontal and parallel to the depositional strike, and σ_3 is horizontal and normal to the depositional strike (see Chapter 2).

Dickey et al. (1968) suggested the mechanism to be associated with gravitational loading and subsequent dewatering of fluid (water) in the vertical direction. As compaction progresses, the vertical permeability decreases drastically causing the horizontal permeability in the shales to provide pathway for fluid movement parallel to the bedding planes of the shales. If growth faulting occurs while abundant water is still present in the shales, the pathways to fluid migration parallel to bedding planes will be cut off by the fault planes. As a result, the fluid will begin to support the load due to overburden and overpressure will result. If growth faulting occurs when most of the fluid has been expelled and shales were already compacted, the abnormal pressure would be much lower or even hydrostatic. This phenomenon is well documented in the literature (Fertl, 1976; Dickey et al., 1968) by data from wells drilled into such formations when the well crosses a fault plane. Quite often, the well encounters a high pressure zone and then, after crossing a fault plane, enters a different fault block where the pore pressures are normal or hydrostatic. Drilling in such environments could be challenging.

Shale diapirism and mud volcanoes have been known to cause high pore pressure (Fertl, 1976). This happens due to density inversion including a material of low shear strength. Such conditions are found in delta areas of the major rivers, such as the Mississippi, Niger, Nile and Amazon (Gretner, 1969). In these environments, small shale diapirs are formed by shales having high porosity (high fluid content) and low shear strength that have been rapidly loaded by sands (high permeability). In this case large-scale mud volcanoes are formed. These represent an overpressure phenomenon caused by an intrusion at depth of mud or a mixture of mud and solid rocks. In Azerbaijan, this phenomenon has been pegged as the major cause of overpressure where fluid pressure gradients approaching 0.90 psi/ft (0.208 kg cm^{-2} m^{-1}) were observed. Mud volcanoes are also known to occur in the episodic phases such as in the Niger Delta area. A very similar phenomenon occurs in the subsurface due to salt tectonics. For example, as salt domes grow, they drag sedimentary layers along with them often causing overpressure in the sediments, as salt provides a barrier to fluid flow. (However, dragging of sediments by salt tectonics can cause fractures and this may provide pathways for fluids to escape, leading to normal pore pressure.) This may provide a habitat for hydrocarbons.

3.4   Chemical Diagenesis as a Geopressure Mechanism

Previously we discussed the release of compaction water from shale and its inability to escape the host rock due to low permeability (shale) as an important mechanism of. geopressure. There also are other sources of water that can cause overpressure. All of these sources are due to chemical diagenesis and the phenomenon is thermal history dependent. Examples of the processes are

- burial metamorphism of shale, e.g., smectite to illite; kaolinite to chlorite; illite to mica transformation and in carbonates, e.g., gypsum to anhydrite transformation; and
- kerogen conversion to oil and oil to gas and thermal cracking of kerogen to gas.

All of these processes are kinetic in nature and involve volumetric changes (pore volume increase) with minimal or no changes in porosity and at the rate that does not permit fluid dissipation out of the system (host rock) due to reduced permeability. Therefore, before discussing specific cases listed above, it would be worthwhile to review briefly the kinetics of chemical diagenesis. This is a relevant subject for geophysicists.

3.4.1   Chemical Kinetics

Chemical diagenesis is the basis of petroleum generation such as kerogen to oil and oil to gas generation, as well as many chemical processes that occur in rocks such as various mineral transformations that alter the rock properties. These processes are controlled by the time and temperature history of rocks. The process is thus kinetic in nature. These reactions of interest for us are typically of the form

$$A \rightarrow B + C$$

where A is a reactant and B and C are products. Most theories of chemical kinetics have evolved from the idea of collision of molecules, either in the gas phase or in solutions, resulting in a chemical reaction. All chemical reactions (reactants to products) are describable by the "rate theory" or "collision theory." Collision theory says that a chemical reaction can only occur between particles when they collide. There is a minimum amount of energy that particles need in order to react with each other. If the colliding particles have less than this minimum energy, then they just bounce off from each other without reacting. This minimum energy is called the *activation energy*. The faster the particles are moving, the more energy they have. Fast moving particles are more likely to react. In the geochemistry literature, there are the following ways to increase the rate of reaction:

1. increasing the temperature,
2. increasing the concentration in solution,
3. increasing the pressure in gases,
4. increasing the surface area of a solid, and
5. using a catalyst (a catalyst is a substance that will change the rate of a reaction; a catalyst is often used to accelerate reaction).

A reaction of the type $A \rightarrow B + C$ is expected to follow the *first-order* rate law. Then the following equation holds:

$$\frac{d[A]}{dt} = -k[A] \tag{3.24}$$

Here the parentheses indicate the *concentration* of the reactant A. Reactions of the type in equation (3.24) are known as the *first-order* chemical reaction. Chemical reactions are temperature dependent. The temperature dependency is usually expressed by the Arrhenius equation (Arrhenius, 1889)

$$k(T) = Ae^{-\frac{E}{RT}} \tag{3.25}$$

or in logarithmic form:

$$\ln k(T) = \ln A - \frac{E}{RT} \qquad (3.26)$$

Here T is the absolute temperature in Kelvin, E is the activation energy of the reaction (in J/mol or cal/mol), and R is the universal gas constant (J mol^{-1} K^{-1} or cal mol^{-1} K^{-1}). The pre-exponential factor A (in equation 3.25) has the dimension of a frequency (t^{-1}). It is therefore sometimes termed as "frequency factor." The factor RT is a measure of the thermal energy of the system at a given temperature T. If this factor is small compared to the activation energy E, the reaction rate constant k is also small. Activation energy can be considered as the height of the energy barrier that must be overcome in going from a metastable state (such as kerogen) to "more stable" products (such as oil). Here "more stable" means having lower free energy. At any given temperature, the reactant molecules will have a distribution of thermal energies. Although most reacting molecules would be close to the average, a few will be very low and a few will be very high. As the temperature is raised, a greater percentage of the molecules will have sufficiently high energy to pass the barrier and convert to products. Below we use these concepts in developing a theory for burial diagenesis of shale.

3.4.2 Burial Metamorphism of Shale

Compaction of shales at deeper depths is dependent upon its loading history that alters shale composition, mineralogy and pore pressure through what is known as the chemical diagenesis or burial metamorphism of shales. It is a complex phenomenon; both initial composition of mud that turns into shale, and temperature and geologic time (burial history) play critical roles. The pore water chemistry is also an integral part of this process. There is an enormous amount of literature on the subject. There are two good sources for further study: Potter et al. (2012) and Bjørlykke (2015).

Chemical diagenesis of shale is related to the change in the chemical composition and migration and evolution of fluids (water), and thus it relates to evolution of geopressure. At increasing temperatures and pressures associated with burial, the sediments change chemically and release water in a complex series of nonequilibrium, irreversible reactions that usually occur gradually over a long period of time and elevated temperature. Although the literature is rich in the area of shale burial diagenesis, it has remained mostly isolated in the clay chemistry and geology disciplines and did not receive the attention it deserves in the geophysics discipline other than some qualitative discussions and hypotheses regarding its importance in evolution of geopressure, for example, by Swarbrick and Osborne (1998) and Bowers (2002). The chemical reaction of importance in our discussion of shale diagenesis is smectite – illite transformation (S/I transformation). The following discussions are based on the pioneering work of Hower et al. (1976), Boles and Franks (1979), Eberl and Hower (1976), Foster (1981), Freed (1979), Weaver (1979), Howard (1981), McCubbin and Patton (1981), Perry (1969), Perry and Hower (1970, 1972), Morton (1985), Hall (1993), and Lahann (2002). Dutta (1986, 1987b) also made contributions to this effort.

Smectite and illite are both members of the micaceous clay family. Smectite is also known as montmorillonite. During burial, temperature increases and studies by X-ray diffraction have shown that smectite content gradually decreases and illite abundance increases. (It requires K-feldspar as a catalytic agent.) In a potassium-rich environment, smectite progresses toward illite through formation of a series of randomly and then regularly interstratified mixed-layer illite/smectite intermediate products (Hower et al., 1976; Morton, 1985). In a magnesium-rich environment, smectite can transform into chlorite through a series of irregular and regular mixed-layer smectite/chlorite transition minerals by fixation of magnesium. Progressive illitization of smectite occurs not only in Gulf of Mexico, USA, but worldwide, although it was first reported in Tertiary clastics province of the USA by Burst (1969). Foster (1981) found the following chemical reaction that converts smectite to illite in a potassium-rich environment:

$$\text{Smectite} + \text{K-feldspar} \rightarrow \text{Illite} + \text{Na-feldspar} + \text{Silica} + \text{Water} \qquad (3.27)$$

As mentioned earlier, K-feldspar is a catalyst for the reaction. This transformation begins at $\sim 55°C$ (Perry and Hower, 1972) and ends at $\sim 150°C$. X-ray diffraction studies on fine-grained samples have shown that illitization occurs with decrease of expandable layers. This decrease results in the migration of the primary mixed-layer peak from 17 Å (smectite) toward 10 Å (illite). In Figure 3.6, modified after Dutta (1987b), we show the percentage of illite in the mixed-layer S/I system as a function of temperature from a deep Gulf of Mexico well. Hower et al. (1976) characterized the diagenetic transformation by an "order" parameter, RN, based on the degree of interstratification of smectite and illite layers in the S/I clay and the illite fraction. They defined RN to indicate that each pair of smectite layers are separated by at least N layers of illite layers. Thus, $R0$ is defined to be the case where smectite and illite layers are in the random phase. Creation of each ordered phase is also associated with the release of bound water from the clay into its pore space. Thus as the diagenetic alteration progresses, not only the smectite fraction decreases from its original fraction and the illite fraction increases, but also the ordering of the layers changes gradually from the random phase $R0$, to a short range ordered phase $R1$, several long range ordered phases $R2$, $R3$ and finally terminating with the 100 percent illite phase. Figure 3.7a shows a schematic representation of a hypothetical 20-layer clay crystallite composed of 70 percent smectite and 30 percent illite that is viewed from its base plane. This composite transforms into an $R1$ phase as shown in Figure 3.7a due to chemical diagenesis. The structural units of the individual smectite (15 Å) and illite (10 Å) components are shown in Figures 3.7b and 3.7c, respectively. Thus, the illitization process definitely leads to increased anisotropy in the system due to increased ordering of layers from the random phase, $R0$ to RN (N > 0). As the result of loss of interlayer water during chemical diagenesis, the clay crystallite collapses from 270 to 200 Å, as shown in Figure 3.7a, expelling water into the pore space of the host shale. If this *extra* water is allowed to move out of the shale's pore system, then pore pressure will not change. However, if the low permeability of shales does not allow this move, pressure increase will occur.

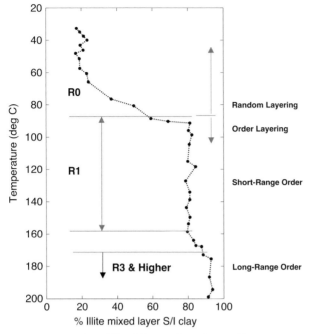

Figure 3.6 Smectite to illite (S/I) transformation profile versus temperature for a shale sample from a well in the on-shore environment of Louisiana. Measurements were made by X-ray diffraction on fine-grained (<5 microns) samples. Figure courtesy of Professor John Hower (personal communication, 1981). The well is overpressured at a depth of ~14,700 ft. Hower has identified several phases of S/I transformation with increased ordering of S/I layers in the mixed-layer clay system.

From the perspective of a geophysical response, the following are the reasons why a study of shale burial diagenesis (chemical digenesis) is important:

1. Release of bound water through smectite-to-illite transformation affects geopressure evolution.
2. Reordering of smectite/illite layers takes place, causing increased anisotropy and effective stress change.
3. Shale compaction (porosity reduction with increased effective stress) trends are significantly affected by temperature and geologic time.
4. Acoustic properties (density, compressibility, velocity, etc.) of shales are significantly affected by the process.
5. Mass transfer from altered shales to nearby sands (reservoir) will affect the reservoir quality of the sand through cementation and secondary porosity.
6. Shale diagenesis provides the link for rock properties trend shifts between deepwater (younger) and shallow water stratigraphy (older). This understanding will enable us to transport rock properties trends from one basin to another.
7. Depth where significant illitization occurs may dictate the "seal" integrity of hydrocarbon targets (prospects) because illitic seals are easier to breach under elevated pore pressure.

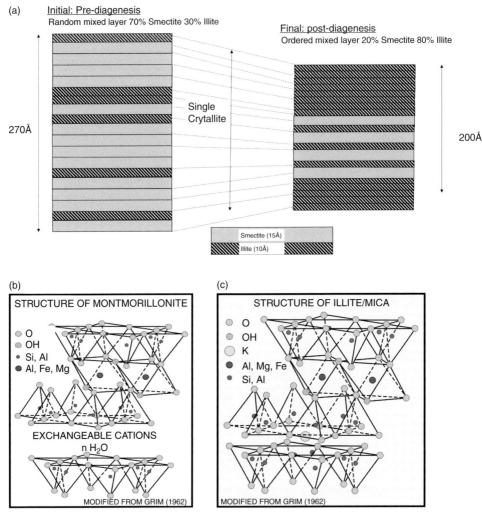

Figure 3.7 (a) Schematic representation of mixed-layer smectite–illite diagenesis in a 20-layer hypothetical crystallite. The smectite component is assumed 15 Å thick and contains two interlayers of water. The loss of water layers due to diagenesis leads to a 10 Å structure, which, when accompanied by other changes (not shown), results in illite. (b) Crystalline structure of illite and (c) smectite. Crystallographic structures after https://pubs.usgs.gov/of/2001/of01-041/html docs/images/monstru.jpg and https://pubs.usgs.gov/of/2001/of01- 041/htmldocs/clays/illite.htm.

Modeling Shale Burial Diagenesis

As mentioned above shale diagenesis is an important component of pore pressure buildup; there is an important interplay between mechanical and chemical reactions, and it influences on when and how high pressure evolves in clastic sedimentary basins. It can also affect pore pressure in carbonate reservoirs, due to occurrences of smectite-rich clays in carbonate deposition environments (Bjørlykke, 2015). Any

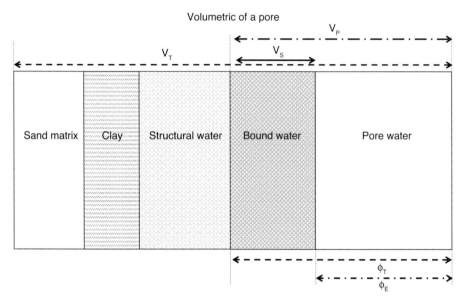

Figure 3.8 A schematic of the distribution of water in shaly sands. Definitions of various porosities, such as total and effective porosity, are also presented.

modeling exercise dealing with shale diagenesis must address the issue of release of water and whether it can move out of the system (host rocks) or not, as well as shale mineralogical changes. Smectite layers in mudrock contain water in three ways: "mobile or intergranular or pore water," "interlayer or bound water," and "structural water." In Figure 3.8, we show schematically the distribution of water in shaly sands. A definition of various porosities, such as total and effective porosities, are also presented; this is important, since various borehole tools may or may not provide the information on the type of porosity it measures. For example, direct density measuring tools yield information on the *total* porosity (*sum* of bound water and intergranular water) whereas porosity measured by neutron log yield data on structural porosity, as it deals with the hydrogen contained in the hydroxyl units of the clay platelets (structural water). During chemical diagenesis, the "bound" water is released in the pore space, and a concomitant redistribution of effective stress happens from the solid grain framework to the pore fluids (rebalancing of overburden load). This results in over-pressure, if the rate of water movement out of the system is not rapid enough to keep up with water accumulation.

We specifically address the numerical modeling of the S/I chemical reaction, which is the most studied reaction. However, the same formulation is equally applicable to other first-order chemical reactions, provided activation energy parameters for these reactions are known (see below). The formulation given here is based on the earlier work by Hower et al. (1976), Foster and Custard (1980), Freed (1981), Bruce (1984), and Dutta (1986, 1987b). Dutta (1986) showed the process to be *thermal history* dependent. The essential elements and assumptions of the model are described below.

Let $N_s(0)$ and $N_s(t)$ be the molar fractions of smectite in the host rock at time $t = 0$ (at the start of the reaction) and at a later time t, respectively. We assume that the smectite–illite reaction can be expressed by the first-order chemical rate theory, namely,

$$\frac{dN_s(t)}{dt} = -kN_s(t) \qquad (3.28)$$

The rate constant k is given by the Arrhenius equation (3.25). Combining equations (3.25) and (3.28) and integrating with respect to t, the smectite fraction at a time t, $N_s(t)$, is given by

$$N_s(t) = N_s(0)e^{-\int_0^t A\exp\left(-\frac{E}{RT}\right)dt} \qquad (3.29)$$

Here E is the activation energy of the Arrhenius model. We define a *diagenetic function* β, first introduced in the compaction model in equation (3.10), to characterize the smectite-to-illite transition as follows:

$$\beta = B_0 N_s(t) + B_1(1 - N_s(t)) \qquad (3.30)$$

Both $N_s(t)$ and $N_s(0)$ are dimensionless fractions. However, the vertical effective stress in equation (3.10) has a dimension and a physical range of values. Hence, we needed to introduce the additional constants B_0 and B_1 in equation (3.30). These should be treated as calibration constants. We assume $N_s(0)$ to be unity. For the activation energy and the frequency factor of equation (3.26) Hower et al. (1976) proposed a value of $E = 19.6 \pm 3.6$ kcal/mol and $A = 2s^{-1}$ for conversion of synthetic Beidellite to a mixed-layer S/I in laboratory experiments. Dutta (1987b) obtained a value of $E = 19.3 \pm 0.7$ Kcal/mole and $A = 0.4 \times 10^5$ (year)$^{-1}$ by using X-ray diffraction data on the conversion of smectite to illite from the data of Hower et al. (1976) and Freed (1981). Examples of the data fit for extracting the activation energy and preexponential factors are given in Dutta (1987b, Figure 9) and Dutta (2002b, Figure 2). These values were further confirmed by using more data from basins outside the Gulf of Mexico (Dutta et al., 2014; Dutta, 2016).

In the traditional Arrhenius model, the temperature T is the only variable that controls the rate of chemical reaction. However, this is not adequate in geology and geochemistry where the chemical processes are significantly affected by geologic time t. In the above model, geologic time is included explicitly through a model for T as follows. For the sake of simplicity, we assume that the present day geothermal gradient α did not change during the paleo-time, and T_0 is the reference temperature (sea-bed temperature, for example). Then the temperature T at a given depth $Z(t)$ of the compacting sedimentary column is

$$T(t) = T_0 + \alpha Z(t) \qquad (3.31a)$$
$$Z(t) = \gamma t \qquad (3.31b)$$

and thus,

$$T(t) = T_0 + \alpha\gamma Z(t) \qquad (3.31c)$$

Here γ is the burial rate between time $t = 0$ and t, α is thermal gradient, and these are assumed to be constant in this time interval. In actual numerical applications, we should approximate the burial rates by piece-wise linear functions in smaller time steps (layers) with variable burial rates within each layer. In this way, the temperature T in equation (3.28) *depends* on geologic time. (We assumed that the term $\alpha\gamma$ is independent of geologic time.) The term $\alpha\gamma$ in equation (3.31c) has the dimension of temperature/time, and it is termed the *rate of heating*. Therefore, the extent of the diagenetic reaction is controlled by the time duration of the material (smectite) staying in a given temperature window, and *not by temperature alone*. Lahan et al. (2002) also came to the same conclusion. However, he favored a kinetic model that used a second-order rate theory instead of the first-order rate theory used in the present model.

The diagenetic function is the "modulating" factor that affects the effective stress versus void ratio of shale (equation (3.10)) and was first introduced by Dutta (1986). The void ratio ε, is defined as the ratio of the pore volume of a rock to that of the solid rock (equation (3.11)). The generalized compaction model given in equation (3.10) indicates that as the effective pressure in the sediment increases, the porosity decreases for a *fixed* value of β, namely, a fixed diagenetic state. This is the so-called *Type-1 or the mechanical compaction process*. However, the model also shows another possible compaction process in which the porosity change is negligible but the effective pressure changes significantly due to alteration of shale- mineralogy. This is associated with the changing value of β as shown in Figure 3.9a. Figure 3.9b shows the nature of the water-release curve for this alternate compaction process. This is the so-called *Type II or chemical diagenesis process* (Dutta, 1986) and *"fluid expansion" process* as discussed by Bowers (1995). However, in the model presented here, the transition from mechanical compaction to chemical diagenesis is *not* abrupt. It is a gradual process dictated by the shape of the diagenesis function which in turn is controlled by the activation energy and the thermal history of the sediments. Initial smectite abundance has a significant impact on pore water freshening because it controls both the total amount of bound water available for release and the magnitude of the porosity correction that defines the pore water volume. The model assumes the availability of potassium feldspar as a catalytic agent.

The modeling curves of Figure 3.9a showing a plot of the diagenetic function is appropriate for a sedimentary basin that has undergone the typical subsidence of the Miocene age with a constant thermal gradient of 1.20°F/100 ft. The first derivative of the curve in Figure 3.9a is the amount of bound water released as the result of diagenesis, and is shown in Figure 3.9b. The temperature, at which the peak water release occurs, i.e., $dN_s(t)/dt = 0$ in Figure 3.9b, is given by,

$$E/RT_m = \ln{(ART_m^2)}/\gamma\alpha E \qquad (3.32)$$

This equation can be solved iteratively. Using the values of E and A given earlier, equation (3.32) yields $T_m = 98°C$ for a burial rate of 0.5 ft/1000 years; $T_m = 107°C$ for a burial rate of 1.0 ft/1000 years, and $T_m = 117°C$ for a burial rate of 2.0 ft/1000 years,

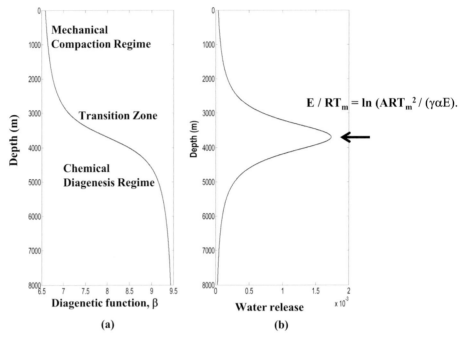

Figure 3.9 (a) A plot of the diagenetic function β for a fixed burial rate of 1000 ft/million years and a fixed thermal gradient of 0.012°F/ft. (b) First derivative of (a) showing the nature of the water-release curve due to S/I transition.

respectively. Thus, the peak water release due to chemical diagenesis shifts deeper and deeper with increase in burial rate.

Lopez et al. (2004) documented a relationship for shale compaction that included explicit dependence on temperature and shale mineralogy. In terms of effective stress, this is expressed as:

$$\sigma = \sigma_0 e^{-\frac{\lambda(T)\varepsilon}{CEC}} \tag{3.33}$$

The authors defined the function $\lambda(T)$ as

$$\lambda(T) = a + bT \tag{3.34}$$

Here ε is the void ratio, T is temperature, and a and b are calibration coefficients, and CEC is the cation exchange capacity. The cation exchange capacity (CEC) is a measure of the shales' ability to hold positively charged ions. The clay minerals in shales have negatively charged sites on their surfaces which adsorb and hold positively charged ions (cations) by electrostatic force. Thus, it is a parameter that indicates shale mineralogical variations (smectite to illite, for example). This parameter defines the amount of cations capable of exchange and is usually expressed in millequivalent weight per 100 g of rock (meq/100 g). As noted by Lopez et al. (2004), the form of the compaction model in equation (3.30) was first proposed in the Shell literature during the early 1980s by Sims and Dutta (see also Dutta, 2002b; Wilhelm, 1998) and it is referred to within Shell as the Sims–Dutta model (as reported by Lopez et al., 2004, p. 54). In Figure 3.10a, modified after Lopez et al. (2004), X-ray diffraction data on the cuttings from a deep well in the Ursa area (deepwater Gulf of Mexico, USA) are presented. This shows clearly the extent of the smectite-to-illite (S/I) transformation. The authors used the *CEC* model using a weighted average of the known *CEC* values for the shale minerals present, and thus accounting for variations in relative abundance with depth. For the sake of simplicity, in Figure 3.10a they approximated the *CEC* function as a step function, going from *CEC* = 22 to *CEC* = 19, at a depth of 17,500 ft. This depth corresponds to a temperature of 175°F (79.4°C). Lopez et al. (2004) further noted that the measured transition in the CEC values from Mars/Ursa areas also corresponded to marked increase in overpressure in Mars and Ursa wells – both from authors' modeling and check shot velocity data. These results are shown in Figures 3.10b and 3.10c.

Figure 3.11(a) shows the variations of the illite fraction, $(1-N_s(t))$, of equation (3.29) for the indicated constant burial rates at a fixed thermal gradient of 1.2°F/100 ft as reported by Dutta (2016). Figure 3.11b shows the variation with respect to different thermal gradients for a fixed burial rate of 1 ft/1000 years. These figures clearly show the kinetic nature of the S/I transformation – older rocks and rocks with higher temperatures begin the S/I transition at shallower depths, while the opposite is true for younger rocks and lower temperature regimes. Thus, *the depths at which the S/I transition starts and ends can vary significantly – by as much as 10 kft.*

The model described here could be further enhanced by a more detailed approach for paleo-temperature reconstruction, such as in the traditional basin modeling algorithms. This is discussed in Chapter 10. Al-Kawai (2018) also addressed this issue on a real data set from Gulf of Mexico (Thunder Horse field of BP) and showed that the effect of thermal transients could significantly affect the depth profile of S/I transformation. Petmecky et al. (2009) carried out a 3D basin modeling approach for pore (and effective) pressure analysis and obtained velocity in a basin scale by inverting an unspecified transform that related effective pressure to velocity and temperature. However, the authors did not use the S/I conversion model in their approach. An integrated work of this type is highly desirable not only for pore pressure modeling but also velocity modeling for depth imaging as discussed in Dutta et al. (2014, 2015a, 2015b) and Petmecky et al. (2009). The kinetic model discussed here is similar to the kerogen conversion model as discussed by Tissot and Welte (1978).

Further Examples of Modeling Results on Profiles of Chemical Diagenesis of Shales

As noted above, chemical compaction of shale due to S/I transformation should be predictive from time-temperature evolution of sediments under gravitational loading. In Figure 3.12 we show a comparison of the predicted S/I diagenetic depth profile (solid curve) with the X-ray derived smectite proportions (in dashed curves) in mixed-layer S/I clay in a well in Harris County, Texas, USA. The data are from Hower et al. (1976) for two clay size fractions, < 0.1 μm and 0.2–0.5 μm. The well was drilled to a total depth of

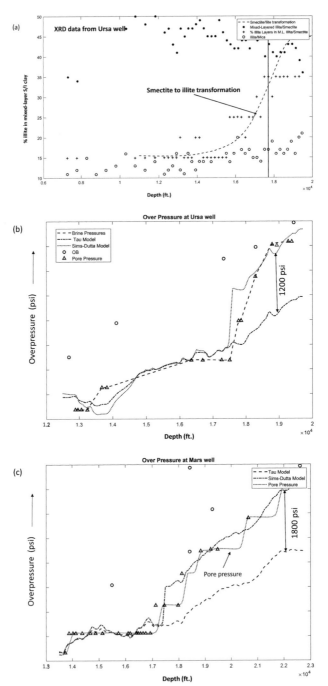

Figure 3.10 (a) X-ray diffraction data on cuttings extracted from a well at Ursa show the onset of the smectite-to-illite transformation in the clay minerals as a function of depth. A model for the cation exchange capacity (CEC) can then be built (inset) to investigate the consequences of this effect on the modeled pore pressures using the Sims–Dutta model. Modeled overpressures as a function of depth, at (b) Ursa and (c) Mars, using the Tau (proprietary Shell model) and Sims–Dutta (Shell model) models for velocity-to-vertical effective stress transformation, as compared to extrapolated brine pressures at the wells. Note that the vertical scales for overpressure are not noted on the figures. The large increase in overpressure at the Yellow event may be explained by the onset of the smectite-to-illite transformation (shown in Figure 3.10a), as described by the Sims–Dutta model. The tau model required ad hoc recalibration at this depth to explain the effect. Modified after Lopez et al. (2004).

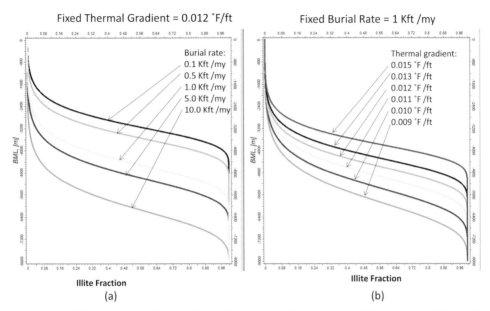

Figure 3.11 A representation of illite fraction generated due to chemical diagenesis for (a) a fixed geothermal gradient and varying sedimentation rates and (b) a fixed sedimentation rate and varying geothermal gradient. These models were generated using the steady state heat flow assumption (no transients).

18,945 ft and it penetrated Miocene and Oligocene rocks. The shale samples on which the measurements were made were taken from the intervals in the Lower Miocene (approximate age of 20 million years) to the Upper Oligocene (approximate age of 30 million years). The well encountered the boundary between Anahuac and Frio formations at a depth of 7620 ft, and the Frio formation extended to the entire depth of the well. The age-depth data shown in Figure 3.12a were supplied by Hower (private communication), and used in the model calculation shown in Figure 3.12b along with the thermal gradient data of Figure 3.12c, also provided by Hower (private communication). The agreement between the X-ray measurements of smectite fraction and the prediction based on the kinetic model of chemical compaction is fairly good.

Another example is shown Figure 3.13 for a well drilled by the DOE, USA – Pleasant Bayou Well No. 1, for which complete X-ray clay mineralogy data as well as other data such as well logs and stratigraphic and paleo-data were available. This is an on-shore well located in the Brazoria Fairway, Brazoria County, Texas. The data were reported by Freed (1979). The author used the method proposed by Hower et al. (1976) to acquire and analyze the data. For each sample, the clay samples were < 2 μm. In Figure 3.13a, we show the predicted shale diagenetic profile and compare with the actual X-ray diffraction based data from the cuttings from this well. The age-depth data and the thermal gradient data are shown in Figures 3.13b and 3.13c, respectively. The geothermal gradients are based on the equilibrated bottom-hole temperature measurements. The prediction from the theory with the data is in reasonable agreement for the S/

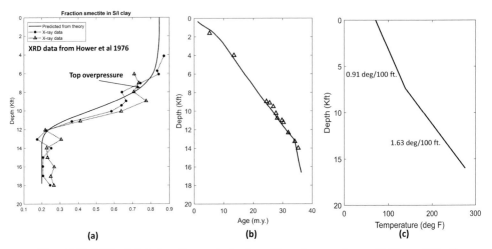

Figure 3.12 (a) Predicted clay diagenesis depth profile and comparison with X-ray diffraction data; the data are from Hower et al. (1976). (b) Age versus depth data and (c) temperature versus depth data used in the modeling (continuous curve of Figure 3.12a). Modified after Dutta (1987b, Figure 9).

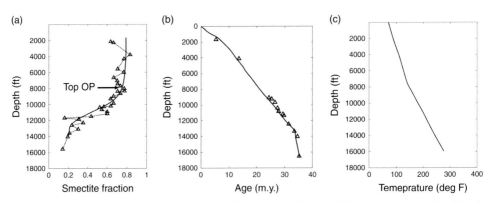

Figure 3.13 Predicted diagenetic depth profile for an on-shore well in Texas as reported by Freed (1979). (a) Modeling and prediction of the diagenetic depth profile (continuous line). (b) Age versus depth profile. (c) Thermal gradient data used in the calculations.

I depth profile. The mid-point of the transition occurs at a depth of ~ 11,000 ft. This is consistent with the data. Freed (1979) also reported a marked change in the ordering of the smectite and illite layers in the mixed-layer S/I system – layering was $R0$ order for samples from depths less than 11,500 ft and was $R1$ order for samples from depths > 11,750 ft.

Since the model of chemical compaction discussed here is kinetic, we expect to have a significant effect of reaction time (varying burial rates) on the diagenetic depth profile. Thus, older rocks which had a longer time to alter at a given temperature would proceed further with alteration (smectite-to-illite conversion)

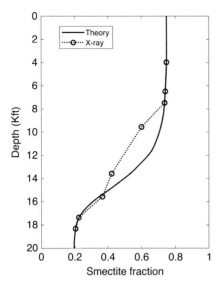

Figure 3.14 Predicted S/I diagenetic depth profile for an offshore well from Louisiana that contains younger rocks (Plio–Miocene). The data are from Perry (1969). Note that the midpoint of the S/I transition is at deeper depths as predicted by the kinetic model. This profile is different from that shown in Figure 3.12 for a well that penetrated much older rocks.

than the younger rocks. Perry (1969) was the first to observe this effect in two wells from the Gulf of Mexico, USA. His data from a well that penetrated the Plio–Miocene sections, referred to as well-C in his publication, are presented in Figure 3.14 along with the model predictions. The profile for this well is very different from the well in Figure 3.12a which penetrated much older rocks – the mid-point transition in well-C (Figure 3.14) occurs at a deeper depth than that in the well represented in Figure 3.12a – a well that penetrated much *older* rocks. Thus, as we progress from the Texas–Louisiana coastline toward deeper waters, the stratigraphy gets younger and buried deeper, and so does the S/I transformation profile versus depth. This is consistent with the profiles for the Mars/Ursa wells shown by Lopez et al. (2004) in Figure 3.10. The mid-point of transition is at ~17,000 ft for that well.

It should be clear from these examples that the data and models, such as those discussed above are very powerful tools for *predicting* shale mineralogical transformations such as the S/I transformation due to chemical diagenesis. We will see later that models of this type can be used to *predict* acoustic properties of shales such as the density, the velocity and pore pressure profile from burial history analyses. These predictions can help us to create synthetic seismograms to tie with reflection seismic data and put realistic constraints on various subsurface models based on seismic data such as pore pressure and imaging. This will be discussed in Chapters 6 and 10. In addition, we think that S/I profiles can also be used as paleo-thermometer in basin modeling endeavors.

3.5   Kerogen Conversion and Hydrocarbon Generation as Mechanisms of Geopressure

As early as 1978, Meissner (1978), Spencer (1987), and Sweeney et al. (1987) recognized hydrocarbon generation as an important mechanism of geopressure in the Williston Basin (Bakken Formation). This is lately considered as a prolific area for hydrocarbon production (unconventional resources). The cause of buildup of high pore pressure in the organic carbon rich reservoirs (high kerogen concentration) is certainly related to volume changes that occur in petroliferous rocks due to kerogen conversion to oil and then to gas (Welte et al., 1997). See also Bethke (1989), Luo and Vasseur (1996), Lee and Williams (2000), and Bredehoeft et al. (1994). These authors also found that the volume change depends on the type of kerogen sources and the density and volume of the petroleum products generated during maturation. It is controlled by the time-temperature history of the rocks.

For high-permeability rocks the excess pressure does not become large as generation proceeds, because the excess fluid volume is removed by fluid flow. For low-permeability rocks, the pore pressure could increase to the fracture pressure, leading to increasing permeability by generating fractures through which the excess oil or gas can escape; therefore, fracturing by oil or gas generation is a possible mechanism for migration starting from initially low-permeability source rocks. These rocks have a permeability of the order of 0.01 μd (10^{-20}m^2) or less. Excess pore fluid pressures are generated when the rate of volume increase, due to the transformation of higher-density organic matter (such as kerogen) to less dense fluids (oil or gas), is faster than the rate of volume loss by the flow. Only two reaction rates are needed for the conversions, a low-temperature reaction rate for the kerogen/oil conversion ($E \sim 24$ Kcal/mol; $A \sim 10^{14}$/my) and a high-temperature reaction rate for oil/gas conversion ($E \sim 52$Kcal/mol, $A \sim 5.5 \times 10^{26}$/my), according to Berg and Gangi (1999). High pressures are more readily developed by gas generation than by oil generation because of the much lower density of gas (Barker, 1990). Thermal cracking of oil to gas in a closed reservoir can cause a pressure large enough to fracture the rock, even at relatively low fractions of oil/gas conversion.

3.5.1   Model for Hydrocarbon Generation as a Contribution to Geopressure

Both Berg and Gangi (1999) and Hansom and Lee (2005) showed that the process of hydrocarbon generation – kerogen to oil and then oil to gas, can be modeled by the kinetic theory as presented earlier for smectite – illite conversion. Both processes create volumetric changes within the rock pore volume as a light material (kerogen) is converted to a lighter material (oil) and then to an even lighter material (gas). The transformation rate of hydrocarbon generation depends largely on the ambient temperature and the time window available for the host rock to mature. The product of the two quantities is known as the "heating rate." It is an important parameter that has a significant effect on the kerogen conversion. The rate of fluid flow out of the host rock is governed by the host rock permeability, the viscosity of the fluid and the excess pressure gradient. As the sediments are buried, the ambient temperature increases and the rate of transformation to hydrocarbon increases. However, as the compaction progresses, the permeability decreases considerably causing the fluid flow out of the host rock to be retarded. If the rate of retardation of outflow of fluid is smaller compared to its production rate, overpressure will occur. The process is thus complex and requires not only mass-balance but also details of the basin evolution including geological phenomena such as heat flow, uplift, erosion, and faulting.

If the permeability of the host rock is large, then the excess volume changes and the accompanied pore pressure increase due to hydrocarbon generation would be accommodated by the fast out-flux of the fluid. In this case, excess pressuring will not occur. However, for low-permeability host rocks, such as organic rich shale (e.g., Bakken shale of Williston Basin and Kimmeridge shale of North Sea), the fluid flow is retarded and causes excess pressure. For these rocks, the pore pressure could increase to as high as the fracture pressure. This would cause microfracturing to occur and lead to pathways for fluid (oil or gas) migration. These rocks have permeabilities of the order of a few nanodarcy.

In the following, we will consider in particular the rate of kerogen-to-oil conversion that is kinetic in nature as pointed out earlier. The time rate of change of kerogen to oil, at a given time, t, can be modeled by the first-order chemical rate theory as

$$\frac{dM_k(t)}{dt} = -K_k(t)M_k(t) \tag{3.35}$$

or by integrating:

$$M_k(t) = M_k(0)e^{-\int_0^t K_k(t)dt} \tag{3.36}$$

Here $K_k(t)$ is the reaction rate for kerogen conversion to oil, $M_k(t)$ is the mass of convertible kerogen at time t and $M_k(0)$ is its initial mass. The fraction of kerogen $X_k(t)$ converted to oil is

$$X_k(t) = \frac{M_k(0) - M_k(t)}{M_k(0)} \tag{3.37}$$

Therefore,

$$X_k(t) = 1 - e^{-\int_0^t K_k(t)dt} \tag{3.38}$$

Similarly, the fraction of oil $X_0(t)$ converted to gas is

$$X_0(t) = 1 - e^{-\int_0^t K_o(t)dt} \tag{3.39}$$

where K_o is the rate of gas generation. For oil generation, the rate of reaction is given by the Arrhenius equation,

$$K_k(t) = A_k e^{-\frac{E_k}{RT(t)}} \tag{3.40}$$

where E_k is the activation energy for kerogen-to-oil transformation, A_k is the associated pre-exponential factor, R is the gas constant, and $T(t)$ is the time-varying temperature in degrees Kelvin. The kinetic parameters in the Arrhenius equation depend on the types of kerogen and oil in the source or the host rock. A similar definition applies for oil-gas generation involving pre-exponential factor A_o and activation energy, E_o, namely,

$$K_o(t) = A_o e^{-\frac{E_o}{RT(t)}} \tag{3.41}$$

To estimate the amount of excess pressure generated by hydrocarbon generation, we need to couple the kinetic models described above to a basin evolution model that predicts the temperature and pressure distribution in sedimentary basins. Hansom and Lee (2005) proposed the following hydro-geologic model in 2D:

$$\varphi\beta\frac{\partial P}{\partial t} = \frac{\partial}{\partial x}\left(\frac{\kappa_x}{\mu}\frac{\partial P}{\partial x}\right) + \frac{\partial}{\partial z}\left\{\frac{\kappa_z}{\mu}\left(\frac{\partial P}{\partial z} - \rho g\right)\right\} - \frac{1}{1-\varphi}\frac{\partial\varphi}{\partial t} + \varphi\alpha\frac{\partial T}{\partial t} + Q_k + Q_o \tag{3.42}$$

Here, ϕ is porosity, β is compressibility, P is pore pressure, α is geothermal gradient, κ_x and κ_z are permeabilities in the x and z directions, μ is viscosity, and Q_k and Q_o are source terms associated with kerogen-to-oil and oil-to-gas conversion. The first two terms on the right-hand side of the above equation describe the flow of fluid (Darcy flow) away from the source rock beds. The third term describes compaction due to porosity reduction and the fourth term describes aquathermal expansion. The source term, Q_k for kerogen-to-oil conversion is described by

$$Q_k = \left(\frac{\rho_k}{\rho_o} - 1\right)\frac{dX_o(t)}{dt} \tag{3.43}$$

where X_o is the kerogen's capacity for generating oil and ρ_k and ρ_o are kerogen and oil densities, respectively. Similarly, the source term, Q_o, for oil-to-gas conversion can be described by

$$Q_o = \left(\frac{\rho_o}{\rho_g} - 1\right)\frac{dX_g(t)}{dt} \tag{3.44}$$

where ρ_o and ρ_g are the densities of oil and gas, respectively. We note that the extent of gas generation is given by X_g, which is the oil's capacity for generating gas. Gas and oil densities can be calculated using the equation-of-state approach as suggested by

Hansom and Lee (2005). Alternatively, one could use the Batzle and Wang (1992) model as discussed in Appendix A.

Hansom and Lee (2005) solved equation (3.42) for oil and gas generation numerically in 2D including disequilibrium compaction, aquathermal pressuring and hydrocarbon generation as geopressure mechanisms. The numerical models simulated the deposition and compaction of a 10 km thick shale at the uniform rate of 1 mm a^{-1} over a period of 10 Ma. The permeability of shale was taken to be 10^{-6} d. The basal heat flow (needed for temperature evolution in the model) was set to be a constant value of 1 HFU (or 41.8 milli Watts per meter squared). The pre-exponential factor A_k and the activation energy E_k of oil generation from kerogen were taken to be $6.15 \times 10^{16}\,h^{-1}$ and 218 KJ mol^{-1}, respectively. The values of A_o and E_o of gas generation from oil were set arbitrarily to $4 \times 10^9\,h^{-1}$ and 200 $KJmol^{-1}$, respectively. Figure 3.15a shows how the calculated oil and gas (methane) generation (as a volume fraction of the capacity) at the deepest nodal point in the shale evolves with time. The deepest source bed generated oil between 6 and 7 Ma and subsequently reached the gas window between 5 and 4 Ma. The simulation result showed that the overpressure increased significantly due to oil and gas generation in this low-permeability rock. Oil and gas generation together caused excess pressure up to about 40 percent of that generated by compaction only. Figure 3.15b shows the depth range for oil (3–4 km in this model) and gas (6–7 km in this model) generation and corresponding overpressures at the end of sediment deposition. Figure 3.15 shows that overpressure increases significantly over the entire range of oil and gas windows in time.

Studies such as the one by Hansom and Lee (2005) indicate that kinetic parameters, local and regional heat flow, permeability of source rocks, and sedimentation rates have profound influence on the timing, duration, and depth of oil and gas generation and theireffect on geopressure. This in turn impacts significantly the spatial and temporal

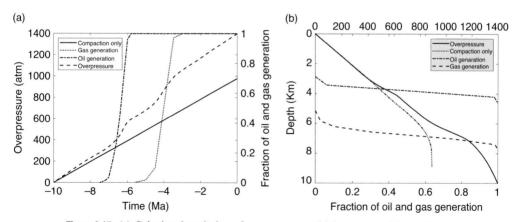

Figure 3.15 (a) Calculated evolution of pore pressure (thick curve) and thermal maturity at the deepest nodal point of the basin accepting continuous sedimentation of 10-km-thick shale. The dash-dotted curve shows oil generation, and the thin dash curve shows gas generation. (b) Overpressure (atm) and thermal maturity versus depth profiles after deposition of 10 km shale. Predicted overpressure assuming compaction only is shown as a thin dash curve. Note the high overpressure due to gas generation.

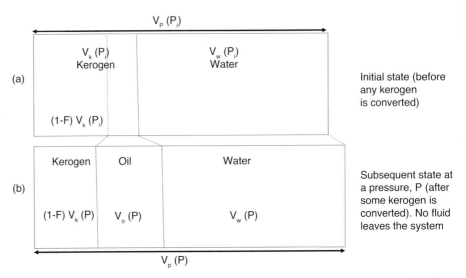

Figure 3.16 Variation of the pore, water, kerogen, and oil volumes with pore pressure. (a) The volume at the initial pore pressure, P_i (before any kerogen is converted into oil). (b) The volume at a subsequent pore pressure, P, when a fraction F of kerogen has been converted to oil (assuming no fluid flow from the pore volume).

distribution of geopressure. However, we note that this mechanism for geopressure is *local* and valid for rocks that have high total organic carbon content (TOC).

Models such as the one discussed above are tied to the basin evolution history that includes thermal history and chemical kinetics of hydrocarbon generation. This is a complex model and simulations require access to commercial 3D basin history reconstruction programs in the market. An approximate model has been proposed by Berg and Gangi (1999). They proposed an analytic expression for pressure changes associated with hydrocarbon generation. The conceptual model is shown in Figure 3.16 (modified after Berg and Gangi, 1999).

Their assumptions are (1) pore pressure changes are associated with volumetric changes within the rock due to kerogen-to-oil and oil-to-gas conversion, (2) permeability of source rocks are very small so that pore pressure build up due to hydrocarbon generation is not dissipated by fluid flow, (3) the stress state is isotropic so that the source rock fails when the pore pressure reaches the overburden pressure (an upper limit), (4) the properties of the rock, organic matter, and fluids remain constant during hydrocarbon generation, and (5) only two sets of kinetic parameters are needed for kerogen-to-oil conversion – a low-temperature reaction rate for the process of kerogen conversion to oil and a second set for oil-to-gas conversion that has a much higher rate of reaction. Under these assumptions, they showed that for kerogen-to-oil conversion, the pore volume $V_p(P)$ would depend only on the pressure and the fractional conversion, F,

$$V_p(P) = V_w(P) + V_k(P) + V_o(P) = V_w(P) + (1 - F)V_{k0}(P) + V_o(P) \qquad (3.45)$$

where V_{k0} (P) is the volume of the original kerogen at pressure P. This yields the following expression for the excess pore pressure ΔP for kerogen-to-oil conversion:

$$\Delta P = \frac{vF(D-1)}{(c_w + c_p) + v[(1-F)(c_k + c_p) + FD(c_o + c_p)]} \tag{3.46}$$

where

v = ratio of the convertible kerogen volume to water volume in the pore volume at pressure P
F = the fraction of the initial mass of convertible kerogen that is converted to oil
D = ratio of kerogen-to-oil-density at pressure P_i
c_w = the compressibility of water
c_p = the compressibility of the (nonmineral) pore space
c_k = the compressibility of the convertible kerogen
c_o = the compressibility of the oil

Similar expressions have been derived by Berg and Gangi (1999) for oil-to-gas generation (see Berg and Gangi, 1999, equation (15)). The fraction of kerogen-to-oil conversion, F, depends on stiffness of the rock (through c_w and c_p), initial kerogen/oil density ratio D, and initial kerogen/water ratio v.

A weakness of their model is that it does *not* predict at what depth a given fraction of conversion, F, is expected. For a specific source rock this depth depends on the details of the basin evolution together with its thermal history and the type of kerogen, as noted earlier. The variation of F with depth for kerogen to oil conversion is given by

$$F = 1 - e^{-\int_0^t K_k(t)dt} \tag{3.47}$$

Similarly, for oil-to-gas conversion, F is given by

$$F = 1 - e^{-\int_0^t K_o(t)dt} \tag{3.48}$$

where $K_k(t)$ in equation (3.47) is the oil generation rate and $K_o(t)$ in equation (3.48) is the oil-to-gas generation rate, and both are given by the Arrhenius model, equations (3.39) and (3.40), respectively. Thus the static model needs to be applied in two stages: first compute the fraction, F, of kerogen converted to oil versus depth using equation (3.47) and the fraction of oil converted to gas vs. depth using equation (3.48) and then calculate the excess pressure, ΔP, due to each of these transformations as given in equation (3.46). We give an example from Berg and Gangi (1999) for Austin Chalk. In Figure 3.17, we show their calculated fractions converted to oil and gas, respectively, as a function of depth. In Figure. 3.18, modified after Berg and Gangi (1999), we show the pressure buildup with depth during oil generation for different ratios of initial

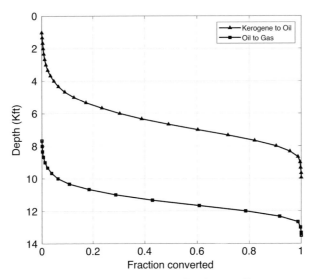

Figure 3.17 Oil and gas generated as a fraction, F, of kerogen converted to oil and a fraction, F, of oil converted to gas versus depth. Modified after Berg and Gangi (1999).

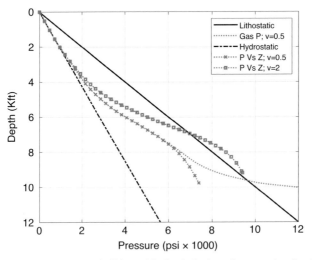

Figure 3.18 Pressure buildup with depth during oil generation for different ratios of initial kerogen to initial water volume as noted. Modified after Berg and Gangi (1999).

kerogen to initial water ($v = 0.5$ and $v = 2.0$). Berg and Gangi (1999) presented a case study for Austin Chalk, Texas and demonstrated not only the overpressure due to oil and gas generation but also effects of the fluid migration pathway due to microfracturing. Microcrack growth due to overpressure from kerogen maturation was also studied by Coussy et al. (1998) for a 2D system, and by Jin et al. (2010) and Fan et al. (2010, 2012) for a 3D penny-shaped crack embedded in a kerogen-shale solid. Using linear elastic fracture mechanics, Yang and Mavko (2018) modified the approach of Berg and Gangi (1999) to mathematically model microcrack growth in source rocks during kerogen maturation. The Yang and Mavko (2018) model calculates microcrack sizes (surface areas, lengths, apertures, and volumes) and the amount of overpressure throughout the maturation process, for both laboratory and geological settings. Compared to laboratory settings, microcracks are much smaller under geologic settings because of the significant overburden stress and stiffer rock frames. Under geologic settings the crack apertures are in the submicron regime with lengths ranging from 100 to 300 μm. As the kerogen matures and creates overpressures, formation of such microcracks connects isolated microscale oil pockets, thus providing primary migration pathways.

3.6   Chemical Diagenesis due to Gypsum-to-Anhydrite Transformation

This is a local overpressure generation mechanism in evaporite sequences. It involves mineralogical changes for gypsum ($CaSO_4 2H_2O$) to anhydrite ($CaSO_4$) and release of substantial amount of bound water in the host rock. This is a source of geopressure. The *reversible* reaction is given by

$$CaSO_4 2H_2O \leftrightarrow CaSO_4 + 2H_2O \qquad (3.49)$$

The chemical reaction starts at ~ 40°–60°C at atmospheric pressures (Bjørlykke, 2015). The process therefore can begin at relatively shallow depths of burial (Jowett et al., 1992). In the North Sea, this mechanism is responsible for generation of very high pore pressure in Permian-age Zechstein carbonate (Mouchet and Mitchell, 1989). This mechanism is considered to contribute to high overpressure in several evaporite formations in the Middle East. Massive amounts of anhydrite occur when salt domes form a caprock. Anhydrite is 1–3 percent of the salt in salt domes and is generally left as a cap at the top of the salt dome when the halite is removed by pore waters. The typical cap rock is a salt, topped by a layer of anhydrite, topped by patches of gypsum, and then topped by a layer of calcite (Walker, 1976). We note that higher temperature and pressure favor the stability of anhydrite. This is because anhydrite has a more compact structure than gypsum. The loss of water resulting from gypsum to anhydrite leads to a volume reduction of ~ 40 percent. The transformation of anhydrite to gypsum leads to a volume increase of 60 percent (Bjørlykke, 2015). The volume increase can change the local tectonic stress, faulting and near-surface deformation. The published literature has few quantitative study on the effect in overpressure.

3.7   Charging through Subsurface Structures (Lateral Transfer of Fluids)

Lateral transfer of fluids through dipping sand or any permeable bodies embedded in overpressured shale has been known to contribute to overpressure in the sand because of the much higher permeability of the sand than that of the surrounding encasing shales. This was discussed by Dickinson (1953), and expanded by Mackenzie et al. (1986), and Traugott and Heppard (1994) who coined the term *centroid effect*. The permeability of shale is extremely low compared to reservoir rocks. Consequently, the pore pressure in reservoirs are usually very different from the adjacent shales. The concept of "centroid" arose from the observation that shale and sand pressures follow different local pore pressure gradients. Consider a reservoir rock (sandstone) buried in overpressured shale. We assume that the pressure in the sandstone has equilibrated to that of shale over a geologic time. The fluid is assumed to be brine. Then we consider the situation where the overpressured shale is asymmetrically loaded, and is undergoing differential compaction due to varying sedimentation rates. This centroid concept arose from the observation that sand pressures must follow the loading of shale that is undergoing differential compaction due to varying sedimentation rates. This causes the sand layer tilt as shown in Figure 3.19a causing fluid to flow updip in the sand. However, the fluid pressure gradient in the sand is in equilibrium with the original pore pressure at the depth of the sand prior to tilting. Because of tilting, the sand acts as a conduit for transfer of pressures to the updip. At depth shale pore pressures exceed the sand pore pressures, but at shallower locations the sand pore pressures exceed shale pore pressures. The *centroid* is the depth at which sand and shale pore pressures are equal. This effect is shown schematically in Figure 3.19b. Then, what will be the pore pressure if we drill wells in location B (updip) and A (downdip) as shown in Figure 3.19c for a tilted sand body truncating against a fault that does not allow fluid to escape? Since there is no fluid flow, the overpressure in the reservoir sand unit remains constant. The fluid pressure at A at a depth Z_A is given by

$$P_A = P_0 + P_h \qquad (3.50)$$

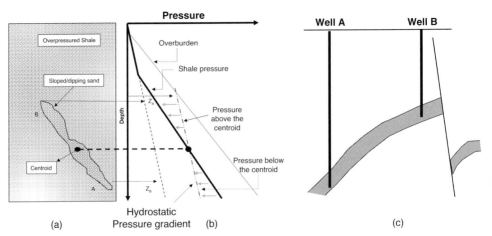

Figure 3.19 Pore pressure transfer in sand embedded in overpressured shale. Modified after Bruce and Bowers, (2002). Possible geologic scenarios tied to these schematics are fluid migration in folded and tilted reservoir rocks.

where P_0 is the overpressure (pressure in excess of the normal pore pressure) and P_h is the normal or hydrostatic pressure. The fluid pressure at B at a depth Z_B is then given by

$$P_B = P_A - (Z_A - Z_B)g\rho_w \qquad (3.51)$$

where ρ_w is the fluid (brine) density.

Let us put some numbers. Let us assume that prior to tilting, the brine-filled reservoir sand (pressure gradient of 0.465 psi/ft) was buried at a depth of 9000 ft below the sea-bed (mudline) and was in equilibrium with the adjacent shale with a pore pressure gradient of 0.8 psi/ft (15.42 ppg). The excess or overpressure in the sand is 3015 psi. {0.8 (psi/ft) × 9000 ft – 0.465 (psi/ft) × 9000 ft}. This sand is now asymmetrically tilted with a relief of 3000 feet. Therefore, the updip location of the sand is now at 6000 ft below mudline, while the downdip is at 12,000 feet below mudline. The pore pressure at the updip location is now 5845 psi {0.465 (psi/ft) × 6000 ft + 3015 psi = 5805 psi}. This equates to a pressure gradient of 0.97 psi/ft (18.69 ppg). Thus, the mud weight to drill this hypothetical sand at the updip location will be 18.69 ppg, as compared to pretilting mud weight of 15.42 ppg. This highlights the structural effects on pore pressure computation. It is for this reason, drillers prefer not to drill exactly at the crest of a structure. It can cause a blowout. The effect could be even more serious when the buoyancy effect due to hydrocarbon is considered as explained below.

Equation (3.52) shows a model for estimation of reservoir pressure at any point in the sand body (Stump et al., 1998) due to centroid effect:

$$\Delta P = \left[\left(\frac{\beta}{\beta + \beta_f(1 - \phi)}\right)\rho_b - \rho_f\right]g\int_0^L Z(x)dx \qquad (3.52)$$

Here, ΔP is the amount of overpressure within the sand, β is matrix compressibility, β_f is fluid compressibility, ϕ is porosity, ρ_b is bulk density, ρ_f is fluid density, g is acceleration due to gravity, Z is depth, and L is sand relief. Through a combination of seismic inversion, attribute analysis, and 3D visualization techniques, the sand bodies and their properties can be characterized, thus allowing an estimation of possible excess pressure due to structural relief. Figure 3.20, modified after Dutta and Khazanehdari (2006), shows two sand bodies mapped using acoustic impedance and velocity data. The pore pressure was computed using equation (3.52). Panel 1 shows the pore pressure in the background shale. Extracted sand geometry is shown in the panel 2. A computer program estimates the geometry of the individual sand bodies and computes the excess pressure due to structural relief as discussed in the text (panel 3). The pressure cube that is more representative for a background shale lithology (first panel) is then corrected according to the estimated pressure for each individual sand body and it is shown in the fourth panel. The color scales are indicated for the fluid pressure gradients in pounds per gallon or ppg. Once the sands are delineated and the excess pressure due to structural relief is corrected, the pressure variation due to fluid buoyancy (oil, gas) can also be easily modeled as discussed below by allowing for fluid density (oil or gas) changes. It

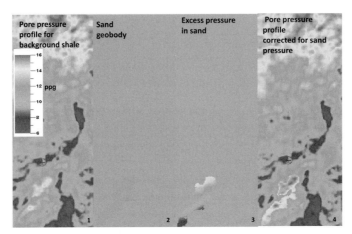

Figure 3.20 Pore pressure changes in sands with a relief that is embedded in overpressured shale. The first panel shows the background pressure in shale. The second panel shows the sand geometry as extracted from the acoustic impedance inversion of 3D seismic data. The third panel shows the excess pore pressure in sand. The fourth panel shows the total pore pressure. After Dutta and Khazanehdari (2006). The range of pore pressure is from 6-16 ppg as noted. (A black and white version of this figure will appear in some formats. For the color version, please refer to the plate section.)

should be noted that the lateral transfer of pressure is not a primary mechanism itself for creating overpressures. Nonetheless, this process can exert a strong influence on many of the pore pressure profiles encountered in the subsurface, and may mask recognition of the underlying causal mechanism.

3.8   Hydrocarbon Buoyancy as a Cause of Overpressure

We stated in Chapter 1 that hydrocarbon column heights can result in substantial overpressure at the top of the reservoir. This is due to the buoyancy effect of the hydrocarbon (it is lighter than brine). The effect is further magnified by the structural relief as discussed above. Buoyancy makes a body (or a drop of immiscible fluid) rise when submerged in a fluid of a higher density. Where two immiscible fluids (oil and water) are put together, they create a buoyancy pressure that is a function of the density difference between the two fluids and the length of the lighter fluid column. The buoyancy pressure, ΔP, is the excess pressure in a confined reservoir due to the density difference between hydrocarbon (lighter density) and brine. Thus,

$$\Delta P = (\rho_w - \rho_h)g\Delta Z \tag{3.53}$$

or equivalently expressed as

$$\Delta P = (grad_w - grad_h)g\Delta Z \tag{3.54}$$

where ρ_w and ρ_h are the brine density and the hydrocarbon density, respectively. ΔZ denotes the height of the hydrocarbon column above the base of the column, g is the acceleration due to gravity, and $grad_w$ and $grad_h$ are the pressure gradient of water and hydrocarbon, respectively. Below we give some fluid densities and their equivalent pressure gradients.

Fluid type	Fluid density (g/cc)	Pressure gradient (psi/ft)
Water	1.00–1.15	0.433–0.498
Oil	0.58–0.90	0.250–0.390
Gas	0.16–0.35	0.070–0.150

In Figure 3.21, we illustrate the effect of buoyant forces by an example in a situation where a hydrocarbon-filled reservoir of column height ΔZ lies atop on a water-filled reservoir. The system is encased by shale so that fluids do not flow out of the system. Below the hydrocarbon/water contact, pore pressures follow a hydrostatic trend – offset from normal hydrostatic pressure, if the brine-filled reservoir pressure is above hydrostatic. Above the hydrocarbon/water contact, pressures follow a slope that depends on the hydrocarbon density. This slope may be 0.07–0.15 psi/ft for gas and 0.25–0.39 psi/ft for oil. Because hydrocarbons are lighter than water, the amount of the overpressure in the hydrocarbon column increases with elevation above the hydrocarbon/water contact. This extra "boost"

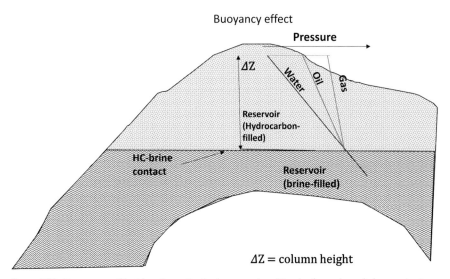

Figure 3.21 Buoyancy effect in a hypothetical reservoir with a hydrocarbon–brine contact. Buoyancy force increases since the hydrocarbons are lighter (lower density than brine). Thus, the buoyancy pressure is largest for gas-filled reservoir.

Reservoir pressures in tilted sand beds (without and with hydrocarbon)

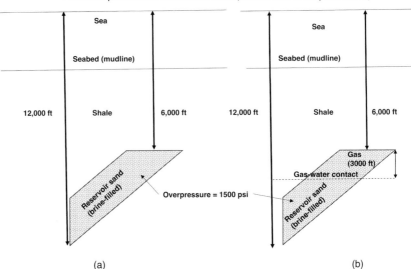

Figure 3.22 Reservoir pressures in a hypothetical tilted sand bed. (a) Brine-filled reservoir. (b) Reservoir with a gas–brine contact. The gas column height is assumed to be 3000 ft.

in overpressures is the *buoyancy effect*. These are also applied to the connate water, and this is the process by which water is expelled downward from the reservoir as the petroleum accumulates. Once the irreducible water saturation is reached, no further water expulsion takes place, and the water pressure increases with that of the adjacent petroleum.

The buoyancy pressure concept is easy to understand from the basic physics. However, its importance in the context of drilling oil and gas wells is often underestimated. We explain this point further with an example in Figure 3.22.

Example

As shown in Figure 3.22a, we assume that a dipping sand reservoir is filled with water (brine) and the water pressure gradient is 0.465 psi/ft. The updip portion of the sand reservoir is at the depth of 6000 ft from the top of sea while the downdip portion of the reservoir is at a depth of 12,000 ft. Furthermore, we assume that the sand reservoir is encased by a very low-permeability rock, such as shale and the sand reservoir is overpressured to 1.500 psi. Figure 3.22b is similar to Figure 3.20a except that a gas-water contact exists at a depth of 9000 ft. Thus, the gas column height is 3000 ft. We want to compute the following in cases (a) and (b) at the top and base of the sand reservoir: (a) pore pressure in psi, (b) pore pressure gradient in psi/ft, and pore pressure gradient in equivalent mud weight, i.e., in pounds per gallon or ppg.

Calculations

Case (a)

Brine case

Pore pressure at 12,000' = normal pressure + overpressure
= (0.465 psi/ft × 12,000 ft) + 1,500 psi
= 5580 psi + 1500 psi
= 7080 psi

Pore pressure gradient to 12,000' = 7080 psi/12,000 ft
= 0.590 psi/ft
= 11.37 ppg

Pore pressure at 6000' = normal pressure + overpressure
= (0.465 psi/ft × 6000 ft) + 1500 psi
= 2790 psi + 1500 psi
= 4290 psi

Pore pressure gradient to 6000' = 4290 psi/6000 ft
= 0.715 psi/ft
= 13.78 ppg

Case (b)

Gas case

Pore pressure at 6000' = normal pressure + overpressure + buoyancy pressure
= (0.465 psi/ft × 6000 ft) + 1500 psi
+ ((0.465 psi/ft – 0.10 psi/ft) × 3000 ft)
= 2790 psi + 1500 psi + 1095 psi
= 5385 psi

Pore pressure gradient to 6000' = 5385 psi/6000 ft
= 0.8975 psi/ft
= 17.29 ppg

Pore pressure at 12,000 ft = 7080 psi
Pore pressure gradient to 12,000 ft = 0.590 psi/ft = 11.37 ppg

This examples illustrates that (1) pore pressure is affected by the dip of a permeable layer – also known as the "relief" – and that (2) the buoyancy pressure contributes significantly to the pressure gradient and its effect could be sizable, especially for gas. It is for these reasons that drillers typically do not drill for targets at the crest of a structure, as chances of blowouts are higher, especially if the reservoir is over-pressured and gas is encountered at the crest. High-relief structures at shallow depths probably were responsible for blowouts of early days' rotary drilling. In deep water depths (water depth > 500 m), we often encounter dipping sand beds at shallow depths (~ 140 m) below seabeds that are slightly overpressured and sediments are hardly morphing into rocks. With some "relief" and especially with gas, numerous blowouts were reported in the Gulf of Mexico, USA in these environments. This is known in the industry as drilling hazards associated with *shallow-water-flow* (*SWF*) *sands*. We shall discuss this in detail in Chapter 11.

3.9 Hydraulic Head as a Cause of Overpressure (Erosion/Uplift, Elevation Related to Datum)

Sometime this effect is called fossil pressure in the literature (Fertl, 1976). Fossil pressure is the term often associated with formations with uplift and subsequent erosion – the degree of overpressure is associated with the amount of erosion. The concept of fossil pressure is best explained by an example in Figure 3.23. Figure 3.23a shows a normally pressured sand (pressure gradient of 0.465 psi/ft) that was buried at a depth of 5000 ft and then sealed. To reach this sand by drill bit would, of course, require a "mud" weight of 8.952 ppg (0.465 psi/ft × 19.25= 8.952 ppg). This sealed sand was subsequently uplifted by 2000 ft and a part of the overburden was eroded – its current depth is 3000 ft as shown in Figure 3.23b. This sand is now overpressured by the amount of 930 psi (i.e., 0.465 psi/ft × 5000 ft – 0.465 psi/ft × 3000 ft = 930 psi). To reach this sand by a drill bit, we require an equivalent "mud" weight of 0.775 psi/ft ((0.465 psi/ft × 3000 ft + 930 psi)/3000 ft) or 14.919 ppg (0.775 psi/ft x 19.25). This is due to datum change. Overpressure can also result from a hydraulic head in an adjacent highland area, where there is hydraulic continuity into the subsurface. An example is shown in Figure 3.24. It shows that normal compaction and abnormally high pore pressure situations can arise when the well elevation is lower than the outcrop elevation. For the example shown in this figure, the pore pressure at the well site is 6045 psi (41.69 MPa) (0.465 psi/ft × 13,000 ft). This leads to a pore pressure gradient of 6045 psi/10,000 (ft) or 0.504 psi/ft (0.013 MPa/m or 9.702 ppg). This is substantially overpressured solely due to datum. The reservoir is still normally compacted.

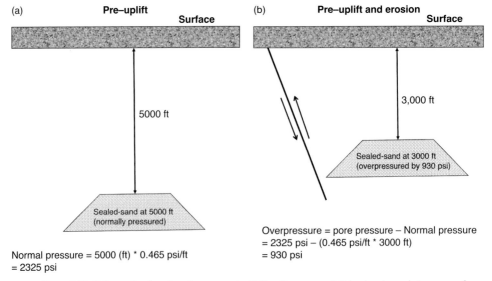

Figure 3.23 Schematic showing the concept of "fossil pressure." It is simply maintenance of overpressure after erosion and uplift. The datum change causes excess pore pressure gradient.

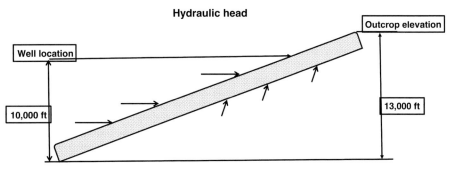

Figure 3.24 A schematic of pressure gradient change due to elevation change (hydraulic head).

3.10   Aquathermal Pressuring as a Mechanism of Geopressure

Aquathermal pressuring is based on the pressure, temperature, and density data for water as reported by Kennedy and Holser (1966). The pressure of water heated in a closed vessel (constant volume) will rise by about 125 psi/°F. Based on this observation, Barker (1972) proposed it as a possible source of geopressure. He noted that as rocks are buried, they are heated as the temperature rises with depth in the earth due to heat produced by decay of radioactive elements, and heat flowing upward through the crust from the mantle. Because heating causes expansion of pore fluid at the depth in a confined and relatively low-permeability and low-compressibility rock, expanding pore fluid (brine) will lead to increased pore pressure (assuming that the seal is prefect and the expanding fluid does not escape). Barker (1972) quantified the change of the fluid pressure ΔP as a direct temperature variation, ΔT, by the following expression:

$$\Delta P = \frac{\left(\alpha_w - \alpha_s\right)}{\beta_p + \beta_w - \beta_s} \Delta T \tag{3.55}$$

where α_s and α_w are, respectively, the thermal expansion coefficients of solids and water; β_p, β_w and β_s are the compressibilities of the pore volume, of water and skeletal structure of the rocks, respectively; and ΔT is the temperature change. Thus, an increase in temperature will lead to increase in pore pressure, provided $\beta_p + \beta_w > \beta_s$ (which is usually the case). Barker showed that for typical geopressured environments, the aquathermal mechanism could lead to an increase of 1000 psi for a temperature increase of $\sim 8°F$. However, for the mechanism to be viable, he assumed that

1. there must be a *perfect* seal over geologic time;
2. there must be a *constant* pore volume maintained; and
3. in a overpressure zone, development of a perfect seal must have occurred at a *lower* temperature and therefore, at shallower depths than at the present depth and temperature.

All of these assumptions are questionable and the literature is quite critical (Chapman, 1994a; Magara, 1975; Dianes, 1982a; Luo and Vasseur, 1996). It is now regarded that aquathermal pressuring is a *minor* cause of overpressure.

3.11   Osmotic Pressure as a Source of Geopressure

Osmosis is the movement of water from an area of low concentration of solute to an area of higher concentration of solute. A solute is atoms, ions, or molecules dissolved in a liquid such as sodium chloride (NaCl) in water. The rate of osmosis is determined by the total number of particles dissolved in the solution. The more particles dissolved, the faster becomes the rate of osmosis. This situation occurs for sediments near salt domes, for example. If a membrane is present, water will flow to the area with the highest concentration of solute. Osmotic pressure is the pressure created by water moving across a membrane due to osmosis. The more the water moves across the membrane, the higher is the osmotic pressure. The movement of water will stop when the chemical potentials across the membrane become the same. Osmotic pressure is given by (Lewis, 1908)

$$\Delta P_{os} = VMRT \tag{3.56}$$

where ΔP_{os} is osmotic pressure difference, V is Van 't Hoff's factor, M is the molar concentration of solution (mol/L), R is universal gas constant, and T is temperature in Kelvin. Van 't Hoff's factor of a solute is determined by whether or not a solute stays as it is or breaks apart in water. Some solutes break apart and form ions or charged atoms in water. For example, NaCl will form sodium (Na+) and Chlorine (Cl^-) ions in water. The Van 't Hoff's factor of NaCl is two because it breaks into two ions. That osmotic pressure will develop across a shale membrane is not disputed. However, there is little evidence that this is an important cause of overpressure. In fact, Osborne and Swarbrick (1997) calculated an upper bound to the magnitude of overpressure from osmosis to be about 4 MPa (580 psi) for the average shale of North Sea. Neuzil (2000) conducted experiments in a shallow borehole to measure osmotic pressure and calculated its upper limit as 2900 psi in low-porosity smectite-rich sales. Thus, one can rule out osmosis as a viable source of overpressuring. Mouchet and Mitchell (1989) also noted, "It seems that the capacity for osmosis to generate abnormal pressure is limited to special cases such as sharply contrasting salinity, proximity to salt-domes, and lenticular series. In most instances of abnormal pressure, the role of osmosis is difficult to prove and must be thought of as minor." We agree with this assessment.

3.12   Summary

In this chapter we discussed various sources of geopressure. Most of the mechanisms are operative in clastics (sands and shales) or clastics encased in low porosity carbonates or carbonate muds. For shale, the most important mechanism of over-pressure is mechanical compaction followed by chemical diagenesis associated with mineralogical changes within the shale and further water release at depth. For sands, lateral transfer of fluid associated with structural relief is very important. Pore pressure in sands can be lower than, equal to, or higher than the pore pressure in the encasing shales. The excess pressure in sands depends on the structural relief – higher the relief, higher is the excess pressure. In anticlinal structures, the excess pore pressure can cause hydraulic fracturing and therefore, equivalent mud weights must be adjusted by the drillers prior to spudding a well. Furthermore, in a sand reservoir with relief, the buoyancy effect due to hydrocarbon is a very important source of "local overpressure" and must be a part of the well trajectory and mud weight planning. Aquathermal and osmotic effects causing geopressure are minor causes. For basins with significant tectonic activity, overpressure due to tectonic stresses (lateral compression) could be the most important source of geopressure, after mechanical and chemical diagenesis. Most of our knowledge to date has come from drilling wells in tectonically active areas. Quantitative effects associated with this phenomenon, however, is very difficult to predict at present.

4 Quantitative Geopressure Analysis Methods

4.1 Introduction

In this chapter we shall discuss quantitative methods for analysis of geopressure. There are numerous techniques available for this. In Table 4.1 we list the most common and useful techniques for predicting and detecting geopressure at all stages of operation – exploration to drilling and postdrilling. During the exploration phase, regional geology and seismic methods are the most commonly used methods. Although gravity and magnetic methods are available in the industry for geologic applications such as mapping basement and salt bodies, we could not find much in the literature that documents using these methods for geopressure analysis.

Regional geology-based analyses are usually carried out to assess the "big picture" – establish a regional framework before locating local structures and lithostratigrahy, establish possible migration pathways of fluids, locating barriers to fluid movement and hence the existence of "closed" systems or pressure barriers, assess potential hydrocarbon traps and qualitatively identify possible risky zones or formations for drilling. In Figure 1.11 we showed a map of the global distribution of abnormally high pore pressure. A good summary of the habitat of geopressure is given by Law and Spenser (1998). This reference has a very good summary of selected attributes of abnormally pressured regions of the world including age of abnormally pressured rocks, depositional systems, structural setting, depth to top of abnormal pressure, range of pressure gradients, temperature to top of abnormal pressure, thermal maturity of abnormal pressure, fluid pressure phase, associated hydrocarbon accumulation type, and cause(s) (Law and Spenser, 1998, Table 1). A summary of the literature indicates that there is a *strong* correlation between abnormal pressures and hydrocarbon accumulation (traps). Below we show a summary of our observations:

➢ Majority of hydrocarbon fields are between hydrostatic gradients (~ 0.46 psi/ft) and ~ 0.85 psi/ft.
➢ Threshold for known economic hydrocarbon accumulation is approximately 0.85 psi/ft (19.125 kPa/m).
➢ $\sim 95\%$ of commercial oil/gas fields are between 0.55 to 0.75 psi/ft (12.375 to 16.875 kPa/m).
➢ "Hard" or "super high" pressured fields are discovered (> 0.85 psi/ft) but these often contain smaller reservoirs, smaller "fetch" area and deplete fast.

Table 4.1 Techniques for prediction of geopressures during various stages

Stage	Sources of data	Pore pressure indicator	Comments
Before drilling (rank wildcat)	Surface geophysical detection	Seismic velocity, gravity	Gravity data not always available
	Basin modeling	Various	Hydrodynamics of basin affects results
	Geological interpretation	Various	Analogues as guide
Before drilling (with well)	Seismic/well logs	Velocity, density, resistivity, conductivity	Logs need environmental corrections; Electrical logs need thermal/salinity corrections
While drilling	Drilling parameters	Penetration rate, D-exponent, MWD/LWD logs (sonic, density, resistivity, conductivity)	
	Drilling mud parameters	Mud gas cuttings, pressure kicks, Flow-line temperature, pit-level and total pit volume, hole fill-up, mud flow rate	
	Shale cutting parameters	Bulk density, shale formation factor, volume, shape, size of cuttings, sand/shale ratio	
	Correlation between new and existing wells	Various drilling data	
	Seismic while drilling	Velocity from drill bit as a source	AKA "seismic while drilling," new seismic data acquired
		Repeated velocity modeling/migration	AKA "seismic guided drilling" technology (no new seismic data acquired as opposed to seismic while drilling); it is actually repeated updates of a small volume, data not available while drill bit is going down
After drilling	Basin modeling		Emerging technology; Turnaround time is an issue
	Subsurface geophysical detection	Vertical seismic profiling, check shot and various logs	
	Well logging	Sonic, resistivity, density, neutron, and downhole gravity	
When well is tested	Measure and monitor pore pressure in short zones	Repeat formation tester, drill stem test, pressure bombs, shut-in pressure	Directly measured pressure data; pressure measured in permeable zones only

➤ Unconventional gas and oil accumulations are known to occur in abnormally high pore pressure basins such as in Bakken shales of Williston Basin (Meissner et al., 1978) and other basins of the North America (e.g., Alberta Basin of Canada [Masters, 1979, 1984]; Green River Basin of Wyoming, Colorado, and Utah [Law et al., 1989; McPeek, 1981; Law, 1984; Spencer, 1987]; and Niger Delta, Nigeria [Chukwu, 1991]).

➤ "Subsalt and sub-basalt traps" are complex and difficult to predict (salt or basalt movement may or may not create "pressure bleed-off" path).

➤ Subsalt and sub-basalt pore pressures are difficult to image seismically due to poor illumination; however, major discovered fields are rarely in "hard" overpressured formations.

➤ Carbonate reservoirs with high pore pressure is common (e.g., Middle East) but these are hard to predict using conventional seismic or petrophysical well log–based methods.

➤ Occurrences of overpressured aquifer sands in young Tertiary clastics are well known; these sands do not have commercial potential but they do pose drilling challenges (geohazard).

It is worth noting that drilling costs go up very high in "hard" pressured formations (> $ 125 million and often $ 150–200 million per well in deepwater environment). Regional or semiregional geologic studies are thus very important for establishing the pressure regime and hydrocarbon habitat in abnormally pressured environments. In untested or virgin areas, this is often the most useful method for predicting high pressure formations. Either seismic velocity analysis or 3D basin modeling techniques or both simultaneously could be used for this purpose. In any case, either of the approach must use some kind of model that relates an attribute such as velocity or porosity to pore pressure or effective stress and overburden stress. In this chapter we shall discuss various methods to accomplish that. In Table 4.2 we provide a list of methods used for pore pressure analysis.

Table 4.2 Methods for quantitative analysis of geopressure

1. Direct calibration methods
   (a) Pennebaker
   (b) Hottman and Johnson
   (c) Eberhart-Phillips et al.
2. Horizontal method
   (a) Ratio method
   (b) Modified horizontal method (Eaton's method)
3. Vertical or equivalent depth method
4. Effective stress methods
   (a) Bowers
   (b) Dutta

These methods can be and have been used at all stages of geopressure analysis. The tools (probes) used with these methods are many – wireline logs (resistivity, sonic, density/neutron, gamma ray), tools while drilling (drilling rate and d-exponents), measurements while drilling tools (resistivity, sonic, gamma ray, and density/neutron) and various seismic tools (check shot, vertical seismic profile or VSP, and surface seismic data). The only direct methods to measure pore pressure are drill stem tests and wireline formation tests; however, these are all after-drilling tools and the data are collected from permeable formations only.

While reviewing these methods below, we note that each of these methods has their intrinsic strengths and weaknesses. There is no holy grail. Best practitioners use a host of methods, realizing that the "tools" to drive these methods have their own limitations such as the depth of penetration of formations and whether these tools are pre-drill, during-the-drill, or post-drill. For these reasons, quite often a host of methods are used at the well site during the drilling phase.

Before reviewing the various analysis methods as outlined in Table 4.2, we shall discuss below briefly the nature of normal compaction trends and describe the general characteristics of under compacted zones. For pressure prediction in shale, compaction is an important phenomenon and it provides a basis for many commonly used methods.

4.2   Normal Compaction Trends and Characteristics of Undercompacted Zones

In order to evaluate geopressure quantitatively, most methods utilize the concept of Normal Compaction Trend (NCT) to establish a background trend for reference purpose. This was first applied to describe high pore pressure in shales since trends for these are easier to establish, especially if the formations are massive such as in the Miocene sections of the Tertiary Clastics in the Gulf of Mexico. High pore pressures in these basins are mostly due to undercompaction as discussed in Chapter 3 – the greater the undercompaction, the higher the porosity, and the greater the percentage of the load carried by the rock matrix (and lower the vertical effective stress). Compaction represents a reduction in porosity and increase in vertical effective stress with increasing depth as we saw in Chapter 3. An example of several compaction trends for mudrock and shale were shown in Figure 3.3. Typically these trends are nearly straight lines when logarithm of porosity is plotted against depth as in the case of Athy's model as shown in Figure 3.3. Such trends are very useful prior to carrying out pore pressure analysis in a particular part of a basin. However, slopes of these trends are highly variable – these depend on mineralogical content of rocks, non-argillaceous mineral content such as quartz, carbonates, etc., as well as diagenetic alteration of rocks with burial. Hence care must be taken to interpret these trends.

In Chapter 3 we have seen the importance of fluid movement in low permeability rocks due to compaction. In case of abnormally high pressured rocks, fluids will tend to move from high pressured areas toward formations under hydrostatic conditions. If there is a seal between the two zones, there will be a lack of fluid communication and it will show up as an abrupt change in pressure conditions. This type of pressure transition zone is known to occur at the base of the sandier section in the Gulf of Mexico deep shelf gas plays; it represents particular drilling challenges in terms of well control and casing design. However, if the seal is not perfect, escaping fluids will bring about a gradual buildup of pore pressure as compared to the upper part of the high pressure sequence. Such an interval will exhibit a *transition* zone. This is often the case in the deepwater clastics. The thickness of the transition zone is controlled by mineral composition of the sediments, especially its permeability, drainage conditions (dictated by the sedimentation rate), permeability of rocks in the transition zone and the geologic time available for fluid drainage (e.g., due to hiatus in sedimentation). It is easier to detect abnormal pressure if the transition between the different pressure regimes are gradual. This implies that the thicker the transition zone, the easier it will be for the detection of overpressured zones. Some examples of relatively simple and more complex transition zones are shown in Figures 4.1a and 4.1b, respectively.

The existence of a transition zone implies changes in rock property, especially permeability. It may be caused by changes in depositional environment such as from purely clay-rich sediments to sandy shales or shaly sands or carbonates. However, under favorable temperature and pressure conditions, diagenetic changes in rocks (shales) can also cause transition zones to develop due to chemical diagenesis such as smectite-to-illite transformation as discussed in Chapter 3. Drilling has shown that multiple transition zones are also possible. These pose significant hazard to drilling.

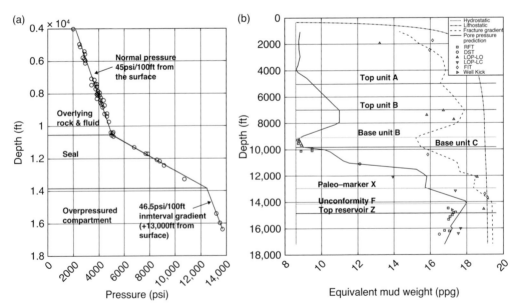

Figure 4.1 An example of (a) relatively simple transition zone from Cook Inlet (modified after Demings, 1996); (LSU Geophysics handout) showing a sharp transition and (b) a more complex transition zone from N. Sea for a high-pressure well (modified after Swarbrick et al., 2005).

The locations of pressure transition zones and their gradients are controlled by (a) the mechanism creating the abnormal pressure, (b) the rock properties within the transition zone, and (c) the time period over which the high pressure gradients have been operating. The character of many pressure transition zones will change dynamically through time as a result of the evolution of fluid movement and pressure development and bleed-off of the basin in which they are found. We must remember that over geologic time, all transition zones exhibit properties of a dynamic system – fluids continue to migrate and leak; there is no such thing as a *perfect seal* over geologic time.

4.3   Methods to Predict Geopressure

4.3.1   Direct Calibration Methods

These methods rely on *direct* calibration of a physical variable such as velocity, resistivity, or bulk density of rocks to pore pressure, either using analytic curves or charts. The underlying assumptions behind these methods are: (1) abnormally high pore pressures are generated due to compaction disequilibrium, (2) normal compaction trends are well-established, (3) lithology of rocks are known, and (4) the rocks or formations where a prediction is to be made is identical or very similar to the rocks where the calibration charts or curves are generated. The methodology consists of the following steps: (1) for a defined attribute such as porosity or velocity, establish a "normal" trend with increasing depth (especially, for shales where the trend is easier to establish), (2) measure the "deviation" of the attribute from the normal trend and then (3) "calibrate" that departure from the normal trend with pressure data from well(s). This approach is based on the observation that a geopressured formation exhibits several of the following properties when compared with a normally pressured section at the same depth: (1) higher porosities, (2) lower bulk densities, (3) lower effective pressures, (4) higher temperatures, and (5) lower velocities. The most well-known methods under this prediction scheme are

1. Pennebaker's method (overlay method),
2. Hottman-Johnson's method (normal compaction trend line method), and
3. Eberhart-Phillips's method (regression method).

Pennebaker's Method (Overlay Method)
Pennebaker (1968) developed an approach for predicting pore pressure using seismic or well log velocity (sonic) data. In this method, first, a velocity to pore pressure gradient (in ppg) chart is created from well logs to convert velocity to pore pressure gradient as a function of depth. Then the velocity data from an area where a prediction is to be carried out is overlain. Ensuing pore pressures gradients are then read (again in depth) directly to predict it. In Figure 4.2a we show such a chart that Pennebaker developed for South Texas along the Rio Grande River from Maverick County to Offshore from Cameroon County. The indicated lines show vertical interval velocity

trends for indicated pore pressure gradients in ppg. The vertical axis is depth in semi-log scale. The assumption is made that mechanical compaction is the only mechanism and further the mathematical form for compaction is similar to that formulated by Athy (1930a). The interval velocity is assumed to be varying with depth in an exponential manner as was verified by numerous wells in the study area. The key point to note is that Pennebaker's method needs to establish a trend for normal compaction – then abnormal pressure can be related to degree of departure from the normal pressure base line as shown in Figure 4.2b. The Pennebaker's procedure is fairly simple and very straightforward. This is perhaps one of the earliest versions of a rock physics template for pore pressure prediction. However, there is a caveat. There could be – and there always is – wide variations in the (normal compaction) trends. There are multitudes of reasons why that is the case. Pennebaker (1968) identified two major causes: a change in lithology amongst rocks of same age group and change in age of rocks of similar lithology (shale). The trend shifts to lower transit time (higher velocity) due to lithology change is shown in Figures 4.2c for calcareous sands and in Figure 4.2d for calcareous shales. These rocks are of the same age group. Figure 4.3 shows the effect due to age of rocks of similar lithology. There is a shift to higher transit time (lower velocity) as one moves from onshore to offshore – with older sediments, to more recently deposited, less compacted sediments closer to the Gulf of Mexico. These shifts in the trend are sizable and require experienced analysts to develop and use these charts. Further, the calibration charts being tied to depth are not transportable from one basin to another or even from one part to another part of the same basin although they are very valuable, if properly calibrated. Once the normal compaction trend is established that accounts for age and lithology of rocks, one can then overlay interval traveltime profile(s) at a location from seismic data to predict pore pressure as shown in Figure 4.2b.

Subsequent to Pennebaker's classic paper, many authors addressed issues associated with establishing normal compaction trends. An example is given by Reynolds (1970) and is shown in Figure 4.4. It clearly depicts the effect of lithology on velocity data – in this case, seismic interval velocity as used by Reynolds (1970). Thus, without detailed knowledge of well site geology (lithology, age), overpressure trends can be significantly distorted. This means that any velocity data must be interpreted and guided by geology to account for expected subsurface lithology variations. We will discuss this in greater detail later when we discuss how to handle pore pressure variations due to lithology changes within shale due to chemical diagenesis (Chapters 5 and 6) and carbonates.

Hottman and Johnson's Method

In a classic paper Hottmann and Johnson (1965), described a method to detect over-pressured rocks using sonic and resistivity logs from wells in Miocene and Oligocene age formations from offshore Louisiana. For acoustic logs (and also for resistivity logs), the authors created a direct calibration curve between the logarithm of $\delta(\Delta t)$ the departure of the sonic transit time, Δt, from the normal compaction trend when plotted versus depth and pressure gradient (in ppg), to predict overpressure in shales. An

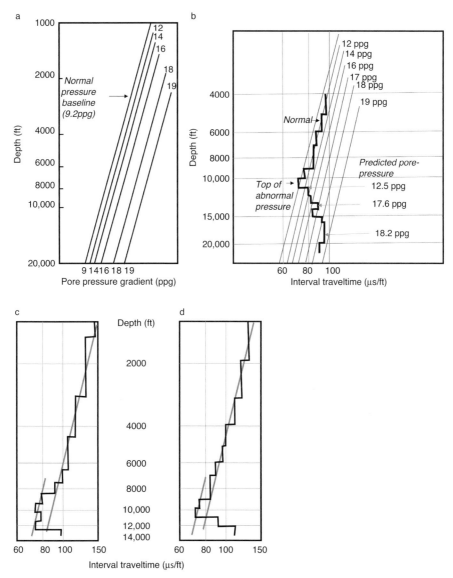

Figure 4.2 An example of the overlay method of Pennebaker (1968). (a) Abnormal pressure versus depth trends as related to the degree of departure from normal baseline pore pressure. (b) Interval traveltime computed from seismic data at a specific well location showing the onset of overpressure below a depth of 10,000 ft. Examples showing shifts in normally pressured baselines for calcareous (c) sands and (d) shales.

example of the calibration curve is shown in Figure 4.5 for sonic logs. The data were taken from 16 wells in offshore Louisiana. We should note that the fluid pressure gradients are measured in sands adjacent to pressured shales; an assumption has been made that the sands and shales are in pressure equilibrium. This may or may not be the case due to possible lateral pressure communication between shales and adjacent sands. The authors did not account for this effect in their measurements. Hottman and Johnson's data are shown in Table 4.3.

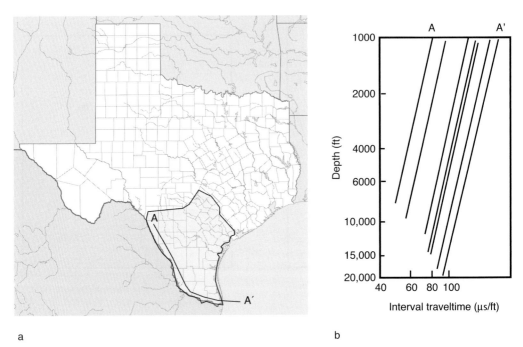

a   b

Figure 4.3 Trend of interval traveltime versus depth in South Texas along the Rio Grande River in Maverick County to offshore from Cameron County.

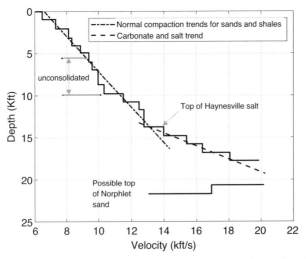

Figure 4.4 Interpreted velocity analysis for a prospect in southern Mississippi showing the effect due to lithology change. Without knowledge of well site lithology, geopressured trends can be distorted and misinterpreted. Modified after Reynolds et al. (1971).

A curve fit to the data yields the following equation:

with
$$P/Z(psi/ft) = a + b\ln(\delta(\Delta t)) \tag{4.1}$$

Table 4.3 Fluid pressure gradient in psi/ft and sonic
traveltime departure from normal trend $\delta(\Delta t)$ in μsec/ft

Fluid pressure gradient	Sonic departure time
(Psi/ft)	$\delta(\Delta t)$ *(μ sec/ft)*
0.87	22
0.62	9
0.82	21
0.84	27
0.86	27
0.73	13
0.54	4
0.86	30
0.56	7
0.84	23
0.92	33
0.6	5
0.88	32
0.9	38
0.91	39
0.81	21
0.97	56
0.78	18

Source: After Hottman and Johnson (1965, Table 1)

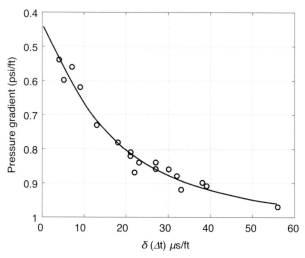

Figure 4.5 Pressure calibration plot of Hottman–Johnson. Plotted are the fluid pressure gradients
(psi/ft) measured from 16 offshore Gulf of Mexico wells versus interval transit time departure
from the NCT at the same depths where pressure gradients are measured. We should note that the
fluid pressure gradients are measured in sands adjacent to pressured shales; an assumption has
been made that the sands and shales are in pressure equilibrium. This may or may not be the case
due to possible lateral pressure communication between shales and adjacent sands. The authors
did not account for this effect in their measurements.

$$a = 0.281 \; b = 0.174.$$

We note that Hottmann and Johnson (1965) implicitly assumed that the normal compaction trend, Δt_n, is given by a relation of the type

$$\Delta t_n = \Delta t_0 e^{-Z/\lambda} \tag{4.2}$$

and so the variable $\delta(\Delta t)$ is given by

$$\delta(\Delta t) = \Delta t - \Delta t_0 e^{-Z/\lambda} \tag{4.3}$$

where $\delta(\Delta t)$ is the interval transit time departure from that of the normal trend, Δt_0 is the transit time at $Z = 0$, λ is the slope of the of the transit time data determined from the normally pressured section of the sonic log, and Z is depth in meters. Kan and Swan (2001) compiled data from several basins of varying ages and expressed their results in terms of the following quadratic equation

$$P(Z)/Z = P_n(Z)/Z + C_1\delta(\Delta t) + C_2(\delta(\Delta t))^2 \tag{4.4}$$

where $P(Z)/Z$ is the pore pressure gradient (in Pa/m) at depth Z (in meters), and $P_n(Z)/Z$ is the pore pressure gradient (in Pa/m) under hydrostatic pressure condition at the same depth. Their results are reproduced in Table 4.4.

In Figure 4.6, we show pressure calibration charts from several Tertiary Clastics basins (modified after Kan and Sicking, 1994). Plotted in the horizontal axis is $\delta(\Delta t)$ in µs/ft and in the vertical axis is fluid pressure gradient in psi/ft. We note substantial differences in these charts; this is expected as the compaction properties of sediments from respective basins are very different due to geology. We again remind the readers that such charts are based on the assumption that mechanical compaction is the predominant mechanism for overpressure. Therefore, usage of charts such as these may lead to erroneous estimates of pore pressure when other geopressure mechanisms – other than mechanical compaction, begin to play an important role.

Hottmann and Johnson (1965) clearly pointed out in their paper that the pressure data were from sands adjacent to (overpressured) shales and that an assumption was

Table 4.4 Regression coefficients C_1 and C_2 for $\delta(\Delta t)$ as shown in equation (4.4)

Basin and age	C_1 (Pa/µs)	C_2 (Pa.m/µs^2)
Gulf of Mexico Miocene	143	−0.42
Gulf of Mexico Pliocene	67.6	−0.1
Gulf of Mexico Pleistocene	42.7	−0.28
North Sea	13.1	0.168
Alaska	36.5	0.48
Northwest Australia	42.4	0.22
South China Sea	32.8	1.54

Note: Pore pressure gradients are expressed in Pa/m, the depth in m, and the interval transit time Δt in µs/m.

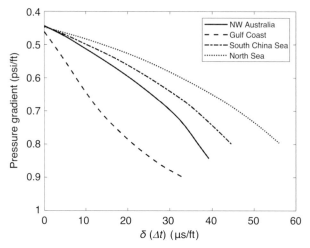

Figure 4.6 Acoustic log transit time departure versus fluid pressure gradient for several basins. Modified after Kan and Sicking (1994).

made that the shales were in pressure equilibrium with adjacent sands – a critical assumption that was missed by the subsequent users and practitioners of their technique. This is an important point to remember: *Most pore pressure prediction techniques are based on analysis of shales, either using seismic, sonic or resistivity logs; however, the calibration charts are created based on pressure measurements in adjacent sands with the assumption that these are in pressure equilibrium with shales. That may not be the case due to fluid transfer as discussed in* Chapter 3 – *sand pressure may be lower, equal or greater than the shales depending on the relief of the sand and its clay content.*

In summary, the basic assumptions in the Hottmann and Johnson's as well as Pennebaker's methods are as follows:

1. Mechanical compaction is the only source of overpressure.
2. Porosity of shales decreases with compaction (and therefore, with depth) in a systematic fashion. Consequently, the interval transit times of shales also change systematically with depth.
3. There is no uplift or erosion in the basin. In other words, the effective stress that a rock (shale) experiences today is the maximum effective stress that it has experienced so far. There are no effects due to effective stress reversal or unloading or hysteresis.
4. Shale mineralogical composition remains the same throughout the burial. Thus, the analysis based on a single compaction trend line could be inadequate for other types of shales such as calcareous, silty, and hard or limey-shales.
5. All thermal effects known to affect compaction of shales are ignored.
6. The mathematical form of the normal compaction trend is not based on any principle of rock physics. For example, it does not address the end-point of the

normal compaction trend of a shale when its porosity exponentially approaches zero.

Hottmann and Johnson's technique is still one of the two most widely used techniques in the petroleum industry – the other being Eaton's method (see below). The strength of Hottman and Johnson's method is that it is very simple to use and the methodology is applicable to shales in a wide range of clastic basins. The technique is also applicable to other attributes such as density, resistivity and conductivity logs as well as drilling penetration rate logs for shales. However, there are several weaknesses of the method:

1. It is an empirical method and therefore, area dependent. It depends heavily on "local" calibration and therefore, is not transportable from basin to basin. Thus the method should be applied with great caution in areas with poor or no well control.

2. Establishing "normal compaction" trend is very critical. The authors noted that attribute analysis must be made on "thick and clean" shales. But they did not specify how thick is "thick" and how clean is "clean." Some authors (e.g., Eaton, 1976) suggested that shales be at least 20 feet thick. Any error in the slope and intercept plot of the "normal compaction" trend could significantly alter the predicted magnitude of overpressure, especially at deeper depths.

3. As noted earlier, the relation applies strictly to overpressure caused by undercompaction and not to other mechanisms such as those that are affected by thermal history of basins such as aquathermal pressuring, smectite to illite metamorphism, kerogen maturation and late-stage fluid charging.

4. The method requires multiwell database, both for normal compaction trend analysis and pore pressure calibration. Computations based on a single-well analysis could be unreliable. Additionally, for this reason the method may not be suitable in rank-wildcat areas or exploration areas with limited or no well control. However, for appraisal and development work, this method is very reliable.

5. The pressure calibration curves, being empirical in nature, require an extensive and reliable pressure data base such as those from Repeat Formation Tester (RFT) or Drill Stem Test (DST) measurements. The shapes of these curves (see Figure 4.6) require that pressure data be made available from shallow to deeper parts of the calibration wells. Quite often this is not possible as shallow pressure data are rarely collected during operations. For deeper parts, the data are collected only for "reservoirs with pay" both due to economic and safety criteria. Because of these limitations, the quality of the calibration curve is severely hindered. The limited pressure data may not provide a reliable guidance to the exact shape of this curve. This type of situation is quite common, especially, during the exploration phase, and limits the applicability of the technique severely.

6. The quality of the calibration charts (for both acoustic and resistivity logs) as provided by Hottmann and Johnson (1965) in their publication are exceptionally good. However, this is not typical. In Figure 4.7, modified after Mouchet and Mitchell (1989), we show a more typical case. The authors chose to represent the

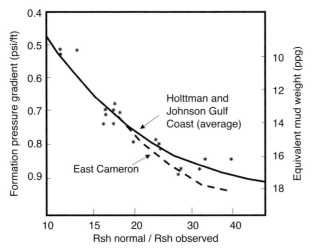

Figure 4.7 Pressure calibration charts for resistivity data. Modified after Mouchet and Mitchell (1989).

ratio of $R_{sh,normal}/R_{sh,observed}$ as the variable related to pore pressure gradient. Note the paucity and scatter of data at high pressure gradient. Charts such as these could result in a pore pressure uncertainty by as much as 25% at the high end of the pressure gradient.

A comment on the construction of normal compaction trend is worth noting. There is a growing awareness of the importance of temperature on the rock properties as the oil and gas industry has drilled multitudes of wells that penetrate formations that are known as high-pressure and high-temperature (HPHT) wells. It suggests that both the slope and the intercept of normal compaction trend data (say, for sonic transit times or resistivity trends) are affected by increasing temperatures and that undercompaction, therefore, may not be operative at these temperatures (Dutta, 1986; Dutta et al., 2014; Bowers, 1995). Chemical diagenesis may be the operating mechanism at these temperatures thus invalidating the single normal compaction trend concept with depth with a fixed slope and intercept when plotted in semi-logarithm space. As we shall see below, the well-known Eaton's method also has this deficiency. Methods based on effective stress do not suffer from this deficiency (Bowers, 1995; Dutta, 1987b; Dutta et al., 2014).

Eberhart-Phillips et al. Method (Regression Method)

This method by Eberhard-Phillips et al. (1989) is based on a curve fit to a set of laboratory data on compressional wave velocity V_p of some clean and shaly sandstone under controlled pore pressure and confining pressure conditions. The authors suggested the following relation:

$$V_p = 5.77 - 6.94\phi - 1.73\sqrt{C} + 0.446(\sigma - e^{-16.7\sigma}) \qquad (4.5)$$

where ϕ is porosity (fraction), σ is effective stress (kbar), and C is clay content (fraction). Here V_p is expressed in km/s. This is a simple relation that has been used for predicting effective stress given a set of measurements on velocity, clay content and porosity much like what Gardner et al. (1974) suggested earlier. The data used by Eberhart-Phillips et al. (1989) were obtained by measurements on consolidated rocks, similar to that by Gardner et al. (1974) and hence the use of the relation is limited. Certainly, it may not be valid for soft-clastics, especially at low effective stress and thus, the model must be re-calibrated. To predict pore pressure P, we would require a separate estimate of overburden stress, S, since $P = S - \sigma$. Thus a recommended use of this empirical method is to recalibrate the coefficients with petrophysical well logs and core data before application.

4.3.2   Horizontal Methods

Ratio Method
This method has been applied with some success for sonic, density and resistivity logs. In this method, pore pressure is calculated assuming that pore pressure is the product of the normal pressure multiplied by the ratio of the measured value to the normal value for the same depth. For sonic logs, this implies

$$P = P_n \left(\frac{\Delta t}{\Delta t_n} \right) \qquad (4.6)$$

and for resistivity (R) logs,

$$P = P_n \left(\frac{R_n}{R} \right) \qquad (4.7)$$

Here, the symbol n, stands for the normal trend value when extrapolated. As noted by Mouchet and Mitchell (1989), it is totally empirical and the results are not always satisfactory. Calibration of this method requires knowing precisely the appropriate normal values of each parameter such as the sonic traveltime or resistivity. We note that in this method, one does not use either vertical effective stress or overburden stress explicitly and thus it can result in erroneous values of calculated pore pressure – for example, the calculated value at a depth could be larger than the overburden stress at that depth. This would be unphysical. Mouchet and Mitchell (1989) noted that this method is more useful for drilling-exponent approach to calculating pore pressure.

Modified Horizontal Method (Eaton's Method)
This is one of the most widely used methodology (Eaton, 1970, 1972, 1975, 1976). It is also based on the analysis of Normal Compaction Trend (NCT). The essential steps are very similar to what we discussed earlier in the context of Pennebaker and Hottmann

and Johnson's methods. Eaton's method (Eaton, 1975) is usually written in terms of interval transit time (sonic log data) as follows:

$$P = S - (S - P_n)\left(\frac{\Delta t_n}{\Delta t}\right)^3 \tag{4.8}$$

or alternately, in terms of velocity as,

$$P = S - (S - P_n)\left(\frac{V}{V_n}\right)^3 \tag{4.9}$$

Here P is the actual pore pressure, S is the vertical overburden stress, P_n, is the normal (hydrostatic) pressure, Δt is the measured interval transit time corresponding to measured velocity, V and Δt_n, and V_n are the corresponding normal compaction interval transit time and velocity, respectively. This implies that if $V = V_n$, i.e., the rock is normally compacted, then $P = P_n$. Conversely, if $V < V_n$, i.e., the measured velocity is lower than the normal velocity, $P > P_n$, which means the rock is over-pressured. Eaton's model uses overburden gradient explicitly and thus it allows for effects such as air gap and water depth. In terms of effective pressure, σ, Eaton's equation (equation (4.9)) can be written as

$$\sigma = \sigma_n\left(\frac{V}{V_n}\right)^3 \tag{4.10a}$$

or alternatively

$$\sigma = \sigma_n\left(\frac{\Delta t_n}{\Delta t}\right)^3 \tag{4.10b}$$

where σ_n is the effective pressure under normal pressure conditions. It is given by

$$\sigma_n = S - P_n \tag{4.10c}$$

Thus Eaton's method appears to be an effective stress based method. Eaton's method may not be the most accurate method for complex geological settings, but it is often used as a standard against which all other pressure prediction models are compared. Eaton's method is also used widely with resistivity, conductivity and drilling-exponent data. These relations are given below:

$$P = S - (S - P_n)\left(\frac{R}{R_n}\right)^{1.2} \quad \text{Resistivity} \tag{4.11}$$

$$P = S - (S - P_n)\left(\frac{C_n}{C_n}\right)^{1.2} \quad \text{Conductivity} \tag{4.12}$$

$$P = S - (S - P_n)\left(\frac{D_c}{D_{cn}}\right)^{1.2} \quad \text{"D"– exponent} \tag{4.13}$$

Eaton exponents, namely, 3.0 for sonic and 1.2 for others, are a *mystery*. We could not find any theoretical justification for this in any of Eaton's publications. Most practitioners of this technique believe that this exponent should be treated as a free parameter – the value being fixed by calibration well(s). Sarker and Batzle (2008) even suggested that this exponent should vary with depth. Bowers (1995) also raised the issue. Eaton's exponent is a measure of the sensitivity of the sonic velocities to effective stress (Ebrom et al., 2003). For most cases in the Gulf of Mexico, an Eaton exponent of 3 for P-wave velocities and 2 for S-wave velocities is appropriate (Ebrom et al., 2003). A larger value of the *exponent* indicates insensitivity of the velocities to changes in effective stress. However, a single value of the *exponent* is often insufficient to predict pore pressures for the entire section, and this number needs to be varied with depth depending on the degree of overpressure in the subsurface. Users should be aware of this issue.

Additionally, as with the method of Hottmann and Johnson, Eaton's method is very sensitive to the choice of the normal compaction trend versus depth. Most of the error occurs due to lack of velocity data at shallow depths. A poor choice at shallow depths can have a large impact on the reliability of predicted pressure at deeper depths. Bowers (1995) also pointed out that there is no justification of why a NCT of an attribute (sonic or resistivity, for example) should be a straight line when the semi-log of the attribute is plotted versus depth. From rock physics point of view, there is little justification for this.

We have found that Eaton's method can be and has been seriously misused as the method suffers from the same weaknesses as outlined above for Hottmann and Johnson's method. While it is reasonable to expect that shales may, and often, do vary in lithology, an accounting of this effect has turned the technique into a "curve-fitting exercise" – varying the slope and intercept of the normal compaction trend line to fit a set of pressure data points at deeper depths. One can obviously see the problem in the example shown in Figure 4.8. The derived NCT (solid line) obviously does not fit the data; it has been adjusted to fit a set of pressure data at deeper depths (pressure data are not shown). An alternate NCT model as shown by the dashed line in Figure 4.8 apparently fits the NCT data well. However, using this new NCT line would increase the pore pressure at larger depth (say, at 4500 ft). So which pressure should we believe from the sonic log? The basic problem is that the so-called NCT in Eaton's model has no basis – it is neither based on theory nor on a model that describes the velocity of a rock (shale) under normal pore pressure condition. It is for all practical purpose a subjective decision. Quite often the intercept of the so-called NCT does not obey the rock/fluid velocity at the near surface (zero depth, for example). There is another bad practice that has propagated in the commercial business of pore pressure – the use of "multiple normal compaction trends" when one trend line does not fit the data! These "multiple normal compaction trends" usually have no physical basis. So users need to be very aware of this pitfall. On the other hand, with calibration and with the understanding that the well to be drilled is a good analogue of a well that has been used as a calibration well, this method can give good results. It is also very easy to use.

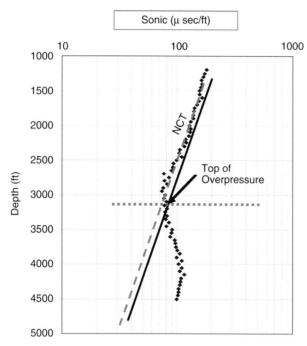

Figure 4.8 Problem with the Eaton method. The derivation of the NCT is often subjective; in this example the NCT has been adjusted to match the pore pressure at deeper depths (beyond 3500 ft) neglecting the shale transit times at shallower depths. An alternate NCT (long-dashed line) fits the shallower data better but would yield much higher pore pressure at deeper depths. So which trend should we use? What if there is no well?

Eaton's method is applicable to a wide range of geologic environments provided care is taken to calibrate the model. Simple equations such as the Eaton model are always preferable over graphical methods such as Pennebaker or calibration chart approach of Hottman and Johnson. The method is applicable to all velocities – VSP, check shot, sonic logs, and surface seismic.

4.3.3   Vertical or Equivalent Depth Method

The methods discussed so far have one common feature – mechanical compaction disequilibrium is assumed to be the only cause of overpressure. The equivalent depth method is an extension of that assumption. It assumes that the same shale with equal physical properties at different depths will have equal vertical effective stress, namely, the difference between the total vertical overburden stress and the pore pressure. This is illustrated in Figure 4.9 for interval transit time, Δt. Every point A in an undercompacted or overpressured zone is associated with a normally compacted point B. The depth of point B (Z_B) is called the *equivalent* depth. The fluid contained within the pores of the shale at A has been subjected to all the lithostatic load in the course of the burial from Z_B to Z_A. Using Terzaghi's law, the vertical effective stresses at points A and B are given by

$$\sigma_B = S_B - P_B \tag{4.14}$$

$$\sigma_A = S_A - P_A \tag{4.15}$$

where S and P denote vertical overburden stress and pore pressure, respectively. However, at the equivalent depth, we have

$$\sigma_A = \sigma_B \tag{4.16}$$

This then implies

$$P_A = P_B + \Delta S_{AB} \tag{4.17}$$

where

$$\Delta S_{AB} = S_A - S_B \tag{4.18}$$

Thus in Figure 4.9, interval transit time must have an equivalent depth where vertical projection from a point inside the reversal intersects the plotted data at a common point above the reversal on the normal compaction trend. This statement applies to any attribute that depends on porosity such as bulk density, velocity, resistivity and conductivity (after proper temperature correction). For the equivalent depth method to work we must have the same rock properties at two depths. As we shall see below this is not always the case as mechanical compaction disequilibrium or undercompaction gives way to chemical diagenesis causing high overpressure that cannot be explained by undercompaction mechanism alone. An example of where the equivalent depth method works and where it does not is shown in Figure 4.10. This example is taken from Bowers (1995). Others have also reported evidence similar to this. For example, Magara (1975) found that the Equivalent Depth method under predicted the pore pressure data of

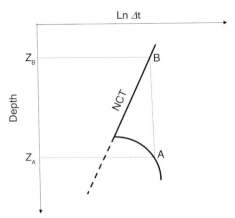

Figure 4.9 Illustration of how the equivalent depth method for pore pressure prediction works. Both points labeled A (overpressured interval) and B (normally pressured interval) have the same interval transit time.

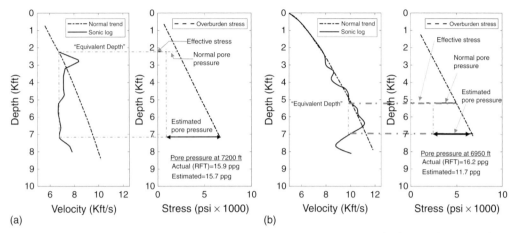

Figure 4.10 An example of the case where the equivalent depth method (a) works and (b) fails. Modified after Bowers (1995).

Hottman and Johnson (1965). Bowers (1995) also showed the same to be true of the data presented by Hottman and Johnson (1965).

A natural consequence of equivalent depth concept as explained above suggests that for shales, increasing velocity with burial (increased vertical overburden stress and compaction) will result in higher bulk density. This is always the case at shallow burial depths where fluid expulsion from the shale keeps in stride with the burial rate. If pore pressure is elevated over the hydrostatic condition due to rapid burial (compaction disequilibrium), velocity reversal will happen as the burial continues (vertical overburden stress increases with increased depth but vertical effective stress decreases) along with decrease in bulk density. This will be evident in a cross-plot of bulk density and velocity or interval transit time of shales. Such relationships are observed routinely from borehole data (sonic, density, and resistivity logs). Thus undercompaction will necessarily create a reversal in *both* velocity and density data with increasing pore pressure. However, if there is an additional fluid source at depth (e.g., any mechanism that depends on temperature and causes the shales to further dewater), then a reversal in velocity *may not* accompany a change in porosity, and hence, bulk density. This was also noted by Bowers (1995). Thus bulk density logs can be used to discern whether compaction disequilibrium works at all depths. In most cases, as burial continues, mechanisms other than compaction disequilibrium begin to contribute to geopressure at deeper depths. Most of these mechanisms are associated with elevated temperature and overburden stress.

4.3.4   Effective Stress Methods

In our view, effective stress based methods for pore pressure prediction are more fundamental than those which use a totally empirical formula such as in the Eberhart-Phillips et al. (1989). Most of the pressure prediction methods using effective stress are based on a three-step process. These are: a compaction model (relations between porosity and effective stress), an attribute relation (a relation between porosity and a directly measured attribute such as velocity or slowness, resistivity or conductivity), and an estimate of overburden stress. The key point to note is that porosity is eliminated from the final formulation by using measured attributes such as velocity and resistivity. This yields effective stress from the measured attribute (velocity or resistivity). Terzaghi's law facilitates the conversion between effective stress and pore pressure by subtracting effective stress from estimated overburden stress. We shall discuss these steps in detail below.

Compaction Models (Porosity versus Effective Stress Relation)

We know that mechanical compaction causes porosity to decrease with increasing vertical effective stress as defined in Terzaghi's law. Vertical effective stress is the stress acting on the solid framework of the rock-fluid system. Overpressure results when the fluid flow in a low permeable system (shale) is retarded and the outflow cannot keep up with the rate at which overburden builds up. At deeper depths (higher temperature), chemical diagenesis becomes very important – this is when pore fluid pressure increases without much change in the porosity of the host sediment (shale). When this happens, porosity becomes more or less independent of effective stress. Typically, for this to happen, there is usually an extra source of water, for example, shale alteration due to burial diagenesis and/or aquathermal pressuring.

There are many compaction models discussed in the literature for mechanical compaction. (We discussed some of these in Chapter 3.) Some of the well-known ones are listed below:

$$\text{Athy-type models}: \phi = \phi_0 e^{-K\sigma} \tag{4.19}$$
$$\text{Terzaghi (1923)}: \phi = 1 - \phi_0 \sigma^{c/4.606} \tag{4.20}$$
$$\text{Palciauskas and Domenico (1989)}: \phi = 1 - \phi_0 e^{-\lambda\sigma} \tag{4.21}$$
$$\text{Holbrook (2004)}: \sigma = \sigma_{\max}(1 - \phi)^\alpha \tag{4.22}$$
$$\text{Dutta (1987b)}: \sigma = \sigma_0 e^{-\varepsilon\beta(T)} \tag{4.23}$$

Here T is temperature, $\beta(T)$ is a function of temperature that is implicitly dependent on time, and K, c, λ, $\sigma_{\max}$, α, σ_0 and ϕ_0 are lithology dependent constants. Void ratio is denoted by ε; it is the ratio of the pore volume to rock volume. We like these models because the parameters included here are fundamental in nature and depth does not appear explicitly. Depth is a poor substitute for geologic variables such as effective stress and temperature. Out of these models, the first four are strictly valid for mechanical compaction only; the fifth model is valid for *both* mechanical compaction and compaction associated with chemical diagenesis. Let us examine this further in terms of porosity versus effective stress relations.

Examining the compaction models as given above, we note that the porosity-effective stress relations, with the exception of model in equation (4.23), are single-valued functions. All Athy-type relations assume the existence of a *virgin* compaction curve that is tied to a single-lithology model. This is also the fundamental assumption in Bowers's model (Bowers, 1995). In reality, shales do have a composite mineral composition – each component has its unique compaction curve and that when effective stress reverses (decreases), the final compaction curve could be different in destination than its starting point. This phenomenon is known as chemical diagenesis. Goulty (1998) termed this as *chemical compaction*. (We prefer the term chemical *diagenesis*.) In geology the phenomenon of this type is known as burial metamorphism of shale in the context of shale alteration and dewatering (Hower et al., 1976).

There is another possibility for chemical diagenesis; one that occurs in a *closed* system (shale) after it has undergone some undercompaction. In this case a secondary pressure generation mechanism takes place and the rock cannot stay on the virgin compaction curve because of this fluid source (fluid expansion). Examples are aquathermal pressuring and osmotic phenomenon. In this case the mechanical compaction process is arrested, porosity change is minimal, the effective stress deceases due to increase in fluid pressure and the rock begins to display a form of hysteresis behavior in porosity-effective stress domain. As we shall see later, this behavior is more pronounced in the velocity-effective stress domain (Bowers, 1994).

Dutta (1986, 1987b) accommodated chemical diagenesis models of the type just discussed by suggesting that the compaction space – porosity reduction by increasing effective stress, is actually a *four-dimensional* space that involves *temperature and geologic time* as two additional variables besides *porosity and effective stress*. This is the basis of Dutta's compaction model as shown in equation (4.23). The diagenetic function β, is a function of temperature and time. The shape of this function was discussed in Chapter 3 in the context of smectite to illite transition (Figure 3.9).

Attribute Relation (Attribute versus Porosity Relation)

Earlier we discussed some relations between porosity and effective stress. Porosity is a fundamental quantity. However, it is also very difficult to measure *in situ*. Most measurements deal with measuring some *other attributes* that are correlated to porosity. Velocity and density are two attributes that depend directly on porosity and it is why borehole acoustic and density logs are used for pore pressure detection and analysis. So is the resistivity. Here we will point out some well-known relations between porosity and velocity and then show how we can relate velocity to effective stress.

A fundamental concept that relates velocity to porosity is that of the critical porosity model postulated by Marion et al. (1992) (see Figure 2.27). They showed the experimental data that is the basis for their model – relation between compressional velocity and porosity in sand-shale mixtures. Figure 2.27 shows the cut-off between Wood's equation (suspension model) and the transformation to rock (load bearing entity) at porosities of 38% to 50% and is defined as the critical porosity. Velocity trend at porosities higher than the critical porosity is dominated by the Wood's equation and that at porosities smaller than the critical porosity, is dominated by a load-bearing porosity-velocity relation such as the one labelled as Voigt average and described by Nur et al. (1991). The increase of velocity with decreasing porosity is due to compaction.

The literature is full of relationships between porosity (and bulk density) and velocity of a given rock type (Nur and Wang, 1989). Table 4.5 summarizes some of the widely used relations for sandstones, shaley sandstone and mudrock. Most of these relations have been validated by laboratory measurements (at ultrasonic frequencies).

Some authors (Han et al., 1986; Castagna et al., 1985; Eberhart-Phillips et al., 1989) addressed shaly sand issues and studied the effect of clay content on sandstone velocities. However, very little is published about such relationships for shales, where geopressure occurs. For sonic logs, the Pickett-type equation (Pickett, 1963) is frequently used:

$$\Delta t = A_0 + B_0 \phi \tag{4.24}$$

where Δt is the sonic interval transit time (slowness) (usually expressed in μs/ft), and A_0 and B_0 are lithology-dependent constants. Another commonly used relationship is that of Gardner et al. (1974). This is usually expressed as

$$\rho_b = C V^D \tag{4.25}$$

where ρ_b is bulk density, V is interval velocity of P-waves, and C and D are lithology dependent constants. A curve fit to the data of Gardner et al. (1974) yields C = 0.23 and D = 0.25, if velocity is expressed in ft/s and the bulk density in gm/cc. Note that the well-known Wyllie time-average equation (Wyllie et al., 1956; Gardner et al., 1974) relating porosity to transit time, namely,

$$\Delta t = \phi \Delta t_f + (1 - \phi) \Delta t_m \tag{4.26}$$

where Δt_f and Δt_m are the inverse of velocity or slowness through the fluid and solid phases (grains of the rock matrix), respectively, is a special case of the Pickett-type equation. It suggests that the total traveltime through a rock is the sum of the traveltime through fluid and the traveltime through the solid rock. This is a heuristic assumption. Issler (1992) published a very useful porosity-sonic transit time equation for shales and mudstones based on logs and core measurements. It is an extension of a similar relationship published earlier by Raiga-Clemenceau et al. (1988). Issler's (1992) equation is

$$\Delta t = \Delta t_m (1 - \phi)^{-x} \tag{4.27}$$

Table 4.5 Velocity–porosity–clay content relationships for sandstones and shaley sandstones

Wyllie's time average equation (Wyllie et al., 1956)

$$\frac{1}{V_P} = \frac{\phi}{V_f} + \frac{1-\phi}{V_M}$$

Raymer-Hunt–Gardner relation (Raymer et al., 1980)

$$V_P = (1-\phi)^2 V_M + \phi V_f \qquad\qquad 0 < \phi < 0.37$$

$$\frac{1}{V_P^2} = \frac{\rho\phi}{\rho_f V_P^2} + \frac{\rho(1-\phi)}{\rho_M V_M} \qquad\qquad \phi > 0.47$$

$$\frac{1}{V_P} = \frac{0.47-\phi}{0.10}\frac{1}{V_P(\phi=0.37)} + \frac{\phi-0.37}{0.10}\frac{1}{V_P(\phi=0.47)}$$

where $\boxed{V_P(\phi=0.37)}$ is calculated from the low-porosity equation at $(\phi=0.37)$

and $\boxed{V_P(\phi=0.47)}$ is calculated from the high-porosity equation at $\phi=0.47$

Tosaya (1982)

$$V_P(\text{km/s}) = 5.8 - 8.6\phi - 2.4C$$
$$V_S(\text{km/s}) = 3.7 - 6.3\phi - 2.1C \qquad\qquad C \text{ is clay volume fraction}$$

Castagna et al. (1985) (for shaley sands of Frio formation)

$$V_P(\text{km/s}) = 5.81 - 9.42\phi - 2.21C \qquad C \text{ is clay content}$$
$$V_S(\text{km/s}) = 3.89 - 7.07\phi - 2.04C$$

For mudrock (clastic silicate rocks composed primarily of clay and silt-sized particles)
$$V_P(\text{km/s}) = 1.36 + 1.16V_S(\text{km/s})$$

Han et al. (1986) (for shaley-sandstones)

$$V_P(\text{km/s}) = 5.59 - 6.93\phi - 2.18C$$
$$V_S(\text{km/s}) = 3.52 - 4.91\phi - 1.89C \qquad\qquad \text{at 40 Mpa}$$

For clean sandstones
$$V_P(\text{km/s}) = 6.08 - 8.6\phi$$
$$V_S(\text{km/s}) = 4.06 - 6.28\phi \qquad\qquad \text{at 40 Mpa}$$

Eberhart-Phillips et al. (1989) (regression to data obtained by Han et al., 1986)
$$V_P(\text{km/s}) = 5.77 - 6.94\phi - 1.73\sqrt{C} + 0.446(P_e - 1.0e^{-16.7Pe})$$

where x is termed as *acoustic formation factor* and it depends on lithology. This is very similar to Archie's resistivity equation. According to Archie's model, the resistivity R of a fluid filled porous rock can be expressed in terms of resistivity of brine R_w, and formation factor, F as

$$R = FR_w = a\phi^{-m}R_w;$$
$$F = a\phi^{-m} \tag{4.28}$$

Here, coefficients a, and m are lithology dependent (Archie, 1942). Equation (4.27) is similar to equation (4.28) in form.

We again note that out of all the relations discussed so far in the section that relate porosity to velocity or its inverse, only two relations have been stated by the authors to be explicitly valid for shales – that of Issler (1992) and Gardner et al. (1974). Clearly we need more active research in this area since we may not assume that relations valid for sandstones and shaly sands are equally applicable for shales with variable matrix composition. In addition, most laboratory measurements are made on consolidated rocks; it is not clear whether these relations can be extrapolated to their respective soft-rock end-members or to higher pore pressure (low effective pressure) states since rock properties are affected by their diagenetic states and acoustic properties (velocities) are impacted by hysteresis (Gardner et al., 1974). In addition, there needs to be validation of laboratory measurements by borehole data, especially under geopressured conditions for shales. Barring this, petrophysical well logs are the best source of data for shales.

Relations between Velocity and Effective Stress

Since porosity is not *directly* measurable *in situ* by any tool, it is desirable to eliminate it in any relation between velocity and effective stress. The same argument applies to relations between resistivity and effective stress. The Terzaghi's law facilitates the conversion between effective stress and pore pressure by subtracting effective stress from estimated overburden stress. We shall illustrate the procedure by some concrete examples below. This approach leads to a systematic procedure or recipe by which the resulting model can be made depth independent and thus become "transportable" across basins. Calibration is still required as the dimensions of velocity and effective stress are different and contain empirical coefficients.

An example of this approach can be illustrated by using Athy-type models as shown in equation (4.19). This relates porosity to effective stress. If we use Issler's relation for the attribute law for interval transit time or slowness (equation (4.27)), we can eliminate porosity and obtain the following relation between interval transit time and effective stress:

$$\sigma = 1/K \ln[\phi_0 \Delta\tau^\alpha / (\Delta\tau^\alpha - 1)] \tag{4.29}$$

where $\Delta\tau = \Delta t / \Delta t_m$ and $\alpha = 1/x$. Recall that both K and α are lithology dependent coefficients. Many such relations are possible based on the type of compaction model and the attribute relations used. Of course, these relations have empirical coefficients and therefore, need to be calibrated as noted earlier.

Dutta's Method

In Chapter 3 we discussed mechanical compaction and chemical diagenesis involving shale. The latter includes overpressure mechanisms that depend on temperature and time such as aquathermal pressuring, uplift/erosion and effective stress reversal, fluid source at deeper depths in a confined system such as smectite dehydration and conversion of smectite to illite, kaolinite to illite and illite to muscovite and mica. Some of these mechanisms (chemical diagenesis) change the matrix of the host rocks (shales) – examples are all chemical reactions that deal with burial diagenesis of clay (smectite to illite; kaolinite to illite, illite to muscovite and mica)- while others do not (aquathermal pressuring, smectite dehydration at lower temperature and higher pressure). Chemical diagenesis not only alters the matrix properties of the host rocks such as the density, and bulk and shear moduli (Dutta, 2002b) but also systematically alter the rearrangement of the minerals in the clay platelets as evidenced by long-range ordering in X-ray diffraction measurements (Hower et al., 1976; Freed, 1979). The alteration of rock matrix in burial diagenesis is permanent and irreversible and there is no conclusive evidence that the bound water released from the clay platelets during metamorphism has different viscosity or density other than the usual changes of water properties due to elevated temperature and pressure.

Earlier researchers (e.g., Hottmann and Johnson, 1965) were aware of the importance of mechanisms other than mechanical compaction disequilibrium for overpressure in shales although they did not explicitly state it. As Bowers (1995) noted correctly, Hottmann and Johnson (1965) did include the fluid expansion effect in their cross-plot of fluid pressure gradient and sonic log-derived interval transit time departure from the normal compaction trend $\delta(\Delta t)$. The onset of chemical compaction due to burial diagenesis in shales is neither abrupt nor unpredictable – it is gradual and the phenomenon is kinetic and dictated mainly by the thermal history of the hosting shales. In some cases the chemical reaction needs a catalytic agent such as potassium from the decomposition of potassium-feldspar. Dutta (1986, 1987b) was the first to develop a quantitative and predictive model for the S/I transformation (smectite to illite transformation is designated as S/I transformation throughout the text) based on the geochemical work of Hower et al. (1976) and Freed (1979). This was discussed in detail in Chapter 3. The pioneering work of Bowers (1995) showed that fluid sources at depth are additional sources of water that cause very high overpressure in shales. He further validated this observation by using data from outside of the Gulf of Mexico.

The generalized compaction model proposed by Dutta (1986, 1987b; Dutta et al.,2011; Dutta et al., 2014) that includes both mechanical compaction and chemical diagenesis is given below:

$$\sigma = \sigma_0 e^{-\beta(T(t))\varepsilon} \tag{4.30}$$

where the void ratio, ε, is defined as the ratio of the pore volume of a rock to that of the solid rock, namely,

$$\varepsilon = \frac{\phi}{(1 - \phi)} \tag{4.31}$$

Eliminating porosity between equations (4.27), (4.30), and (4.31), the interval transit time or slowness (inverse of velocity) is given by

$$\Delta t = \Delta t_m (1 + \beta^{-1} \ln(\sigma_0/\sigma))^x \qquad (4.32)$$

where β is the thermal history dependent burial diagenesis function, σ is vertical effective stress, σ_0 is the vertical effective stress required to reduce porosity to zero, x is a lithology dependent coefficient (for shale it is ~ 2.13), and the suffix m denotes the matrix material. This is known as the Dutta model (Dutta, 2002b; Dutta et al., 2014) that relates acoustic slowness (inverse of velocity) to effective stress for various lithologies (characterized by the parameter x) and various diagenetic states of clay undergoing S/I transformation characterized by the parameter β. We recall that the parameter x is the lithology coefficient (*aka* acoustic formation factor) in Issler's equation (equation (4.27)). In Table 4.6 we provide some suggested values of x and. Δt_m

In the model given above for velocity, the lithology-dependent parameters such as Δt_m and x (in equation (4.32)) need to be related to the diagenetic state of the shale. The effect of diagenesis is to gradually alter the matrix properties of shales as the shale metamorphism progresses. Therefore, to account for this effect in equation (4.32), the terms containing Δt_m and x, using the diagenetic function $N_s(t)$ of equation (3.29), need to be modified to

$$\langle \Delta t_m(t) \rangle = \Delta t_{m,s} N_s(t) + \Delta t_{m,i}(1 - N_s(t)) \qquad (4.33)$$

$$\langle x \rangle = x_s N_s(t) + x_i(1 - N_s(t)) \qquad (4.34)$$

Suffixes s and i are used to denote smectite and illite, respectively. Thus $\Delta t_{m,s}$ and $\Delta t_{m,i}$ denote, respectively, the slowness for smectite and illite *grains*. A similar definition holds for the lithology exponents, x_s and x_i, in equation (4.34). From the authors' experience, the typical values for x_s and x_i are, respectively, 2.17 for smectite and 2.20 for illite. In the earlier publication (Dutta et al., 2014), values of 67.056 µs/ft and 2.19 were used as average values for $\Delta t_m(t)$ and $\langle x \rangle$ respectively. In principle σ_0 also should be redefined by using the diagenetic function as done in equations (4.33) and (4.34).

Table 4.6 Recommended values for the parameters in the Dutta model

Lithology	Coefficient, x	$\Delta t_m (\mu s/ft)$
Sandstone	1.60–1.80	54.0–55.0
Shale	2.13–2.20	67.0–69.0
Dolomite	1.90–2.10	43.0–44.0
Calcite	1.70–1.80	47.0–48.0
Smectite	2.15–2.17	68.45*
Illite	2.00–2.20	?

* Measurements based on clays (Wang et al., 2001).

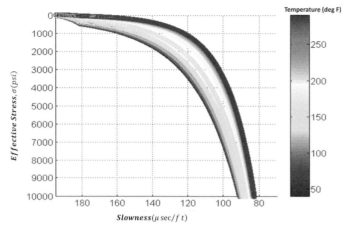

Figure 4.11 A plot of effective stress (psi) versus interval transit time (μs/ft) of a "generic shale" of Miocene-age from the Dutta model (equation (4.32)) for indicated temperatures in color. As noted in the text, the model incorporates both mechanical compaction and chemical diagenesis. Often in the literature, chemical diagenesis is referred to as chemical compaction (e.g., Liu et al., 2018). (A black and white version of this figure will appear in some formats. For the color version, please refer to the plate section.)

However, since reliable measured value of σ_0 for the end-member illite was not found, default value for smectite was used instead for all computations presented in the literature. Carefully controlled experiments for determining the compaction curves for pure clays containing smectite and illite (and possibly, other clay minerals) will be very useful for understanding the end-member shale properties.

A plot of equation (4.32) is shown in Figure 4.11 for a "generic shale" – the color code indicates temperature. The burial rate has been assumed to be constant at 1 ft/ thousand years and is assumed to be appropriate for a Miocene age shale. Again we see that a given state of shale (transit time or velocity) is dictated by variables such as effective stress, temperature, and time (thermal history) – and not depth. How a shale got to a final depth (today) totally depends on its genesis – initial clay parameters and its burial path in time-temperature domain. Furthermore, normal compaction in this model is dictated by its effective stress at normally pressured condition σ_n and the time-temperature history. It is unlikely that when logarithm of transit time is plotted versus depth (as in the Eaton model) it will be a straight line.

For a given diagenetic state of shale, one can also *predict* the expected range of its velocity at a given depth for different pore pressure states using the model discussed here. Figure 4.12a shows such a plot of expected shale-velocity versus depth (off shore) corresponding to the diagenetic function shown in Figure 4.12b. The parameters for this model are chosen to be appropriate for an average Miocene-age shales in the Gulf of Mexico with a constant thermal gradient of 1.2°F/100 ft and the seabed temperature of 40°F. Each curve in Figure 4.12a corresponds to the indicated pore pressure gradient in ppg. The right-most curve is valid for the hydrostatic pore pressure

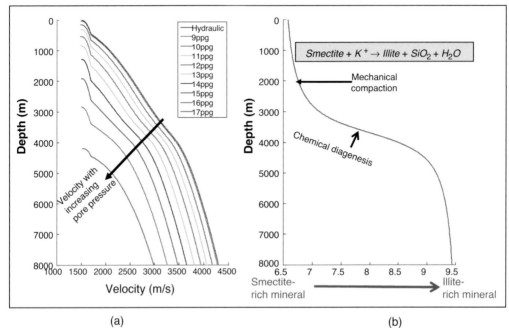

Figure 4.12 (a) Predicted rock velocity versus depth in the Dutta model for various pore pressure gradients (in ppg) as indicated. (b) The diagenetic function β that was used in the velocity plot. The parameters are appropriate for generic smectitic shale of average Miocene age that has undergone burial under an assumed thermal gradient of 0.012°F per foot. (A black and white version of this figure will appear in some formats. For the color version, please refer to the plate section.)

condition, and it is the *predicted Normal Compaction Curve* from the model that is *rock-physics compliant*. Imprints of diagenetic changes are clearly seen in these curves – these are not straight line curves in semi-log space as assumed by many models. All of these curves are *completely predictive* by simply using the geologically controlled parameters such as those used in the basin modeling algorithms. The gradient discontinuity of the curves at shallower depths in Figure 4.12a are due to rocks compacting from a suspension stage to a load bearing stage (Nur et al., 1991). Critical porosity is the porosity above which the rock can exist only as a suspension. In sandstones the critical porosity is ~ 40%, that is the porosity of a random close pack of well-sorted rounded quartz grains. This pack is often the starting point for the formation of consolidated sandstones. We are assuming that this definition also applies for fine-grained argillaceous sediments such as clays although it is not established experimentally. In Table 4.7 we show typical values of critical porosity for various natural rocks (Nur et al., 1998).

In Figure 4.13a we show a plot of velocity versus effective stress of equation (4.32) for varying values of the diagenetic function, β, starting from 6 to 15. Superimposed on

Table 4.7 Critical porosity of natural rocks

	Porosity (%)
Sandstones	40
Limestones	40
Dolomites	40
Pumice	80
Chalks	65
Rock salt	40
Cracked igneous rocks	20
Oceanic basalts	20

Source: After Nur et al. (1998)

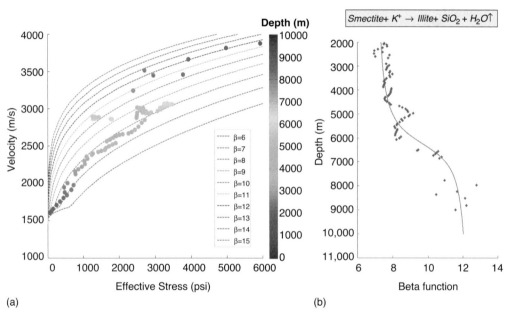

(a)   (b)

Figure 4.13 (a) A plot of predicted velocity versus effective stress of shale for various discrete values of the diagenetic function β as indicated (see equation (4.32)). Superimposed on this are the data from a well in the Gulf of Mexico. (b) A plot of the diagenetic function that was used in the construction of the plot on the left. (A black and white version of this figure will appear in some formats. For the color version, please refer to the plate section.)

this figure are the data from a well – the corresponding diagenetic function for that well is shown in Figure 4.13b that was reconstructed using the age-depth and temperature-depth relations for that well. Referring back to Figure 4.13a, we note that the velocity increase with the effective stress is at first rapid, but becomes gradual until an approximately constant terminal velocity is attained for a fixed high value of effective stress for a given β. For applications to deepwater sediments, note that the nonlinear portions of these curves at low effective stresses are the most relevant to high pore pressures (corresponding to low effective stresses) at deeper depths. Furthermore, shallow water flow (SWF) zones (shallow hazards) occur for very low effective stresses – about 100 to 300 psi is very typical. This is also in the nonlinear portion of the curves of Figure 4.13a. The shape of curves in Figure 4.13a are similar (for a fixed value of β) to what Tosaya (1982) observed from laboratory measurements of granular material of almost constant porosity (see Figure 3.4) and prior to that, by Wyllie et al. (1956).

Effect of Chemical Diagenesis on Shale Density-Slowness Trend in the Dutta-Model

Chemical diagenesis in shale has a distinct effect on its density. Dutta (1997a) first proposed a shale classification scheme based on burial diagenesis of shale due to smectite to illite (S/I) transformation. This classification scheme has been slightly modified as more data were analyzed and reproduced in Figure 4.14. (The original model was developed by Dutta and discussed in a BP Exploration Report No. HO 97.0004, Western Hemisphere. This report was released by BP on May 8, 2000.) The figure shows a cross-plot of density versus transit time of shales and sands (clean as well as "dirty" or shaly sands) from a number of wells in the Gulf of Mexico. The shale classification scheme based on its diagenetic state is given below:

Shale classification scheme	General characterization
Eo-diagenesis (Early)	<25% illite in S/I clay; mainly loss of water; hydrostatic to moderate pore pressure gradients (~9–14 ppg); ≤125°F; random layering in S/I clay.
Meso-diagenesis (Middle)	25–75% illitic beds in S/I clay; short-range ordering in S/I clay; loss of bound water; gain of interlayer potassium; silicon replaced by aluminum and release of Mg^{+2}, Fe^{+2}, Ca^{+2}; ~125°–225°F; may be associated with generation of hydrocarbon if kerogen is present; moderate to high pore pressure gradients (~14–17 ppg).
Telo-diagenesis (Late)	> 75% illitic beds in S/I clay; long-range ordering of illitic beds in S/I clay; >225°F; may be associated with HC cracking and gas generation if kerogen is present; pore pressure gradient greater than 17 ppg; cementation of adjacent reservoir rocks highly possible.

The pre- and post-diagenetic trends are labelled as EO (early) – and Telo (late)-diagenesis in Figure 4.14. Slightly modified trends of this type were also published by Dutta (2002a). This is shown in Figure 4.15. Figure 4.15a shows the X-ray diffraction data of Freed (1979) and Figure 4.15b shows the cross-plot of bulk density versus acoustic transit time of shale for the same well (DOE Well No. 1 in Pleasant Bayou, Texas). Katahara (2006) published a cross-plot of density and interval transit time data validating the trends of Dutta (2002a). This is shown in Figure 4.16. In Figure 4.17, we show an example from North Sea as reported by Swarbrick et al. (2014). These end-member trends for smectitic and illitic shales are believed to be valid worldwide provided no *re-worked* illitic shales are deposited over smectitic shales (as in Cook Inlet, Rocky Mountains, and similar tectonically active basins).

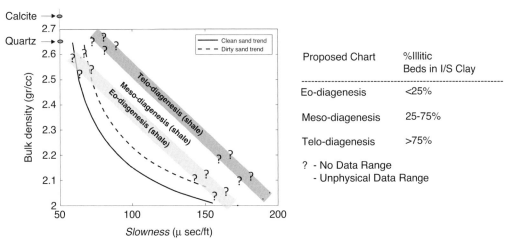

Figure 4.14 Shale classification scheme based on the chemical diagenesis of smectitic shales and their characteristics. See text for details.

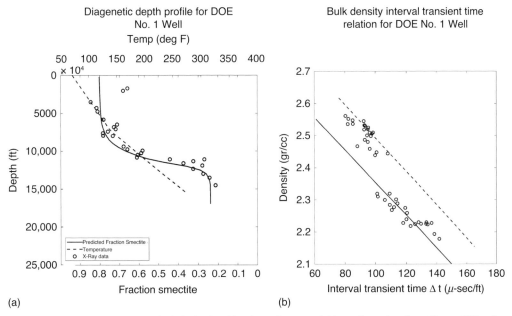

(a)

(b)

Figure 4.15 A modified shale classification scheme and (a) confirmation from X-ray diffraction data and (b) well logs from Pleasant Bayou well as reported by Freed (1979).

Examples given above show only end-member trends. In a recent study, Liu et al. (2016) documented the *complete transition* from smectite-to-illite for a deep well in the Gulf of Mexico. Their analysis and the data are shown in Figure 4.18. The end-member trends (labelled smectitic and illitic trends) in this figure are based on analyses of many well logs from all over the world. It is not derived from the analysis of data

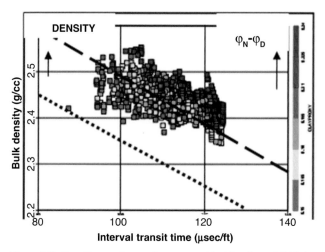

Figure 4.16 Postdiagenetic trend as reported by Katahara (2006) based on data analysis based on sonic, density, and neutron porosity logs. Sonic density versus interval transit time showing neutron porosity minus density porosity. The dotted line is the Eo-diagenetic (smectite-rich) trend of Dutta (2002a), and the dashed line is the Telo-diagenetic (illite-rich) trend of Katahara (2006). Similar patterns are seen across the Gulf of Mexico shelf in shales that are both highly compacted and highly overpressured. Data are from the Gulf of Mexico.

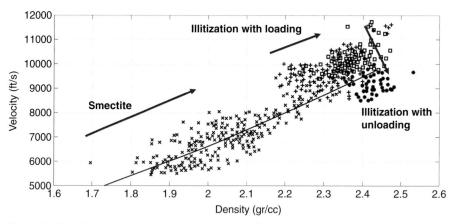

Figure 4.17 Postdiagenetic trend as reported by Swarbrick et al. (2014) based on data from sonic and density logs. Data are from the North Sea.

from this particular well. The well log data shown in Figure 4.18 are from a *specific* deep well in the Gulf of Mexico that underwent the complete transition. The rocks are of Miocene-Oligocene ages. The color codes indicate depth ranges where the shales are found. The pink curve that is overlain on the well data is the *predicted* evolutionary path from the model for this particular well using the Dutta model. It is *not* a curve fit to the data. In a series of snapshots cut-offs at indicated depths we show in Figures 4.19a–4.19h, how the shales are progressing in this two-dimensional (density and transit time) space starting from depths below 6000 ft (where only mechanical compaction

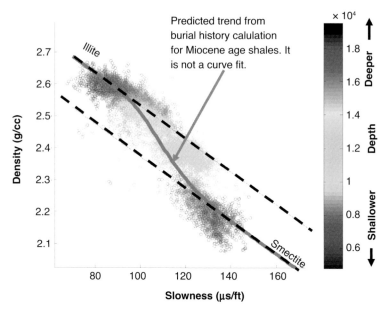

Figure 4.18 Generalized relationship between velocity and bulk density of shale from a deep well in the Gulf of Mexico (Liu et al., 2016). The pink curve is the predicted relation for generic Miocene-age shales from the Dutta model. From Dutta (2016). (A black and white version of this figure will appear in some formats. For the color version, please refer to the plate section.)

occurs) to the final depth at 20,000 ft where shales are completely metamorphosed to illitic shales due to chemical diagenesis. These snapshots show the entire *evolution* of the bulk density and transit time of shales for this well. The pink curve is the *prediction* based on burial history analysis as outlined in Chapter 3. The model and the data are in agreement although, as noted above, the model parameters were not adjusted for this well. Trends such as these suggest that it is possible to predict the acoustic properties of shales (under the assumptions as outlined earlier) based on *shale burial history analysis before a well is drilled*. This is a powerful statement! For example, the acoustic properties of a Pliocene age shale could be extrapolated to Miocene – Oligocene age based on assumed thermal history. This allows the possibility of construction of synthetic seismograms using thermal history of rocks *before* a well is drilled.

In summary, the diagenetic trend of bulk density of shales in the Dutta method is of the form

$$\rho = \rho_s N_s(t) + \rho_i(1 - N_s(t)) \tag{4.35}$$

where

$$\rho_s = a_s + b_s \Delta t \tag{4.36}$$

and,

$$\rho_i = c_i + d_i \Delta t \tag{4.37}$$

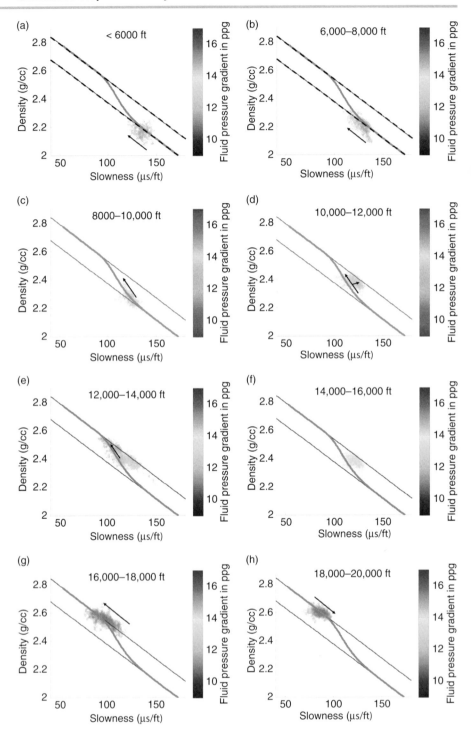

Figure 4.19 These figures (a) – (h) show the systematic metamorphism of smectite to illite and its imprint on the acoustic properties of shales shown in Figure 4.18. Shown are the density and transit times from the Miocene–Oligocene well every 2000 ft (a) starting at less than 6000 ft and (h) ending at 20,000 ft. The pink curve is the predicted relation for generic Miocene-age shales from the Dutta model. (A black and white version of this figure will appear in some formats. For the color version, please refer to the plate section.)

Here, suffixes s and i denote smectite and illite, respectively; ρ_s is the end member density versus slowness trend of smectitic – shales (or the prediagenetic end member), and ρ_i is the end-member trend for the illitic – shales (the so-called post diagenetic end member), respectively. Δt is the interval transit time or slowness, and $N_s(t)$ is the smectite fraction at time t, and a, b, c, and d denote calibration coefficients. Typical values for these coefficients are $a_s = 2.8854$, $b_s = -0.00513$, $c_i = 3.0501$, $d_i = -0.00518$ following the convention shown in Figure 4.18, namely, bulk density is increasing upward and transit time is increasing to the right. The above relations are valid for porosities less than the critical porosity. For values higher than the critical porosity, the sediment has no shear strength and one should use a model for velocity and density that is appropriate for a suspension model – a phase composed of clay particles and water. We note that the smectitic and illitic end-member trends depend on the mineral composition of the shale upon deposition.

It is clear that the density and transit time cross-plot of the type discussed can be used to demonstrate the onset of fluid expansion and *chemical compaction* – a term coined by Goulty et al. (2016). (We refer to this as chemical diagenesis.) This is illustrated in Figure 4.20. This figure shows a heuristic burial history model that predicts the evolutionary relation between bulk density and slowness of shales with increasing effective stress for purely mechanical compaction for smectitic or pre-diagenetic trend (1), unloading or decreasing effective stress (2), mixed mechanical compaction and chemical diagenesis (3), purely chemical diagenesis (4) and, post-diagenetic or illitic trend (5). In our terminology, the entire process of "unloading" and "fluid expansion" as used by Bowers (1995) is termed as *chemical diagenesis*. The key point to note is that the onset of the chemical diagenesis will vary not only from basin to basin, but may also vary within a basin; the pressure buildup in the host shale will be

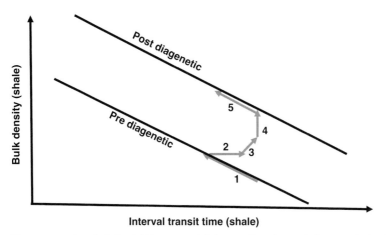

Figure 4.20 A heuristic burial history model that predicts the evolutionary relation between bulk density and slowness with increasing effective stress for purely mechanical compaction for smectitic or prediagenetic trend (1), unloading or decreasing effective stress (2), mixed mechanical compaction and chemical diagenesis (3), purely chemical diagenesis (4), and postdiagenetic or illitic trend (5).

dictated by how a particular basin has evolved and how the permeability of the host sediments facilitated the migration of the excess water released from the host rocks into the pore space. Thus, any basin modeling algorithm used for analysis of evolution of sedimentary basins will yield incorrect results if the diagenetic process within host-shales are not included.

We note that equation (4.32) will yield the slowness for normal compaction $\Delta t_{n,dutta}$ if the effective stress is computed using hydrostatic fluid pressure condition. The NCT-trend thus derived is supported by the rock physics considerations – it is *not* curve-fitted to a particular well data, and therefore, it is predictable from burial history simulation and can be used to carry out the traditional Eaton – type approach for pore pressure prediction using the traditional-Eaton equation expressed in terms of effective stress (σ), if so desired. Thus, one could use the following approach for Eaton-type model with a newly defined NCT for shales:

$$\sigma = \sigma_n \left(\frac{\Delta t_{n,dutta}}{\Delta t}\right)^{\gamma} \tag{4.38a}$$

$$\Delta t_{n,dutta} = \Delta t_m \left[1 + \frac{1}{\beta_n} \ln\left(\frac{\sigma_0}{\sigma_n}\right)\right]^x ; \beta_n = B_0 \approx 6 - 7 \tag{4.38b}$$

$$\sigma_n = S - P_{hyd} = S - P_{hyd,grad}(psi/ft) \times Z(ft) \tag{4.38c}$$

where σ_n is the effective stress (psi) under normal or hydrostatic conditions, Δt is slowness (μs/ft), $\Delta t_{n,dutta}$ is the normal compaction trend (slowness) in the Dutta model, γ is the traditional Eaton exponent (typically 3.0, although it could be a variable exponent that depends on depth), Δt_m is the slowness of the rock matrix mineral grains, (typically 68 μs/ft), σ_0 is the stress at which porosity approaches zero (typically, 26,500 psi), β_n is the dimensionless prediagenetic function (equal to B_0 which is typically within 6~7), x is the Issler exponent (typically, ~1.97–2.13 for shale), P_{hyd} is the pore pressure (psi) under hydrostatic pressure condition, $P_{hyd,grad}$ is the pore pressure gradient (psi/ft) under hydrostatic pressure condition (typically, 0.465 psi/ft), S is overburden stress (psi), and Z is depth (ft). The NCT given by equation (4.38b) will no longer be a straight line in a semi-log space of slowness and depth, as in the traditional Eaton model. This modification to the traditional Eaton approach will remove the guesswork in constructing a NCT while still using the traditional Eaton method, since it is *predicted* by the diagenetic function appropriate for the burial conditions of the sediment under assumed hydrostatic pore pressure conditions. This approach may also eliminate the need for arbitrary NCT trend shifts often used in the traditional Eaton approach. However, it will not eliminate the need, if any, for adjustable or depth-dependent Eaton exponents.

Bowers's Method

In a series of interesting papers, Bowers (1994, 1995) documented the importance of temperature on velocity – effective stress relations for shales. As temperature increases with burial there is a density change of pore water which causes "fluid expansion" – a term coined by Bowers (1994, 1995). This will necessarily increase the pore pressure due to aquathermal pressuring (Barker, 1972). If the pore fluid is hydrocarbon, density reduction with increasing temperature could be sizable, especially when liquid hydrocarbon is converted to gas. If the sedimentary rock contains kerogen, then the thermal cracking of kerogen to hydrocarbon would also contribute to increased pore pressure (Barker, 1990; Meissner, 1978). As density reduction progresses with increasing temperature, vertical effective pressure would decrease (reverse) as compared to the case for lower temperature. However, Luo and Vasseur (1996) and Swarbrick et al. (2002) showed the aquathermal pressuring effect to be small (< 500 psi with ultra-low permeability rocks).

Bowers (1994, 1995) also addressed the phenomena of effective stress *reversal* (along with reversal of any attribute dependent on effective stress such as velocity, density and resistivity) beside aquathermal pressuring due to increased temperature. For example, fluid source at depth due to S/I transformation would be a part of this discussion. He introduced the term *"unloading"* to describe effective stress reversal during uplift and aquathermal pressuring or other fluid source in a closed system such as chemical diagenesis due to S/I conversion in the host shale. Thus, he implicitly assumed that S/I transformation caused pore pressure to increase and effective vertical stress to reduce due to fluid release in a confined system (host shale). However, he did not include the effect of mineral compositional changes in the relation between velocity and effective stress in contradiction to the necessary feature of this transformation (Hower et al., 1976; Freed, 1979). The release of bound water from the smectite transformation in the low-permeability environment causes high pore pressure. Attendant vertical effective stress reversal is always associated with load transfer from the host rock matrix to pore fluid resulting in increased pore fluid pressure. Since velocity of a rock/fluid system is mostly affected by the grain to grain contact, a load transfer will result in weaker contacts and hence reduction (reversal) in velocity, and resistivity. The density reversal may not happen at that juncture (when velocity reversal happens) due to the fact that porosity is already very low at those deeper strata and a further decrease in porosity results in insignificant change in density. The compaction is thus not reversible. The situation here can be termed *overcompacted* – it is however, associated with much *higher* pore pressure, and *lower* velocity than the case for *undercompacted* case. In this method, therefore, one must cope with the problem of determining the paleo-maximum stress experienced by the shale to determine the *unloading path* in the velocity versus vertical effective stress domain.

The depth at which *stress unloading or load transfer* will start to take over is not predicted by this method without assistance from petrophysical well logs, such as density and sonic. Bowers (1995) did offer a clue to determining when the unloading will occur. He observed that if equivalent depth method for pressure analysis in shales at higher temperatures (> 100°C) were used, it would necessarily result in under prediction of pore pressure. He suggested that when fluid expansion causes high overpressure, it causes unloading, and therefore elastic rebound. So, the search for high overpressure is basically a search for rebound in his method. He further noted that bulk properties undergo less rebound than transport properties. This means that sonic

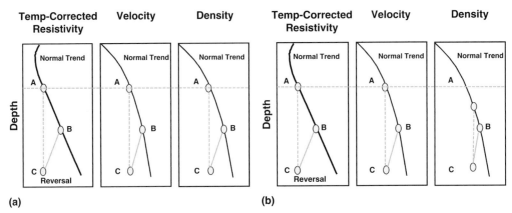

Figure 4.21 (a) Detecting reversals without high overpressure. Vertical projections of temperature-corrected resistivity, sonic, and density from a point inside the reversal intersects the log at a common point above the reversal. (b) Detecting reversal with high overpressure. Vertical projections of temperature-corrected resistivity, sonic, and density from a point inside the reversal do not intersect the logs at a common point above the reversal; the density log is crossed at a deeper depth. Modified after Bowers (1995).

velocity and resistivity data in highly overpressured intervals undergo larger reversals than bulk density measurements. This suggests a simple way to predict when unloading happens- *look for the depth at which the equivalent depth method fails* as shown in Figure 4.21. Vertical projections of sonic velocity, porosity and density from a point inside the reversal do not intersect the logs at a common point above the reversal; the porosity and density logs are crossed at a deeper depth than the velocity log. An example of a high pressure well showing this feature is shown in Figure 4.22. The fourth track of the figure compares pore pressure measurements (RFTs) and mud weights used to drill the well with pore pressures computed from the sonic velocities using the Equivalent Depth Method. The sonic velocity and resistivity logs undergo reversals not seen by the density log, and Equivalent Depth Method underestimates pore pressure.

Bowers (1995) proposed a quantitative model for accounting of fluid expansion and unloading effects that cause reversal in effective stress, when using velocity data, with the following equation:

$$V = V_0 + a[\sigma_{max}(\sigma/\sigma_{max})^{1/u}]^b \qquad (4.39)$$

where the reference velocity, V_0, is taken as 5000 ft/s, and a and b are empirical and locally calibrated constants, σ is vertical effective stress and u is termed as the elastoplastic term. The unloading parameter u is a measure of how plastic the sediment is (see Figure 4.23). In that figure, $u = 1$ implies no permanent deformation because the unloading curve reduces to the virgin curve and $u = \infty$ corresponds to completely irreversible deformation, since $V = V_{max}$ for all values of effective stress less than σ_{max}.

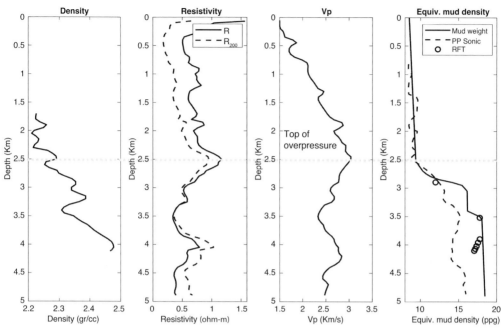

Figure 4.22 High-pressure well example in Bowers method. Sonic and resistivity logs undergo reversals not seen by the density log. Pore pressures are underestimated when undercompaction is assumed to be the cause of overpressure (Equivalent Depth Solution). Curve labeled R is raw resistivity data; curve R_{200} is resistivity data corrected to a common temperature of 200°F (93° C). Modified after Bowers (1995).

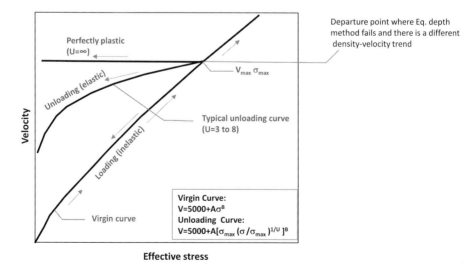

Figure 4.23 The meaning of the unloading parameter U in Bowers's model.

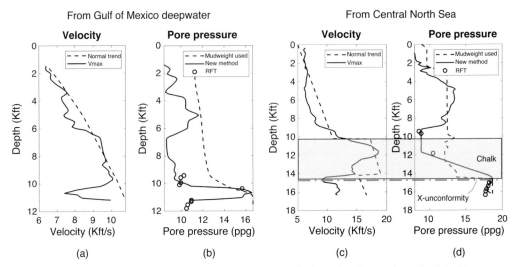

Figure 4.24 An example of the pore pressure prediction using Bowers's method. (a–b) Deepwater, Gulf of Mexico. (c–d) North Sea. Here we show velocity log and predicted pore pressure from the log using the unloading model of Bowers (1994, 1995). These examples are from Bowers (1994, 1995).

According to Bowers, u typically ranges between 3 and 8. Essentially, coefficients a, b, V_{max} and u should be treated as locally calibrated constants.

In Figures 4.24a and 4.24b we give an example of Bowers method for pore pressure estimation in deepwater Gulf of Mexico and in Figures 4.24c and 4.24d from Central North Sea (modified after Bowers, 1995). The Gulf of Mexico example is based on data from a well drilled in nearly 1400 ft of water in the Gulf of Mexico. Figure 4.24a is a plot of the sonic log data, and a normal trend line (labelled as *virgin curve* by Bowers) analytically computed from the following equation:

$$V = V_0 + a\sigma^b \tag{4.40}$$

There are a number of small velocity reversals above 9700 ft, and one major reversal between 9700 ft and TD. Bowers noted that because all of the shallower reversals are very *weak*, they were not considered to be due to fluid expansion. (This is a subjective decision.) Within the *large* reversal, the velocity at first drops at a rate similar to that in the smaller reversals. However, at 10200 ft, the slope steepens significantly. This slope break was interpreted by Bowers to be the onset of fluid expansion overpressure. Therefore, in the unloading relation, V_{max} was assumed to be the velocity at 10,200 ft, not the peak velocity at 9700 ft. Where the velocities near TD are above the value at 10,200 ft, the virgin curve was used to compute effective stresses. The computed solution were based on $u = 3.13$, $a = 28.3711$, and $b = 0.6207$. Figure 4.24b shows the predicted pore pressure data. Open circles are pore pressures determined from RFT measurements. The dotted line shows the mud weights that were used, while the solid line is the estimated pore pressure. It can be seen that the Bowers's method is able to track both the rise in pressure, and the pressure regression. The example from Central North Sea (Figures 4.24c and 4.24d) also confirms these observations. Similar patterns are seen across the Gulf of Mexico shelf area shales as well as North Sea, Malay Basin, India and other parts of the world that are highly compacted and highly overpressured.

Pressures as high as the one shown in the above example cannot be explained by the compaction disequilibrium method alone such as the ones postulated by Eaton. Bowers suggested an easy but heuristic fix: If the Eaton pore pressure estimates are too low, Eaton- equation can be adjusted to yield lower effective stresses at deeper depths. One way is to increase V_n by shifting (and rotating, if needed) the normal trend line to fit the measured or estimated pressure data. Another is to use a variable Eaton exponent for a fixed NCT. Both of these approaches have considerable shortcomings. We shall examine this further. In Figure 4.25a we give an example based on the data presented earlier in Figure 4.8. That figure is reproduced in Figure 4.25a; however, we displayed two alternate NCTs labelled A and B that also fit the shale traveltime data (labelled D). (We noted earlier that the NCT that fit the measured pore pressure data – not shown, but *not* the traveltimes, is labelled C.). In Table 4.8a, we show computed pore pressure results for these *alternate* realizations of NCTs at the TD of the well (4500 ft). We have used the standard Eaton exponent of 3 for these computations. The predicted pressures show significant variations – from 18.45 to 18.99 ppg. These variations are *larger* than the driller's typical *safety* margin at TD. It also raises a question: Should we use a NCT that fits the shale traveltime data or measured pore pressure data? (Ideally, the NCT should fit *both* the shale travel and pore pressure data.) What should we do if the measured pore pressure data are not available?

In Table 4.8b we show the sensitivity of the *Eaton exponent* on computed pore pressure for a fixed NCT. Again, in Figure 4.25b we show the same figure as shown earlier in Figure 4.8, except the trend line A has been removed. For computations, we use the NCT labelled B that fits the shale traveltime data (but not the pore pressure

Table 4.8a Effect of multiple realization of NCT (as shown in Figure 4.25a) on predicted pore pressure using the standard Eaton method (Eaton exponent of 3)

Normal compaction trend (NCT)	A	B	C
Overburden[a] (psi)	4500	4500	4500
Hydrostatic pressure[b] (psi)	2092	2092	2092
Sonic traveltime at TD, $\Delta t (\mu s/ft)$	100	100	
NCT value at TD, $\Delta t_n (\mu s/ft)$	30	40	43
$\left(\dfrac{\Delta t_n}{\Delta t}\right)^3$	0.027	0.064	0.0795
Pore pressure (psi)	4434.98	4345.89	4308.56
Pore pressure gradient (ppg)	18.99	18.61	18.45

Note. The quantitative values are computed at TD (4500 ft).
[a] We have assumed overburden gradient to be 1 psi/ft for the sake of ease in computation. [b] We have assumed hydrostatic gradient to be 0.465 psi/ft.

Table 4.8b Effect of variable Eaton exponents (for a fixed NCT as shown in Figure 4.25b) on computed pore pressure using the Eaton model

Eaton exponent	1.0	*3.0*	2.0
Overburden[a] (psi)	4500	*4500*	4500
Hydrostatic pressure (psi)	2092	*2092*	2092
Sonic traveltime at TD $\Delta t(\mu s/ft)$	100	*100*	100
NCT value of B at TD $\Delta t_n(\mu s/ft)$	40	*40*	40
$\left(\dfrac{\Delta t_n}{\Delta t}\right)^3$	0.40	*0.064*	0.16
Pore pressure gradient at TD (ppg)	15.145	*18.61*	17.62

Note. The quantitative values are computed at TD (4500 ft). The computations used are based on NCT labeled "B" in Figure 4.25b. [a]We have assumed overburden gradient to be 1psi/ft for the sake of ease in computation. [b]We have assumed hydrostatic gradient to be 0.465 psi/ft.

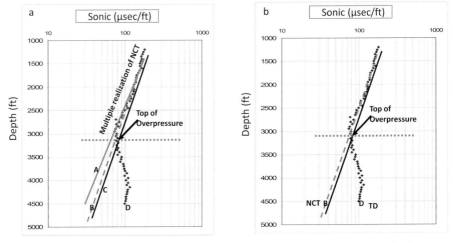

Figure 4.25 Sensitivity test of the Eaton model. (a) Effect of multiple realizations of NCTs labeled A, B, C (for a fixed Eaton exponent of 3) on Eaton-derived pore pressure. See Table 4.8a for computed results. (b) Effect of variable Eaton exponent (from 1 to 3) on computed pore pressure at TD using NCT labeled B. See Table 4.8b for computed results. Without any pore pressure data, it becomes very difficult and subjective to create a reliable model.

data) and vary the Eaton exponent from 1 to 3. The sample computations in Table 4.8b show a marked variation in the computed pore pressure at TD (4500 ft) – from 15.145 to 18.61 ppg. These exercises show the uncertainty on the construction of Eaton model both in terms of NCT and Eaton exponent. Thus Bowers is correct in pointing out that this uncertainty allows the user to use the original Eaton–virgin curve relation to be transformed into an unloading curve. Most users of the Eaton method found the traditional Eaton parameters with a single trend line to predict pore pressure unreliable at deeper depths where unloading effect becomes dominant. Sarker and Batzle (2008) went as far as even suggesting that the Eaton exponent should be regarded as a continuously varying parameter dependent on depth. The fix suggested by Bowers is practical but highly empirical and data dependent, and hence, could be subjective. *Eaton did not give any rationale for using an exponent of 3.* We have also found on numerous occasions that the best-fit NCT that fits the shale traveltime data do not always match the pore pressure data at deeper depths.

Next we discuss the strengths and weaknesses of the Bowers's method. The strength of the method is that it recognizes the deficiencies of the well-known compaction disequilibrium mechanism as the sole source of overpressure in shales and offers a solution that recognizes the importance of an additional mechanism – *fluid expansion* in a closed system. It also honors the well-known hysteresis effect in rocks. The solution proposed by Bowers is practical and easy to implement. It is better than the traditional Eaton-method. The weakness of the method is that the recognition of the onset of the depth at which fluid expansion mechanism will be dominant requires availability of sonic, density (or resistivity) log and pressure data from deep well(s). This may not be always possible, especially, in exploration projects in basins with no or limited deep well control. The method is therefore a *post-drill technique* just as the Eaton method. Furthermore, the method assumes that all shales can be characterized by their own *single* compaction trend in the velocity and vertical effective stress when there is no unloading. In other words, shale compaction must necessarily possess a single *virgin curve*. This is not possible if shale mineralogy changes. Furthermore, there is no reason to believe that all unloading curves *must* terminate at the predetermined virgin curve as postulated by Bowers (see Figure 4.23).

Comparison of Bowers and Dutta Models: Stress Unloading and S/I Transformation

In Figure 4.26 we show a comparison of the two models. Salient features of the both models are tabulated in Table 4.9. Figure 4.26a shows the velocity versus vertical effective stress in the Bowers model and a comparison with the corresponding relations in the Dutta model is shown in Figure 4.26b. The vertical axes in both figures display the same variables. The Dutta model uses a third variable – temperature, while the fourth variable – time is hidden (assumed in this plot to be appropriate for Miocene age rocks). In the Dutta model, the trends are bounded by two extremes – a smectitic or pre-diagenetic shale trend at low temperature and an illitic or post-diagenetic shale trend at higher temperature while in between we have a continuously varying compositional change of shale properties that depend on the given ratio of smectite to illite at a given temperature. The differences between the Bowers and Dutta models are: (a) there is a single *inelastic loading* curve in Bowers model while the Dutta model has no such limitation because the loading curves depend on thermal history, (b) *elastic unloading* in the Bowers model assumes that the illitic shales have the same compaction curve as the smectitic shales – that is the illitic shales would fall on the same "virgin" curve as shown in Figure 4.26a. In the Dutta model (Figure 4.26b), the velocity – effective stress relation for illite is different from that of the smectitic shales. This means there is no elastic loading in this model as the mineral transformation from smectite to illite is irreversible. Furthermore, the relation between velocity and interval transit time in the Dutta model is a multivalued function that depend

Table 4.9 A summary comparison of Bowers's and Dutta's models

Bowers's model
- Recognizes the importance of thermal mechanisms
- Proposes simple velocity to effective stress that includes unloading/fluid expansion mechanisms
- Assumes a single "virgin" compaction model; depth prediction at which unloading occurs requires density logs
- Ignores changes associated with shale mineral composition
- Needs an independent measure of when the method is operable (density log, for example)
- Needs calibration

Dutta's model
- Complex model – dependent on thermal history of basins and kinetics of chemical diagenesis
- Accepts variable shale mineral composition with burial history
- Properties are determined by effective stress, temperature, geologic time
- Provides an easy link to basin history simulation
- Onset of diagenesis is predictive
- Shale acoustic properties are predictable prior to a well being drilled
- Rock parameters need calibration

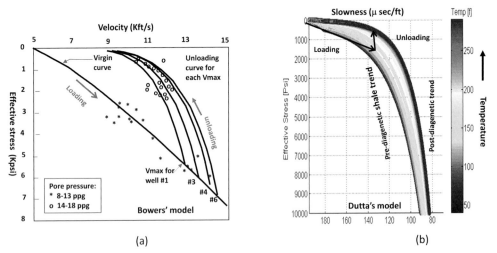

(a)   (b)

Figure 4.26 A graphical comparison of the (a) Bowers's and the (b) Dutta's models. See text for a discussion. (A black and white version of this figure will appear in some formats. For the color version, please refer to the plate section.)

on the mineral composition of the shales as denoted by the diagenesis parameter β in equation (4.32). The main point is that in the Dutta-model, the S/I transformation is always "inelastic." For Bowers's model to be valid for S/I transformation, the process must occur without change in the shale mineral composition. That is not the case. Bowers's model should work for cases where we know the paleo-maximum vertical effective stress. How do we know this prior to drilling? We do not. Dutta model is generally more valid and does not require knowledge of paleo-maximum vertical effective stress – it is a part of the burial history model.

4.4 Pore Pressure Prediction in Carbonates (and Other Competent Rocks) Where Common Shale-Based Techniques Do Not Work

4.4.1 Problem Statement

High pore pressure in carbonates are well known (Kumar et al., 2010; Atashbari and Tingay, 2012; Green et al., 2016). Some noted that the pressures from reservoirs in the Middle East can vary dramatically from relatively benign, such as in the Arab C and D formations (in Saudi Arabia), to highly overpressured, in Kuwait (19.5 ppg EMW), Oman/Kuwait in the Upper Jurassic and Gotnia formation in Sabriyah Field, Kuwait. The methods and models that are available in the literature for pore pressure prediction works well for clastic (mostly shale) but not for nonclastic such as carbonates. Porosity loss in carbonate sediments is mostly due to chemical alteration associated with porosity changes as opposed to shale where porosity loss is mainly due to mechanical compaction. Chemical processes that contribute to carbonate porosity changes are mainly pressure solution and subsequent transport and deposition of mineral constituents. The processes are complicated further by the complex chemical interaction between fluids present within the cracks and pores and the mineral composition of solid material. Thus the porosity and effective stress are *not* related in carbonates as opposed to shales. This is the main reason why conventional usage of velocity versus effective stress approach that is so successful in the clastics does not work for carbonates. This is shown in Figure 4.27. Velocity is sensitive to effective stress in shale (Figure 4.27a) but not so for water-saturated limestone (Figure 4.27b). This loss

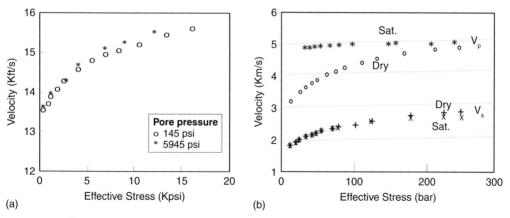

Figure 4.27 Comparison of P- velocity for (a) shale (modified after Tosaya, 1982) and P- and S-wave velocities for (b) limestone (modified after Avseth et al., 2005). 1 bar = 14.5 psi and 1 Kft/s = 304.8 m/s

of sensitivity of velocity to effective stress is common for all carbonates and other competent rocks. It is the main reason why the methods that we discussed so far do not provide useful prediction for pore pressure in carbonates – *a high pore pressure would be associated with high velocity* (in contrast to shale where high pore pressure is associated with higher porosity and lower velocity).

Some authors have used the pore pressure in embedded shales within carbonate interval as an indicator of pore pressure in carbonates (Huffman, 2002). However, this assumption is not valid in general and a predictive model based on this assumption may produce erroneous results. Any model for pore pressure prediction in carbonates must deal with a key reason: *we must identify key parameters linking the P-wave velocity response to the pore-pressure change.* This will include (1) developing new rock physics-based models and (2) a way to extract high-frequency variations from velocity measurements (logs, seismic) and relate these to pore pressure. Recently some progress has been made in this area. Below we discuss the current status in this area and provide example applications.

4.4.2   Rock Models for Pore Pressure in Carbonates

Carbonate rocks are stiffer and contain overpressure without any associated influence on porosity. So we need models that can focus on the fluid pressure component of the rock/fluid system. The recent literature considers two such models which we believe to be promising: (1) Pore compressibility based model and (2) the Biot model.

Pore Compressibility–Based Model

Compressibility (inverse of bulk modulus) method, as a way to compute pore pressure in carbonates, was first discussed comprehensively by Atashbari and Tingay (2012) and later, modified by Azadpour et al. (2015). The models proposed by these authors use rock porosity and compressibility to calculate the pore pressure. Any change in pore spaces due to overpressure is a function of bulk and pore volume compressibility. Atashbari and Tingay (2012) used bulk and pore volume compressibilities as parameters based on relations between the two by Zimmerman (1991) to calculate the pore pressure as given below:

$$P = \left(\frac{(1 - \phi)C_b \sigma_{eff}}{(1 - \phi)C_b - \phi C_p} \right)^{\gamma} \tag{4.41}$$

Here P, is pore pressure, ϕ is fractional porosity, C_b is bulk compressibility (psi^{-1}), C_p is pore compressibility (psi^{-1}), σ_{eff} (psi) is the effective stress defined by the authors as the vertical overburden stress minus hydrostatic pressure and γ is an empirical constant ranging in value from 0.9 to 1.0. (Note that when γ is not equal to one, the dimensions for equation 4.41 do not match and the equation should be viewed as empirical functional form, much like Gardner's equation which relates velocity to density. Atashbari and Tingay (2012) used 0.97 in Figure 4.28.) Compressibility values are usually obtained by core analysis (static measurements). The results based on

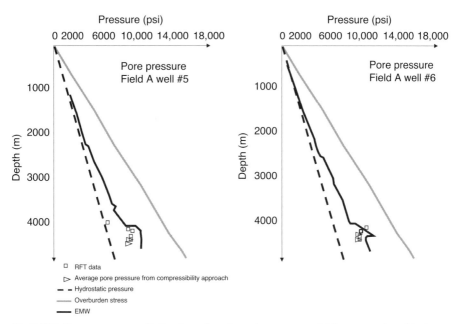

Figure 4.28 Pore pressure prediction in carbonates using compressibility approach and test on data from two wells in Iran. Modified after Atashbari and Tingay (2012). The value for the exponent γ in equation (4.41) was 0.97.

compressibility method are therefore, discrete values, since the core data are available only in some specific depth intervals. An example of this application is shown in Figure 4.28 for two wells (limestone) in Iran. The model yielded reasonable values of pore pressure as compared with RFT data in the reservoir interval.

Thus, a continuous pore pressure log in carbonates can be generated by using geophysical well logs to create compressibility logs needed for the model in equation (4.41). Pore compressibility for each of the zone can be calculated based on the well log data and using the relation between pore volume compressibility and rock bulk volume compressibility (Khatchikian, 1996; Suman, 2009). In order to do this, first Gassmann's equation (Chapter 2) is used to transform the sonic velocities to *dry* conditions using reservoir parameters evaluated through log analysis. Then, the velocities are converted to bulk rock and pore volume compressibility which in turn depend on matrix properties and porosity. As this compressibility is obtained from sonic log velocities, the pore compressibility thus derived is *dynamic* and therefore, *static* corrections need to be applied (Khatchikian, 1996). This is discussed in Chapter 12. This is a promising approach for pore pressure prediction in carbonates. However, the drawback is that some static measurements (core based) are usually needed to calibrate the model (dynamic model) and such measurements are seldom made in reservoir rocks outside of the zone of interest.

Biot Model for Pore Fluid Component of Velocity

Biot (1962) discussed a model for bulk and shear modulus of fluid-filled porous elastic rocks by treating the rock framework and fluid as two interpenetrating elastic media.

We discussed this in Chapter 2. Yu et al. (2014) used the Biot theory to isolate a component related only to pore fluid in the rock physics model for velocity. The goal is to assess the magnitude of this component and relate it *directly* to the frequency content of the velocity log (either from sonic or seismic inversion). In the Biot model, a rock is described by a porous framework characterized by bulk modulus, K_{fr}, shear modulus μ_{fr}, grain density of ρ_s and porosity ϕ. When saturated with a fluid of bulk modulus, K_f, and density, ρ_f, the velocity of the saturated rock V_p and its bulk density ρ_b are given by (Yu et al., 2014)

$$V_p{}^2 = V_{p,dry}{}^2 + \Delta \tag{4.42}$$

where

$$V^2{}_{p,dry} = \frac{K_{fr} + \dfrac{4}{3}\mu_{fr}}{(1-\phi)\rho_s} \tag{4.43}$$

$$\Delta = \frac{K_f}{\rho_b \phi} \frac{(1 - K_{fr}/K_s)^2}{1 + \dfrac{K_f(1 - \phi - \dfrac{K_{fr}}{K_s})}{\phi K_s}} - \frac{\phi \rho_f}{\rho_b} \frac{K_{fr} + \dfrac{4}{3}\mu_{fr}}{(1-\phi)\rho_s} \tag{4.44}$$

and

$$\rho_b = (1-\phi)\rho_s + \phi\rho_f \tag{4.45}$$

Yu et al. (2014) thus decomposed the velocity into *two* components – one from the fluid part denoted by Δ and the other from the rock framework part. The fluid component part of the model was further related to certain "high-frequency" components of a sonic log through what the authors termed as *wavelet transformation method* (see below).

Wavelet Transformation to Extract High-Frequency Variation

Yu et al. (2014) showed that the small-scale fluctuations of the P-wave velocity (sonic log) were caused by pore pressure change in carbonate rocks and the large-scale fluctuations of the P-wave velocity depended on the rock framework. Thus, the pore pressure in overpressured formation can be identified by the high-frequency detail of *wavelet transformation* of P-wave velocity. So what is the wavelet transformation method? Mathematically, the wavelet transformation of a signal is the cross-correlation of a signal with a wavelet of varying central frequency. Thus it is a technique to decompose a signal to identify its frequency distribution through time. Yu et al. (2014) applied this technique to relate to pore pressure measurements in carbonates. Their results for the wavelet transformation method are shown in Figure 4.29. They chose that (high) frequency decomposition of sonic log that showed the "*highest amplitude fluctuation*" (panel cd4) in Figure 4.29. The *lowest component* (panel ca4) was correlated to the *rock framework velocity*. The total velocity of the sonic log is thus a superposition of several high frequency components (cd1–cd4)

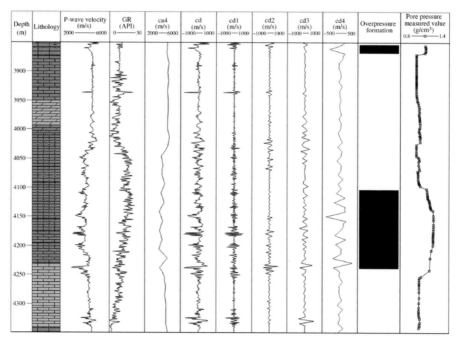

Figure 4.29 P-wave velocity decomposition chart using the wavelet transformation and overpressure formation chart based on the measured data in TZ-A well (Tarim basin) as shown by Yu et al. (2014). The black zones show the overpressure formation, ca4, the low-frequency approximation of the fourth-layer number, denotes the contribution from the rock framework to the P-wave velocity; cd1, cd2, cd3, and cd4 are the high-frequency details; and cd is the net contribution from the pore fluid to the P-wave velocity.

superimposed on a single low frequency component (ca4). The advantage of this decomposition is that one can relate the high amplitude fluctuation (cd4) *directly* to pore pressure in the carbonate section. This is shown by the authors in the last panel of Figure 4.29. It clearly demonstrates that the approach is very powerful as it can pick up *small-scale fluctuations* (due to pore fluid pressure) that are dominated overall by a low frequency trend – in this case, rock framework velocity. A velocity (e.g., velocity labelled cd4) versus effective stress (overburden minus pore pressure) model can be developed based on this approach. We think that this approach can be used to detect high pore pressure in other high velocity rocks also such as various types of basement rocks and evaporite sequences.

High-Frequency Velocity from Seismic Impedance Inversion

Some authors (Huffman, 2002; Sengupta et al., 2011) have advocated the use of high fidelity seismic velocity based on inversion of seismic data as a tool for pressure prediction in carbonates. Others (Green et al., 2016) believe that local calibration based on rock compressibility measurements on core data coupled with detailed geologic modeling and analysis to isolate the details of the carbonate formation and pressure measurements along with seismic-inversion based velocity correlations, is a viable approach to pressure prediction in carbonates. In our opinion, this type of integration of various sub-disciplines is promising for pore pressure prediction in carbonates. Below we discuss the results of seismic inversion based approach for carbonates. We will defer the details of the seismic inversion techniques to Chapter 5.

Seismic Inversion Approach to Pore Pressure in Carbonates
The use of surface seismic data for amplitude inversion is well documented; it is frequently used for impedance (defined as the product of velocity and density) analyses for rock/fluid identification. However, this technique can also be used for pore-pressure analysis (as shown in Chapter 5) as it can yield much higher resolution in velocity required for carbonates and many drilling applications in both carbonates and clastics. Seismic inversion is excellent at deriving impedance contrasts across layered interfaces within the reflection data, which reveal short-scale (high-frequency, several meters) variations of impedance; however, it is not able to determine the large-scale (low-frequency, hundreds of meters) vertical trends. This is provided by the conventional velocity analysis. Sengupta et al. (2011) used 3D seismic inversion to create a 3D geomechanical model and showed that pore pressure in carbonates (Sabriyah Oil Field, Kuwait) can be estimated using the calibrated high frequency velocity model in a salt-anhydrite sequence. In the example application (Sengupta et al., 2011) the pore pressure was very high and the seismic impedance inversion method delivered good results. This case study is discussed in Chapter 12.

4.5   Measurement of Pore Pressure

There are many commercial tools available to measure *in situ* pore pressure. These tools use pressure transient analysis techniques to estimate pore pressure (much like the heat diffusion measurement to obtain temperature information). In the petroleum industry, these measurements are recorded under the category of procedures called *"well testing."* A well test is a period of time during which the rate and/or pressure of a well is recorded to estimate well or reservoir properties, to prove reservoir productivity, or to obtain general reservoir management data. Thus, these techniques are designed for obtaining data from *permeable or reservoir formations* only. As such they provide little useful data on pore pressure in shale unless the geologic formations allow for pressure equilibrium between the reservoir and the encasing shale. This is, however, seldom the case. So the following discussion on tools is a very brief summary on the measurements of *reservoir* pore pressure. This is intended for layman as the subject is very complex and a thorough discussion is beyond the scope of this book.

4.5.1   RFT and MDT Tools

Repeat Formation Tester (RFT) and *Modular Dynamic Formation Tester* (MDT) are a class of *formation tester* tools that were introduced to the petroleum industry as early

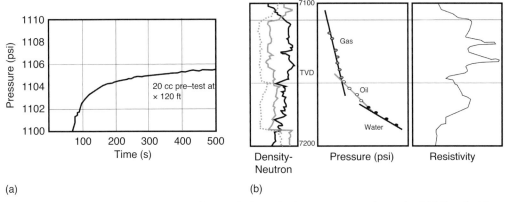

Figure 4.30 (a) Pressure transient analysis to obtain pore pressure data using MDT tool with a pretest chamber. (b) An example of how pressure data are used to identify fluid types in a reservoir fluid (density change causes slope change in reservoir pressure versus depth plot) and establish fluid contact points. The lateral shifts, if any, of the pressure versus depth plots indicate whether reservoirs are in communication or there is a barrier in between. Such data are useful to establish the extent of the fluid contact using seismic data.

as 1975. These tools were designed to collect fluid samples from formation (oil, water, gas) and measure pressure from multiple depths. MDT is now viewed as a replacement of RFT tool due to its accuracy. These tools can be used, either in cased or open-hole. These tools are typically run to the desired depth. It is necessary to seal the tool against the side of the wellbore. If the test is in casing, a perforation charge is fired to establish communication between the well and formation. A small volume of formation fluid is then produced and collected either into the tool or through the tool to the wellbore. In Figure 4.30a we show a result of pressure transient analysis that shows the equilibrated reservoir pressure after transients die out (Kuchuk et al., 1996). In Figure 4.30b we show how formation pressure data are used to identify fluid types (differentiated by fluid densities) at any given depth in a reservoir and to locate fluid contacts. In most of the RFT/MDT tools, there are capabilities to make pressure transient analysis in both vertical and horizontal directions. This allows one to estimate permeability anisotropy from pressure transient analysis. These tools are accurate and reliable. They are, however, limited in practice by the borehole diameter – 6 inches to 14.75 inches (Mouchet and Mitchell, 1989).

4.5.2   DST Tool

A *Drill Stem Test* (DST) is a special tool that is mounted at the *end* of a drill string. It is usually run with a gamma ray tool for making sure that it is clamped at the right depth. The tool isolates a formation or a part of the formation and stimulates and flows a formation to determine the type of fluids present and the rate at which they can be produced. The rate of flow is dependent on the permeability of the reservoir rocks in the formation. Besides making pressure measurements, the tool also makes temperature measurements. It is typically a part of the assembly known in the industry as *bottomhole assembly* (*BHA*). A basic drill stem test BHA consist of a packer or packers, which act as an expanding plug to be used to isolate sections of the well for the testing process, valves that may be opened or closed from the surface during the test, and recorders used to document pressure during the test. In addition to packers, a downhole valve is used to open and close the formation to measure reservoir characteristics such as pressure and temperature which are charted on downhole recorders within the BHA. During normal well drilling, mud is pumped through the drill stem and out of the drill bit. In a drill stem test, the drill bit is removed and replaced with the DST tool and devices are inflated above and below the section to be tested. As noted above, these devices are known as packers and are used to make a seal between the borehole wall and the drill pipe, isolating the region of interest. A valve is opened, reducing the pressure in the drill stem to surface pressure, causing fluid to flow out of the packed-off formation and up to the surface. Since measurements are carried out from the rig, it is the *preferred* tool for measurements in tight formations as the time period over which the measurements are made is much longer that the MDT/RFT tools. However, the measured pressure is the average pressure of the formation layer that has been isolated.

4.5.3   Comments on Downhole Pressure Measurements and Fluid Flow Relaxation Time of a Low-Permeability Rock

As noted above the downhole pressure measuring tools yield information on *permeable* formation only (reservoir). For low-permeability rocks such as shale, pressure transient analysis such as the ones by MDT/RFT/DST tools fails because pressure pulse transmission could take very *long* time and that rules out any *in situ* measurements. We may ask just how long is long?

To answer this question, we *simulate* a pressure pulse *transient* analysis test in a low permeability medium (sample size ~ 1 inch) whose permeability-porosity relation is given by a power law that depends on void ratio, ε (ratio of pore volume to rock volume) as shown in Figure 4.31. The simulation set up is shown in Figure 4.31a. The simulation is performed on a 1 inch long shale sample. The porosity is 0.18. The permeability is 1.73×10^{-6} mD. It is jacketed with a confining pressure P_0 of 4000 psi and it is held constant throughout. Initially the pore pressure p_0 is held at 100 psi throughout. At time $t > 0$, the pore pressure is raised to 1100 psi at one end (Figure 4.31b) and held constant thereafter. This creates a pressure pulse that travels through the sample. The sample begins to compact as fluid leaves the sample over time. Note that the surfaces at both ends are moving. In a computer simulation we are asked to measure the volume of fluid per unit area through time and determine the time required for pore pressure to equilibrate. This setup is very similar to what Brace et al. (1968) devised for permeability measurement in granite. The question is: How long must one wait before a steady-state flow is established? This will obviously be a function primarily of permeability. The mathematical description of the model is given below.

The approach taken by Brace et al. (1968) is valid but there are two restrictive assumptions. They neglected the consolidation of sample as the pressure pulse traverses and the compaction diffusivity, defined as $-\dfrac{\kappa}{\eta}\dfrac{\partial\sigma}{\partial\varepsilon}$ where κ is permeability of the rock, η is viscosity of fluid (water), σ is effective stress and ε is void ratio $\left(\dfrac{\phi}{1-\phi}\right)$, is assumed constant. The first assumption implies that the rock frame is rigid. This may be a valid assumption for low permeability tight-sandstone or granite but not for shales. The second assumption allowed Brace et al. (1968) to use a simple diffusion equation to model pressure pulse propagation. In the experiment we just described, we need to deal with moving boundaries at both ends of the sample and the time-dependent compaction of the sample.

We assume that the origin of the coordinate system is attached at one end of the sample at all times so that the displacement of the solid particles at that point is zero. The basic equations for flow and compaction are obtained by following the steps:

(a) Darcy's law

$$Q = -\frac{\kappa}{\eta}\frac{\partial p}{\partial z} \tag{4.46}$$

(b) Relative flux of fluid

$$Q = (V_w - V_r)\phi \tag{4.47}$$

(c) Equation of state of solid

$$\rho_r = \text{Constant} \tag{4.48}$$

(d) Equation of state of water

$$\rho_w = \text{Constant} \tag{4.49}$$

(e) Equation of continuity for solid

$$\frac{\partial[\rho_r V_r(1-\phi)]}{\partial z} = -\frac{\partial[\rho_r(1-\phi)]}{\partial t} \tag{4.50}$$

(f) Equation of continuity for water

$$\frac{\partial(\rho_w V_w \phi)}{\partial z} = -\frac{\partial(\rho_w \phi)}{\partial t} \tag{4.51}$$

Here Q is the volume of fluid passing per unit area per unit time with respect to the solid, p is pore pressure, V_r and V_w are the velocity of the rock grains and that of the water, respectively, and ρ_w and ρ_r are the densities of water and solid particles of the sample, respectively. From the above, we get

$$\frac{\partial(Q + V_r)}{\partial z} = 0 \tag{4.52}$$

Using equations (4.50) and (4.51) and noting that the speed of the solid particles V_r at $z = 0$ is 0, we obtain the compaction equation

$$\frac{\partial[Q(1-\phi)]}{\partial z} + Q_0 \frac{\partial \phi}{\partial z} = -\frac{\partial \phi}{\partial t} \tag{4.53}$$

where

$$Q_0 = [V_w \phi]_{z=0} \tag{4.54}$$

is the flow of the fluid at the end of the sample where the origin of the coordinate system is located. From Terzaghi's law, the effective stress for this experiment is

$$\sigma = P_0 - p(z,t) \tag{4.55}$$

And combining this with the Darcy's law, we get

$$Q = \frac{\kappa}{\eta} \frac{\partial \sigma}{\partial z} = \frac{\kappa}{\eta} \frac{\partial \sigma}{\partial \varepsilon} \frac{\partial \varepsilon}{\partial z} = -D \frac{\partial \varepsilon}{\partial z} \tag{4.56}$$

The compaction diffusivity of the system is defined as

$$D = -\frac{\kappa}{\eta} \frac{\partial \sigma}{\partial \varepsilon} \tag{4.57}$$

Thus D is not constant as it depends on porosity and the effective stress both of which are functions of t and z. On combining the above equations we get the following compaction equation in terms of the void ratio ε:

$$(1+\varepsilon)^2 \frac{\partial}{\partial z} \left(\frac{D}{1+\varepsilon} \frac{\partial \varepsilon}{\partial z} \right) - Q_0 \frac{\partial \varepsilon}{\partial z} = \frac{\partial \varepsilon}{\partial t} \tag{4.58}$$

We still have moving boundaries to deal with. We can handle this by transferring to the Lagrangian system of coordinate transformations from (z, t) to (z', t') such that

$$z'(z,t) = \int_0^z \left[\frac{1}{1+\varepsilon(z,t)} \right] dz \tag{4.59a}$$

$$t' = t \tag{4.59b}$$

Using this new coordinate system, the compaction equation (4.58) reduces to

$$\frac{\partial}{\partial z'} \left[\frac{D}{1+\varepsilon} \frac{\partial \varepsilon}{\partial z'} \right] = \frac{\partial \varepsilon}{\partial t'} \tag{4.60}$$

Equation (4.60) needs to be solved with the initial and boundary conditions

$$\varepsilon = \varepsilon_0; t' = 0, z' = 0 \tag{4.61a}$$

$$\varepsilon = \varepsilon_1; t' > 0, z' = 0 \tag{4.61b}$$

The constitutive relation between the void ratio and the effective stress is given by a relation of the form introduced earlier

$$\varepsilon = \beta \ln(\sigma/\sigma_0) \tag{4.62}$$

where β and σ_0 are constants. (We ignore any effect of temperature in the model.) The variation of permeability with porosity (and void ratio) is given by $\kappa = m\varepsilon^n$ ($m = 0.0034$, $n = 5.0$). This completes the mathematical description of the 1D model of the test to simulate the basic physics behind a MDT tool for pore pressure measurement using pressure transient technique. In Figure 4.31c we show computed fluid flow from each end of the sample as a function of time. It shows that it takes over *four days* to reach a steady state and not several minutes as shown in the field example of Figure 4.30. That is a long time for fluid flow in a one inch sample (shale)! It is for this reason the RFT/MDT/DST tools cannot provide pore pressure data from shale or shaly formations. This is a challenge that the industry faces – *we can't get reliable pressure measurements in shales using downhole tools where we need it the most*. Currently all pressure data for shales are either *inferred* from measurements on adjacent permeable beds (sands) or from *kicks*.

4.6   Leak-Off Test, Extended Leak-Off Test, and Fracture Gradient

In Chapter 2 we discussed the basics of continuum mechanics and its extension to poroelastic medium (rocks). The concept of principal stress was also discussed. Figure 2.24 showed the orientation of three principal stresses at any point in the earth. These stresses have certain orientations and magnitudes and directly contribute to fracturing of rocks. For drilled vertical wells, S_v, S_h, and S_H denote the three principal stresses. The vertical stress S_v is due to the combined weight of the overlying rocks and fluids (overburden stress) and it is obtained by integrating density logs. Due to Poisson's ratio effect, the vertical stress due to overburden will tend to spread and expand the underlying rocks in the horizontal lateral directions. This lateral movement is counteracted by the surrounding rock material and causes the horizontal lateral stresses S_H and S_h, known as maximum and minimum horizontal stresses, respectively, to develop. We note that this is true for extensional basins. Such is not the case for compressional or strike-slip basins (see Chapter 13). Naturally occurring earthquakes also affect the stress directions and their magnitudes.

It is erroneous to assume that the two horizontal stresses are equal. As noted, magnitudes of horizontal stresses are affected by Poisson's ratio of surrounding rocks – the higher the Poisson's ratio, the higher the horizontal stresses. We also remind that the estimates of Poisson's ratios for various rocks as measured from static experiments are *not* the same as those obtained from dynamic measurements such as from borehole logs and seismic data. However, a thorough discussion of this important subject is beyond the scope of this book and we refer the readers to the excellent book

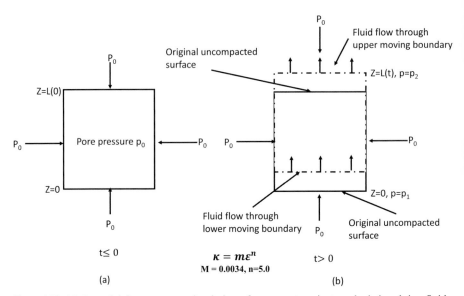

Figure 4.31 (a) A model for computer simulation of pressure transient analysis involving fluid flow through a shale sample of 1-inch length, porosity of 0.18, and permeability of 1.733×10^{-6} mD. The sample is jacketed at a confining pressure P_0 (4000 psi) at all times. Initially the pore pressure in the sample is p_0 (100 psi). (b) The experiment begins with raising the pore pressure at one end to 1100 psi and held constant thereafter. This creates a pressure pulse that travels through the sample.

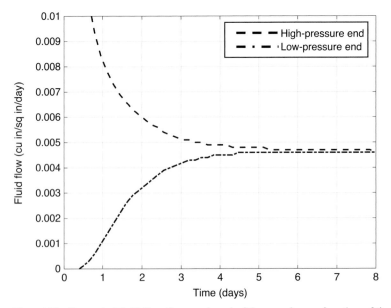

Figure 4.31c Computed fluid flow from each end of the sample as a function of time. The steady state is reached after three days. The permeability of the sample κ is low (1.73×10^{-6} mD).

by Zoback on reservoir geomechanics (Zoback, 2007). Avasthi et al. (2000) proposed the following equation to estimate minimum horizontal stress

$$S_h = (v/(1-v))(S-\alpha P) + \beta P \qquad (4.63a)$$

and Li and Burdy (2010) proposed a model to predict S_H namely,

$$S_H = \frac{(S_h - P)}{v} - S + 2P \qquad (4.63b)$$

where v is Poisson's ratio, P is pore pressure, and α is Biot's consolidation constant. We note that XLOT tests are the direct way to estimate horizontal stresses. The two horizontal stresses are equal only when these are related to the overburden stress only. However, this seldom is the case, as for example, in tectonically active areas due to faulting or mountain building. Direct measurements of lateral stresses are difficult and expensive in boreholes; these are usually parameterized based on equations of the type as shown above. Hydraulic fracture testing is the most effective technique for obtaining estimates of horizontal stresses in borehole environment. However, due to practical considerations such as economy of the project and the considerable amount of time that it takes to make measurements in *open* boreholes (a dangerous time in the life of a borehole!), often only a limited set of data points are measured (Hudson and Harrison, 1997).

4.6.1   Safe Mud Window, Casing Program, Lost Circulation, Leak-Off Tests

For drilling applications pore pressure prediction is critical to well construction and planning. This involves designing a safe casing program and planning mud weights that are within a safe limit – loo low mud weight can cause invasion of fluids from the formation into the borehole ("kicks") and may cause wellbore collapse. Too high mud weight can cause loss of borehole fluids into the formation by fracturing the formation surrounding the borehole ("lost circulation"), cause decrease in the rate of penetration leading to differentially stuck pipe and resulting in formation damage. In addition a decision has to be made whether to drill with water based or oil based mud – the latter is very expensive. This also impacts the "drill bit" program (roller cone versus PDC bit).

Safe Mud Window and Casing Program

In Figure 4.32a we show a simple graphical relationship between casing depth, pore pressure gradient and fracture gradient. The casing program is typically designed from the bottom up. It is a common practice that mud weight is usually kept slightly higher than the true formation pressure gradient by approximately 0.2 to 0.3 ppg. This is known as the *trip-margin*. A similar safety margin is also established by drillers by keeping the mud weight slightly *less* than the fracture gradient by ~ 0.2 to 0.3 ppg. This is known as the *kick-margin*. While drilling a well every attempt is made to *keep the mud weight within the window between*

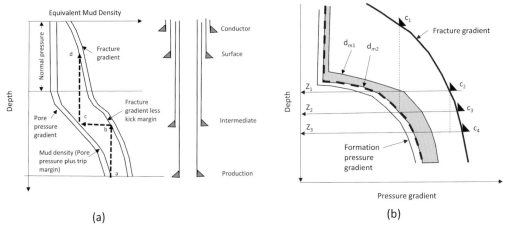

Figure 4.32 (a) Shows how fracture gradient affects the optimization of casing points. Note the safe mud window established by the driller – the mud weight is kept slightly higher than the pore pressure gradient and slightly lower than the fracture pressure gradient. (b) Consequence of excessive mud weight on the position of the casing. Mud weight d_{m1} after setting casing at C_1 will require setting further casings C_2 and C_4, whereas a mud weight of d_{m2} only requires casing C_3 to be set.

trip-margin and kick-margin. This is known as the *safe mud window.* This window is established by the driller *prior* to drilling and it impacts the casing strings and the casing program as shown in Figure 4.32a. Any error in pre-drill estimation of pore pressure and fracture gradients can lead to running out of casing strings and thus unable to reach the target depth. Therefore, an accurate estimate of pore pressure is critical as is the establishing a *safe mud window* for safety. Figure 4.32b shows the consequences of excessive mud weight on the position of the casing (after Mouchet and Mitchell, 1989). In this example the use of mud weight d_{m1} after setting casing at C_1 will require setting further casings C_2 and C_4 whereas a mud weight of d_{m2} only requires casing C_3 to be set. Clearly it shows how important it is to predict pore pressure accurately – the safety margin (safe mud window) dictates the casing setting depths, and whether a well can be drilled safely to deeper target depths without running out of casings.

Differential Drill Pipe Sticking and Lost Circulation

This is a serious situation caused usually by *underbalanced* drilling. This happens when the equivalent mud weight is substantially *less* than the pore pressure (the difference is known as differential pressure and not to be confused with differential stress as used in the Terzaghi principle). In this case while the drill bit approaches the stuck-point, the borehole starts caving in and the torque on the drill string increases simultaneously. (Loss of borehole is the usual outcome.) This condition may require sidetracking the borehole. The seriousness of the differential sticking depends on several other factors such as how long the pipe remains motionless against the formation, the slickness of filter cake (mud penetration to the borehole), and the permeability of the formation.

Lost circulation is typically caused by excessive mud weight that causes mud to overcome the fracture gradient and leak into the formation – a situation known as *overbalanced* drilling. This is an expensive proposition especially when drilling mud is of a synthetic blend. Occasionally, circulation of the drilling mud can be re-established by reducing the mud weight. However, this practice is not common. A large overbalanced drilling, on the other hand, often causes the casing to be set too high to cure lost circulation. This might create unsafe drilling conditions as well as running out of casing prior to reaching target.

Only *rarely* a well is drilled *deliberately underbalanced* when the mud weight is *less* than the formation pressure. This situation arises when the gap between the formation pressure gradient and the fracture pressure gradient is too small to define a safe mud window. The practice dictates drilling with mud weight *slightly higher* than the fracture gradient and when the well *kicks*, the driller would lower the mud weight to control the well. This cycle is repeated multiple times until the target depth is reached. (In the industry this practice is jokingly known as *drilling for the kicks*!) Some of the very high pressure wells in the N. Sea have been drilled that way. Only a few experienced drillers attempt this. Nonetheless, the practice is very dangerous and there are cases when underground blowouts have happened due to misjudgment.

Leak-Off (LOT) and Extended Leak-Off Tests (XLOT)

Fracture gradient measurement is quite straight forward for land operation because it is not affected by water column. However, the fracture gradient will be reduced in deepwater environment. This is because water is less dense than rocks. Increasing water depth will reduce both the overburden pressure and the formation fracture pressure. Therefore, offshore wells will have smaller safety margin between mud weight and fracture pressure than land wells. Furthermore, at the same water depth, the fracture pressure at the shallower section will be decreased more than the deeper depth. What's more, particularly at a shallow depth where the average overburden is greatly reduced by water column, more casing strings will be required to reach the planned casing depth.

Fracture gradient of a formation is determined typically by hydraulic fracturing. This is a *direct* method. Hydraulic fracturing occurs when pore pressure is high enough to induce sufficiently high stress field which will cause rock to fracture. The geometry of the stress field depends on the geometry of the borehole or cavity in which the pressure is applied and the geometry and material properties of the surrounding rocks. This is a common technique used to measure *in situ* component of the stress field when *small* volume of fluid is pumped into the formation in a test known as "*leak off test*" (LOT). The LOT technique is frequently used in the oil industry to assess the "fracture gradient" of the formation (i.e., the maximum borehole pressure that can be applied without

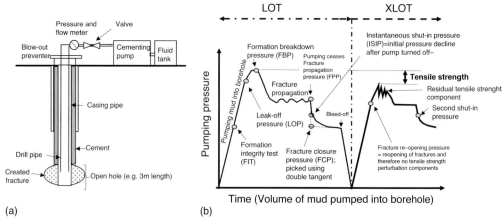

(a)   (b)

Figure 4.33 (a) Schematic of a borehole configuration during leak-off test (LOT) and extended leak-off test (XLOT). Modified after Yamamoto (2003). (b) Idealized relation between pumping pressure and volume of injected fluid during LOT and XLOT. Modified after White et al. (2002).

mud loss) and to determine optimal drilling parameters such as mud density that can be used in the open section of the borehole after setting a casing (Kunze and Steiger, 1991). To stimulate tight reservoirs and measure *in situ* horizontal stress, drillers pump *large* amount of fluid volumes into the formation for *longer* period of time. This is known as the *extended leak off test* (XLOT).

In both LOT and XLOT procedures, the pumping pressure tests are carried out immediately below newly set casing in a borehole as shown in Figure 4.33a. It is similar to other pumping pressure tests known in the industry as pressure integrity test, formation integrity test, or casing shoe integrity test. Each of these tests has a different target pumping pressure. While the LOT procedures are relatively simple, an XLOT is a more complex test with extended pressurizing procedures. To carry out LOT or XLOT after setting casing and cementing, a short length (several meters) of extra open hole is drilled below the casing shoe. The casing shoe is then pressurized by drilling fluid. The pressure at the casing shoe is equal to the sum of the hydrostatic pressure of the drilling fluid column and the shipboard pumping pressure. Figure 4.33b shows an idealized pumping pressure curve for XLOT (White et al., 2002). As the fluid is pumped in, the pressure in the borehole increases, and the *leak-off pressure* (LOP) is reached. This occurs when fluid begins to diffuse into the formation at a more rapid rate as the rock begins to dilate (Figure 4.33b). Generally, a LOT is a test that finishes immediately after LOP is reached. We discussed this in Chapter 2. During an XLOT acquisition program, pumping continues beyond the LOP point until the pressure reaches *formation breakdown pressure* (FBP). This creates some new fractures in the borehole wall. Pumping is then continued for a few more minutes, or until several hundred liters of fluid have been injected, to ensure stable fracture propagation into the undisturbed rock formation. The pumping pressure then stabilizes to an approximately constant level, which is called the *fracture propagation pressure* (FPP). When FPP is reached pumping is ceased. This is known as *shut-in*. The *instantaneous shut-in pressure* (ISIP) is defined as the point where the pressure decrease after shut-in deviates from a straight line. The most important pressure parameter from XLOT test is the *fracture closure pressure* (FCP). As the name suggests, this occurs when the newly created fractures close again. FCP is determined by the intersection of two tangents to the pressure versus mud volume curve (Figure 4.33b). As Yamamoto (2003) pointed out, the value of FCP represents the minimum principal stress. This is because the stress in the formation and the pressure of fluid that remains in the fractures have reached a state of mechanical equilibrium. Since these tests are carried out in open borehole sections, care must be exercised for safety, especially in high pressure wells.

4.6.2   Empirical Models for Fracture Gradient Determination

Fracture pressure is defined as the pressure at which fractures would open in a formation. Fracture gradient is defined by the ratio of fracture pressure and the depth. In the process of well planning and drilling, an understanding of fracture gradient and its relationship to mud weights (related to pore pressure) is critical to ensure safe drilling and well operations. It impacts the casing setting depths, and the casing and the mud program. From the well control stand point, the fracture gradient directly effects how much influx volume can be successfully contained in the wellbore. If the wellbore pressure is over the fracture pressure, formations would be fractured and this situation will result in loss of drilling fluid into formations. Additionally, it might lead to well control problems. *Thus the scheduled mud densities in any one stage should not exceed the lowest expected fracture gradient in the open hole.*

When LOT or XLOT data are not available, we need to rely on various empirical relationships that use stress analysis techniques for fracture gradient prediction. There are numerous such relationships (Hubbert and Willis, 1957; Haimson and Fairhurst, 1967; Matthews and Kelly, 1967; Eaton, 1969; Anderson et al., 1973; Althaus, 1997; Pilkington, 1978; Daines, 1982b; Breckels and van Eekelen, 1982; Constant and Bourgoyne, 1988; Aadnoy and Larson, 1989; Wojtanowicz et al., 2000; Barker and Meeks, 2003; Fredrich et al., 2007; Wessling et al., 2009; Keaney et al., 2010; Zhang, 2011; Oriji and Ogbonna, 2012). Each of these methods have inherent assumptions, many are empirical in nature and we encourage the ardent readers to consult the bibliography. Below we discuss some of the most popular methods.

Hubbert and Wills's Method

The earliest discussions of fracture gradient model appeared in Hubbert and Willis (1957). It is valid for tectonically passive areas. The authors assumed that the fracture in rocks occurs when the applied fluid pressure exceeds the sum of the minimum effective stress (which is the difference between the overburden stress and the pore pressure) and the formation pressure as shown below

$$FP = \sigma'_h + P = S_h \tag{4.64}$$

Assuming that the effective minimum stress is horizontal and has a value of one-third of the effective overburden stress, namely,

$$\sigma'_h = (S_h - P)/3 \tag{4.65}$$

they obtained the following well-known expression for fracture pressure,

$$FP = (S - P)/3 + P \tag{4.66}$$

or

$$FP = (S + 2P)/3 \tag{4.67}$$

The fracture gradient is obtained by dividing the fracture pressure by depth. It should be noted that the Biot consolidation coefficient is usually assumed to be 1 in fracture gradient calculation in the oil and gas industry; therefore, it is not considered in the related equations. There is, however, some debate about this in industry (Sarkar and Batzle, 2008). Hubbert and Willis are sometime credited with the following equation for fracture pressure, namely,

$$FP = (v/(1 - v))(S - P) + P \tag{4.68}$$

where v is the Poisson's ratio of the rock. This equation is not given in their paper, but taking $v = 0.25$, this formula reduces to equation (4.66). From experience, we found that the Hubbert and Willis (1957) model over predicts values for fracture pressure in geopressured formations.

Matthews and Kelly's Method
Matthews and Kelly (1967) introduced a variable matrix coefficient, K_i, and suggested the following expression for fracture pressure:

$$FP = K_i(S - P) + P \tag{4.69a}$$

where the effective stress coefficient (also known as the matrix stress coefficient) is defined by

$$K_i = \frac{\sigma_h}{S} \tag{4.69b}$$

The matrix stress coefficient, K_i, is a *depth-dependent function* that is calibrated with local fracture gradient measurements as shown in Figure 4.34. The depth axis in this figure is the *equivalent normal depth* as defined earlier in this chapter. Therefore, it is doubtful whether this type of empirical curve would be applicable at deeper depths where we would expect various thermal phenomenon to be

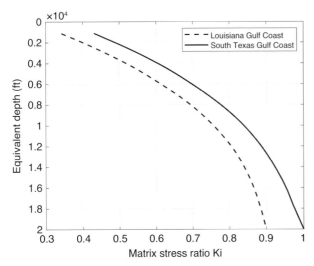

Figure 4.34 An example of the matrix stress ratio, K_i, for the Matthews and Kelly (1967) method for fracture pressure. Modified after Nguyen (2013).

operative in generating abnormally high pore pressure. This is because the equivalent depth could not be calculated in that situation. In their paper, Matthews and Kelly (1967) obtained K_i from the fracture initiation pressures. Therefore, this fracture gradient is higher than the fracture extension gradient (the minimum stress gradient). Values of K_i is higher for the sands of South Texas Gulf Coast which are more clayey than the offshore sands of Louisiana. Thus one needs to be careful in using these curves as these are not universal and local calibration is a must.

Eaton's Method

Eaton (1969) assumed that rock deformation was elastic and replaced the matrix stress coefficient in equation (4.69) by a term that depends on the Poisson's ratio, v

$$FP = (v/(1-v))\sigma + P \qquad (4.70)$$

where σ is vertical effective stress defined as $\sigma = S - P$. In this method lithology plays an important role through the Poisson's ratio term in equation (4.70). However, we note that this Poisson's ratio really refers to values from *static* measurements as opposed to *dynamic* measurements, as for example, through velocity of P- and S-waves as given below:

$$v = \frac{0.5(V_P/V_S)^2 - 1}{(V_P/V_S)^2 - 1} \qquad (4.71)$$

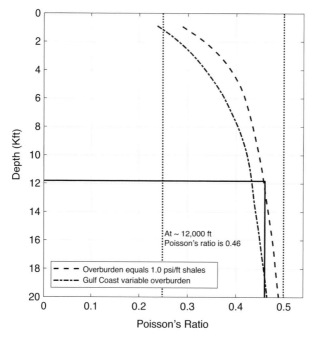

Figure 4.35 Values of (so-called) Poisson's ratio in Eaton's model. Modified after Nguyen (2013).

This has created some confusion in the application of the Eaton method in both geoscience and engineering communities (Zhang and Yin, 2017). To obtain v common practice in the engineering community is to use the sonic and shear logs from borehole measurements and the geoscience community the conventional P-wave seismic data in conjunction with the shear data from shear-logs. Both practices are error prone due to unknown amount of velocity dispersion associated with the dynamic measurements. The correct data should be obtained from static measurements of Poisson's ratio. In Figure 4.35 we show a plot of Poisson's ratio versus depth as recommended by Eaton (1969) who back calculated these values from regional fracture pressure and pore pressure data using equation (4.70). These are *not* measured Poison's ratios.

Daines's Method

Daines (1982b) superposed a horizontal tectonic stress, σ_t onto Eaton's equation for fracture pressure estimation. This additional term has been characterized as the required minimum pressure within the borehole to hold open and extend an existing fracture. Daines's equation for fracture stress can be written in the following form:

$$FP = (v/(1 - v))\sigma + P + \sigma_t \tag{4.72}$$

Table 4.10 Recommended values for Poisson's ratio

Lithology		Poisson's ratio
Clay		0.17–0.50
Conglomerate		0.2
Dolomite		0.21
Limestone micritic		0.28
Limestone sparitic		0.31
Limestone porous		0.2
Limestone Fossiliferous		0.09
Limestone Argillaceous		0.17
Sandstone Coarse		0.05–0.10
Sandstone Medium		0.06
	fine	0.03
	shaly	0.24
	fossiliferous	0.01
Shale	calcareous	0.14
	dolomitic	0.28
	siliceous	0.12
	silty	0.17
	sandy	0.12
Siltstone		0.08

Source: Modified after Daines (1982b) and Mouchet and Mitchell (1989)

where *FP* is the fracture pressure; and σ_t is the superposed horizontal tectonic stress. The value of σ_t is usually estimated from the leak-off tests. In the original formulation this is assumed to be constant for the entire well. Zhang and Yin (2017) claim that

$$\sigma_t = C\sigma \tag{4.73}$$

is a fair approximation to tectonic stress where C is a constant. In that case, Daines's equation can be written as

$$FP = (C + v/(1 - v))\sigma + P \tag{4.74}$$

Daines (1982b) suggested some values for the Poisson's ratio in equation (4.72). These are given in Table 4.10. It appears that these values are not from dynamic measurements such as seismic or velocity logs but rather like Eaton's recommendations (back calculated from field measurements of fracture pressure and pore pressure data) and so these should be used with caution. We suggest that these values should be calibrated locally with leak off test data, whenever possible.

Zhang (2011) suggested some upper and lower bounds on fracture pressure (neglecting any temperature effect). According to him, the maximum fracture pressure can be approximated by

$$FP_{max} = 2S_h - P \qquad (4.75)$$

where

$$S_h = (v/(1-v))\sigma + P \qquad (4.76)$$

Equation (4.75) can be used as the upper bound of fracture pressure (or gradient), and Eaton's method can be used as the lower bound of fracture pressure (or gradient) (Zhang et al., 2008; Zhang, 2011). The average of the lower bound and upper bound of fracture pressures can be used as the most likely fracture pressure (Zhang, 2011):

$$FP_{avg} = (1.5v/(1-v))\sigma + P \qquad (4.77)$$

where FP_{avg} is the most likely fracture pressure.

4.6.3   Summary

In summary, fracture pressure estimation prior to drilling a well is a key requirement for well planning and drilling safety. In order to prevent "kicks" during drilling it is a common safety practice to keep the mud weight slightly higher than the mud weight required to balance the true formation pore pressure at a given level. When drilling through a high formation pore pressure zone we need to increase the mud density to maintain equilibrium for the newly drilled formation. But this might have an undesired consequence: the mud pressure would be increased throughout the entire open hole, including in front of previously drilled weak zones. This typically causes loss of mud and poses a serious drilling safety issue. It is for this reason one must attempt to avoid the situation by restricting mud weight to a value below the fracture pressure at the level in question. We must remember that drillers obtain Leak-off Test (LOT) just below a previously run casing shoe prior to drilling ahead, since this is likely to be the weakest point in the next drilling phase. So the trick is to use a mud weight that will neither break the formation at the current level nor the weakest part of the formation already drilled. Evaluating fracture pressure requires estimate of S_h – the minimum component of in situ stress. In most cases (in the absence of extra tectonic stress component), the minimum stress is horizontal and the following model is valid:

$$S_h = K\sigma + P \qquad (4.78)$$

where K is the ratio of the effective horizontal to effective vertical stress. The discussions that we had previously in this section regarding models for fracture pressure or fracture gradient have been from consideration of the value of this coefficient. Various empirical models try to estimate this. In our opinion, the best estimate is provided by either actual measurements from LOT or XLOT or values obtained from analogue wells.

Any model that uses the Poisson's ratio method to estimate K must be dealt with carefully before using because of the inherent ambiguity on the actual measurements of this quantity. We referred to this ambiguity as the difference between static and dynamic values of Poisson's ratios. This difference is accentuated by the presence of low strain amplitude but frequency dependence of Poisson's ratio in dynamic measurements and the presence of high strain amplitudes in static (frequency independent) measurements. These differences could be large and, if not reconciled, can create serious safety issues while drilling. We will also address this issue further in Chapter 12.

4.7   Subsalt Pore Pressure and Fracture Pressure

In Figure 4.36a we show a schematic of many challenges the drillers face while drilling through and around salt (Perez et al., 2008). These challenges are mostly due to salt's tendency to move through the sediments. These movements leave an imprint on pore pressure as indicated and provide habitat for hydrocarbon entrapment (good news!) and affect imaging and drilling adversely (bad news!). Poor subsalt velocities are known to affect seismic imaging and this impacts accurate positioning of well trajectory and prohibits estimation of reliable pore pressure. With the advancement of high quality seismic data acquisition techniques such as ultra-long offsets and continuous azimuth coverage and exponential growth in compute power for data processing, some progress has been made in the area of subsalt pressure analysis. We shall discuss some of the advanced seismic velocity analysis techniques to address this in Chapter 6. Here we show some examples. In Figure 4.36b, we show an example from the Green Canyon, Gulf of Mexico where seismic data was acquired with ultra-long offset and multiple boats. The high quality seismic data provided better image as compared to the legacy data and enabled reliable estimation of subsalt pore pressure as verified with sonic log, and mud weights (see figure 4.37). We note that the *crest* of the anticline coincides with velocity *lows* and *higher* pore pressure. Below the salt, the well encountered pore pressure regressions as shown in Figure 4.37 and it was abandoned due to safety concerns. The solid lines (left panel) represent velocity versus depth and pore pressure (middle panel) as predicted by a rock physics model (rock physics template at the well location) as discussed in Chapter 6.

Another example from a different well in the GC-area is shown in Figure 4.38. It is also a subsalt well. It is known that salt movement causes development of a "rubble zone" beneath the salt. The rubble zone usually consists of a series of shale stringers that are embedded in poorly consolidated sands. The zone could be overpressured at the entry point because of gas pockets under the salt (known in the industry as the "attic gas" – it occurs through migration of gas from below through dipping permeable beds). Since salt has lower density compared to the sediments, it creates higher pore pressure just underneath the salt and drillers face a situation where a formation needs to be drilled with a very narrow gap between pore pressure and fracture pressure. This causes lost circulation, kicks and stuck

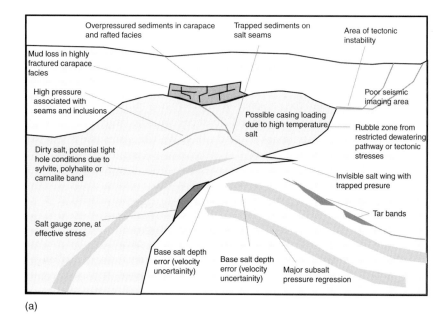

(a)

(b)

Figure 4.36 (a) A schematic of the effect of salt movement on stratigraphic changes creating pockets of overpressured sediments and posing challenges to velocity analysis, imaging and drilling. (b) An improvement in technology for velocity model building (rock physics guided tomography) and high-quality seismic data provided better subsalt image as compared to the legacy image shown on the left side. Note that the crest of the anticline coincides with low velocity (and therefore, higher pore pressure). The velocity scale is from 5000 ft/s to 16000 ft/s. After B. Deo, private communication, 2004, based on oral presentation by Bhaskar Deo at the SEG / Beijing 2014 International Geophysical Conference & Exposition, Beijing, China, April 21–24. (A black and white version of this figure will appear in some formats. For the color version, please refer to the plate section.)

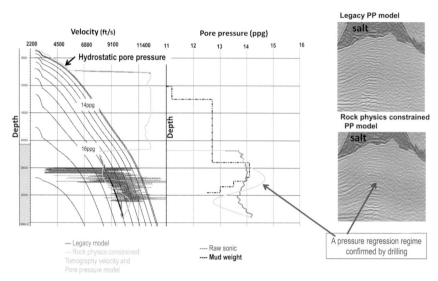

Figure 4.37 Subsalt velocity (tomography) and pore pressure constrained by rock physics during velocity modeling and comparison with sonic log, legacy tomography (unconstrained), and mud weight from drilling (Green Canyon area). Seismic data were ultra-long offset. Solid lines represent velocity versus depth for various pore pressure gradients from a rock model as discussed in Chapter 6. The first panel represents velocity from 2200 ft/s to 11400 ft/s in step of 2300 ft/s. The second panel is pore pressure in ppg starting from 11 ppg to 16 ppg in step of 1 ppg. After B. Deo, private communication, 2004, based on oral presentation by Bhaskar Deo at the SEG / Beijing 2014 International Geophysical Conference & Exposition, Beijing, China, April 21–24. (A black and white version of this figure will appear in some formats. For the color version, please refer to the plate section.)

pipe and in dire situations, even blow outs. Pre-drill prediction of such events is extremely difficult.

Fracture pressure modeling for subsalt wells is very challenging especially when a drilling project requires penetrating thick salt formations as in the Gulf of Mexico and other petroleum basins (e.g., Egypt). As noted earlier, large uncertainty in fracture pressure is a direct result of unreliable subsalt pore pressure. A major contributor to this is salt creep and poses a major challenge for borehole stability (Zhang et al., 2008); therefore, a heavier mud weight is used to control salt creep. For example, mud weight in subsalt wells can be as high as 80%–90% of the total vertical overburden stress. This high mud weight would definitely require a higher fracture gradient in the salt formation. Zhang and Yin (2017) reported LOT and FIT data in salt formations in 15 wells in the Gulf of Mexico (10 in the Mississippi Canyon and 5 in the Green Canyon). Their results showed that the LOT and FIT pressures in most salt formations were larger than the overburden stress by as much as 1000 psi. The main reason for this is thought to be salt creep although incorrect measurements of LOT in subsalt wells cannot be ruled out. Zhang and Yin (2017) also commented that their data fit the Matthews and Kelly (1967) model for fracture gradient estimation but only after making the matrix stress coefficient, K, in that model a depth-dependent variable (see Zhang and Yin, 2017,

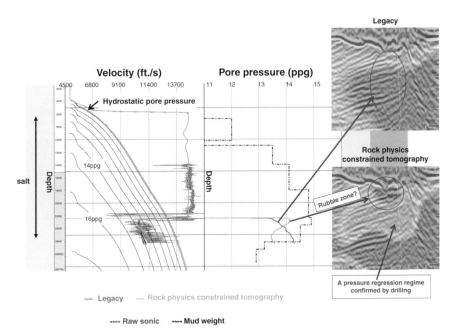

Figure 4.38 Subsalt velocity (from tomography) and pore pressure constrained by rock physics during velocity modeling and comparison with sonic log, legacy tomography (unconstrained), and mud weight from a well in the Green Canyon area. Solid lines represent velocity versus depth for various pore pressure gradients from a rock model as discussed in Chapter 6. Seismic data used were ultra-long offset. The figures on the right show pore pressure from legacy (top) and rock physics constrained tomography model (lower) for pore pressure showing a possible "rubble zone" below salt that has higher pore pressure, The left panel is velocity (ft/s) versus depth (ft) starting from 4500 ft/s to 13700 ft/s in step of 2300 ft/s. The middle panel is pore pressure in ppg, starting from 11 ppg to 15 ppg in step of 1 ppg. After B. Deo, private communication, 2004, based on oral presentation by Bhaskar Deo at theSEG / Beijing 2014 International Geophysical Conference & Exposition, Beijing, China, April 21–24. (A black and white version of this figure will appear in some formats. For the color version, please refer to the plate section.)

Figure 5). However, the data fit to obtain a depth-dependent matrix stress coefficient is not very convincing due to a large scatter of data points.

To summarize, subsalt drilling is very challenging due to poor seismic illumination and salt creep as well as narrow gap between pore pressure and fracture pressure that results in mud loss, stuck pipe and even, blowout. This is where fit-for-purpose technology can help mitigate risk. Techniques such as the ones discussed in Chapter 6 may prove useful.

4.8   Overburden Stress Evaluation

By definition, the overburden or lithostatic stress (also known as confining stress or total vertical stress), S, at any depth, z, is that pressure which results from the *combined* weight of the rock matrix and the fluids in the pore space overlying the formation of interest as well as the weight of the static water column (density, ρ_w and height, z_w) if in an offshore environment and the air pressure, S_{air}. This is written as

$$S = S_{air} + g \int_0^{z_w} \rho_w \, dz + g \int_{z_w}^z \rho_b \, dz \qquad (4.79a)$$

and

$$\rho_b = (1 - \phi)\rho_r + \phi\rho_f \qquad (4.79b)$$

and for mixture of water and hydrocarbon

$$\rho_f = f_w\rho_w + (1 - f_w)\rho_{hc} \qquad (4.79c)$$

Here ρ_b is the bulk density, ρ_w is water density, ρ_r is rock grain density, ρ_f is the fluid density, ϕ is the fractional porosity, S_{air} is atmospheric pressure, g is the acceleration due to gravity, f_w is volume fraction of water, and ρ_{hc} is the density of hydrocarbon. In deriving the above equation it is assumed that acceleration due to gravity g is a constant over z. In reality, g is strictly not a constant but a function of z and should appear inside the integral in equation (4.79a). But since g varies little over depths which are a very small fraction of the Earth's radius, it is placed outside the integral in practice for most near-surface applications which require an assessment of lithostatic pressure. In deep-earth geophysics/geodynamics, g varies significantly over depth and it may not be assumed to be constant.

Overburden stress is an important quantity that is essential for quantitative estimation of pore pressure and fracture pressure. Because of sediment compaction and the subsequent increase in bulk density with increasing depth, the overburden stress increases with depth. The oft used assumption of 1 psi/ft for overburden stress gradient in the literature (Fertl, 1976; Mouchet and Mitchell, 1989) may not be valid everywhere. For offshore sediments, there is a zone of unconsolidated sediments below seabed (~300 m). In this zone, the fractional porosity is larger than or equal to the critical porosity (Marion et al., 1992) and the sediments behave more like suspension than rock that can bear some shear stress. This must be accounted for in the density model used in equation (4.79a). Although the computation of overburden stress is well defined, it is quite often misused. In the early days the drilling community stayed away from the process of integration as required in equation (4.79a) and they resorted instead to discrete summation over large intervals to keep computations simple (see, e.g., Mouchet and Mitchell, 1989, pp. 203–207). This practice could lead to erroneous results and is not recommended.

4.8.1   Direct Method for Overburden Stress Computation

There are two approaches to quantitative estimation of overburden stress: direct and indirect. The direct method is to use the recorded density log in the integral in equation (4.79a) together with estimates of water density (for offshore) as a function of temperature, pressure and salinity as given in Appendix A. This is the best practice. However, there is an issue here that is worth noting. Overburden stress estimation is a 3D problem.

Integrating bulk density logs is essentially a 1D method. How to reconcile the difference? There are two ways to reconcile the difference. The first approach is to use the velocity data from 3D seismic and convert it to density by using one of the techniques described below. The other is to use the 3D density data, if available, from airborne gravity surveys (Jorgensen and Kisabeth, 2000). The authors proposed using joint inversion of several sets of data to constrain the density in 3D. This type of survey can yield lateral changes in the density due to tectonic stress imprints related to salt movement. It can also delineate the boundaries between salt and sediments. However, the resolution of the data is limited and the density extraction from inversion would yield blocky pseudo-logs of density. Thus density logs are the best available tool today for direct overburden estimation.

The industry has access to downhole gravity meter tools. Borehole gravity measurements (BHGM) respond passively to the bulk density of a large rock volume surrounding the borehole that is orders of magnitude larger than the volume sampled by logs or cores. It measures the density difference between two depth intervals – it is not a continuous density measuring tool. A series of precise gravity measurements are collected at discrete intervals by stopping and reading the gravimeter at preselected borehole depths. Gravity measurements are unaffected by casing or formation damage caused by drilling (Jageler, 1976; Beyer, 1987). Nind and Macqueen (2013) present an interesting case study of density measurements using this tool. Nonetheless, BHGM is not a common source of information for bulk density measurements because unlike other wireline tools; it is not a continuous measuring tool.

Below we show a best-practice workflow for estimating overburden stress from datum to depth of interest in a continuous manner. It is not always possible to obtain direct estimates of overburden stress continuously from sea level to total depth or TD. This is mostly due to unavailability of continuous density log. Beside, even when the log is made available, there may be sections where the data are deemed unusable. In these cases, we need to resort to indirect empirical methods. Some of these models are discussed in Section 4.8.2.

Best practice for continuous overburden stress evaluation

Sea level to sea bed	Depends on salinity of water column (1.01 to 1.05 g/cc)
	Use regression models as given in Appendix A
Sea bed to top of logged interval	Density log from offset well
	Density-Transit time relations plus seismic velocities
	Regional trends (e.g., Traugott model)
Top of logged interval to total depth (TD)	Density log
	Density log from offset well
	Density-transit time relations plus seismic velocities
Intervals of poor quality	Density log from offset well
Density log	Density-transit time relations plus seismic velocities
Frontier areas or deeper intervals where no well data are available	Density-transit time relations plus seismic velocities
	Regional trends

4.8.2　Indirect Methods for Overburden Stress Computation

In general, some logs – in particular density logs, are not recorded in the upper parts of wells because the borehole diameter is too large and the walls are often too rugose to provide usable data. In situations like these indirect methods are commonly employed. These methods use velocity data from either seismic or sonic logs to estimate bulk density of formations. This enables the creation of overburden stress estimates prior to drilling a well. Below we provide some recipes for computing bulk density from velocity information. All of these models should be calibrated with density logs, whenever possible.

Methods That Use Depth as a Variable

A literature search revealed a host of methods under this category. Some of the more popular ones are described below.

Amoco Method

Many drillers use depth as a parameter for estimating bulk density. A useful empirical relationship is given below for average bulk density of a formation as a function of depth that is independent of lithology:

$$\rho_b = 1.953 + \frac{(Z/3125)^{0.6}}{8.345} \tag{4.80}$$

where Z is the depth below sea-bottom expressed in feet and the bulk density ρ_b in g/cm^3. This relationship is anecdotal and attributed to Amoco-drillers and supposedly, valid for deepwater sediments as well. We have no reference for this.

Traugott Method

Traugott (1997) proposed a depth-dependent model for overburden stress *gradient* (OBG) in ppg (1 psi/ft = 19.25 ppg = 2.31 g/cc.) using the following

$$OBG = [8.5z_w + \rho_{avg}z_{mbl}]/z \tag{4.81}$$

where the true vertical depth z (also referred as the measured depth) is referenced to the derrick floor, i.e., referenced to the top of the mud column, z_{mbl} is the depth below mudline and z_w is the water depth. All depths are expressed in feet (1 meter = 3.281 feet). Thus $z = (z_{mbl} + z_w + A)$ where A is the air gap or the height of the derrick above the floor. In the above expression ρ_{avg} is the average density based on regional density measurements (expressed in ppg) and given by

$$\rho_{avg} = 16.3 + (z_{bml}/3125)^{0.6} \tag{4.82}$$

As an example, if z_w and A are 2000 feet and 50 feet, respectively, the overburden gradient is 13.7 ppg, if z is 5200 feet and 16.7 ppg, if z is 12300 feet. Traugott-model is a simple algebraic equation and requires no other data and yields overburden gradient based on depth below mudline. The model yields reasonably good overburden stress although it overestimates bulk density at mudline slightly. Furthermore, it assumes that the water density is fixed at 1.018 g/cc.

Dutta Method
Dutta examined density and sonic logs from hundreds of offshore wells from many basins throughout the world and found the following simple algebraic relationship for overburden stress S as a function of depth to be useful, especially in rank-wildcat areas:

$$S = Az_{bml}^2 + Bz_{bml} + Cz_w + P_{air} \qquad (4.83)$$

where S is overburden stress in psi, z_{bml} is depth in feet as measured below sea-bottom, z_w is water column height in feet and P_{air} is the atmospheric pressure. (It is the pressure exerted by the weight of the atmosphere, which at sea level has a mean value of roughly 14.6959 pounds per square inch.) The values of the calibration coefficients are $A = 0.5432 \times 10^{-5}$; $B = 0.8783$; $C = 0.455$.

Zamora Method
The Zamora (1989) method also uses depth as a variable. However, it adds a correction factor that depends on age of rocks as shown for overburden gradient in ppg:

$$OBG = [8.5\ z_w + (8.03 + 0.232A)\ z^{1.075}]/z \qquad (4.84)$$

The parameter A has values tied to age as shown below

Holocene – Pliocene: 0–5;
Miocene-Oligocene: 5–9;
Eocene-Paleocene: 9–10;
Cretaceous-Triassic: 10–11;
Permian-older: 11–14

Depth-Dependent Model with Compaction
In this approach a compaction model that depends on depth such as the one by Athy (1930a) is used to estimate porosity and then bulk density. From equation (4.79b)

$$\phi = \frac{\rho_r - \rho_b}{\rho_r - \rho_f} \qquad (4.85)$$

Using Athy's relation with mudline porosity of ϕ_0, compaction coefficient λ, and depth below mudline z_{bml}, namely

$$\phi = \phi_0 e^{-\lambda z_{bml}} \qquad (4.86)$$

we get

$$\lambda = \frac{\ln(\frac{\phi_0}{\phi})}{z_{bml}} \qquad (4.87)$$

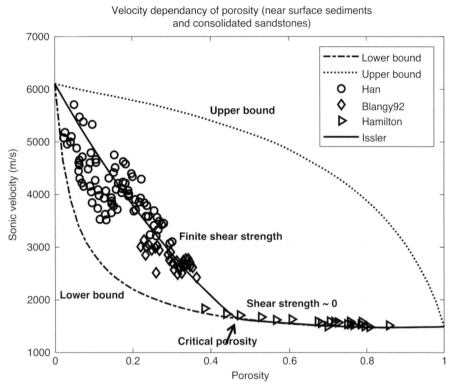

Figure 4.39 Velocity versus porosity relation from a compilation of data as indicated. Shown are the relations from the Issler equation as discussed in the text and the lower and upper bounds. The velocity for porosity higher than the critical porosity is given by the Wood's equation (Wood, 1941) and is valid for suspension of solid particles in a fluid; therefore, the mixture has no shear strength.

If z is the measured depth from Kelly bushing, and z_w is the water depth, we have $z_{bml} = z - z_w$, and after some algebraic manipulation, the overburden stress is given by

$$S = P_{air} + \rho_w g z_w + \rho_r g z_{bml} - \frac{g(\rho_r - \rho_f)\phi_0}{\lambda}\left(1 - e^{-\lambda z_{bml}}\right) \qquad (4.88)$$

where ρ_w is the density of sea water. Assuming the average matrix density to be 2.6 g/cc and the fluid density ρ_f to be 1.074 g/cc, porosities could be computed from a given porosity versus depth profile. Zhang (2011) carried out a regression on the porosity data available to them versus depth and obtained a mud line porosity of 0.4088 and a compaction coefficient λ, of 8.56091E-05 ft^{-1}. Although the above relation for overburden appears to be a complex relation, we note that the key parameter is the porosity that has been derived from a compilation of compaction curves. The rest is simply an algebraic manipulation.

Barker and Wood Method

Barker and Wood (1997) examined LOT data from 70 wells in the Gulf of Mexico and derived cumulative average bulk density from these data. They suggested the following expression for the overburden gradient in ppg

$$OBG = (8.55z_w + 5.3z_{bml}^{1.1356})/z \qquad (4.89)$$

where sea water density is taken to be 8.55 ppg, z_{bml} is depth below mudline, z is total vertical depth measured from Kelly bushing (ft) and z_w is water depth (ft). All depths are expressed in feet. They neglected to add the term dealing with atmospheric pressure.

Methods That Use Velocity as Input Data

Estimation of bulk density from velocity (sonic log or seismic) is essentially an exercise in picking an appropriate relationship between velocity and porosity or density. This has to be done carefully as these relations depend on various factors such as consolidation state of rocks and how well defined the lithologies are. In Figure 4.39 we show a typical plot of velocity versus porosity of offshore sediments. The data in Figure 4.39 are a collection from near surface sediments in offshore environment and those from measurements by Han (1986) on consolidated rocks. The near surface sediments are characterized by very low shear strength. We note that the point where the lower bound deviates from the data on consolidated rocks is known in the literature as the *critical porosity* (Marion et al., 1992). The dashed line for porosity smaller than the critical porosity is based on the model proposed by Issler (1992).

For porosity larger than the critical porosity, the model is appropriate for suspension based on Wood's model (Wood, 1941)

$$V_p = \{[\phi\beta_f + (1-\phi)\beta_r][\phi\rho_f + (1-\phi)\rho_r]\}^{-0.5} \qquad (4.90)$$

Table 4.11 Coefficients of the power law form curve fit (equation (4.91)) to the velocity–density data of Gardner et al. (1974) (A) and coefficients of the curve fit (equation (4.92)) to the velocity–density data of Gardner et al. (1974) using a quadratic equation (B)

(A)				(B)		
Coefficients of equation (4.91)			V_P range (Km/s)	Coefficients of equation (4.92)		
Lithology	*d*	*f*		a	b	c
Shale	1.75	0.265	1.5 – 5.0	−0.0261	0.373	1.458
Sandstone	1.66	0.261	1.5 – 6.0	−0.0115	0.26	1.515
Limestone	1.36	0.386	3.5 – 6.4	−0.0296	0.461	0.963
Dolomite	1.74	0.252	4.5 – 7.1	−0.0235	0.390	1.242
Anhydrite	2.19	0.160	4.6 – 7.4	−0.0203	0.321	1.732

Note: Part (B) yields a better representation of the data.
Source: Reprinted after Mavko et al. (2009)

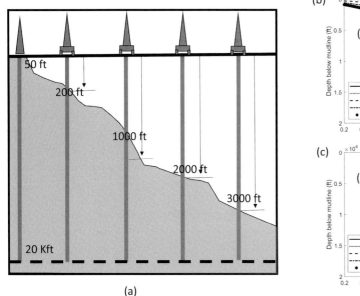

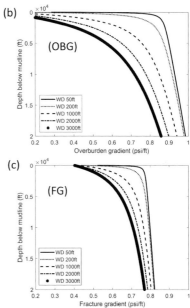

Figure 4.40 Effect of water depth on overburden and fracture pressure gradients. (a) A schematic of a well to be drilled at indicated water depths. (b) A plot of the overburden pressure gradient in psi/ft. The decrease of the gradient with increasing water depth is clear – a result explained by the fact that for a given depth as the water column increases we are replacing denser material with lighter material (brine). In (c) we show the offshore fracture pressure gradient as predicted using the model of Eaton (equation (4.96)) as a function of water depth. All gradients are measured from KB.

where V_p is the P-wave velocity of a suspension of solid particles (density ρ_r and compressibility β_r) in a fluid (volume fraction of fluid is the porosity ϕ, density ρ_f, and compressibility β_f). A note of caution: there are situations where velocity may not directly yield porosity since there are many factors that affect velocity but not porosity. An example is the case for very low porosity rocks where velocity may be affected by closure of cracks and not porosity. Gardner et al. (1974) suggested a very useful empirical relation between P-wave velocity and bulk density

$$\rho_b = dV_P^f \tag{4.91}$$

where coefficients d and f are lithology dependent and are given in Table 4.11a (taken from Mavko et al., 2009). The power law curve fit is to the data of Gardner et al. (1974). The velocity ranges are also listed in the table. Table 4.11b (also taken from Mavko et al., 2009) shows the various coefficients of equation (4.92) after curve fit using a quadratic fit to the same data of Gardner et al. (1974),

$$\rho_b = aV_P^2 + bV_P + c \tag{4.92}$$

We note that the coefficients in above models are lithology dependent as it should be. However, Gardner et al. (1974) also suggested a useful relation between bulk density and velocity of P-wave that represents an *average* between various rock types as shown below:

$$\rho_b \sim 1.741 V_P^{0.25} \tag{4.93}$$

where V_P is in km/s and ρ_b is in g/cm^3, or

$$\rho_b \sim 0.23 V_P^{0.25} \tag{4.94}$$

where V_P is in ft/s and ρ_b is in g/cm^3. The lithology independent relation shown in equation (4.93) or equation (4.94) is most widely used by the geophysicists. However, we recommend using the lithology-dependent equations, whenever possible.

A widely used but fundamentally *incorrect* model of porosity estimation for consolidated rocks from sonic log and hence, to compute bulk density and overburden stress, is the Wyllie's time-average equation (Wyllie et al., 1956)

$$\Delta t = \Delta t_r (1 - \phi) + \phi \Delta t_f \tag{4.95}$$

where

Δt = transit time from sonic log in $\mu sec/ft$
Δt_r = transit time of the matrix (grain) material in $\mu sec/ft$
Δt_f = transit time of the fluid in $\mu sec/ft$
ϕ = fractional porosity

For water, a typical value of Δt_f is 200 µs/ft. The relationship shown in equation (4.95) assumes that the total traveltime through a fluid-filled rock is a sum of the traveltime through the rock grains and that through the fluid component. It should not be used for young Tertiary clastics such as in the Gulf of Mexico. There are many more appropriate relationships between porosity and velocity (see, e.g., Mavko et al., 2009) and these are generally valid for wider range of rock types (except shale) than the Wyllie's time-average in equation (4.95). All these relations work well when properly calibrated to local data.

For seismic use, any physical model that relates porosity to velocity may be used to estimate bulk density and overburden pressure. Another alternative is to use the Gardner et al. (1974) model as given in equation (4.94). Seismic data have the distinct advantage that the data are nearly always available, and furthermore, it covers the near surface material as opposed to the log-based data. For this reason, seismic data can be used to infill the overburden stress in the shallow depths as well as any other depths where log values are missing or unreliable. In this way, a continuous depth-dependent overburden model can be generated for pore pressure and fracture pressure determination in 3D.

If no seismic data are available for a well, and there is no log data from an analogue well, bulk densities may be estimated on the rig-site by measuring density on cuttings. This data is always available when a well is drilled. The other alternative is to use any of the depth – dependent relations of the type discussed above (equations (4.80)–(4.89)). However, this should be the last option as the model needs to be calibrated.

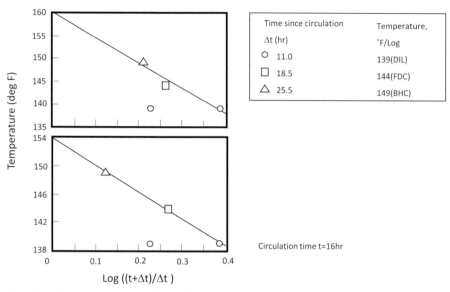

Figure 4.41 Temperature versus depth from borehole data. Horner-plot method for correcting BHT data based on a circulation time of 16 hours and noted times-since-circulation at the inset. The method requires at least three temperature measurements. Circulation time was 16 hours in this example. Modified after Peters and Nelson (2009).

4.9   Effect of Water Depth on Overburden and Fracture Pressure Gradients

Over the last several decades exploration for hydrocarbons has taken the petroleum industry to deepwater. This activity has heightened the industry's exposure to several drilling related challenges. Earlier we talked about some basic concepts regarding casing and mud programs. The gap between the overburden and the fracture stress and its depth dependency is a key parameter that impacts drilling in this regard. It is obvious from discussions in Section 4.8 that as water depth increases, the overburden stress is systematically reduced because water is less dense than rocks. To illustrate this quantitatively we consider the schematic of wells to be drilled under varying water depths as indicated in Figure 4.40a. In the example calculations, the true vertical depth is fixed at 20 Kft and the water depth varies from 0 to 3000 ft. We use the depth-dependent model for overburden stress (S) as given in equation (4.83) and the following for the fracture pressure gradient (FPG in psi/ft) (Eaton's model)

$$FPG = \frac{v}{1-v}\frac{S-P}{z} + \frac{P}{z} \tag{4.96}$$

where z is the true vertical depth (ft) measured from Kelly bushing and v is Poisson's ratio. We have assumed the fluid pressure gradient to be 0.7 psi/ft and have taken an average Poisson's ratio value of 0.3. Figure 4.41b shows the overburden stress gradient in psi/ft, and Figure 4.41c shows the fracture pressure gradient in psi/ft. These figures clearly illustrate the significant decrease in both the overburden stress and fracture pressure gradients with increasing water depth.

The model discussed above points to an important consequence in the shallow part of the stratigraphy below seabed in deepwater environment. The gap between the fracture pressure and the overburden pressure decreases as the water depth increases resulting in possible extra casing to be set to prevent lost circulation. If drill bit encounters overpressured formations in this situation, one may have to deal with the possibility of a blowout. This is known as the shallow-water-flow (SWF) problem in the industry and has caused substantial loss of resources in the Gulf of Mexico and elsewhere (Ostermeier et al., 2000). This phenomenon and some possible remedies are discussed in detail in Chapter 11.

4.10   Temperature Evaluation (Direct and Indirect Methods)

4.10.1   Introduction

Earlier in this chapter we discussed the importance of thermal effects (formation temperature and its history) on pore pressure models (Bowers, 1994; Dutta, 1986, 2016) in the context of burial metamorphism of shale and release of bound water. Formation temperature is also an important parameter in analysis of resistivity logs, as well as for the detection of fluid movement, and in geochemical modelling of formations and the maturity of hydrocarbons. The sister discipline of basin modeling treats the subject well (Hantschel and Kauerauf, 2009) because of the importance of petroleum systems modeling for hydrocarbon exploration and production. However, the subject has been treated with benign neglect in the exploration geophysics literature. Yet thermal effects on rock/fluid properties are very important and it affects pore pressure in significant ways (Swarbrick et al., 2002; Osborne and Swarbrick, 1997). In this section, we shall discuss some key aspects of how to obtain estimates of subsurface temperature for modeling thermal processes in the context of quantitative modeling of geopressure. The procedure can easily be applied to correct resistivity and conductivity logs for geopressure analysis.

The most direct way of obtaining continuous subsurface temperature data is via direct borehole temperature measurement logs. The temperature log results from a tool for measuring the borehole temperature of the *fluid* in the borehole that is in *equilibrium* with the formation. (This is very important.) Although temperature sensors are attached to every logging tool combination that is run in a well for the measurement of the maximum temperature (assumed to be at the bottom of the well), tools that can *continuously* measure temperature as the tool travels down the well are rare. Oil companies do not consider using it to be an essential part of their logging program due not only to reasons just mentioned earlier but also the requirement that the temperature of the fluids in the borehole be in equilibrium with the formation for reliable data – a process that may take from many tens of hours to days and thus adds considerably to the cost of the well. The equilibrium condition holds only in shut-in wells or production wells. It is for these reasons the geologists and geochemists use indirect methods. Some of the common techniques are discussed below.

4.10.2   Methods for Estimation of Formation Temperature

Horner Method for Direct Measurement in Borehole

Log-headers contain useful information regarding maximum bottom-hole temperature (BHT) measurement. This is always available. However, these measurements do not truly reflect the formation temperature as the borehole fluid may not be in equilibrium with the formation – a process that may take days as mentioned. The drilling mud is circulated during drilling and prior to inserting the wireline tool, and this drilling mud is cold compared to the formation. The cold drilling fluid invades the formation and cools it down very efficiently via heat convection. During circulation of drilling fluid the temperature of the borehole reaches an equilibrium defined by the cooling effect of the drilling fluid and the heating effect of the formation. When the circulation of the drilling mud stops (e. g., in preparation for the insertion of a wireline tool), the borehole gradually regains the true formation temperature, because the large mass of formation around the borehole heats the drilling fluid up to its ambient temperature. This process is slow because it occurs via mainly heat conduction which is less efficient than heat convection. It is for this reason various methods have been adopted in the past to correct the logged BHT to real formation temperature (Hermanrud et al., 1990; Hunt, 1996). The most common method is the Horner-plot. The Horner-plot method proposed by Dowdle and Cobb (1975) gives a reliable static formation temperature in regions of low geothermal gradient, and can be expressed by

$$T_m = T_f - \log[(t + \Delta t)/\Delta t] \qquad (4.97)$$

where T_m is the measured bottom-hole temperature, T_f, is the true formation temperature, t is the circulation time, and Δt is the time since circulation or the elapsed time after circulation has ceased. This plot is a straight line that intersects $(t + \Delta t)/\Delta t = 1$ at the formation temperature.

The Horner-plot method gives a reliable static formation temperature in regions of *low geothermal gradient less than or equal to 4°C per 100 m* (Sasaki, 1987). Horner-corrected BHT data must include a minimum of *three* logging runs that record time and temperature for each run. For the Horner method to work, it is important that the well not be circulated during logging. Also, it must not be circulated between the several logging runs of a Horner suite. An example of how to use Horner-plot is given in Figure 4.41 taken from Peters and Nelson (2009). The data that were used to create Horner plot are also shown in Figure 4.41 Based on published comparisons of DST and Horner-corrected BHT data from the same depths, the standard deviation of corrected bottom-hole temperatures is about ±8°C (±14°F) as documented by Peters and Nelson (2009). Some studies show that corrected data may still be systematically biased lower than true formation temperature. BHT data are an important source to reduce uncertainty and needs to be considered when carrying out quantitative geopressure evaluation that include thermal effects. Continuous temperature logs are also useful to detect the presence of overpressured zones, where the hot over-pressured fluids escape into the borehole and are noted by a rise in the measured temperature. It is for this reason drillers routinely measure the mud temperature on the rig-site. However, the accuracy of measured temperature data is not known and hence this indicator can be used only in a qualitative sense.

Velocity Method for Indirect Estimation of Formation Temperature

Several authors have advocated the use of seismic and well log–based velocity data to estimate subsurface temperature (Houbolt and Wells, 1980; Alam and Pilger, 1986; Brigaud et al., 1990). With seismic data, it is possible to estimate temperature in 3D. These methods traditionally use assumptions of *steady-state* heat flow in sedimentary basins and relate thermal conductivity to velocity to obtain relationships between temperature and traveltime. We note that the P-wave velocity of rocks is controlled by the same fundamental properties (e.g., porosity, mineralogy, texture, and pore fluid) as thermal conductivity. Hence, rescaling the seismic P-wave velocity profile will approximately reflect the variation in thermal conductivity of a sedimentary sequence. Of notable interest is the paper by Houbolt and Wells (1980). The basic elements of his model are:

$$q = -K(\frac{dT}{dx})$$ (4.98)

$$\frac{V_P}{K} = cons$$ (4.99)

$$K = \frac{1}{(aT + b)}$$ (4.100)

Equation (4.98) is the Fourier heat flow model, where q is the heat flow, T is the temperature, x denotes depth (positive downward and hence the negative sign for heat flow), a and b are constants and K denotes thermal conductivity. Equations (4.99) and (4.100) are the basic assumptions that thermal conductivity at a reference temperature is proportional to P-wave velocity (equation (4.99); Karl, 1965) and that thermal conductivity is inversely proportional to temperature (equation (4.100); Clark, 1966; Schartz and Simmons, 1972). By combining equations (4.98)–(4.100) and expressing seismic velocity in terms of traveltime t, Houbolt and Wells (1980) obtained the following relationship between temperature and traveltime:

$$T_j = T_0 e^{aq(t_j - t_0)} + c(e^{aq(t_j - t_0)} - 1)$$ (4.101)

where $c = b/a$

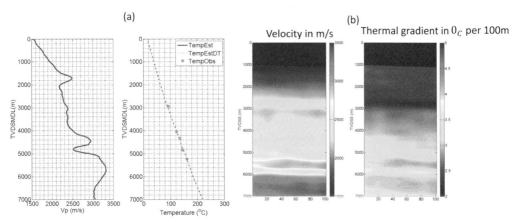

Figure 4.42 An example of prediction of temperature from velocity based on a steady state model of heat flow as discussed in the text. (a) Predicted temperature using a calibrated seismic velocity data that is shown on the left panel. The color dots on the right panel represent BHT data that was corrected for circulation effects using Horner's method. The dashed line shows predicted temperature versus depth profile. TVDSS denoted total vertical depth below sub-sea level. (b) Predicted thermal gradient data in 3D using seismic velocity. Left panel shows seismic velocity (m/s) with X-axis from 0 to 100 seismic traces with 50 m spacing (a total of 5 km). The target well is around the center trace #50. Y-axis is the true vertical depth in meters below mean-sea (TVDSS in m). The right panel shows estimated temperature gradient (in centigrade/100 m). The color bar is the estimated temperature gradient (degrees centigrade per 100 m). (A black and white version of this figure will appear in some formats. For the color version, please refer to the plate section.)

Here T_0 and t_0 are temperature and traveltime, respectively, at top and T_j, t_j are the temperature and traveltime, respectively, at the bottom of a depth interval labelled by the index j. From studies on several shut-in wells, Houbolt and Wells (1980) obtained global estimates of parameters a and c. These are $a = 1.039$ and $c = 80.031$. Heat flow is a sensitive parameter in the model and it was determined by bottom-hole temperature measurement in a borehole (calibration).

Equation (4.101) can be used to estimate subsurface temperature estimate at any depth from a known surface temperature and known traveltime (typically 0 s) and traveltime to the depth point of interest. An example of the application is given in Figure 4.42 for a well (offshore) for which velocity (left panel of Figure 4.42a) and BHT data as shown by red dots on the temperature versus depth plot of the right panel of Figure 4.42a were available. The BHT data were corrected for circulation effect using Horner's method. The 3D seismic velocity data is shown on the left panel of Figure 4.42b. The seismic data covered a cube of 5 × 5 × 5 km. The seismic velocity data was obtained using CIP tomography and the velocity inversion process was guided by rock physics as discussed in Chapter 6. The well was located around trace number 50 on the seismic velocity section shown on the left panel of Figure 4.42b. The right panel of Figure 4.42b shows a cross section of predicted geothermal gradient using the seismic velocity. This example shows that the predicted results are in excellent agreement with measured data for this mini-basin.

Alam and Pilger (1986) also present an interesting case study using this approach in South Louisiana. They correlated the variations of the predicted geothermal gradient (thermal anomalies) with growth faults and top of overpressured zone in the area of study. They speculated that the thermal anomalies coincided with the fluid (hydrocarbon) migration. This possibility is intriguing. However, it needs more analysis and validation.

In a recent paper, Duffaut et al. (2018) present an approach for temperature prediction from velocity data that is similar to that of Houbolt and Wells (1980). However, they replaced equation (4.99) with a velocity that explicitly depended on clay fraction – this is desirable. The authors suggest the following relation that relates temperature to velocity

$$T(x) = T_0 + \frac{q}{a_0} \int\limits_0^x \frac{dx'}{[1 - V_{cl}(x')]V_p(x')} \tag{4.102}$$

where, a_0 is a constant, q is heat flow, and V_{cl} denotes clay fraction (obtained from gamma-ray log). However they neglected the temperature dependency of thermal conductivity (equation (4.100)). So as depth increases, their thermal conductivity will remain the same. In reality, it should decrease with increasing depth. The error due to this could be significant at deeper depths.

The preceding discussions on the prediction of the thermal field from seismic velocity assumes that the heat flow in the sedimentary basin could be described by the steady-state assumption. This is not a valid assumption – and certainly not true for the Tertiary clastics of the Gulf of Mexico and similar basins. In addition, most of the velocity based approaches do not apply when salt and carbonates are present. These lithologies have high values of thermal conductivity and the relations are not applicable for these lithologies. Duffaut et al. (2018, p. 384) claim that their model (equation (4.102)) "performs reasonably well for other lithologies (diatomites, carbonates, salt, and basalt), assuming typically $V_{cl} < 0.2$." It is interesting but more studies are needed to validate the claim.

Basin modeling is an alternate technique that can be used for prediction of thermal fields. It is discussed in Chapter 10. Here most thermal transients due to sedimentation rate variations and fluid flow and pore pressure are included along with a variety of lithologies such as salt and carbonates. Effect of thermal transients in younger basins could be significant. An example is presented by Al-Kawai (2018) in Figure 4.43. The study area is shown in Figure 4.43a along with a seismic line and two wells with bottom-hole temperature data. Figures 4.43b and 4.43c show the temperature differences at the two well locations between two solutions at each location – a steady-state solution and the one that includes all *thermal transients due to variable sedimentation rates and overpressure generation*. The thermal transient model was based on a basin

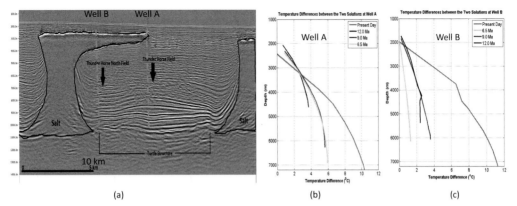

Figure 4.43 Effect of thermal transients on present-day temperature and its effect on chemical diagenesis. Vertical axes in (a) - (c) are depth in meters (a) Seismic line showing the turtle structure (right). (b) Basin modeling based calculations of temperature. The figures show the temperature differences between that predicted by a steady-state heat flow assumption and a model that includes *all* thermal transients at multiple time steps, including present day as indicated in (b) and (c). After Al-Kawai (2018). (A black and white version of this figure will appear in some formats. For the color version, please refer to the plate section.)

modeling approach (Hantschel and Kauerauf, 2009). The differences between the two solutions are significant (as large as 10°C at TD). This suggests that a combination of basin modeling and seismic approach may yield reliable estimate of temperature. However, more research and case studies are needed.

4.11   Summary

In this chapter we discussed various methods to quantify pore pressure, fracture pressure and overburden stress using a host of geophysical methods such as velocity, density and resistivity from well logs and seismic data. Most of these methods are based on approaches that work for shales. In shales, it is easier to define a trend based on normal compaction although defining such trends are occasionally complicated by the occurrence of shallow overpressure. Extending shale based methods to other lithologies is not an easy task. As noted earlier, in reservoir rocks such as sandstones, the pore pressure could be lower than, equal to, or higher than the adjacent shales. Some promising methods for carbonates are emerging and we discussed some of these in this chapter. The 3D structure of subsurface formations and the stress distribution influence the pore pressure regime. Occurrence of higher than hydrostatic pore pressure is a result of porosity reduction with burial. However, it is a complex process. It involves mineralogical changes within rocks that are a function of the thermal history of the rocks as well as the complex chemical interaction between the pore water and the minerals. While our understanding of the methods to predict pore pressure quantitatively is much better in clastics than in carbonates and other hard rocks (such as low porosity basement-type rocks), we still have a long way to go when we aspire to address the complete 3D models. Some of these issues are discussed in Chapters 13 and 14.

A major issue pertains to calibration of models as discussed in this chapter. This is not a simple matter. The quality of the calibration data from wellbore measurements are not easy to assess as the data (such as mud weight) have a large uncertainty that is directly related to drilling practices in an area. Drilling is a messy operation – of course, safety is of paramount importance here and second to none, including "useful data" collection. Naturally. This precludes using equivalent mud weight as a quantitative indicator of pore pressure. (It is at best an upper limit of true pore pressure.) Yet, most of the literature is full of this practice! (This includes many of our own publications!) Downhole pore pressure measurements are neither routine nor in preponderance. Furthermore, these are always made in permeable formations only and very rarely yield the correct subsurface pore pressure in low-permeability rocks such as shales. So the calibration data are lacking where we need it the most! We need a downhole tool that can measure pore pressure in low-permeability rocks.

Occasionally calibration points are based either on the occurrence of *kicks* (in which case the pore pressure in the permeable formation producing the kick must be higher than the equivalent mud weight and lower than the kill mud weight), or on observations of *instabilities* in shales (a subject that we did not discuss in this book as there are excellent books on the subject) (see, e.g., Aadnoy and Looyeh, 2010). In the former case, it is *assumed* (often erroneously) that the pore pressures in shales and the adjacent permeable formations are the same. In the latter case, the assumption is that instabilities occur when the mud weight has fallen below the pore pressure. In fact, wellbore instabilities that are due to compressive breakouts can occur at pressures that are *higher or lower* than the pore pressure. Therefore, the assumption that collapse begins to occur at a mud weight equal to the pore pressure can result either in an overestimate or an underestimate of pore pressure. If neither occurs, pore pressure is assumed to be less than the equivalent mud weight. It is an erroneous assumption.

With the exception of the Dutta-model, most of the methods require that the rock obeys a single, monotonic, compaction-induced trend, and that no other effects are operating. In reality, active chemical processes can increase cementation, leading to increased stiffness (higher velocities, and higher densities), which can mask high pore pressure, and increased resistance to further compaction, which can lead to erroneous prediction of the onset of overpressure. This is why we do not have an acceptable method for pore pressure prediction in high velocity rocks such as carbonates, although as noted some progress is being made. We discussed how elevated temperatures lead to a transformation of the predominant shale mineral during the compaction process. For example, an increase in temperature transforms a water-bearing smectite to a relatively water-free (and denser) illite. This transformation occurs over a range of temperatures near 100°C, but they can vary with fluid chemistry and burial conditions; furthermore, the depth at which this temperature is exceeded varies from basin to basin due to variable heat flow. Subsequent diagenetic processes such as cementation due to hot-water transport and subsequent reprecipitation inhibits reliable pore pressure prediction.

We briefly touched upon the subject of temperature estimation using models that relate temperature to velocity or traveltime data. These sorts of models are very useful as it makes it easier to deal with more complex geopressure mechanisms that require temperature or thermal history of rocks as input.

5 Seismic Methods to Predict and Detect Geopressure

5.1   Introduction

That velocity is a very useful tool for pore pressure prediction prior to drilling a well has been known for a long time (Bilgeri and Ademeno, 1982; Hottman and Johnson, 1965; Pennebaker, 1968; Reynolds, 1970; Reynolds et al., 1971; Dutta, 1987a, 1987b). The recognition was based on the fact that overpressured formations exhibit some of the following diagnostics when compared with the same formation had it normally compacted at that depth: (1) lower effective stress, (2) lower velocity, (3) lower bulk density, (4) higher porosity, and (5) higher temperature.

In Chapter 3 we discussed various mechanisms for overpressure. Some of these mechanisms – compaction disequilibrium, for example, yields a direct relationship between velocity and effective stress (and therefore, to pore pressure through Terzaghi's principle). Some other mechanisms yield a multivalued relation between effective stress and velocity (Bowers, 1994). Since the dimension of effective stress (MLT^{-2}) is different from that for velocity (LT^{-1}), it is obvious that any relation between effective stress and velocity must involve "empirical" coefficients with dimensions (MT^{-1}) that must be determined and calibrated. This complicates pore pressure prediction using velocity as a tool. A failure may be attributed to either an inappropriate model, a bad velocity estimate used as a probe or inappropriate data from field measurements. Another observation is that one needs an accurate velocity trend to denote the onset and magnitude of overpressure. This is further complicated by the spatial scale (or frequency content) of the velocity field as a tool for overpressure determination. It is a limitation, especially, when seismic velocity is used.

In this chapter, we shall identify various sources of velocity data – *checkshot, VSP, well logs, laboratory and seismic measurements*. The goal is to obtain velocity variations in 3D that not only reflects the subsurface structures in depth but also conveys the expected range of velocity variations that is compliant with rock physics principles, structural geology, and stratigraphy of formations and geopressure. This is not an easy task as the process is compounded by how the data are acquired, processed, and interpreted. We note that all velocity information are *inferred* from data – be it from core, borehole, or surface seismic. Velocity estimation from core data is fairly simple as length over traveltime defines the velocity in a controlled environment. Borehole logs such as sonic logs are very useful but are affected by conditions of the borehole when the data are acquired. Seismic velocity data are not affected by borehole but are neither unambiguous nor unique due to acquisition geometry and analysis techniques employed to estimate velocity. Furthermore, how the seismic data are acquired dictates whether the data are applicable for pore pressure prediction in a given geologic area. For example, seismic data acquired using narrow azimuth and limited offset are typically not applicable to such areas as subsalt. Besides, seismic reflection data are always acquired and processed with the primary goal of enhancement of geological structural interpretation and as such may not be suitable for pore pressure prediction without additional processing, interpretation and calibration. It should be remembered *that not all velocity models are created equal*. Some are not suitable for geopressure prediction and detection. This is the single most reason why earlier applications of seismic velocity failed to predict correct pressure regime in sedimentary basins.

In this chapter we shall discuss various ways to obtain velocity data for pore pressure analysis and point out how to establish a link between seismic traces that are recorded in "space and time" and the "space and depth" that is required by the drilling community. This chapter should be viewed as a short tutorial on the subject and is intended for use by a broad community of geophysicists, geologists, petrophysicists, and drillers who are not necessarily specialists on velocity analysis. We would like to raise an awareness of various techniques for velocity analysis so that we can appreciate whether the velocity data intended for pressure prediction are actually usable and would provide reliable prediction. We also provide some guidelines for best-practices. Since seismic data are noninvasive predictive tool for geopressure analysis, an investment in understanding its strengths and weaknesses should be worth the cost. This understanding is *essential* for the pressure prediction community and may even prevent "failure" of a pressure prediction project that uses seismic data as a predictive tool!

A high-level workflow for pore pressure prediction using seismic velocity is shown in Figure 5.1. It is basically a four step process: The first step includes a seismic data processing workflow ("processing") to enhance S/N and enable imaging, and is followed by a velocity analysis and imaging loop step where 3D velocity and 3D seismic image are iteratively obtained through a velocity model building and imaging loop. Throughout the velocity model building process additional information may be injected so the model can honor not only seismic observations but also checkshot, marker and/or well log data ("interpretation"). Once a suitable velocity model in depth is derived, in the next step of the workflow a model that links velocity to pore pressure ("transform") and a set of input parameters associated with the model are chosen based on the geological model. Lastly, a calibration step based on wellbore data ("calibration") concludes the four step workflow.

There are uncertainties associated with each step and the reliability of the predicted pore pressure is impacted by a composite uncertainty – not necessarily a linear one as the error propagation throughout the workflow is nonlinear. Since seismic velocity is the driver of the workflow, the accuracy of predicted pore pressure is mainly impacted by the velocity field. This velocity field must reflect changes with pore pressure. For this to happen, the seismic model building step must deal with separating *imaging velocity* from the velocity that is close to *rock velocity* – a differentiation that is often

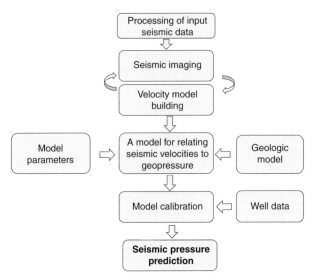

Figure 5.1 A high-level workflow for pore pressure prediction using seismic data. See text for details.

missed by the seismic data processors. Rock velocity refers to the velocity of wave propagation in a piece of rock (Al-Chalabi, 1994; Al-Chalabi and Rosenkranz, 2002) and is related to the elastic constants of the rock. On the other hand, seismic measurements of velocity are derived (inverted) from an average traveltime over the horizontal distance through which the seismic energy travels. Hence, the two velocities may *not* be the same. The process of extraction of *rock velocity* from *seismic velocity* is neither unique nor devoid of pitfalls. Simple tie to well velocity at a given location is the most common procedure in the seismic community; however, it is just a part of the story but not the complete story. What about the depth beyond where the well ends? What about the space away from the well or between the wells? The rest of this chapter deals with some of the steps of the workflow outlined in Figure 5.1 and especially, concentrates on the characteristics of seismic velocity-depth functions that *mimics* the rock velocity and how to accurately determine these from surface seismic data for geopressure prediction.

5.2   Measurements of Velocity

5.2.1   Introduction

Predrill geopressure prediction always involves application of seismic-based *interval velocity* analysis at depths, thanks to the pioneering works by Hottmann and Johnson (1965) and Pennebaker (1968) who set the stage for the applications of velocity to geopressure prediction. None of these pioneers were seismic analysts – Hottmann and Johnson were petrophysicists and Pennebaker was a drilling engineer. Since then, the broader seismic community took notice of these works and promoted some of the practices that are still in use. However, the jargon created by seismic experts were not made transparent to the drilling community. Consequently, they shied away to a great extent from this then emerging technology. The knowledge dissemination gap thus created continues even today and it is getting wider by the rapid deployment of various high-end seismic analysis techniques. Cross-fertilization of interdisciplinary activities did not happen. Below, we shall address many facets of seismic velocities with the goal of making some of the jargons palatable to our sister communities (petrophysicists and engineers) not specializing in seismic tools, beginning with *direct* measurements of velocity (required for calibration) and ending with high-end seismic analysis methods for high-frequency velocity modeling appropriate for geopressure quantification.

5.2.2   Laboratory Measurements

When a mechanical force is applied to a material, it will experience a change in volume and/or shape. If the deformation is elastic, it implies that once the applied force is removed, the material will return to its original volume and shape. As a result, elastic deformation conserves energy. Elastic deformation is commonly described in terms of stress and strain. In Chapter 2 we discussed these in detail. From the perspective of a rock physicist, velocity is a well-defined and unique concept – it is the velocity with which elastic energy propagates through a piece of rock of finite dimension. It can be measured *directly* using laboratory equipment. For elastic, isotropic, and homogeneous rocks, the rock velocity for compressional or P-wave V_p and shear or S-wave V_s are given by

$$V_p = \sqrt{\frac{K + \frac{4}{3}\mu}{\rho_b}} \tag{5.1a}$$

$$V_s = \sqrt{\frac{\mu}{\rho_b}} \tag{5.1b}$$

and

$$\rho_b = \phi\rho_f + (1 - \phi)\rho_r \tag{5.2}$$

where ρ_b is bulk density, ϕ is porosity, ρ_f is fluid density, ρ_r is the grain density of the solid grains, and K and μ denote the bulk modulus and the shear modulus of the material, respectively. In the laboratory setting these velocities are measured directly on a sample of about 5–6 cm in length and ~ 2 cm in diameter by sending an acoustic pulse in the ultrasonic frequency range and measuring the arrival time as shown in Figure 5.2a. The sample is usually under a uniform confining stress (simulating overburden stress) while the pore fluid (oil, brine, or gas) can be maintained at a different pressure. The velocity is obtained by dividing the length of the sample by the transit time of the pulse. The velocity thus obtained is a function of the rock and the fluid properties as well as the stress state of the sample. The velocity thus determined is *dynamic* – it is frequency dependent and measurements are made under very *low* strain amplitude. Thus elastic

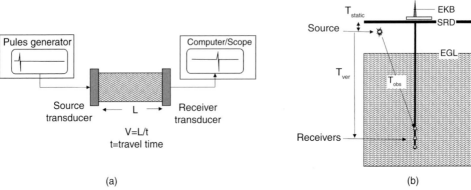

(a)

Laboratory measurement of velocity

(b)

Checkshot geometry – straight well

Figure 5.2 Measurement of velocity in (a) laboratory and (b) field using checkshot that shows the geometry for an offshore environment. For deviated wells, we need to make ray path corrections as discussed in the text. EKB: elevation Kelly Bushing; SRD: subsea rotating device; EGL: elevation ground level.

constants determined from these measurements are known as *dynamic* elastic constants. A *direct* measurement of elastic constants, for example, applying stress on a sample and measuring the resulting strains is known as the *static* elastic constants. The two are not the same – the static measurements are frequency independent but the measurements employ very large strain amplitudes. Geomechanical models describing borehole stability require static elastic constants and pore pressure estimates. Gardner et al. (1974) showed that the dynamic velocity measurements (core, log, seismic) yields velocity data that are *directly* related to the *effective stress* defined as the difference between the confining stress and the stress on the pore fluid. The subject of relating static measurements of elastic constants to dynamic measurements is not an easy one. The relations are mostly empirical and usually restricted to reservoir intervals. We address some of these relations in Chapter 12.

The ratio of the P-velocity to S-velocity is

$$\frac{V_P}{V_s} = \sqrt{\frac{K}{\mu} + \frac{4}{3}} = \sqrt{\frac{2(1 - \sigma)}{1 - 2\sigma}} \tag{5.3}$$

where σ is the (dynamic) Poisson's ratio. Both σ and V_P/V_s are diagnostics of rock and fluid properties as well as the effective stress and pore pressure. S-waves propagate through materials more slowly than P-waves. In addition, S-waves cannot propagate through fluids, as fluids do not support shear particle motion (transverse to wave propagation direction). In Chapter 2 (Figure 2.32b) we showed the sensitivity of Poisson's ratio and P-impedance (a product of velocity of P-wave and bulk density) to pore pressure for various pore fluids (brine, oil, gas). This plays a fundamental role on the extraction of Poisson's ratio from seismic inversion for pore pressure analysis. These extracted parameters from seismic inversion are more detailed (higher frequency than the conventional seismic velocity analysis) and can be related to the effective stress and hence, pore pressure, provided an independent assessment of overburden stress is made separately.

For pore pressure applications, there is a vast literature on the use of conventional P-velocity. We shall address some of these. However, very little is done with S-velocity. In our opinion, this has to do with the high cost and complexity of multi-component data acquisition and processing, either using borehole seismic or surface seismic data. Both of these require additional multicomponent sources and receivers. However, *combined* use of P- and S-wave velocity data can discriminate between pore pressure and fluid content such as gas. Both of these cause lowering of P-velocity. However, S-wave velocity is lowered by elevated pore pressure in brine-filled rock but not so much by change of fluid content (except for a slight increase in velocity due to change in bulk density – hydrocarbon replacing brine). (See Figure 2.32.)

5.2.3   Velocity from Checkshot Survey

Checkshot measurements belong to a class of measurements known as *velocity survey*. The purpose of a velocity survey is to produce a down-going seismic wavelet at the surface near a well and then to measure the time required for that wavelet to travel to a known depth where a seismic receiver is positioned in the well. The borehole receiver is locked successively at several different depth levels, and the vertical traveltime to each level is measured (Anstey and Geyer, 1987). There are various types of borehole measurements of velocity. These include checkshot and vertical seismic profile or VSP of various geometries. Checkshots are direct transmission measurements at seismic wavelengths of traveltime from a surface source to a receiver at a known depth. These data are most widely used to produce time–depth relationship which is then used to map the vertical well in depth into two-way seismic traveltime. The time–depth velocity curve can be inverted into interval velocity (e.g., Lizarralde and Swift, 1999) and used to calibrate sonic and seismic velocities. These are essential for reliable pore pressure estimates. In Figure 5.2b, we show a typical checkshot geometry for a vertical well. The receiver is a geophone anchored at a predetermined depth in the vertical well. This eliminates the time–depth uncertainty as the depth is known. The measured traveltime and depth can be inverted to interval velocity or used as additional information to constrain seismic tomography as will be discussed later.

Traveltime measurements from checkshot surveys as shown in Figure 5.2b are geometry dependent. This means checkshot data from deviated wells must be corrected for geometry. If one has a velocity model (e.g., layered $V(z)$) then ray tracing is used to "verticalize" the traveltime such that $T_{\text{vertical}} = f(T_{\text{observed}}, h, z, V(z))$. When velocity is constant under straight ray approximation the relationship is simplified to a simple geometrical correction such that

$$T_{\text{vertical}} = T_{\text{observed}} \cos\{\tan^{-1}(h/z)\} \tag{5.4}$$

This correction is strictly valid in 1D (i.e., $V = V(z)$). When the horizontal offset h is large or when complex settings are expected, this must be corrected for a 3D geometry for practical use. Since the frequency content of the source in a checkshot survey is very similar to that for surface seismic data, it is used to calibrate seismic data since there is very little ambiguity in the location of the receiver (depth in a borehole where the geophone is anchored). Attention, however, must be paid for analyzing checkshot data at the shallow part of the borehole as the data are often affected by the borehole condition. For land data, the checkshot times are often affected by near-surface conditions (known as "static" effects in the seismic jargon). This static shift (based on prestack time imaging methods) could be as large as 100 ms. This must be taken out of the checkshot based time–depth pairs in the shallow section by performing depth domain migration of the image. Failure to do so will result in erroneous tie of the seismic and checkshot times (and hence, the velocity for pore pressure prediction using seismic and zero-offset VSP). It is advisable to get hold of raw checkshot times and process the data rather than obtaining a list of "processed" checkshot times. This is a best practice. In this case, "trust but verify" is a good mantra.

5.2.4   Velocity from Vertical Seismic Profile (VSP)

Like checkshot surveys, VSP is a measurement where the seismic source (air gun for offshore and dynamite or vibroseis for onshore) is at the surface but the receivers (geophones) are lowered down into a borehole. The major difference between a VSP and a checkshot survey is that VSP data are recorded at much *smaller* spatial sampling intervals than checkshots and the receivers are placed at *regular* intervals (see, e.g., Balch and Lee, 1984; Hardage, 1985). The authors noted that specifically, the vertical distance between successive VSP traces should not exceed one-half of λ_{min}, where λ_{min} is the shortest wavelength contained in the recorded VSP wave field.

When a seismic wave field is recorded with this small spatial sampling interval, several processing techniques can be used to separate the down-going and up-going wave fields (Hardage, 1985). Once the up-going wave field is isolated from the more dominant down-going wave field, the up-going reflection events can be properly analyzed and interpreted and used to produce an improved image of the subsurface. The main advantage of this type of measurement is that we know at which depth the receiver is, while from surface seismic data we have to infer it from the time and the seismic velocity (which may be inaccurate). Furthermore, seismic velocity and imaging analysis are based on traveltime data recorded in *two-way* travel by the wavelet. This causes ambiguity in extracted depths. These differences are further exacerbated by various algorithms used for migration of seismic data that assist in locating depths of migrated images. Excellent discussions of seismic processing is contained in Yilmaz (2001).

There are many types of VSP acquisition techniques: Zero-offset (coincident shot and receivers) as shown in Figure 5.3a, offset VSP shown in Figure 5.3b where the source is at a large distance from the location of the borehole, and walk away VSP in which the acquisition is made by moving the source away from the borehole and

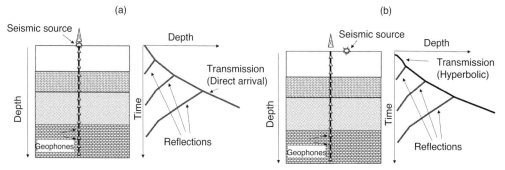

Figure 5.3 Vertical seismic profile (VSP) geometry. (a) Zero-offset geometry where the source and well location with receivers are mostly coincident. (b) Offset-geometry where there is a large lateral distance between the source and the well with receivers.

keeping the receiver at a fixed depth in the borehole. In a layered earth, the reflection points associated with a zero-offset VSP occur close to the vertical line passing through the source and receiver coordinates. Thus, the image made from these data will illuminate the subsurface in only a narrow vertical corridor passing through the receiver location. However, if there is structural dip, the reflection points associated with a zero-offset VSP can occur at significant horizontal distances from the vertical line passing through the source and receiver. When properly processed, such data can produce high-resolution images extending from the receiver position to the farthest reflection point coordinate. Interval velocities from VSP are very useful for pore pressure analysis. It is usually better than velocities obtained from inversion of surface seismic data (tomography, for example).

Zero-offset VSP yields higher accuracy vertical compressional velocity as the dense time–depth data pairs stabilize the inversion to interval velocity (Lizarralde and Swift, 1999). In *offset* VSPs we can obtain *both* compressional and shear velocity information. In addition to velocity information in VSP, processed VSP also provides a seismic trace that ties waveform changes in time to formation boundaries in depth. This is useful information to calibrate seismic data. Pore pressure based on VSP data is usually very reliable – more so than the surface seismic or sonic logs since the VSP data have higher resolution (higher frequency content that overlaps with the frequency content of the seismic data) and contains low frequency information as well that is lacking in the sonic logs.

5.2.5   Velocity from Sonic Logs in Borehole

Velocity from sonic logs are very useful for quantitative analysis of geopressure. Sonic tools (monopole and dipole) are designed to measure the wave velocities of the formation surrounding the borehole. In essence, the sonic tool can be considered as a miniature seismic refraction experiment carried out within the (assumed) cylindrical borehole. There are various types of sonic tools available in the industry such as the Borehole Compensated tool (BHC) and the Long-Spaced Sonic tool (LSS). The sonic

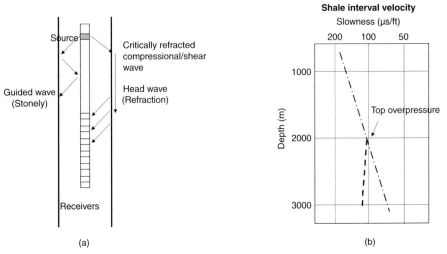

Figure 5.4 Schematic of (a) Borehole Compensated (BHC) Sonic tool showing wavefront arrivals and (b) an example of interpreted sonic data for geopressure detection. Deviation from the trend below depth marked *top overpressure* where velocity is lowered is indicative of geopressure.

tool is centered in the hole by means of centralizers, and contains one or more sources and receivers as shown in Figure 5.4a. A source radiates acoustic energy, which is transmitted into the borehole fluid. When the wavefront impinges on the borehole wall, a refracted compressional wave is generated. If formation shear velocity is higher than the acoustic velocity of the fluid, a refracted shear wave will also be generated. The refracted waves travel along the borehole wall, reradiating energy into the fluid. Energy arrives at receivers on the logging tool at a time that is linearly proportional to their distance from the source. Thus, formation elastic-wave velocities can be determined by measuring the difference in arrival times, Δt, at two receivers a known distance apart. The measurements are expressed in units of microseconds per foot or per meter. The velocity is calculated by dividing the transmitter-receiver distance (fixed) by the traveltime Δt, from the transmitter to the receiver. To compensate for the variations in the drilling mud thickness, there are actually two receivers, one near and one far. This is because the traveltime within the drilling mud will be common for both, so the traveltime within the formation is given by $\Delta t = t_{far} - t_{near}$ where t_{far} is traveltime to the far receiver and t_{near} is traveltime to the near receiver.

Quite often it is necessary to compensate for tool tilt and variations in the borehole diameter. In this case both up-down and down-up arrays can be used and an average traveltime is calculated (Sheriff and Geldart, 1995). In the early days, the sonic tools operated at about 30 KHz. Newer tools use lower frequency, between 3 and 5 KHz. Acoustic log source types fall into three categories: monopole, dipole, or quadrupole. In Figure 5.4b we show an example of using velocity data from LSS tool (monopole) for pore pressure analysis. (Deviation from the trend below depth marked *top overpressure* where velocity is lowered is indicative of overpressure.)

Some cautionary notes on sonic tools are worth notice. Velocities from sonic logs are very useful although the information on velocity is obtained at shorter wavelengths – much shorter than the VSP and surface seismic but much larger than the core samples on which direct laboratory measurements are carried out. In sonic logs, one actually records the refracted signals from the walls of the borehole and thus, it is affected by the conditions of the borehole such as borehole rugosity and mud penetration into the formation during the drilling process. An additional way in which the sonic log tool can be altered is increasing or decreasing the separation between the source and receivers (LSS tool). This gives deeper penetration and overcomes the problem of low velocity zones posed by borehole wall damage. Thus velocity data from sonic logs needs to be corrected for borehole conditions including mud penetration into formation through permeable pathways. This is known as *environmental correction*. No such corrections are needed for either VSP or surface seismic data.

One common problem with the sonic tool is cycle skipping: a low signal level, such as that occurring in large holes, gas prone intervals and soft formations, can cause the far detectors to trigger on the second or later arrivals, causing the recorded Δt to be too high (velocity too low). Therefore, the accuracy of sonic logs should always be checked prior to using the recorded traveltimes for pore pressure analysis. It is our experience that often regular- and long-spaced log measurements conflict, and this should be taken into account when there are disagreements between seismic data and sonic log data. Other considerations include that the resolution of sonic logs is several feet whereas the resolution of seismic reflection data is on the scale of hundred feet or so and the two methods use significantly different frequency ranges. This dependence of velocity on frequency is known as *dispersion* effect. It is for these reasons sonic logs *must* be corrected by checkshot data prior to using the data for well-seismic tie and definitely, pore pressure calculations. This is imperative. In order to investigate how the varying size of a borehole has affected a particular sonic log, the results should be plotted against those of a caliper log (a log that measures the borehole radii and informs the user about damage to the borehole). In general, data from any sonic logging tool must be accompanied by a caliper log for quality control. This is an important requirement. To summarize, sonic log-based traveltimes should be corrected for environmental corrections, dispersion and calibrated with checkshot data prior to using the velocity data for geopressure applications. We interject a comment based on our experience: while the petrophysicists are well aware of this and practice the procedures carefully, the seismic community often lagged the inquisitiveness to query the data quality of sonic logs, especially when using such data from unknown or undocumented sources. (Simply overlaying sonic velocity data on seismic velocity data is not a good practice!)

In the context of reliable velocity extraction from sonic and VSP measurements for pore pressure analysis we need to assess the reliability of the data and the scale of measurements. There are several reasons why VSP interval velocities may be different from sonic logs (Stewart, 1984). These include the volume of rock sampled by each method, instrumental biases, and wave propagation effects such as anisotropy, short-path multiples, and velocity dispersion (dependence of velocity on frequency). There is some interesting physics behind some of these potential discrepancies between VSP and sonic logs. However, a discussion on the subject is beyond the scope of this book and we refer the readers to the excellent book by Hardage (1985). Because of the larger rock volume of investigation, most of which may be considered to be undisturbed by the presence of the borehole, VSP is often considered to be more reflective of the true formation properties. Thus, the calibration of sonic logs to VSP is also of importance prior to using sonic logs for pore pressure calculations. Often the pore pressure estimates from sonic log is accepted as *ground truth* against which all other measurements are benchmarked. This may not be a good practice unless sonic logs are assessed for quality. A necessary requirement is to use only checkshot calibrated sonic log derived velocities for pore pressure applications. In our experience this best practice is seldom followed due to various reasons, a chief one being unavailability of checkshot data. It must be remembered that the sonic log gives a one-way traveltime, and the seismic technique gives a two-way traveltime. Besides, we will see below that the extent of rocks sampled by the two techniques are quite different and corrections employed to account for differences such as intrinsic rock anisotropy (variation of velocity with the direction of propagation of waves) could be different.

5.3   Seismic Velocity from Traveltime Analysis and Anisotropy

5.3.1   Introduction

Seismic velocity is an ill-defined term. Seismologists have derived a large number of different types of velocity such as average, interval, root-mean-square (rms), instantaneous, phase, group, normal moveout (NMO), Dix, stacking, and migration velocities such as tomography, and velocities from inversion of seismic traces and gathers such as poststack impedance, prestack and full-waveform inversion (FWI). A very good discussion of basics of seismic velocity is provided by Yilmaz (1987, 2001) and a list and discussions of various velocities are provided by Bell (2002, see Table 1). These references indicate that seismic velocity is full of jargon and needs some elaboration. Below we discuss some of these in the context of application to geopressure. The purpose of this short tutorial is that the user should be aware of the various types of seismic velocities and their usefulness and limitations for pressure prediction.

5.3.2   Velocity from Normal Moveout (NMO) of Seismic Gathers

Normal Moveout (NMO) Analysis

In reflection seismology, the key parameters are: source to receiver distance (offset), x, and traveltime from shot to receiver, t, as shown in Figure 5.5a, for a single *horizontal* reflector and in Figure 5.5b for multiple horizontal reflectors. At a given midpoint location in Figure 5.5a, the traveltime along a ray path from source to the depth point and then from the depth point to the receiver is $t(x)$. It is easy to show (using Pythagorean Theorem) that the following relation between $t(x)$ and x holds:

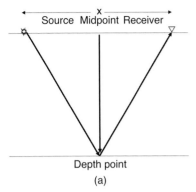

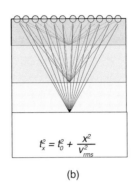

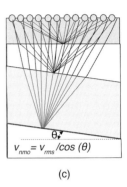

(a)   (b)   (c)

Figure 5.5 Schematics of the basics of seismic velocity analysis. (a) Shows the concept of normal moveout (NMO). (b) Shows normal moveout in a flat layered earth model. (c) Normal moveout in a dipping layered earth model.

$$t^2(x) = t^2(0) + \frac{x^2}{V^2} \tag{5.5}$$

where V is the velocity (assumed to be constant) of the layer (medium) above the reflector, and $t(0)$ is twice the traveltime along the vertical path. Equation (5.5) describes a *hyperbola* in the plane of two-way time versus offset. The difference between the two-way time at a given offset $t(x)$ and the two-way zero-offset time $t(0)$ is called Normal Moveout or NMO, and the velocity that determines the NMO is called the *Normal-Moveout Velocity* or V_{NMO}.

The above formulation for traveltime is valid for a single horizontally flat reflector with a constant velocity. For a horizontally stratified earth (Figure 5.5b), Taner and Koehler (1969) derived the following equation for the traveltime:

$$t^2(x) = C_0 + C_1 x^2 + C_2 x^4 + \dots \tag{5.6}$$

where $C_0 = t^2(0)$, $C_1 = \dfrac{1}{V^2_{RMS}}$, and C_2, C_3, ... are complicated functions that depend on layer thicknesses (Δt_1, Δt_2, ...) and *interval* velocities (V_1, V_2, ...). Here V_{RMS} is the root-mean-square velocity down to the reflector on which depth point D is situated, and it is defined by

$$V^2_{RMS} = \frac{\sum\limits_{i=1}^{i=N} V_i^2 t_i}{\sum\limits_{i=1}^{N} t_i} \tag{5.7}$$

where V_i is the velocity of the ith layer, and t_i is the two-way, zero-offset traveltime of the ith layer. The summations are over all layers from the surface to the reflector.

A further approximation to the traveltime equation given in equation (5.7) can be realized by assuming that the offset is small compared to the depth (known as "small-spread approximation"). In this case, the series in equation (5.6) can be truncated as follows:

$$t^2(x) = t^2(0) + \frac{x^2}{V_{RMS}^2} \tag{5.8}$$

This may be compared with the well-known NMO equation for small offsets:

$$t^2(x) = t^2(0) + \frac{x^2}{V_{NMO}^2} \tag{5.9}$$

Comparing equation (5.8) with equation (5.9), we note that $V_{NMO} \approx V_{RMS}$ for small-spread approximation. For the larger acquisition cables employed these days, higher-order terms in equation (5.6) become important and should be included as a part of the velocity analysis. It is obvious that doing so implies that the moveout or the traveltime equation will no longer be hyperbolic.

Equation (5.8) contains the basis for velocity analysis for a common midpoint gather or CMP gather. This describes a straight line on the t^2 versus x^2 plane. The slope of the line is V_{RMS}^{-2}, and the intercept value at $x = 0$ is $t(0)$. In practice, a least-squares method is used to do the curve fitting. Equation (5.7) yields interval velocities V_i from the RMS velocities. This process is called velocity inversion. The traditional form of the Dix's *interval velocity* equation (Dix, 1955) for the nth layer V_n in terms of the RMS velocities and traveltimes (valid for horizontal layers) is

$$V_n^2 \approx \frac{1}{t_n} \left(V_{RMS,n}^2 \sum_{i=1}^{n} t_i - V_{RMS,n-1}^2 \sum_{i=1}^{n-1} t_i \right) \tag{5.10}$$

Here, $V_{RMS,n}$ is the RMS velocity of the nth layer and t_i is the traveltime of the ith layer. These interval velocities are our best estimates from surface seismic data for the desired rock velocities. However, these need to be calibrated and compensated for anisotropy (see section 5.3.3). There are other limitations as well: accounting for dips and structures, lateral velocity variations, multiples and other seismic noise due to faults, diffractions, and sideswipe. It is important to note that Dix equation employs the following assumptions: (1) layers are horizontally stratified and isotropic, (2) offsets are small so that RMS velocity is a good approximation to the moveout velocity, and (3) the interval velocities are themselves averages over velocities of thinner layers lying below the resolution of the seismic data. Although interval velocities obtained from Dix's inversion is popular, it is also the most restrictive and does not correspond to realistic geology. Its usefulness for pore pressure prediction is also fraught with problems. In early days our nonseismic communities did not grasp the importance of such problems and were disappointed with the results from geopressure analysis.

What if the layers are not flat but dipping as shown in Figure 5.5c? When the reflector is dipping at an angle θ from the horizontal, the traveltime equation in the vertical plane along the dip becomes (Levin, 1971)

$$t^2(x) = t^2(0) + \frac{x^2}{(V/\cos\theta)^2} \tag{5.11}$$

or $V_{NMO}(\theta) = V/\cos\theta$. Again, $t(0)$ is the zero-offset, two-way traveltime. For the dipping layer, the midpoint is no longer a vertical projection of the depth point to the surface as is the case with a flat layer. Thus the terms CDP (Common Depth Point) and CMP (Common Mid-Point) gathers are equivalent *only* for horizontally layered media. When there is subsurface dip (or for laterally varying velocity field) the two are different. Proper stacking of a dipping event requires a velocity that is greater than the velocity of the medium above the reflector. This suggests that a horizontal layer with a high velocity can yield the same moveout as a dipping layer with a low velocity and, hence, can yield the same stacks in the small-spread approximation. This ambiguity can result in a different stacking velocity function, for seemingly similar looking stacks (Levin, 1971). This will also impact its use for geopressure analysis. Modern processing techniques involving *dip moveout* (DMO) usually account for this effect and handles the dip-conflict. Thus, any pressure analysis work should be carried out on *migrated and DMO-processed* seismic data.

Seismic processors and interpreters often work with various forms of seismic gathers such as *common shot and receiver gathers, common-offset gathers*, and *common mid-point gathers*. These are diagrammatically shown in Figure 5.6 under the title "stacking chart." This nomenclature should be understood in the context that many algorithms for

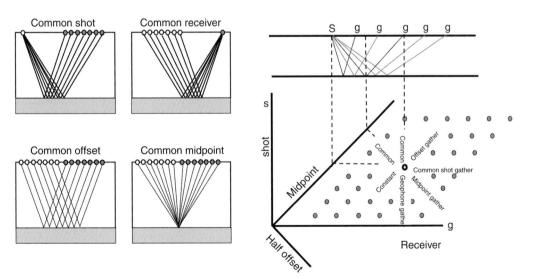

Figure 5.6 Schematic of seismic gathers (left) and stacking chart (right) for 2D configuration as explained in the text.

velocity inversion use some of these types of data sorting for the sake of compliance with the mathematics behind a specific inversion process (AVO, for example).

Basic Time Domain Seismic Velocity Analysis

Normal-moveout analysis is the fundamental basis for determining interval velocity from seismic data. These velocities are used to align reflections in the traces of a CMP gather prior to stacking. This yields the stacking velocity function from equation (5.9) in the (t, x) plane as shown in Figure 5.7. Shown on the left is the typical semblance analysis window. It shows how various velocity functions (horizontal axis) are used to define the *best-fit* hyperbola at a particular traveltime (vertical axis). Also shown are the best-fit rms and the interval velocity functions. The gathers in the middle are displayed for quality control to ascertain whether the moveout velocity flattens the gather at all offsets. These gathers indicate seismic events, which are flattened to estimate the interval velocities shown on the left panel. This is an essential display during velocity analysis; it suggests how flat the events are and how much freedom one has in changing the velocity field – within the signal-to-noise (S/N) ratio in the data, yet keeping the seismic events flattened. This is the standard practice. However, picking velocity functions based on maxima in the semblance analysis will yield *nice looking stacks*, but may not necessarily be the best velocity for geopressure analysis. This criterion tends to yield velocities that increase systematically with increasing depth but may ignore overpressuring which, on the other hand, suggests that the velocity may not always increase with increasing depth. The panels on the right side of Figure 5.7 show velocity fields for adjacent gathers. These displays are also important; they indicate the spatial consistency of the velocity field. Any velocity analysis for geopressure

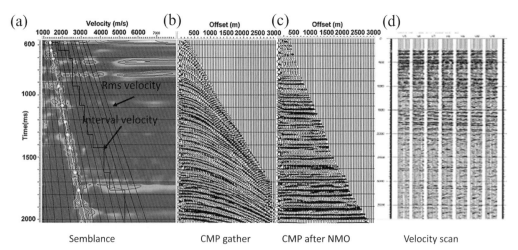

Semblance   CMP gather   CMP after NMO   Velocity scan

Figure 5.7 Basics of velocity analysis. (a) Shows semblance analysis as a guide for best-quality velocity for the seismic gather shown in (b) before NMO and (c) after NMO. (d) Shows velocity scan for multiple gathers. High-speed computers can produce these scans in 3D quickly. This is often the first step in velocity model building using various higher-end velocity inversion techniques.

prediction work must include this step indicating lateral continuity of geology, unless changed by faulting or other geologic features such as pressure compartments. However, such geologic events must have a seismic signature in many gathers, and not just one; otherwise such signatures are false indicators of geologic features and should be discarded while interpreting the velocities. There is a great deal more on conventional velocity analysis such as Horizon Consistent Velocity Analysis (HCVA), Spatially Continuous Velocity Analysis (SCVA), and Dense Velocity Analysis (DVA) and the readers can learn more from the literature (Yilmaz, 1987, 2001; Bell, 2002; Huffman, 2002; Dutta, 2002a, 2002b; Toldi, 1985; Mao et al., 2000).

The interval velocity estimation from seismic stacking velocity analysis has limitations mainly because of the S/N ratio of the gathers, muting, the spread length used for data acquisition, the stacking fold, the choice of coherency measure, true departure from hyperbolic moveout (e.g., anisotropy), the time gate length, the bandwidth of the data, and complex structure with spatially variable velocities. In general, as events on a velocity analysis are picked deeper (in time), the quality of the velocity function degrades. This is because one encounters a range of velocity functions, which flatten an event, creating ambiguity and lack of precision in the picked velocity function. In addition, the range of usable offset angles on the seismic gathers decrease as depth increases; this causes ambiguity in the flatness of gathers. (It is very difficult to flatten a gather at deeper depths when one is sampling only a limited range of offsets, mainly due to cable length limitation.)

Although time domain velocity analysis techniques are used less and less for pore pressure prediction – high-end velocity analysis techniques such as tomography and FWI are more popular – it will not go away. There are still a lot of old seismic data around (e.g., onshore) that cannot be utilized for high-end VA techniques because of how they were acquired. However, these can be re-processed and used for pore pressure prediction with well control. Acquiring new seismic data is not always the viable option to the industry. In addition, most of the tomography or FWI methods need a *good starting model* for velocity for further analysis – otherwise the inherent nonuniqueness problem associated with such techniques can lead to wrong results. With this in mind, we provide below some *guidelines* on how to use the data from conventional time domain velocity analysis for pore pressure prediction.

Guide to Time Domain Velocity Analysis for Pore Pressure Prediction

Quite often, velocities are misused for pressure prediction; stacking velocities need much conditioning before being used for this purpose. Before work begins, a clear understanding of the purpose of the work must be defined: Are the velocities needed for a "regional" understanding of pressure (typically, a grid of 100 × 100 mile)? A detailed image of subsurface pressure at the prospect scale (3 × 3 mile)? At a reservoir scale (100–200 ft layer thickness)? At a wellbore scale (several feet layer thickness at target depth, say, 10,000 ft)? The care and details employed for each of these scales vary and require integration of a host of data other than the seismic, such as well velocities, well logs, and geology. Velocity analysis is a tedious process and is usually carried out repetitively; and at each step additional data and interpretation are incorporated to improve resolution and accuracy. The steps below are intended to provide a guide to the process.

1. A general understanding of the geology should precede any velocity analysis. Using stacked and interpreted seismic sections does this best. All available wells must be posted at appropriate locations along with the key geologic horizons.

2. Conventionally processed stacking velocities are usually unsuitable for pressure prediction work, since they are created for imaging and may have very little to do with the rock velocity. Calibration is an important step.

3. The seismic gathers must be available for quality control and examined for S/N analysis.

4. The processing flow must be clearly identified on the seismic section. The velocities must be processed for de-multiple, DMO and prestack migration (DMO is not necessary if Kirchhoff migration is used). For areas with salt, such as in the deepwater Gulf of Mexico, one must apply appropriate salt mask and mute.

5. The stacking velocity, RMS velocity or the interval velocities given at the top of a stacked seismic section should not be used for pressure prediction work, without looking at the gathers and semblance or velocity spectrum plots for quality assurance.

6. Detailed velocity analysis is best done on a workstation. Essential steps include closely spaced velocity analysis, lateral consistency in the velocity field, smoothing, calibration and interpretation.

7. Usual criterion for picking the velocity (such as semblance maximum on the velocity spectra) may not be suitable for pressure prediction work. Picking that velocity which flattens an event from *near to far* offsets works better. If an event cannot be flattened across its full offset range, then every effort must be made to optimize flattening it as far an offset range as possible (say from near to middle range).

8. Velocity calibration is an essential step in conditioning velocities for pressure prediction. Comparing seismic RMS and interval velocities with those from checkshot surveys does this best. However, if checkshot data are not available, a notion of rock velocity must be used to constrain the velocity field, if possible, from analogue studies. A rock physics template (discussed below) is also a valuable tool to condition the velocity versus depth profile.

9. We should never make a prediction based on a single velocity function, say, at the well location. The velocity field must be checked for lateral and temporal consistency, so that any wild fluctuation or spikes are not present. We recommend that any velocity analysis for pressure prediction work should proceed with examination of at least a dozen (possibly, more) seismic gathers around each site of a calibration or pseudo-well.

10. Velocity picking should be done in several steps of successively increasing detail. Most detailed picking should be event oriented to ensure that no bias is introduced by Dix-calculations for intervals of varying thicknesses.
11. Quality control displays such as NMO-corrected gathers should be used to check accuracy of velocity picks.
12. We should not pick velocities in layers with thickness (in time) less than 50 ms, especially at relatively large depths (say, at 3 s two-way time or more).
13. For a regional scale velocity analysis for pressure prediction work, the velocities should be picked at least in a 1 × 1 km grid or less. Special care must be exercised while interpolating the velocity field. This is because the interpolation process can exaggerate any velocity spikes due to anomalous picks.
14. Velocities must be smoothed by a simple mathematical function, such as a low-order polynomial. Complicated smoothing algorithms, such as splines, may not be worth the trouble they may pose. These functions tend to follow the undulations in the velocity field too faithfully and sometime *create* "geology" when none is present.
15. The conventional velocity analysis as discussed so far uses the Dix model. It assumes a layered earth and ignores ray path bending while flattening a seismic event. This is unphysical. The ray path bending will cause a stack of flat layers to behave as if it were apparently anisotropic, although each layer may be intrinsically isotropic (Yilmaz, 1987). The resultant effect is to cause nonhyperbolic moveout at larger offset ranges. In the jargon of the velocity analysis software packages, this is often referred as the "residual moveout correction." This correction is very important and must be accounted for before any pressure analysis work progresses. This usually requires large-offset data.
16. Velocity structures observed within a spread length must be investigated carefully. These may not be due to geologic variations.
17. Every effort must be made to relate the seismic interval velocities to the *rock velocities*. The velocity field must be constrained by the knowledge of the range of known rock velocities in the area. For example, velocities in excess of 10,000 ft/s in the deepwater Gulf of Mexico must be carefully examined. Velocities from seismic should always be checked whether they are within valid ranges by comparing with a rock model prediction as discussed in Chapter 4. For example, does the velocity predict pore pressure within hydrostatic (*lower bound*) and fracture pressure (*upper bound*)? Any velocity function that is outside this range should be considered suspicious and must be checked and re-picked. Thus, velocity analysis should be guided by a notion of expected rock velocity – not by the quality of the semblance analysis alone. Again, a rock physics template is a valuable tool to guide the process.

18. Any comparison with well log sonic velocities must be done only after the sonic log has been checkshot corrected and filtered to simulate the low-frequency velocities.

19. Usually analysts are trained to pick the *faster* velocities as the depth increases. This may not be a good practice when picking velocities for a quantitative pressure analysis. In fact as pore pressure increases, the velocities do not increase with the depth as rapidly with burial as it would have without any pressure effect. RMS velocities can in fact decrease with increasing depth across a major interval velocity inversion.

20. Special care must be exercised with the layer where velocity has been picked last (namely, at the end of the data set where reliable velocity "picks" can be made). Quite often, the interval velocity is held constant in that layer and then extrapolated to greater depths where there is no actual data to pick velocity. This is a "bad" practice and one should avoid this, since this can result in unphysical pressure values when used by unsuspecting users unfamiliar with the practice.

21. Lateral variations in velocity do occur frequently. This involves encountering low velocity zones in an otherwise high velocity medium or vice versa. This will involve different moveout curves, and hence velocities must account for it by dense velocity picking prior to geopressure estimation.

22. All velocity analyses must be accompanied by some sort of error analysis. Every pick must include a quality assurance procedure (say, 1 for good, 2 for questionable, and 3 for bad, although this is a subjective judgment based on the interpreter's assessment of the S/N of the data). The standard error, χ_{int} in the interval velocity obtained from RMS velocities, χ, when the interval is of thickness h and average depth D is given by

$$\chi_{int} \sim 1.4\chi D/h \tag{5.12}$$

Thus, if we want a precision of 5 percent in the interval velocity of, say, a layer between 1000 m and 1200 m depth, we will need a precision in the RMS velocity at the top and the base of the layer given by

$$\chi = \frac{h\chi_{int}}{1.4D} = 200 \times 5\%/(1.4 \times 1100) = 0.6\% \tag{5.13}$$

Such precision is, although difficult, achievable.

5.3.3   Seismic Velocity Anisotropy

Basics of Velocity Anisotropy

Anisotropy in the context of wave propagation means that velocity depends on the direction of propagation. That rocks are basically anisotropic is unquestionable. Isotropy is an assumption necessitated by the desire to make analysis easier. However, there is a complicating factor when dealing with seismic waves. It is associated with how the data are acquired and the assumption of layered earth geometry behind the technology. A detailed exposition of anisotropy is beyond the scope here; we direct the readers to some excellent articles (Thomsen, 1986; Mavko et al., 2009; Tsavankin, 1997; Al-Khalifa and Tsavankin, 1995; Al-Khalifa et al., 1996; Anderson et al., 1996, 1994; Tsavankin et al., 2003; Fomel and Grechka, 2001). The discussions given here are not intended for specialists in the subject but for those who would like to get exposed to the subject with enough knowledge to communicate with the seismic data processors and analysts intelligently by asking the proper questions so that the velocity conditioning (anisotropy analysis) for pore pressure analysis can be applied with confidence.

Following very closely Fomel and Grechka (2001), we provide here a brief description of seismic anisotropy. In Figure 5.8a we show schematics of several anisotropic configurations that are used in seismic processing. Two major ones are: *Vertical Transverse Isotropy* (VTI) and *Tilted Transverse Isotropy* (TTI). We note that these are called transverse isotropy because there is isotropy in the plane perpendicular to the symmetry axis – the horizontal plane for VTI configuration. The axis of symmetry is an axis of rotational invariance such that if we rotate the formation about the axis, the material is still indistinguishable from what it was before. In a VTI medium the axis is vertical while in a TTI medium the axis is tilted from the vertical. Orthorhombic anisotropy is gaining some traction these days. Horizontal Transverse Isotropy (HTI) is not popular in the industry due to its unphysical nature, except in fractured formations. In a VTI medium, we need *three* parameters to describe the P-wave wavefronts: *vertical velocity and epsilon and delta parameters* of Thomsen (1986). In a recent work, Le (2019) also concluded that for parameterization of VTI medium, the most desirable ones are indeed vertical velocity and Thomsen's epsilon and delta parameters. For TTI media there are two additional parameters related to tilt-angle (Thomsen, 1986).

The seismic velocity analyses discussed in earlier sections yield velocity functions that represent mostly horizontally propagating waves. This velocity function will therefore be *faster* than the well velocity function (vertical velocity); and hence, the predicted pressures, without accounting for anisotropy, will be *lower*. Both VTI and TTI models correct for this. This effect (faster velocity along the beds compared to the vertical) also manifests as *nonhyperbolic* moveout in the *far offsets*, in particular a *pull-up or "hockey stick"* effect in the gathers – the arrivals are unexpectedly early at long offsets.

For transversely isotropic media, Thomsen (1986) introduced a notation for VTI media by replacing the elastic stiffness coefficients with the P- and S-wave velocities along the symmetry axis and three dimensionless anisotropic parameters epsilon (ε), delta (δ), and eta (η). Anisotropy parameter eta (η) is not an independent parameter; it is related to epsilon (ε) and delta (δ). As shown by Tsavankin (1996), the P-wave seismic signatures in VTI media can be conveniently expressed in terms of Thomsen's parameters (ε) and (δ). Deviations of these parameters from zero characterize the relative strength of anisotropy. For small values of these parameters, the weak-anisotropy approximation (Thomsen, 1986; Tsavankin and Thomsen, 1994) yields the following for the group velocity V_g of P-waves in VTI media, as a function of angle ψ from the vertical symmetry axis (Fomel and Grechka, 2001):

$$V_g^2(\psi) = V_z^2(1 + 2\delta \sin^2\psi \cos^2\psi + 2\varepsilon \sin^4\psi) \tag{5.14}$$

where $V_g(0) = V_z$ is the *vertical* P-wave velocity and ε, δ are Thomsen's dimensionless anisotropic parameters which are assumed to be small quantities. For the isotropic case, $\varepsilon = \delta = 0$.

As follows from equation (5.14), the velocity V_x corresponding to ray propagation in the horizontal direction is

$$V_x^2 = V_g^2(\pi/2) = V_z^2(1 + 2\varepsilon) \tag{5.15}$$

Another important quantity is the normal-moveout (NMO) velocity V_{NMO}, which determines the small-offset P-wave reflection moveout in homogeneous VTI media above a horizontal reflector. Its exact expression is (Thomsen, 1986)

$$V_{\mathrm{NMO}}^2 = V_z^2(1 + 2\delta) \tag{5.16}$$

If $\delta = 0$ the normal-moveout velocity is equal to the vertical velocity. Fomel and Grechka (2001) suggested that it is convenient to rewrite equation (5.14) in the following form:

$$V_g^2(\psi) = V_z^2(1 + 2\delta \sin^2\psi + 2\eta \sin^4\psi) \tag{5.17}$$

where

$$\eta = \varepsilon - \delta \tag{5.18}$$

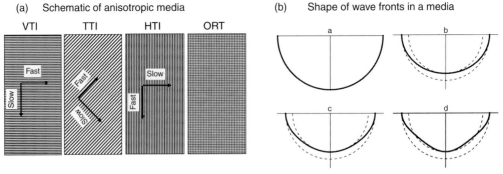

Figure 5.8 (a) Models of various anisotropic media that are used in seismic processing. (b) Shapes of wavefronts in anisotropic media. (upper left) wave front in isotropic media with $\varepsilon = \delta = 0$. (upper right) Elliptically anisotropic media with $\varepsilon = \delta = 0.2$. (lower left) ANNIE model (Schoenberg et al., 1996) with $\varepsilon = 0.2$, $\delta = 0$. (lower right) Anisotropic media with $\varepsilon = 0.2$, $\delta = -0.2$. In each case, solid curves represent the wavefronts. Dashed lines correspond to isotropic wavefronts for the vertical and horizontal velocities. Modified after Fomel and Grechka (2001).

Equation (5.18) is the *weak-anisotropy* approximation for the anellipticity coefficient η introduced by Al-Khalifa and Tsavankin (1995). For elliptic anisotropy, $\varepsilon = \delta$ and $\eta = 0$. Seismic data often indicate $\varepsilon > \delta$, so the anellipticity coefficient η is usually positive. In the linear approximation the anelliptic behavior of velocity is controlled by the difference between the normal moveout and horizontal velocities or, equivalently, by the difference between the anisotropic coefficients ε and δ. Fomel and Grechka (2001) illustrated different types of the group velocities (wave fronts) as shown in Figure 5.8b. Often, an approximation $\varepsilon \cong 2\delta$ is found to be useful (S. Yang, personal communication, 2015) while inverting for these parameters using seismic inversion (see below) as it reduces the dimensionality of the inversion space to just two parameters (vertical velocity and δ).

Determination of Seismic Anisotropy Parameters

As discussed above, for vertical transverse isotropy, three parameters describe the P-wave propagation: vertical velocity V_z, epsilon (ε), and delta (δ). The epsilon and delta parameters are typically determined from *short* and *large-spread* moveout analyses, respectively, and the vertical velocity parameter is usually determined from zero-offset VSP or checkshot data. Core data can be measured to provide anisotropic parameters on small plugs which are often not directly relevant to large seismic surveys. For seismic imaging there are additional methods to estimate ε and δ parameters which are used by the geophysical community:

a. direct determination using "double-scanning" technique (Liu et al., 2016);
b. simultaneous inversion of all anisotropy parameters, including the vertical velocity (Bakulin et al., 2009);
c. rock physics–based technique (Bandyopadhyay, 2009; Bachrach et al., 2011; Bachrach, 2011a, 2011b; Li et al., 2011);
d. slowness polarization analysis of multicomponent VSP (Horne and Leaney, 2000; Tamini et al., 2015).

The double-scanning technique is illustrated by an example in Figures 5.9a–5.9c. It is essentially an iterative or trial-and-error based method that flattens a chosen gather by adjusting ε and δ parameters that have been migrated with a fixed *vertical* velocity. This requires availability of an interactive and user-friendly software platform.

In an alternate procedure, Bakulin et al. (2010b) used simultaneous inversion technique to determine all *three* parameters (vertical velocity, ε, and δ). They used checkshot data to approximate the initial vertical velocity. The inversion was a three-parameter inversion of tomography (see below for a discussion on tomography approach for velocity model building). An example is given in Figure 5.10. We note the oscillatory behavior of the anisotropy parameters ε and δ that mimics the behavior of the vertical velocity. It is not clear why this is so but needs to be understood. We suspect that this is an artifact of the nonuniqueness of the velocity model.

(a)   (b)

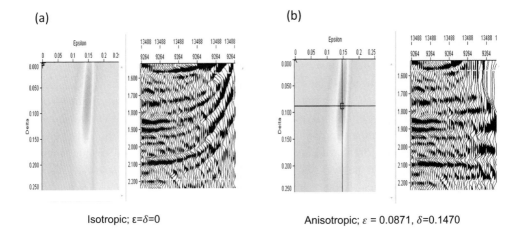

Isotropic; $\varepsilon=\delta=0$ Anisotropic; $\varepsilon = 0.0871$, $\delta=0.1470$

(c)

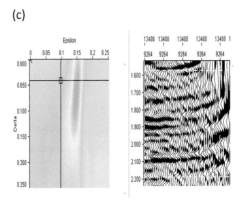

Anisotropic; $\delta = 0.0403$, $\varepsilon =0.0970$

Figure 5.9 Double-scanning technique to estimate anisotropy parameters ε and δ from a seismic gather that has been migrated with a *vertical* velocity (not NMO velocity) from well control. Color bar is a normalized measure of gather flatness (i.e., hotter colors are flatter gathers). Note that different combination of ε and δ can produce similar flatness measure. (A black and white version of this figure will appear in some formats. For the color version, please refer to the plate section.)

The rock physics based method for anisotropy parameter estimation in shale is discussed by Bandyopadhyay (2009), Bachrach et al. (2011), and Li (2014). These authors modeled compacting shales using *microstructure* parameters such as pore aspect ratio, particle alignment and the amount of pore shape deformation using *Differential Effective Medium* (DEM) theory for shales and shaly sands. The advantage of this method over that suggested by Bakulin et al. (2010b) is that the checkshot data are required in that method. However, such data are valid locally and to use that in a large-scale 3D volume as required for Prestack Depth Migration (PSDM) one would need to perform elaborate interpolation and extrapolation. Bachrach et al. (2010) modeled anisotropy in shale

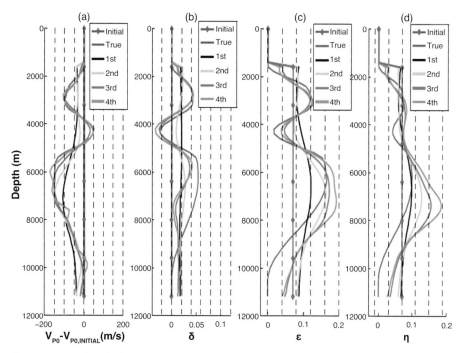

Figure 5.10 Results of a three-parameter VTI tomography using data from vertical checkshot and seismic gathers. Velocity and anisotropy profiles after each iteration are shown together with initial and final models. (a) Update in velocity is shown as a difference between the current velocity at each iteration and initial velocity profile. (b, c) The same for ε and δ. After Bakulin et al. (2010b). (A black and white version of this figure will appear in some formats. For the color version, please refer to the plate section.)

and sandy-shale by including compaction and chemical diagenesis involving smectite to illite transformation and accounted for particle alignment due to this effect and hence, anisotropy. The effect of chemical diagenesis in shale was modeled by a temperature dependent function to approximate the beta-function and it is given below (see equations (3.29) and (3.30) in Chapter 3 and Figure 4.12b):

$$P_s(T) = \frac{1}{2} + \frac{1}{2}\tanh\left(\frac{T - T_T}{2S_T}\right) \tag{5.19}$$

Here, T_T is the temperature where 50 percent of the S/I reaction has occured, and $2S_T$ controls the spread of the reaction. In the model, the chemical diagenesis is captured by changing the mineral elastic stiffness tensor from smectite-rich to illite-rich sediments. Reuss anisotropic average of the elastic tensors according to the temperature dependent ratio is used to achieve representative moduli for the transition. Bachrach et al. (2011) showed using a large checkshot data base from the Gulf of Mexico that the regional trends of the anisotropy parameter derived from such a model compared well with those predicted using checkshot data. In Figure 5.11a, we show the predicted anisotropy versus depth profile for ε and δ parameters that used the tomography-based velocity shown in Figure 5.11b. This approach is very promising. However, it uses parametrization based on microscopic properties of rocks (shale) that are very hard to measure such as the elastic stiffness constants for shales and pore aspect ratio spectrum. In addition, it neglects the burial history dependency of shale properties as a result of chemical diagenesis such as the density and velocity changes of shale matrix and their bulk properties – the authors use only temperature as the sole criterion

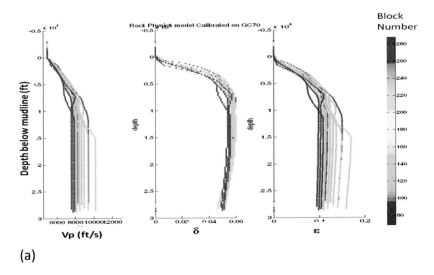

(a)

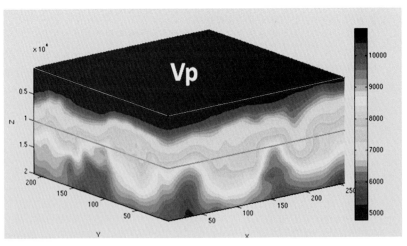

(b)

Figure 5.11 (a) Anisotropy parameter (vertical velocity, ε, and δ) estimation using the rock physics model. Velocity scale is from 6000 to 12000 ft/s in step of 2000 ft/s; δ is from 0 to .06 in step of 0.02 and ε is from 0 to 0.2 in step of 0.1. Modified after Bachrach et al. (2011). (b) The authors used 18 wells from GOM and the seismic velocity data from a tomography model, as shown. (A black and white version of this figure will appear in some formats. For the color version, please refer to the plate section.)

for the onset and extent of the chemical diagenesis of shale. It also requires the X-ray diffraction data for shale chemical diagenesis that is usually not readily available. In an alternate approach, these limitations are managed better by using a constrained rock physics model that uses pore pressure from data or a knowledge of a range of possible pressure variation from rock physics templates (see below).

Slowness polarization analysis uses multi-components VSP to track both velocity (or slowness) along the vertical receiver array and the actual polarization of the seismic wave. This method often provides a good estimation of the VTI parameters as adding vertically-polarized shear wave data reduces the ambiguity in the inversion of the anisotropic parameters (Horne and Leaney, 2000; Tamini et al., 2015).

Examples of Applications to Geopressure Using Conventional Velocity Analysis

In Figure 5.12 we show the effect of data acquired with a large cable on velocity moveout; clearly, large-offset data will yield a better velocity function than those acquired with a cable that yields a smaller offset (30 degrees, for example). The data are from deepwater NW shelf of Australia (Canning Basin) and was acquired with a 6 km cable length. We note that if we restricted ourselves to smaller source-receiver distances (angles less than or equal to $30°$ at the target depth), none of these effects would be detected – all would yield approximately the same velocity functions. This is because one is trying to flatten a hyperbola in the small angle range where it is already close to being flat. The analysis in that case would result in erroneous interval velocity and hence, erroneous pore pressure values. *Thus, we recommend that pressure analysis be performed on a data set containing as large a source-receiver distance as possible.*

It is clear from the above discussions that we should be very careful in using interval velocities derived from stacking velocity analysis for quantitative pore pressure prediction. There are many pitfalls inherent in the technology as discussed; however, many are due to bad practices in the industry – a desire to get quick answers by practitioners not appreciating the assumptions and limitations behind the seismic technology for acquisition, processing, and interpretation of velocity data. We must remember that the purpose of NMO is for imaging and *not* pore pressure analysis. The velocity data *must* be conditioned prior to using it for pressure prediction. On the top of the list is the proper accounting of seismic anisotropy that we just discussed. It is a must- *no seismic velocity data should be used for geopressure analysis without anisotropy corrections.*

An example of the interval velocity is shown in Figure 5.13a overlain with the seismic wiggles. The method employed was conventional velocity analysis (dense picking) with vertical transverse isotropy. The model was calibrated using well data. Migration was carried out in the depth domain. The analysis shows very detailed variations in velocity. It is consistent with the geologic structure. In Figure 5.13b we show the corresponding pore pressure in EMW in units of gm/cc (1 gm/cc = 8.344 ppg). Regions of very high pressure (up to 2 gm/cc or 16.69 ppg) are visible. The well location is shown on the figure.

In Figure 1.14 we showed a 3D cube of pore pressure as an example of the spatially continuous and automated velocity model building procedure (Mao et al., 2000) on a 3D data set in an area that is about 650 square miles in extent. The velocity analysis

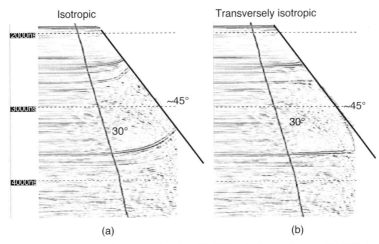

Figure 5.12 (a) A seismic gather showing NMO for the isotropic case. (b) Effect of large offset on velocity moveout on the same gather (transversely isotropic). These data from a deepwater NW Australian shelf were acquired with 6 km cable length. Modified after Dutta (2002b).

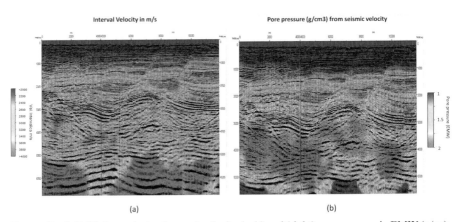

Figure 5.13 (left) High-resolution interval velocity (m/s) and (right) overpressure in EMW (g/cc). Figure courtesy Shuki Ronen, Stanford University. Velocity scale starts from 2000 m/s to 4000 m/s in step of 200 m/s. The overpressure is in SG units starting from 1 to 2. 1 SG unit denotes hydrostatic pore pressure (~ 0.433 psi/ft). (A black and white version of this figure will appear in some formats. For the color version, please refer to the plate section.)

employed used semblance data at *every* CMP. The approach yielded a 3D velocity model in which steep dips and relatively rapid velocity variations are handled easily. That example shows significant clarity of the pore pressure image in 3D, and the high-pressure pockets (greater than 16 ppg mud weight equivalent in red) and velocity reversals are clearly visible.

Another example from Banik et al. (2003) is shown in Figure 5.14. It demonstrates the application of a spatially continuous and automated velocity model building procedure (Mao et al., 2000) to an overpressured area in a deepwater basin, offshore

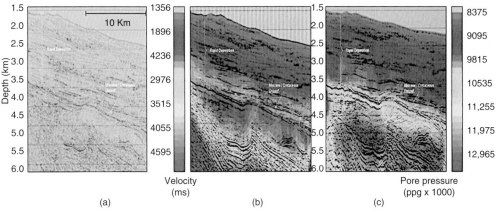

Figure 5.14 Application of spatially continuous velocity analysis and automated velocity model building approach to pore pressure prediction in 3D. (a) A seismic line from a 3D volume, offshore, West Africa (dip line). (b) Prestack Time Migrated (PSTM) image along the dip line. (c) Pore pressure gradient in PPG *multiplied by 1000*. After Banik et al. (2003). (A black and white version of this figure will appear in some formats. For the color version, please refer to the plate section.)

West Africa. Due to the sharp lateral changes in velocity in this area and the absence of deepwater wells, instead of the empirical methods (e.g., Eaton's equation), the Dutta model (Dutta et al., 2001) for velocity to pore pressure transform was used. Figure 5.14a shows a PSTM seismic section along a dip line. Figure 5.14b shows the high-resolution velocity (m/s) for the line in Figure 5.14a. Figure 5.14c shows the pore pressure gradient in ppg (multiplied by 1000). The pressure gradient is seen to increase slowly and gradually in the shallow portion but beyond a prominent geologic horizon, it increases rapidly. Banik et al. (2003) corrected the velocity field for anisotropy as well. They showed that anisotropy correction (5 percent) changed pore pressure by about 1 PPG. Readers are also referred to Chopra and Huffman (2006) for a very good discussion and examples on velocity analysis for pore pressure prediction.

A review of literature showed that properly processed data from conventional time domain velocity analysis can be used for pore pressure prediction provided: (1) the data has long offsets, (2) velocity analysis is performed using *all* the data and not just a small mini-cube around the proposed or an existing well – a bad practice, (3) proper anisotropy corrections are applied to the velocity model, (4) velocities are calibrated, and (5) interpreted to make sure that the model is supported by the local geology. *Caution: Based on our experience we suggest that pore pressure prediction should never be done on a single velocity function.*

5.4 Seismic Velocity from Inversion

So far we have discussed conventional velocity analysis technology – one that is done in time domain and uses semblance-based approaches for flattening seismic gathers to derive interval velocity from stacking velocities. Although techniques such as SCVA

(Spatially Continuous Velocity Analysis) and DVA (Dense Velocity Analysis) facilitate the process through using autopickers and dense velocity picking, these are still semblance-based approaches. Although these are employed routinely for conventional time domain velocity analysis, these have many limitations. We discussed these earlier in this chapter. There are many other techniques, all based on sophisticated inversion procedures, which add value to geopressure analysis work beyond what is possible by the stacking velocity analysis and Dix-interval velocity analysis approaches. These methods use the full bandwidth of the data in the prestack domain and are discussed below.

5.4.1   Seismic Reflection Tomography

Velocity from Seismic Reflection Tomography (aka Common Image Point or CIP Tomography)

Seismic reflection tomography is an *inverse* problem and it is one of the most useful and popular procedure that uses traveltime *inversion* to estimate the subsurface velocity field. Seismic traveltime data are compared to an initial velocity model and the model is then modified until the best possible fit between the model predictions and observed data is found. It can provide velocity fields that relate better to the geologic structures than what is feasible with conventional velocity analysis. In this regard, seismic reflection tomography is similar to medical X-ray imaging using computed tomography (CT scan) in that a computer processes receiver data to produce a 3D image, although CT scans use attenuation instead of traveltime information. Seismic tomography has to deal with the analysis of curved ray paths, which are reflected and refracted within the earth, and potential uncertainty in the location of the reflection points. (CT scans use linear X-rays and a known source.)

Reflection tomography and other inversion techniques are very popular as the cost of computing has come down significantly and machine speed has gone up. Geophysicists are increasingly using velocities from various inversion techniques for geopressure analysis instead of conventional time domain velocity analysis. Tomographic inversion of reflection seismic data forms the basis of all *contemporary* methods of updating velocity models for depth imaging. Tomographic inversion is one of a class of multidimensional inversion methods that provide more detail of the subsurface 3D velocity field associated with 3D geologic structures (Aki and Lee, 1976; Gjoystdal and Ursin, 1981; McMechan, 1983; Bishop et al., 1985; Chiu and Stewart, 1987; Hubral and Krey, 1980). Articles by Jones (2010) and Padina et al. (2006) are good tutorials on the subject and we have drawn from these publications. The mathematics is very involved in this technology and somewhat irrelevant for our purpose. There are excellent commercial packages available in the petroleum industry for velocity model building using reflection tomography. This is a desirable technology to use when reliable velocity fields are required for handling geology with lateral changes in velocity. Geopressure analysis has greatly benefited from this approach.

Among various tomography methods, ray-based reflection tomography and full-waveform inversion based tomography (Tarantola, 1984, 2005; Virieux and Operto,

2009) are the most widely used. Full-waveform inversion tomography is also known as waveform tomography, wave equation tomography, and diffraction tomography. Over the past 10 years, ray-based tomography has evolved as the most valuable tool for velocity model building for depth imaging in complex geologic areas. Reflection tomography (Stork, 1992; Wang et al., 1995; Woodward et al., 2008) replaces the low-resolution layered medium and hyperbolic moveout assumptions of conventional velocity analysis methods with a more accurate ray-trace modeling based approach.

Fundamentals of Ray Tracing Grid Tomography

Pre-stack depth migrations contain redundant images of the earth's subsurface – one image volume for each offset or opening angle for Kirchhoff and wave equation migrations, respectively. The depth variation of reflection events across offset is termed *residual moveout* (RMO). When there is no RMO, the earth model has optimized the data focusing; when there is RMO, we trace rays through the model to determine which parts of the model needs to be changed to flatten the moveout and improve the focusing. We analyze RMO on common-offset image volumes sorted to common image point (CIP) gathers. Standard industry grid tomography generates and solves ray-traced residual-migration equations to produce earth models that flatten RMO on CIP gathers. There are three basic components of depth domain ray tomography, each of which can have its own style. The first component is the data – the representation of the RMO. For ray-based tomography, there are two options: (1) a simple or single parameter analysis, in which the moveout is characterized by a best-fitting parabola or hyperbola, and (2) a complex or multiparameter analysis, in which an image depth is picked independently for many offsets or represented by a higher-order polynomial curve. As computer power has increased and as resolution expectations have grown, the industry standard has moved from single-parameter to multiparameter analysis. Today, we pick single-parameter moveout for preliminary, low-resolution work on a survey, but our later iterations rely heavily on complex multiparameter moveout analysis.

The second component of any ray tomography is the model. Just as with data picks, models can range between two extremes. At one extreme, we have models that can be purely layer based, with properties defined on horizon maps and interpolated linearly between the horizons. Models such as these have relatively few parameters to solve; they are strongly constrained by the user, and they rely heavily on an interpretation. They were probably the standard in the late 1990s and are still useful in areas with poor data quality requiring many a priori constraints. At the other extreme, we have models that can be purely gridded. Today, we mainly use hybrid-gridded models. We divide the model into layers at interfaces in which we have sharp, ray-bending velocity contrasts, such as at the water bottom, at salt boundaries, and at carbonate boundaries. Within layers, we use continuously interpolated property grids. We place surfaces where we believe we can add more detail to the layer boundaries by interpreting seismic images than by analyzing RMO. Property grids within layers are made dense enough to capture the resolution. Surface grids are sampled as finely as interpretable to capture the rugosity of their corresponding geologic structures. Because errors in surfaces are compensated for by erroneous property updates on their underlying grids, poorly determined surfaces restrict the resolution that is attainable beneath them. When high-resolution updates are produced below poorly picked surfaces, they can contain erratic bull's-eye features. For extremely rugose surfaces, the bandlimited nature of the seismic source will result in a ray solution that is unrepresentative of the wave propagation data.

The third component of a ray tomography is made up of the residual-migration equations and their solver. We ray-trace the source and receiver rays corresponding to our depth picks through the model to generate equations that relate changes in the model to changes in the RMO. These are residual-migration equations (Stork, 1992)

$$z'_h = z_h - \sum_i \left(\frac{\partial t}{\partial \alpha_i} \Delta \alpha_i \right) \frac{v}{2 \cos \theta \cos \phi} \tag{5.20}$$

The prime indicates a residually migrated reflector depth, z_h is depth as a function of offset or angle, i indicates a node on the property grid, θ is the angle of incidence for the reflection, ϕ is the reflector dip, v is the effective velocity at the reflector, and $\partial t / \partial \alpha_i$ is the change in traveltime corresponding to a change in property at node i. These are linear tomography equations, in that the $\partial t / \partial \alpha_i$ terms are calculated under the assumption that ray paths do not change as we change the model. Although α_i is most often velocity or slowness, it can be any earth property, such as Thomsen parameters ε or δ (Thomsen, 1986).

Our goal is to find property perturbations $\Delta \alpha$ that minimize the RMO of depth picks residually migrated with a model including those perturbations, i.e., that minimize $z'_h - z'_0$ where z'_h is a non-near-offset pick and z'_0 is a near-offset pick. This is a floating datum criterion, in that all of the picks move to minimize their relative depth errors rather than a static depth error. We have one equation for each non-near-offset pick. The basic tomography update equation is formed by subtracting equation (5.20) for corresponding non-near- and near-offset picks and accumulating the picked depth errors on the right side:

$$\mathbf{L}\Delta\boldsymbol{\alpha} = \Delta\mathbf{z} \tag{5.21}$$

The geometry and background model terms in equation (5.20) are contained in $\mathbf{L}$. Because reflection seismic problems are severely underdetermined, we regularize and stabilize the tomographic problem using preconditioner (row weighting) operator $\mathbf{P}$, smoothing operator $\mathbf{S}$ and column weighting operator $\mathbf{W}$ such that $\Delta \boldsymbol{\alpha} = \mathbf{SW}\Delta \boldsymbol{\alpha}'$. In addition, a damping parameter λ is used. Finally the tomographic solution is obtained by minimizing an objective function $\Phi = \|\mathbf{PLSW}\Delta\boldsymbol{\alpha}' - \mathbf{P}\Delta\mathbf{z}\|_{L_{1.5}} - \lambda^2\|\Delta\boldsymbol{\alpha}'\|$ which is composed of regularized and smooth preconditioned data misfit term together with damping. More details can be found in Woodward et al. (2008). An example is shown in Figure 5.15. On the left we show RMO picks for three azimuths as indicated for six CIP gathers and on the right we show single-parameter correction criteria (γ picks) of RMO for the 2D section (Woodward et al., 2008).

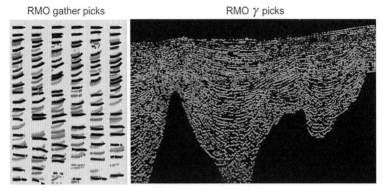

Figure 5.15 (left) Superimposed independent multiparameter RMO picks for three azimuths (red = 30°, black = 90°, and blue = 330°) on six CIP gathers. The picker first fits a line or parabola to the events and then applies a trace-by-trace shift to further match the offsets. (right) Single-parameter correction criteria γ picks of RMO for 2D section QC. Red is for overcorrected gathers, and blue is undercorrected gathers. After Woodward et al. (2008). (A black and white version of this figure will appear in some formats. For the color version, please refer to the plate section.)

The entire workflow for seismic reflection tomography as discussed here is shown schematically in Figure 5.16 (Woodward et al., 2008). We note that the workflow uses anisotropy parameter estimation as well. We discussed this earlier. Most algorithms also impose geologic constraints (typically horizon consistency via steering filters) on the procedure to ensure that the derived model is consistent with geology and interpretable. Blind use of tomography may result in anisotropic velocity fields that will flatten the seismic gather (the sole criterion used by the algorithm) but may not yield a geologically consistent velocity field. This is a very compute intensive process; however, if successfully executed, it yields velocities that are very suitable for pore pressure analysis. There is a caveat, nonetheless. This has to do with the nonuniqueness of the derived velocity model. This may not affect the "image" quality, but it could have a significant impact on the predicted pore pressure and the location (in depth) of the image. We do not want to belabor the point; however, we note that the CIP tomography process uses only two of the three requirements for pore pressure prediction. These are: mathematically converged solution, model compliant with geology and a physically possible model. CIP tomography yields results that are consistent with the first two criteria but the third criterion is often neither tested nor met. A possible solution is discussed in Chapter 6.

Well-Constrained Tomography

To minimize uncertainty associated with the nonuniqueness of a tomography workflow, Bakulin et al. (2009) proposed a method in which they *jointly* inverted the *seismic* and the *well data* locally around the borehole. Their method also allowed for imposing other a priori constraints on the anisotropic velocity model and used all offsets for the determination of Thomsen's parameters – ε and δ. They used the workflow shown in Figure 5.16 (Woodward et al., 2008). In Figure 5.10 we showed an example of this

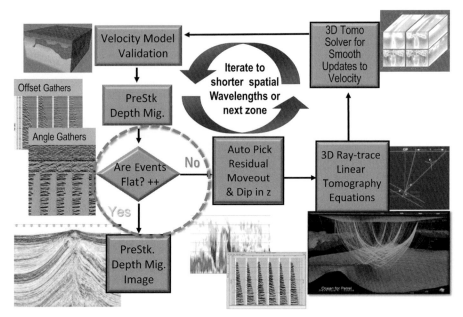

Figure 5.16 A schematic of a workflow for deriving velocity model using reflection seismic tomography (Woodward et al., 2008). (A black and white version of this figure will appear in some formats. For the color version, please refer to the plate section.)

approach for a VTI medium. To manage nonuniqueness of the inversion process, Bakulin et al. (2010b) also accounted for depth variation of reflection events across offset, namely, residual moveout (RMO). Typically, RMO is analyzed on common-offset image volumes sorted to common image point (CIP) gathers. As noted before, this allows us to determine which parts of the velocity model must be updated to flatten the moveout and improve the focusing. RMO analysis is now standard in most industry tomography packages. This accounts for creating a geologically viable velocity model.

Applications to Pore Pressure Prediction Using Velocity from Reflection Tomography

The case study presented here used the workflow shown in Figure 5.16 and was presented by Sayers et al. (2002). This example is located in the deepwater Gulf of Mexico. Existing 2D seismic data were used to predict pore pressure in the region of interest using 2D CIP tomography to refine the velocities. The velocities obtained were then used to derive a predrill pore pressure estimate using the methods of Eaton (1975) and Bowers (1995).

Figure 5.17a shows conventional velocity analysis using semblance-based technique as discussed earlier. Stacking velocities were converted to interval velocities using the Dix equation. The blocky appearance of the velocity field in Figure 5.17a is typical of production-stacking velocities – defined as that appropriate for imaging. This is not adequate for pore pressure analysis. This requires additional smoothing for any meaningful pore pressure analysis on top of smoothing that has already been done into the stacking velocity analysis technique (source-receiver offset related smooth-ing). The interval velocities that result from this process often lack the spatial reso-lution required for pore pressure prediction. Such was the case here. Figure 5.17b shows the interval velocities obtained using CIP reflection tomography. In contrast to the interval velocities obtained from stacking velocities (Figure 5.17a), these velocities provide more details necessary for predrill pore pressure prediction. The authors noted that of interest for drilling in the region of CIP 1200 is the low velocity anomaly at a depth of 2.75 km, which may indicate the presence of overpressure in this area. The spatial extent of this anomaly is very poorly defined based on interval velocities calculated from stacking velocities (see Figure 5.17a). The final prestack depth-migrated section is shown in Figure 5.17c along with the CIP gathers at the drilling location. Figure 5.17d shows Dix-converted interval velocity after highly detailed velocity picking and geologically consistent smoothing. This is a quality control procedure on the tomography. This model is also consistent with the CIP tomography. The overall flatness of the seismic events within the CIP gather indicated that tomog-raphy has produced an acceptable velocity model for imaging which is more accurate than the conventional stacking velocity analysis. The authors noted that the accuracy of the tomographic inversion of surface seismic data was limited by several factors, one of which is the maximum source-receiver offset used in recording the seismic data. Reliable estimates of pore pressure cannot be expected at depths exceeding the maximum source-receiver offset used. This is a good rule of thumb! Figure 5.18e shows predicted pore pressure in ppg for the seismic line using the tomography-based velocity data and Bowers's model (Bowers, 1995) for relating velocity to vertical effective stress:

$$V = V_0 + A\sigma^B \tag{5.22}$$

where V_0 is the velocity of unconsolidated fluid-saturated sediments (taken to be 1500 m/s), and A and B describe the variation in velocity with increasing effective stress. Bowers (1995) obtained the values $A = 4.4567$ and $B = 0.8168$ for a Gulf Coast well studied by Hottman and Johnson (1965) and the values $A = 28.3711$ and $B = 0.6207$ for a deepwater Gulf of Mexico well. Using an estimate of average bulk density $\langle \rho(h) \rangle$, from an empirically based relation of Traugott (see equation (5.23)); Sayers et al. (2002) obtained the pore pressure results shown in Figure 5.18e. The average bulk density used in the model for overburden stress computation is given by:

$$\langle \rho(h) \rangle = 16.3 + (h/3126)^{0.6} \tag{5.23}$$

Here, $\langle \rho(h) \rangle$ is the average sediment density in pounds per gallon (ppg) mud weight equivalent between the seafloor and depth h (in ft) below the seafloor. Recall that a mud weight of 1 ppg is equivalent to a density of 0.1198 g/cm^3 and a pressure gradient of 1.17496 kPa/m. The example presented shows clearly that the velocity analysis using conventional stacking velocity analysis model is usually *not* appropriate for pore pressure prediction. Care must be taken to improve the quality of the velocity analysis,

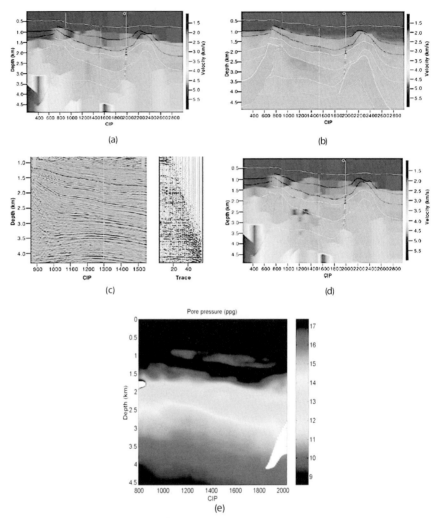

Figure 5.17 Example application (Sayers et al., 2002) of seismic reflection tomography-based velocity to pore pressure prediction. (a) Initial stacking velocity converted to interval velocity using the Dix equation. Note the blocky nature of the velocity due to undersampling and oversmoothing – a typical trait of the stacking velocity analysis. (b) Final interval velocity based on grid-based seismic reflection CIP tomography. The velocity field is much more consistent with geology. Note the velocity anomaly (low) at ~ 2.75 km depth possibly due to high pore pressure. (c) Final predicted depth image using the velocity from (b) and the gather at the well location. The flatness of the gather suggests that the CIP tomography has produced an acceptable solution for depth imaging. (d) Dix-converted interval velocity after highly detailed velocity picking and geologically consistent smoothing. This is a quality control procedure on the tomography. This model is also consistent with the CIP tomography. (e) Predicted pore pressure in ppg using the CIP tomography. Predicted high pressure at ~ 2.75 km was verified with the well data. Pore pressure scale is from 9 ppg to 17ppg in step of 1 ppg. The depths are in km. After Sayers et al. (2002). (A black and white version of this figure will appear in some formats. For the color version, please refer to the plate section.)

for example, by following the guideline presented earlier in this chapter. Tomography is an alternate and a better approach as shown in this example from Sayers et al. (2002). Tomography approach to pore pressure prediction is common these days.

Limitations of CIP Tomography for Pore Pressure Prediction

A tomography approach to building velocity models for pore pressure prediction as illustrated above with the case study has several limitations. In this section we will address some of these. The first issue is the nonuniqueness of the tomography workflow to build a velocity model. This is because the problem is under determined as the surface seismic reflection experiment does not cover all of the model space sufficiently. Furthermore, adding anisotropic parameters makes the number of unknowns very large. Also, as the objective function is nonlinear the final outcome of tomography – a velocity model, is highly dependent on the "initial or starting model" for the iterative approach. Two different starting models will usually yield two different final outputs (velocity). Just which one is appropriate for pore pressure prediction (without any well control) may often not be clear.

The second limitation is that tomography uses *flatness* of seismic gathers, as the most important, and often the sole criterion for velocity modeling workflow as shown in Figure 5.16. While this may be adequate for imaging and *necessary* for pore pressure application, it is usually not *sufficient* for this purpose. Sayers et al. (2002) correctly noted that at depths greater than the offset distance, the predicted velocity field from tomography has the largest uncertainty due to lack of angle range over which the gathers are flattened. To improve the quality of the velocity field in the location where the uncertainty is high we need to impose some physical constraint(s) on the tomography workflow for velocity modeling so that we move toward "earth modeling" rather than "velocity modeling." The two are different. Let us look at a real-life example from a recent field application in the context of pore pressure prediction using velocity from seismic reflection tomography. Figure 5.18 shows two stacked images (panels b and c) using identical seismic data that have been processed in exactly the same way prior to velocity modeling using tomography. The target areas for drilling are denoted on these figures. The seismic image gather on the left (panel a) at the well location has been taken from a 3D cube resulting from a conventional tomography workflow for velocity modeling as shown in Figure 5.16. The image shown in panel b resulted from an "*alternate*" tomography work-flow – *rock physics guided CIP tomography* that used rock physics constraints (to be discussed below) on the conventional tomography workflow. The resulting seismic gather at the predrill location from the new workflow is shown in panel d of Figure 5.18. While the flatness of the seismic image gathers at the well location from both workflows are "acceptable" thus satisfying the gather "flatness" criterion, the image quality is drastically improved in panel c as opposed to the one on panel b from the conventional (unconstrained) workflow. The seismic energy is more focused and shifted both upward and sideways resulting in a better quality image.

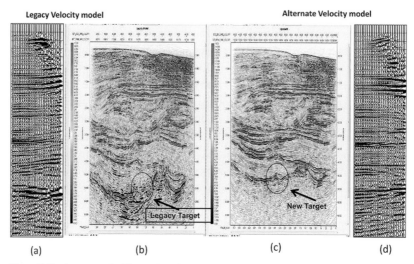

Figure 5.18 An example illustrating the nonuniqueness of reflection tomography. (a) A seismic gather around a well to be drilled. (b) Stacked image using legacy interval velocity volume from CIP tomography. (c) Stacked image from the same seismic data set as used in (b) but using an alternate CIP tomography workflow that is guided by rock physics. This workflow, discussed in Chapter 6, produces a much better image in depth than that shown in (b). (d) The same gather as in (a) flattened with the alternate velocity model. After Dutta et al. (2014). (A black and white version of this figure will appear in some formats. For the color version, please refer to the plate section.)

In this case study there were no well data available in the area for calibration where the well was drilled to verify either of the two models. The well was designed based on the "*alternate* velocity model." So which one turned out to be the correct model? Drilling data from the well verified the "*alternate* velocity model." The lesson? Gather flattening criterion (as used in the legacy model) to build the velocity model (in tomography) is *necessary but not sufficient* for pore pressure prediction. We will discuss in Chapter 6 the basis behind constructing this "*alternate* velocity model" (that includes rock physics constraints) and suggest that the new velocity model building technique for pore pressure analysis imaging is closer to building an earth model.

In some cases CIP tomography workflow generates images in seismic depth that is often substantially off in the zone of interest, causing large uncertainty in exploration targets, seal definition, drilling, and casing point locations and mud weight program. This is clear in Figure 5.18. The target in panel (b) is not well defined. In panel (c) the target is not only well defined but has moved to shallower depths and shifted laterally causing the final well trajectory to be shifted. The errors in seismic velocity such as the one presented here not only cause depth shifts but also cause substantial errors in structure and amplitude expressions which ultimately affect inversion attributes such as impedance and density.

5.4.2   Seismic Velocity from Full-Waveform Inversion

Both time domain velocity analysis and reflection tomography provide estimates of subsurface velocity that are used for geopressure analysis. The reliability of the prediction depends on the quality of the velocity in terms of seismic gather flattening and accounting for geologic variations such as the lateral velocity gradients. The velocities from these techniques depend on traveltime picking and analyses. Full-waveform is an innovative technology for velocity model building and imaging that does not involve any traveltime picking. Instead, it uses the full wave fields recorded at the surface and attempts to model it by sophisticated forward modeling algorithm that includes as much of the kinematics and dynamics as possible during an iterative inversion toward the final earth model (Tarantola, 1984; Virieux and Operto, 2009). While the CIP tomography uses the arrival time information to build a velocity model, FWI uses *both* wavelet shapes and arrival time information. With the advent of high performance computers this technology has progressed rapidly and recent applications to real data have shown promising results for exploration projects as well as time-lapse seismic (Plessix and Perkins, 2010; Sirgue et al., 2009; Vigh et al., 2015). FWI is a multiscale data-fitting method that is well adapted to wide-angle/wide-azimuth acquisition geometries because it uses simultaneously diving and reflected waves. FWI is classically solved with local optimization schemes and is therefore strongly dependent on the starting model definition. This starting model should predict arrival times with errors less than half of the period to overcome the cycle-skipping ambiguity (Virieux and Operto, 2009).

Virieux and Operto (2009) have a good review of FWI. It is an iterative local optimization problem that is generally posed as a linearized least-squares problem; the optimization tries to minimize the difference between the observed and the simulated wave fields. Because this linearized inverse problem is ill-posed, there are always several models to reduce the misfit energy. Prior information can be essential to rule out most of these possible models and to converge toward the most plausible, based on the prior information. Similar to tomography, preconditioning, and/or regularization techniques may alleviate the nonuniqueness of the ill-posed inverse problem. Prior information is generally introduced through regularization in the inverse formulation. For specific applications such as time lapse or pore pressure prediction where other sources of information such as sonic logs, production data, and stratigraphic data are available, they can be incorporated into a prior model so as to ensure robust and consistent results. This can be formulated in the following way in terms of the L2 norm optimization process:

$$\min f(m) = \frac{1}{2} \|F(m) - d\|_2^2 + \lambda \|W(m - m_p)\|_2^2 \qquad (5.24)$$

where F is the forward modeling operator which is typically a finite difference approximation of the seismic wave propagation experiment given an acquisition geometry, d is the recorded data, m denotes the model parameters, m_p is prior information, λ is a regularization weight that balances the contributions from the data and the model terms and W is the confidence field that is also the weighting parameter on the model space.

The other option is to derive the a priori model m_{well} from a well or wells by geological constraints and thus minimize the misfit shown in equation (5.25):

$$\min f(m) = \frac{1}{2}\|F(m) - d\|_2^2 + \lambda\|W_{well}(m - m_{well})\|_2^2 \qquad (5.25)$$

where W_{well} is the confidence field built with and around a well or wells. Constraints in equation (5.25) can also be provided by other priors in lieu of wells such as a *rock physics template* to guide the inversion process to stay within a specified bound (Le, 2019). A typical workflow for FWI is shown in Figure 5.19a. Although most applications to-date are acoustic, the extension to elastic FWI is under way in both industry and academia. The forward modeling operator in FWI is an approximate solution to the wave equation and is much more expensive than the raytracing algorithm used in tomography. Most commonly the acoustic finite difference method is used. Elastic finite difference is much more costly computationally and is still not used in commercial settings in either 2D or 3D.

There is no standard algorithm or procedure for incorporating anisotropy parameters in FWI. The procedures that we discussed earlier in the context of CIP tomography can all be used in FWI. The procedure outlined by Brittan et al. (2013) is particularly easy and found to be useful (Liu et al., 2016). The authors estimated anisotropy in the earth model (Thomsen parameters, delta (δ) and epsilon (ε)) as follows. The delta (δ) field was taken to be zero within the water layer, interpolated from 0 to 5 percent in the top ~150 m (below water bottom) of the dataset and then held constant at 5 percent for the rest of the model. For epsilon (ε), a similar model was utilized – in this case a relatively conservative value of 7.5 percent was used from 150 m to the base of the model. Authors such as Liu et al. (2016) employ $\varepsilon \approx 2\delta$ as the starting model. For FWI, it is a common practice to use a smooth version of the CIP tomography as an initial model. This model should be constrained by available sonic and density logs as much as possible. Large-offset seismic data also adds to the faster convergence of the FWI model to the true model to within the specified tolerance. Convergence to the true model is not guaranteed but a good starting model that has a geologically supported low-frequency trend and is close to the true model usually leads to a good result.

Since FWI is a large-scale L-2 norm optimization process that tries to minimize the difference between the real and the modeled data, it has many limitations. One of the most serious problems is known as "cycle skipping" (Lailly, 1983; Tarantola, 1984; Virieux and Operto, 2009; Pan et al., 2018). This difficulty begins with the inadequacies in the initial model and it is exasperated by the missing low frequencies. One key assumption and criteria in the FWI-inversion scheme is that the modeled and observed waveforms are within *half a wave-cycle* at the lowest frequency to converge iteratively in the right direction. This means the modeled data must have the correct long wavelength structure as much as possible. This criterion is not always met due to large differences between the initial model and the real data. Some authors (Pan et al., 2018) advocated the use of band-passed well log data as a substitute for recovering low frequencies akin to the process of recovering low-frequency data from poststack inversion. Others (Benfield et al., 2017; Kapoor et al., 2013) use a constrained initial model that use diving wave tomography to recover low frequencies. This does not require well data but is highly data driven, requiring multiple passes at the prestack data starting from a low-frequency model such as ~ 3 Hz and increasing the frequency content in steps of, say, 1 Hz up to ~ 7–8 Hz.

An example of the FWI application for imaging is shown in Figure 5.19b (Kapoor et al., 2013). The seismic data are from North Sea (OBC data). The details of the velocity field from the FWI are impressive – the channel in the shallow part of the stratigraphy is clearly delineated by the velocity field as opposed to the panel on the left (VTI tomography) where the channel was *inserted* manually. The central feature is also visible clearly and it is clearly delineated by lateral changes in the velocity.

In a recent paper Benfield et al. (2017) showed an example of the use of FWI for pore pressure evaluation in the East Coast Marine area of Trinidad and Tobago. According to authors, this area is known for its complex geology caused by faulting, shallow gas, mud volcanoes, and highly variable pore pressure – from hydrostatic to very high pressure (~ 15 pounds per gallon or ppg Equivalent Mud Weight). Plumbing pathways or lack thereof across the faults controlled the pressure depletion and sustenance within sand geobodies. This in turn posed significant challenge to drilling as the well calibration of seismic data based prognosis proved to be unreliable. FWI-based application mitigated the challenges for the most part. An example is shown in Figure 5.20. Figure 5.20a shows pore pressure profiles of two offset wells. Both wells penetrated very similar overburden but encountered very different pore pressure profiles, particularly at the sand location at ~ 16,000 ft. The well on the left (Victory-1) indicates that the sand geobodies have hydrostatic pore pressure while the well on the right (Bounty-1) indicates that the sand geobodies have very high pore pressure. This is clearly related to hydrodynamic plumbing of the sand geobodies through the fault system – the sands at the Victory-1 well drains very well through the fault system while sand bodies at the right (Bounty-1 well) do not; hence the pressure in that well at ~ 16,000 feet is very high. The FWI results predicted this behavior clearly through lateral velocity gradient changes in the velocity model. In Figure 20b pore pressure (in green) is the data from the offset well (supported by drilling data, pressure measurements and mud weights) and the red indicates pore pressure from sonic log after the well was drilled. The offset well validated the observations from the FWI model. In Figure 5.20c the authors show an example of comparison between the predicted pore pressure from FWI velocities (dashed curve) and those from tomography (dot and dashed curve) and the observed drilling mudweight data (black block curve). Clearly, FWI-based pore pressures are more consistent with the drilling mudweight than those from the tomography.

The example application of FWI for pore pressure prediction is noteworthy and shows that the velocity model is fit for multiple purpose use. This includes imaging *and* pore pressure, well path trajectory analyses, and, possibly, reservoir characterization.

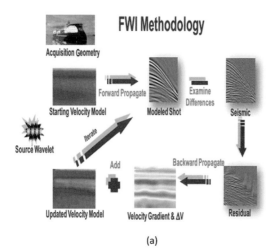

(a)

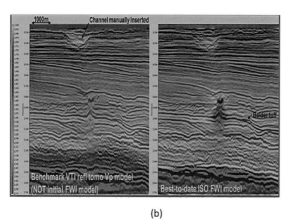

(b)

Figure 5.19 (a) The FWI methodology. (b) An example application of the improvement in velocity field (Kapoor et al., 2013). (left) The benchmark velocity obtained for offshore data (North Sea) using a VTI model. (right) The isotropic–FWI model. The details on the velocity obtained from FWI are better than the one from tomography and better suited for pore pressure prediction. Note that the channel at shallow depths was inserted manually in the left side of panel in (b); the channel was correctly imaged by FWI shown on the right panel of (b). After Benefield et al. (2017). (A black and white version of this figure will appear in some formats. For the color version, please refer to the plate section.)

FWI velocities offer better resolution both vertically and laterally and suggests that it can distinguish between physical geologic processes such as fluid migration due to geopressure from those that are not physical. The industry would greatly benefit by more applications of this type of technology.

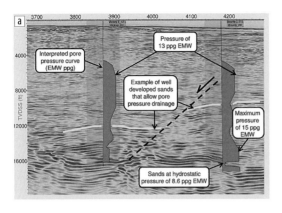

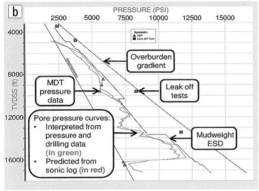

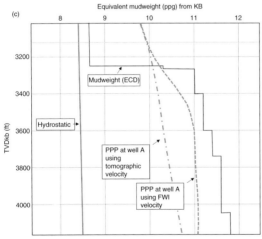

Figure 5.20 Pore pressure analysis (Benfield et al., 2017). (a) Two offset wells, Victory-1 (left-hand side) and Bounty-1 (right-hand side), which penetrate a very similar overburden but have different interpreted pore pressure profiles. (b) The pore pressure interpretation at the Bounty-1 offset well (green line) and the PPP from the sonic log (red line) demonstrate that good calibration of velocities to pore pressure can be made within this basinal geologic setting. It also indicates the drainage of pore pressure through the fault as shown in (a). (c) A comparison of the pore pressures derived from the FWI velocities with those from the tomography. Clearly, the pore pressures derived from FWI velocities are more consistent with the mudweight than the tomography model. After Benefield et al. (2017). (A black and white version of this figure will appear in some formats. For the color version, please refer to the plate section.)

5.4.3   High-Frequency Velocity from Seismic Amplitude Inversion

Introduction

The preceding velocity analysis techniques – conventional, tomography, and FWI – yield a low-frequency background trend for pore pressure. This is important as this is the most stable component of the model that can be obtained from seismic data. However, it has two drawbacks. First, the model assumes that the background lithology is shale and therefore, the predicted pore pressure is identical to the "true" pore pressure. Second, the model will accommodate and smear out any effect on the velocity other than pore pressure. Thus any velocity anomaly such as that due to coal or gas sands, for example, will create a low velocity anomaly that would be counted as contributing to overpressure. That should not be the case. Various seismic inversion techniques can be used to determine high-frequency velocity information from seismic amplitudes. When this is combined with the low-frequency trends from velocity analysis techniques and possibly, FWI, we can obtain a detailed pore pressure profile that is better suited for drilling and well trajectory delineation. In this section we discuss some of the proven amplitude inversion techniques and its use for geopressure analysis.

Poststack Inversion

Poststack techniques are matured techniques and have been used with great success for lithology and fluid discrimination at reservoir scale (Lindseth, 1979; Quijada and Stewart, 2007; Avseth et al., 2005). Several different techniques/methodologies are commonly used to perform acoustic impedance inversion. The prominent ones are recursive, blocky, sparse-spike, stratigraphic and geostatistical inversion (Chopra and Kuhn, 2001). The objective of these techniques is to calculate a band limited impedance from the zero-offset reflectivity. The normal incidence reflection coefficients (R_i) are defined in terms of the impedance (I_i) as

$$R_i = \frac{I_{i+1} - I_i}{I_{i+1} + I_i} \tag{5.26}$$

The above equation can be solved for impedance in a recursive manner, for example. Quijada and Stewart (2007) use the following easy approach. Solving for I_{i+1}, taking the natural logarithm, making an approximation for small R and modeling the seismic trace as a scaled reflectivity R_k, namely,

$$S_k = 2R_k/\gamma \tag{5.27}$$

Equation (5.26) converts to

$$I_{i+1} = I_i \exp\left(\gamma \sum_{k=1}^{i} S_k\right) \tag{5.28}$$

Here γ is a scale factor. The inversion is performed using an approach discussed in Waters (1978). An initial impedance estimate is calculated from the well logs, and the seismic trace is integrated and exponentiated, according to equation (5.28). The Fourier spectrum of the integrated trace is scaled to that of the estimated impedance and a low-pass-filtered impedance is added to the trace. The result is transformed to the time domain, and a new impedance estimate is obtained. The low- and high-cut frequencies depend on the seismic data. The absolute impedance is obtained by adding the low-frequency impedance trend obtained from velocity analysis (high-density conventional velocity analysis, tomography, or FWI). The density model could be supplemented by models such as Gardner et al. (1974). Alternatively, a density model could be created using locally calibrated velocity data with well logs. The low-frequency model for velocity should be calibrated with well data such as checkshot, VSP, or sonic logs.

The next step is to produce a *high-frequency* or *high-resolution* velocity cube. This is done by adding low- and high-frequency velocity and acoustic impedance data from velocity analysis and inversion, respectively. The procedure for pore pressure prediction using seismic inversion as described is given by Dutta and Khazanehdari (2006) and Khazanehdari and Dutta (2006). There are two main steps in this process: (a) Generation of an equivalent density cube from a velocity cube using a rock physics model such as Gardner's equation that is calibrated to local and available well data. Castagna et al. (1993) also give a good overview of various methods for determining density from velocity. We also addressed this in Chapter 4. (b) Combining relative acoustic impedance with the product of velocity and density cubes. The result would be an absolute acoustic impedance (ABSAI) cube containing *both* the high-frequency component from seismic inversion and low-frequency components from seismic velocity analysis. Finally, the high-resolution velocity cube is generated by dividing the ABSAI by a density cube containing both high- and low-frequency components. In the example given below the density cube in 3D is generated from kriging of well density logs and velocity cube. The aim is to extend the range of the hard data (well density) by use of a correlated soft data (velocity). There are a number of kriging techniques that can be deployed, such as collocated co-kriging with trend. Some example applications to pore pressure estimation are given below.

The acoustic impedance volumes so generated have significant advantages. These include increased frequency bandwidth, enhanced resolution and reliability of amplitude interpretation through detuning of seismic data and obtaining a layer property that affords convenience in understanding and interpretation. However, the usefulness of the approach depends on two factors: (1) availability of a robust and reliable low-frequency impedance trend and a viable density model and (2) ability to account for variations in the acoustic impedance due to factors other than pore pressure such as lithology, porosity, and fluid type and its saturation.

Prestack Inversion

Prestack inversion helps in reducing the ambiguities inherent to the acoustic impedance, as it can generate not only compressional but shear information for the rocks under consideration as well as bulk density in some cases. For example, for a gas sand a lowering of compressional velocity and a slight increase in shear velocity is encountered, as compared with a brine-saturated sandstone. Prestack inversion provides this discrimination in such cases. This suggests that Amplitude Versus Offset or AVO technology can be reliably used for pore pressure analysis.

Amplitude variations with offset (AVO) or reflection angle is an important technology that is now matured. Avseth et al. (2005, Chapter 4) contains a very good summary of this technology with several case studies. Briefly, the P-to-P reflection coefficient $R(\theta)$ for a plane P-wave incident at angle θ at an interface between two elastic halfspaces (denoted by suffixes 1 and 2) is approximated by (Aki and Richards, 1980; Shuey, 1985) (see Figure 5.21)

$$R(\theta) \approx R(0) + G \sin^2\theta + F(\tan^2\theta - \sin^2\theta) \tag{5.29}$$

where

$$R(0) = \frac{1}{2}\left(\frac{\Delta V_p}{V_p} + \frac{\Delta\rho}{\rho}\right)$$

$$G = R(0) - \frac{\Delta\rho}{\rho}\left(\frac{1}{2} + \frac{2V_s^2}{V_P^2}\right) - \frac{4V_s^2}{V_p^2}\frac{\Delta V_s}{V_s}$$

$$F = \frac{1}{2}\frac{\Delta V_p}{V_p}$$

with

$$\begin{array}{ll}
\Delta\rho = \rho_2 - \rho_1 & \rho = (\rho_2 - \rho_1)/2 \\
\Delta V_p = V_{p2} - V_{p1} & V_p = (V_{p2} + V_{p1})/2 \\
\Delta V_s = V_{s2} - V_{s1} & V_s = (V_{s2} + V_{s1})/2 \\
\end{array}$$
$$\theta = (\theta_1 + \theta_2)/2 \approx \theta_1$$

Here V_p and V_S denote P- and S-wave velocity, respectively; ρ denotes bulk density and the contrast in material properties is designated by Δ. Equation (5.29) shows the angular range dependency of the reflection coefficient $R(\theta)$. $R(0)$ is the reflection coefficient at normal incidence, G describes the variation at intermediate offsets and is often referred to as the AVO gradient and F dominates the far offsets. For angles less than or equal to 30°, Shuey (1985) suggests that only the first two terms of equation (5.29) are adequate:

$$R(\theta) \approx R(0) + G \sin^2\theta \tag{5.30}$$

The zero-offset reflectivity $R(0)$ is controlled by the contrast in acoustic impedance across an interface. The gradient G depends not only on contrast in P-wave velocity and density but also on the contrast in S-wave velocity across the interface. Thus it is related to Poisson's ratio through V_p/V_s ratio. Thus the gradient values (G) could be related to impedance changes due to gas or oil. We also noted in Chapter 2 (Figures 2.30) that we could obtain pore pressure values from the AVO-gradients. Similar to

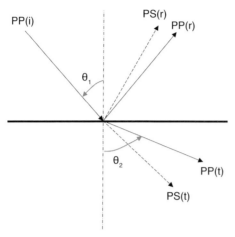

Figure 5.21 Reflections and transmissions at a single interface between two elastic half-space media. The incident plane P-wave is denoted by PP(*i*). All reflected and transmitted waves, denoted by suffixes (r) and (t), respectively, are also shown along with mode converted waves (occurring only at nonzero angle of incidence).

poststack impedance inversion, AVO inversion would require a stable and geologically consistent low-frequency trends of P-wave and S-wave impedances. Availability of shear wave logs for calibration are critical for AVO inversion for pore pressure. We found AVO inversion for P- and S- reflectivities as described by Fatti et al. (1994) to yield reliable results for V_p/V_S and pore pressure (see examples below). The method is valid for incident angles up to 50°.

The AVO derived reflectivities are usually inverted individually to determine rock properties such as lithology, fluids, and pore pressure for the respective rock layers. The accuracy and resolution of rock property estimates depend to a large extent on the inversion method utilized. Below we discuss a *global optimization* approach (Sen and Stoffa, 1991; Mallick, 1995; Ma, 2001; Aleardi and Mazzotti, 2016) that is computer intensive. We have found that these methods produce good results for pore pressure at higher resolution than the velocity analysis techniques discussed so far (Mallick and Dutta, 2002). In these model-driven inversion methods, first, synthetic seismic gathers are generated using an initial subsurface model using a forward model that uses the *elastic* full-wave equation. Thus primary events along with *all multiples* and *mode converted* events are *modeled* as opposed to eliminating them, for example, by preprocessing the seismic data prior to AVO inversion as is the customary practice. These are then compared to real seismic gathers. The model is next modified and synthetic data are updated and compared to the real data again. If after a number of iterations no further improvement is achieved, the updated model is the inversion result. Constraints based on rock physics compliant models are incorporated to reduce the nonuniqueness of the output. These methods utilize a Monte Carlo random simulation approach in the modeling and inversion steps and effectively try to find a *global* minimum without making assumptions about the shape of the objective function and are thus independent of the starting models. The genetic algorithm

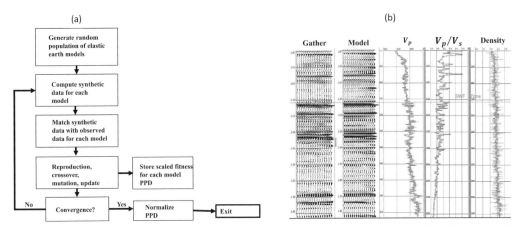

Figure 5.22 Prestack Full-Waveform inversion using Genetic Algorithm modified after Mallick and Dutta (2002). (a) A flowchart detailing the steps involved in the prestack full-waveform inversion of a CMP gather using the GA procedure. (b) Predicted V_P, V_P/V_S, and density from full-waveform and prestack inversion together with uncertainty estimation. The seismic CMP gather and the final solution labeled "synthetic" are also shown at the left using the GA algorithm as discussed in the text and shown in (a). Note the large uncertainty in the estimation of density. This is mainly due to limited offset range of the data.

(GA)-based global optimization approach developed by Mallick (1995) and used by Mallick and Dutta (2002) showed that the extracted V_P/V_S ratios compared very well with pore pressure. In this regard the GA approach is similar to *elastic FWI*; however, instead of a penalty-function and conjugate gradient type of approach in FWI, GA uses a Monte Carlo method that includes multiple possible perturbations of the model (termed reproductions, mutations, and crossover in GA terminology) and searches for global minima that are practically independent of the starting model. In this regard, the inverse problem is solved in a probabilistic framework (Duijndam, 1988; Tarantola, 2005) in which the final solution is represented by the posterior probability distribution (PPD) in the model space. The PPD is approximately represented by the posterior samples after convergence of the GA approach. The entire workflow in the GA approach is shown in Figure 5.22a. An example application is shown in Figure 5.22b. The example of Figure 5.22b is a 1D model and it is intended to simulate P-wave and S-wave sonic logs. The PPD approach yields uncertainties and these are also shown. The density estimation has the largest uncertainty and it is due to the limited range of offsets in the data. The velocity functions (P- and S-waves) shown in Figure 5.22b are ideal for pore pressure applications.

Example Applications of Seismic Inversion for Geopressure Prediction

In this section we present some example applications of seismic inversion to pore pressure prediction. A major advantage of this technology is that it produces *high-resolution pseudo-logs* that mimic the band-passed versions of actual well logs. This brings us closer to exploration drilling applications. Another advantage is that one can discriminate between various lithologies such as carbonate, sandstone, coal and shale and reservoirs with varying fluids (e.g., gas sands) prior to predicting pressure. The following examples elucidate these points.

An example of the application of poststack seismic inversion derived velocity to pore pressure is shown in Figures 5.23 and 5.24. This example is from the Gulf of Mexico, Keathely Canyon Area (KC-291) (Banik et al., 2013). There is a discovery well known as Kaskida in this block. The well penetrated subsalt formations to reach the targets. The 3D seismic data surrounding this well was used for seismic inversion. The low-frequency model was based on conventional but spatially continuous and automated velocity model building process; every trace and time sample was used in the velocity analysis following the procedure by Mao et al. (2000). Figure 5.23a shows the 3D impedance volume from seismic inversion. A comparison of the seismic impedance with that from the petrophysical well logs from the Kaskida well is shown in Figure 5.23b. The comparison is good. It shows that the low-frequency trend was captured well by the velocity model. This high-resolution velocity model was used to generate the 3D volume of pore pressure gradient in ppg, shown in Figure 5.24a. The authors converted the Bowers's model (Bowers, 1994) relating velocity to effective stress, to impedance. This allowed them to use the impedance data directly for pore pressure gradient calculations. Figure 5.24b shows a comparison of the predicted pore pressure gradient at the well location and comparison with the measured values. (These are adjusted mud weights calibrated by RFT data that are not shown.)

The next example is from an offshore area outside of the USA (Khazanehdari and Dutta, 2006). The reservoirs were of the Miocene-age rocks. The purpose of this geopressure modeling exercise was to use the predicted results for well appraisal and help plan the well trajectory to reach the deep targets safely. Since the targets contained various hydrocarbon resources in a structurally complex area with multiple pay zones encased in seals comprising faults and overpressured shales, the project needed to deal with lateral transfer of fluids from adjoining shales and generation of excess pressure due to buoyancy effects associated with hydrocarbons along with undercompaction and burial diagenesis of shales as multiple sources of geopressure. These were discussed in Chapters 3 and 4. This required seismic inversion to understand the reservoir properties and model the effect of hydrocarbons on velocity. A high-density velocity field was generated using an automated velocity picker that generated a spatially continuous stacking velocity based on semblance-style stacking correlations (Mao et al., 2000). Lateral cascaded median smoothing was used to preserve any lateral velocity discontinuities within the data. The method was quality controlled by human intervention at key steps. The results were checked with the semblance analysis procedures and available borehole data (checkshot calibrated sonic logs and VSP). The authors used a sparse-spike-type inversion algorithm that did not require any a priori information to constrain the acoustic solution; hence it is unbiased because it is independently determined on a trace-by-trace basis. The density cube was generated from kriging of well density logs and velocity cube. The aim was to extend the range of the hard data (well density) by use of a correlated soft data (velocity).

By removing the full-frequency density trend from the computed absolute impedance, an estimate of the high-resolution velocity field was obtained that is more suited

(a)

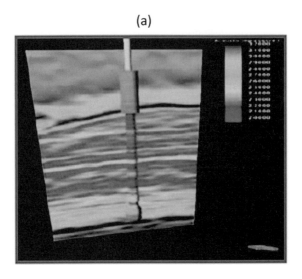

(b)

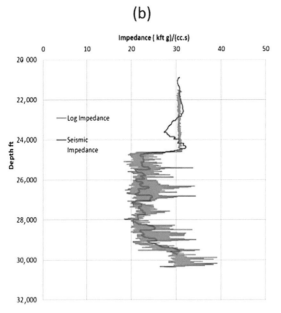

Figure 5.23 (a) An example of high-resolution impedance volume in 3D from seismic data including the overlain impedance trace from the Kaskida well, KC-291 (GOM), and (b) a comparison of the seismic impedance at the Kaskida well (blue) with those derived from the well logs (red). This was a postdrill evaluation study. Modified after Banik et al. (2013). (A black and white version of this figure will appear in some formats. For the color version, please refer to the plate section.)

to drilling applications than those from the low-frequency velocity analyses techniques. In Figure 5.25a we show the absolute acoustic impedance on the left panel. The derived high-resolution velocity cube is shown on the same figure (right panel of

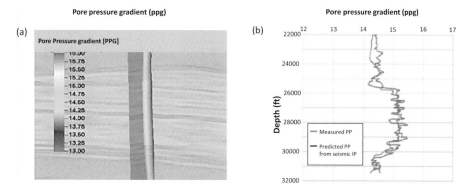

Figure 5.24 An example of predicted pore pressure using poststack seismic inversion method (GOM). Modified after Banik et al. (2013). (left) Pore pressure gradient (Equivalent Mud Weight) in ppg for a seismic data set including the Kaskida well, KC-291 (GOM). (right) A comparison of the pore pressure gradient estimated from the high-resolution impedance inversion labeled Predicted PP from Seismic IP (red) to that estimated from Equivalent Mud Weight (in ppg) and labeled as Measured PP (blue). These are mud weights that is calibrated by RFT data (not shown). (A black and white version of this figure will appear in some formats. For the color version, please refer to the plate section.)

Figure 5.25a. In Figure 5.25b we show the details of the velocity field at the well location. The bottom figure of the panel shows the low-frequency trend and the top panel shows the high-resolution details at the reservoir scale.

The acoustic impedance cube yielded a high-resolution background pore pressure as well as lithology discrimination between sand and shale. Two sand bodies (pay zones) were delineated from surrounding shale and their properties were characterized and visualized by 3D visualization technique as shown in Figure 5.26a. Estimation of excess pressure in these sand bodies due to structural relief was made by following the procedure discussed by Stump et al. (1998) and discussed earlier in Chapter 3. Since the reservoirs had pay, it was required to simulate the velocity for the interval had it been water saturated. This calculation was performed using Gassmann fluid substitution procedure (Gassmann, 1951) as outlined in Chapter 2 and details given by Mavko et al. (2009). After this, aquifer pore pressure was estimated using the lateral pressure transfer mechanism (Stump et al., 1998). Figure 5.26a shows the computed pore pressure gradients in ppg (left panel) for the two sand bodies. We note the left panel of the figure shows the relative pore pressure difference (expressed in fraction) between the sands and the surrounding shale. The right panel shows the corresponding total pressure in sands in ppg. Figure 5.26b shows the composite display of seismically derived pore pressure in ppg (left panel) and the high-resolution interval velocity (right panel) after combining seismic velocity and inversion results. The two well trajectories and the horizontal slice displaying the pore pressure gradients in ppg are also shown in this figure.

(a)   (b)

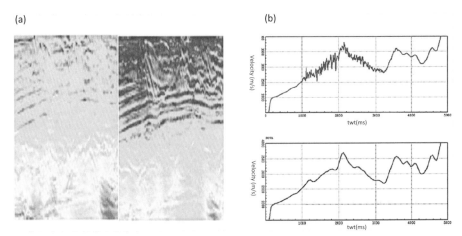

Figure 5.25 High-resolution impedance and velocity from seismic poststack inversion as outlined in the text. (a, left) The absolute acoustic impedance from seismic inversion. (a, right) The derived high-resolution velocity. (b) Comparison between high-resolution velocities (top) with the low-resolution velocity (bottom) at a well location. Velocity scale is from 2000 m/s to 4000 m/s in step of 500 m/s. Two-way time in millisecond (TWT) is from 0 to 5000 ms in step of 1000 ms. After Khazanehdari and Dutta (2006). (A black and white version of this figure will appear in some formats. For the color version, please refer to the plate section.)

(a)   (b)

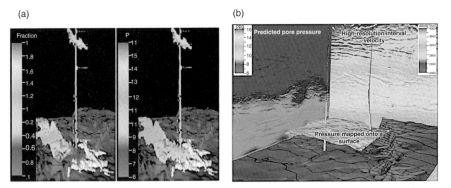

Figure 5.26 Pore pressure gradient in ppg from seismic inversion. (a) Two sand bodies mapped using acoustic impedance and velocity data. (left) The relative change in pressure gradient between the sand the surrounding shale. (right) The total pore pressure gradient in the sands. (b) A composite display of seismically derived pore pressure (left) and the high-resolution interval velocity (right) after combining seismic velocity and inversion results using poststack inversion. The two wells and the horizontal slice displaying the pore pressure gradients in ppg are also shown. The color scale for pore pressure (ppg) is from 6 to 16 in step of 1 ppg. The color scale for velocity (m/s) is from 2000 to 4000 in step of 200 m/s. After Dutta and Khazanehdari (2006). (A black and white version of this figure will appear in some formats. For the color version, please refer to the plate section.)

Figure 5.27 shows a breakdown of pore pressure contributions in the sand bodies due to lateral transfer of aquifer pore pressure and buoyancy effect. Figure 5.27a has four panels. The panel on the left shows pore pressure gradient (ppg) with the assumption that the sand pressure is equal to the surrounding shale pressure. The second panel from left shows in

(a)   (b)

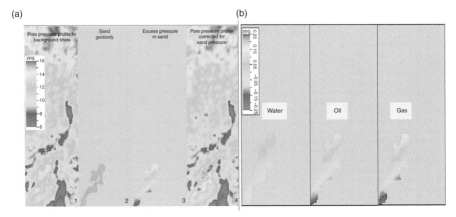

Figure 5.27 Pore pressure contributions in the sand bodies due to lateral transfer of aquifer pore pressure and buoyancy effect. (a) The left panel shows pore pressure gradient (ppg) with the assumption that the sand pressure is equal to the surrounding shale pressure. The second panel from the left shows in detail the sand geometries. The third panel from left shows the excess pore pressure in the sand bodies. The fourth panel from left shows the total pore pressure gradient in sand as compared to the background pore pressure gradient in the surrounding shales. (b) The relative difference between pressure gradient in the inclined sand bodies and the surrounding shale as function of fluid type (water, oil, or gas). After Dutta and Khazanehdari (2006). (A black and white version of this figure will appear in some formats. For the color version, please refer to the plate section.)

detail the sand geometries. The third panel from left shows the excess pore pressure in the sand bodies. The fourth panel from left shows the total pore pressure gradient in sand as compared to the background pore pressure gradient in the surrounding shales. Figure 5.27b shows the relative difference between pressure gradient in the inclined sand bodies and the surrounding shale as function of fluid type (water, oil, or gas).

An example application of the predicted high-frequency information from seismic inversion based on the genetic algorithm (GA) approach for pore pressure prediction is shown in Figure 5.28. This was discussed by Mallick and Dutta (2002). The example is from offshore, GOM area. The area was known as a shallow water flow (SWF) prone area (Mississippi Canyon) with a possibility of disturbance due to geohazards. The well was drilled successfully; however, in the shallow interval a SWF incident was indeed recorded. The goal of the project was to assess if the GA approach to high-resolution velocity could be applied to detect this problem prior to drilling. In Figure 5.22 we showed the predicted P-wave, S-wave, and density at the well location from GA inversion technology. These were used to predict the pore pressure gradient, overburden stress gradient and the fracture gradient (all in ppg) shown in the third panel of Figure 5.28. The Equivalent Mud Weight (EMW) at the SWF level was 9.6 ppg; this is within the range of the prediction from the GA-technology as noted. It should be noted that conventional velocity analysis could not be performed at such shallow depths such as that shown in Figure 5.28; however, inversion did yield useful velocity information at these depths.

The next application shows the results of full-waveform inversion using a *hybrid GA* algorithm in 3D (Mallick and Dutta, 2002; Dutta et al., 2010). This approach was

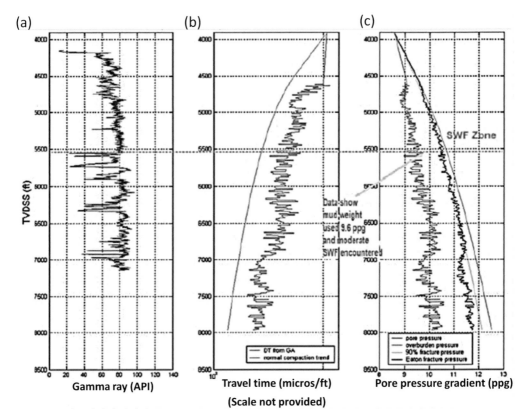

Figure 5.28 Application of high-resolution pore pressure prediction using genetic algorithm (GA) technology for seismic inversion (1D-elastic FWI) as discussed in the text. (a) Gamma-ray curve at the well corresponding to the seismic data in Figure 5.22. (b) High-frequency interval slowness derived from GA based inversion of seismic data shown along with superimposed Normal Compaction Trend (NCT) that is derived from a rock model discussed in Chapter 3 (Dutta model). (c) Predicted pore pressure, overburden pressure, and fracture-pressure gradients in ppg. The fracture-pressure gradient is based on the assumption that the fracture pressure is 90 percent of the overburden pressure.

developed by Mallick (1999) and extended to geohazards (shallow and deep) by Dutta et al. (2010) and Dutta and Mallick (2010). The example presented here is an application in the deepwater Mediterranean Sea, Egypt. This was the first deepwater well drilled in that basin and no well control was available for calibration. The nearest analogue was a well ~100 miles away from the exploration well. The geology is clastics (in the overburden) and carbonates (in the target zone). There were concerns with shallow geohazards as well as high geopressure in the deeper targets. This area of the Mediterranean Sea is known to have extensive chaotic channel systems in the shallower sections, approximately 300 m below the ocean bottom as shown in Figure 5.29a. The channel systems and zones beneath these channels were judged to present high shallow water flow risk during drilling.

Since there was no 3D FWI algorithm available at the time when the project was executed, a hybrid-GA approach was used Mallick (1999). In this approach, the workflow is carried out in two steps. The first step involved using 1D-FWI (elastic) with GA-technology for inversion at 15 locations on the 3D seismic data volume. The pseudo-well locations were chosen after a detailed geologic model building including establishing a stratigraphic framework that identified hazardous zones (both shallow and deep). The inversion provided detailed P-wave, S-wave, and bulk density pseudo-logs at each locations similar to Figure 5.22b. Band-pass versions of these pseudo-wells were used to further build the low-frequency model for subsequent prestack inversion in 3D using 3D-AVO methodology. (Hence the approach was designated as 3D-hybrid elastic FWI inversion.) High-frequency velocities obtained from seismic inversion provided the basis for pore pressure prediction as shown in Figure 5.29a and allowed us to estimate pore- and fracture-pressure gradients in PPG as shown in Figure 5.29b. For pore pressure calculations, further corrections were made due to structural relief of dipping permeable beds as discussed earlier (lateral transfer of fluids). For estimation of the fracture-pressure gradient, V_P/V_S ratios obtained from full-waveform inversion were converted to Poisson's ratio (v)

$$v = \frac{(V_P/V_s)^2 - 2}{2[(V_P/V_s)^2 - 1]} \tag{5.31}$$

and used in the Eaton model for fracture-pressure gradient calculation (Eaton, 1999) after corrections to static values. Similarly, overburden pressure was estimated using bulk densities from the seismic inversion as explained earlier. The results of this study directly impacted the predrill planning and final well positioning. The exploration well was drilled without encountering any geohazard problems (both shallow and deep). The work eliminated the need for a conventional geohazard survey and pilot-hole program. Further, the predicted pressure- and fracture-gradient profiles aided considerably in well planning and drilling in the target area as well.

5.4.4 Pore Pressure Prediction Using Shear Wave Velocity Data

We have seen in Chapter 2 that shear wave velocity (V_s) is an important tool that could be used in the prediction of geopressure. The basis for using shear wave velocity for pore pressure analysis is discussed by Dvorkin et al. (1999a), Dutta (2002b), Dutta et al. (2002c, 2002d), and Ebrom et al. (2003). The key variable is V_P/V_s or Poisson's ratio v defined in equation (5.31). As shown in Figure 2.30, V_p/V_s is strongly related to effective stress σ – decreasing effective stress (increasing pore pressure) causes V_p/V_s to increase rapidly as Poisson's ratio approaches toward its upper limit of 0.5. The fact that this ratio does not change linearly with effective stress suggests that increasing pore pressure has a relatively *stronger* effect on shear wave than on the compressional or P-wave. This is the rock physics basis for using shear wave for geopressure detection and prediction. There is another advantage. We have seen in Chapter 2 that fluid saturation has a significant effect on P-wave but has little effect on S-wave. For example, imaging through gas "chimneys" or "clouds" with P-waves is known to be extremely difficult due to the fact that P-velocities are *reduced* by gas, even with a low

(a)

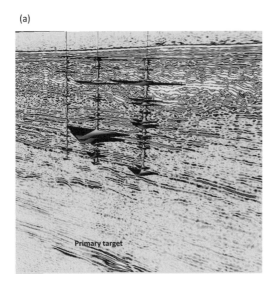

(b)

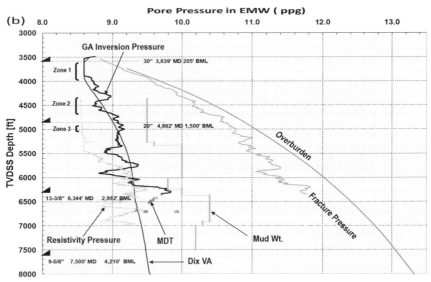

Figure 5.29 Pore pressure prediction using 3D hybrid GA inversion (FWI) of seismic data as discussed in the text. (a) 3D seismic line through three pseudo-wells. The pseudo-well logs were generated using the GA algorithm for 1D-GA for elastic FWI. The magnitudes of the V_P/V_S anomalies are proportional to the size of the diamond displayed. There were 15 pseudo-wells used in the pore pressure prediction shown in (b). (b) Results of the predrill pore pressure prediction (pounds per gallon versus depth) derived using 3D hybrid GA inversion (blue curve) and showing the casing points and mud weights used to drill the well. Note the agreement of the predicted pore pressure with the postdrill modular dynamic tester (MDT) measurements. Also note the narrow mud weight operational window the driller used as per the model prediction. From Dutta et al. (2010). (A black and white version of this figure will appear in some formats. For the color version, please refer to the plate section.)

saturation. Thus using P-waves in this case for pore pressure analysis would yield erroneous result as gas saturation can be confused with high pressure – each reduces P-velocity. Hence shear waves should provide a better tool for detecting geopressure as it can discriminate between gas and overpressure. In addition V_P/V_s can also be used for lithology identification in conjunction with V_P.

For onshore projects, shear wave seismic data can be obtained by either dynamite or vibrator as a seismic source. For offshore environments, both P- and S-wave velocities can be obtained using multicomponent receivers placed at the seafloor (ocean-bottom seismometers or OBS) with air gun as a source. Mode conversion (from P to S) occurs at a target below seabed. V_p/V_s ratio is estimated from PS-converted reflected waveforms of OBS data (MacBeth et al., 1992; Peacock et al., 2009). However, calibration is required and is usually provided by downhole logging of V_s data. Typically event registration method is used to obtain estimates of the interval P- and S-wave velocity ratio (Kao et al., 2010). This is not easy as a proper interpretation and identification of seismic events is needed before events are identified – one needs to know the interface where the seismic events originate from. Proper conditioning of seismic data to analyze PP- and PS-wave velocities is very important to prediction of pore pressure (Kao et al., 2010) as the noise present in the seismic data do not affect the PP and PS events equally – S-wave velocities are more prone to errors than the P-waves. In the Nankai Trough, offshore Japan, Tsuji et al. (2011) used PPS-refracted waves that are converted from the up-going P-waves to up-going S-waves at an interface below the receivers. This was because there was no identifiable PS-converted reflection waveforms for each geological boundary in their study area. (There was no dominant geological boundary within the sedimentary sequence in the study area.)

Most authors who have published applications of OBS data for pore pressure prediction use some modifications to the Eaton method (Eaton, 1975) for the shear wave velocity. Ebrom et al. (2003) suggest using the following equation for S-waves that is similar to what Eaton has suggested for P-waves:

$$\frac{\sigma}{\sigma_n} = \left(\frac{V_{PS}}{V_{PS,n}}\right)^{E_{ps}} \tag{5.32}$$

where V_{PS} is the interval velocity observed from the OBS-processed data (PS) under overpressured conditions, $V_{ps,n}$ is the interval velocity in normally pressured environments, σ is the predicted effective stress, and σ_n is the effective stress in a normally pressured region. E_{ps} is an empirically determined Eaton's PS-wave exponent. For P-waves, Eaton exponent is typically 3.0 (Eaton, 1997). For direct shear wave recording, Ebrom et al. (2003) suggest values in the range of 2.5–2.7 for moderate overpressure (equivalent mudweight in the range of 13 ppg or 0.675 psi/ft) and from 2.0 to 2.6 for high overpressure (equivalent mudweight in the range of 17 ppg or 0.883 psi/ft). For mode conversion such as the case for PS-phase under discussion here, the authors argue that since the process consists of a downgoing P-wave, and an up-coming reflected S-wave, the pressure sensitivity of converted-wave velocities would be intermediate between P-wave and S-wave velocity sensitivities. A depth dependent Eaton exponent is also not ruled out by the authors (Ebrom et al., 2003).

In Figure 5.30 we show an example of the use of OBS data for pore pressure prediction. The data is from the Atlantis Field, deepwater Gulf of Mexico, as reported by Kao et al. (2010). The objective of the project was to estimate pore pressure in the shallow part of the stratigraphy for shallow hazards risk assessment and well head placement. Figure 5.30a shows pore pressure in MPa from a collective output of the six nodes and highlights the three SWF zones with zones 2 and 3 deemed high risk. Computed pressures do show an overall increase in pore pressure over the background. The prediction methodology used equation (5.32). However, the authors used two velocity trends for estimates of the *normally* (hydrostatic pore pressure) compacted sediments – one from Eberhart-Phillips et al. (1989) labeled as EP and another from Hamilton (1976a) labeled as Hamilton. Both models predict pore pressure that is higher than the hydrostatic pore pressure by a significant amount. In Figure 5.30b we show a comparison of the V_p/V_s ratio as predicted from the OBS data with the background Hamilton (1976a) and Eberhart-Phillips et al. (1989) velocities. Note the high value of ~ 7.5 at 100 m below the seafloor. This is consistent with the observations made in Chapter 2 (Figure 2.28) that high pore pressure is associated with anomalously *high* V_p/V_s values over the background (Dvorkin et al., 1999a; Dutta and Mallick, 2010; Dutta et al., 2010).

5.4.5   Pore Pressure Prediction Using Seismic Wave Attenuation

So far we have discussed how seismic velocity can be used for pore pressure prediction. Some authors have suggested that seismic wave attenuation (Q-factor) may be used for this purpose as well (see, e.g., Young et al., 2004; Carcione and Helle, 2002; Salehi and Mannon, 2013). The basis for using Q-factor for pore pressure prediction is substantiated by rock physics (Prasad and Manghnani, 1997; Jones, 1995; Dvorkin and Mavko, 2006; Mavko et al., 2009; Valle et al., 2017) and the effect is physical. As pore pressure increases in a fluid filled porous rock, the grain-to-grain contacts are reduced as compared to the case when the rock was in hydrostatic pore pressure condition. This causes a seismic signal passing through this rock at any frequency to be attenuated more as compared to the case for normally pressured state. The common measures of attenuation are defined and related to one another as follows:

$$\frac{1}{Q} = \frac{M_I}{M_R} = \frac{\Delta f}{f} = \frac{\delta}{\pi} = \frac{\alpha V}{\pi f} = \frac{\alpha(dB/\lambda)}{8.686\pi} \tag{5.33}$$

where

Q = quality factor,
α = attenuation coefficient,
δ = logarithmic decrement,
Δf = resonance width,
M_R and M_I = real and imaginary parts, respectively, of complex modulus M,
V = velocity,
f = frequency,
λ = wavelength.

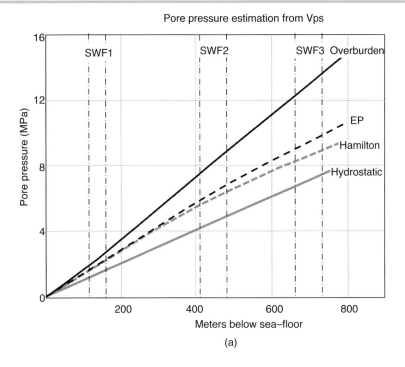

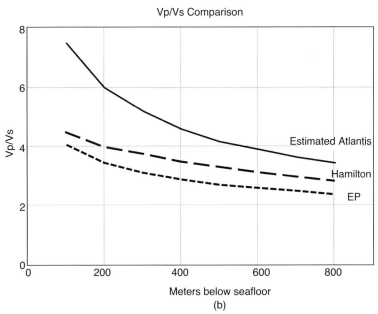

Figure 5.30 (a) Pore pressure prediction using compressional and shear wave data from multicomponent seismic nodes in Atlantis Field, deepwater Gulf of Mexico. (a) Pore pressure in MPa using the modified Eaton model of equation (5.31). Both Eberhart-Phillips (labeled by EP) and Hamilton (labeled by Hamilton) velocities are used as inputs in the pore pressure estimation. Deviation from EP and Hamilton curves suggests the sediments are over hydrostatic pore pressure even as shallow as 100 m below the seafloor. This is indicative of shallow hazards associated with shallow aquifer pressured sediments. (b) Predicted V_p/V_s ratio from OBS data and a comparison with the model predictions using Eberhart-Phillips and Hamilton velocities.

The quality factor Q is the most common parameter used in the seismic industry. It is expressed as

$$Q = \frac{2\pi E}{\Delta E} \quad (5.34)$$

The total seismic energy stored in a cycle is E and the change in energy per cycle is ΔE. *Lower* the value of Q, *higher* is the loss of energy. If we assume that the average frequency corresponds to E, then the change in frequency will correspond to ΔE. So the amplitude decay of a seismic signal can be represented as

$$A(t) = A(0)e^{-\frac{\pi f t}{Q}} \quad (5.35)$$

Since higher frequencies are attenuated faster that the lower frequencies – the effect is more pronounced with lower Q. This effect is labeled in the literature as *seismic frequency based pore pressure prediction*. Typically, Q ~ 30 for weathered sedimentary rocks, ~ 100–200 for brine-filled sedimentary rocks, and ~ 1000 for granite. Multiple studies (Prasad and Manghnani, 1997; Jones, 1995; Dvorkin and Mavko, 2006; Mavko et al., 2009) have shown that when pore pressure is applied to core samples, and a seismic wave is passed through the sample, Q-factor decreases with decreasing effective stress (increasing pore pressure for a fixed confining stress). One study (Siggins et al., 2001) concluded that Q was much more sensitive to pore pressure increase *than* the velocity. Thus with increasing depth of burial the seismic amplitude will decrease (higher frequencies will be attenuated) but the rate of decrease or loss of higher frequencies will be *larger* in the overpressured rocks than the normally pressured rocks at the same depth. An example is shown in Figure 5.31 for various mudweight window curves as noted. The application is from a field in the Mississippi Canyon area of deepwater Gulf of Mexico (Salehi and Mannon, 2013). The field included depleted reservoirs in close proximity to salt intrusions. The velocity-based pore pressure prediction using Bowers's method (Bowers, 1994) from both sonic in green (labeled PPP Bowers DT) and seismic velocity in blue (labeled PP V_{int}) are compared with predicted pore pressure with seismic Q-based method (labeled ppQ). The comparison is interesting. The Q-based method apparently agrees with the data better than the seismic velocity-based approach – the fast velocities within the channel zone apparently yielded lower pore pressure than the Q-based method.

We have no reason to doubt the validity of the results; however, we do raise an issue. It is difficult to assess the quality of seismic velocity used in the project, especially when targets are located within or near salt bodies as these tend to affect the seismic velocity and the image qualities in adverse ways. The fact that the velocity increases rapidly as salt intrusions are approached suggests that velocities might be contaminated and unfit for pressure prediction. So the comparison of the two methods may not be fair. There are several other issues with the Q-based

Figure 1.12

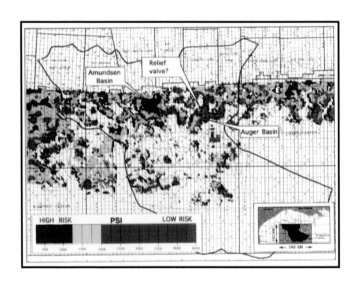

Figure 1.13

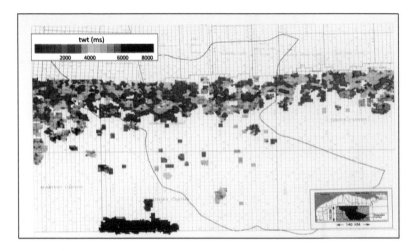

Figure 1.14

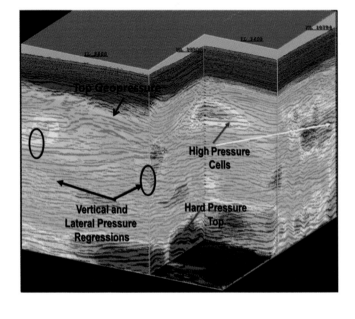

Figure 2.30

Vp / Vs Versus Effective Pressure, Pe

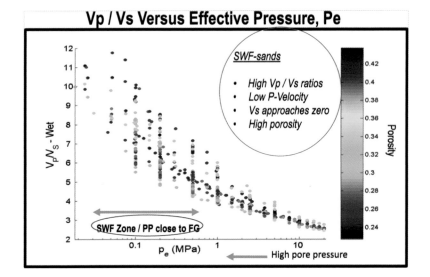

SWF-sands
- High Vp / Vs ratios
- Low P-Velocity
- Vs approaches zero
- High porosity

SWF Zone / PP close to FG

High pore pressure

Figure 3.20

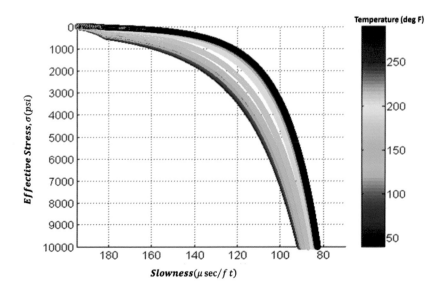

Pore pressure profile for background shale

Sand geobody

Excess pressure in sand

Pore pressure profile corrected for sand pressure

Figure 4.11

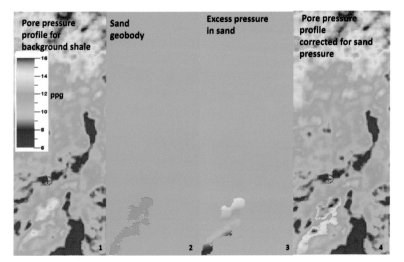

Figure 4.12

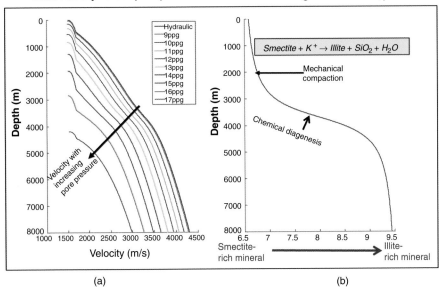

Rock velocity versus pore pressure

Diagenetic funcion, β

(a)

(b)

Figure 4.13

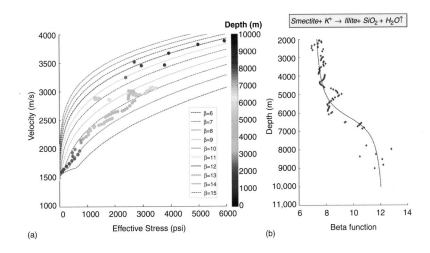

(a)

(b)

Figure 4.18

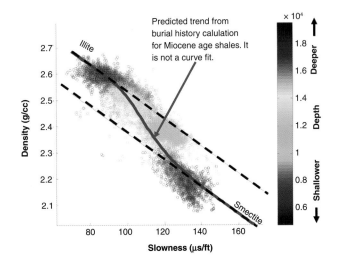

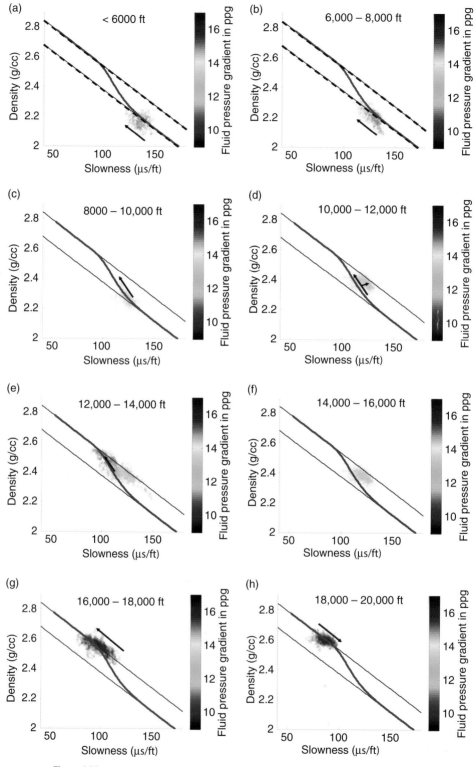

Figure 4.19

Figure 4.26

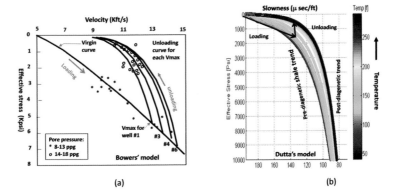

(a)

(b)

Figure 4.36

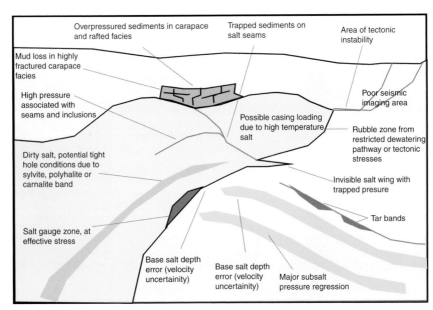

(a)

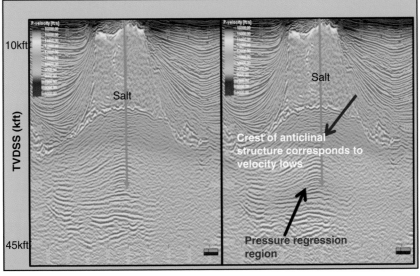

(b)

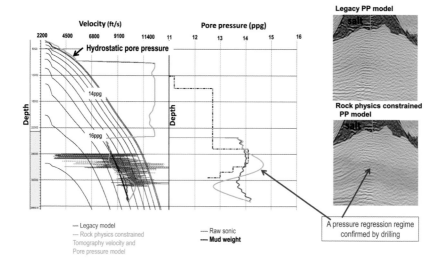

Figure 4.37

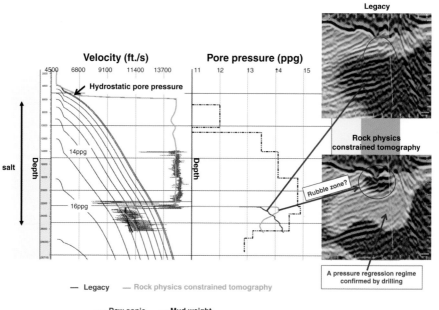

Figure 4.38

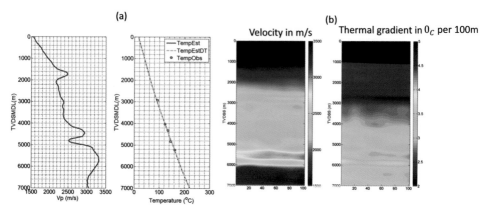

Figure 4.42

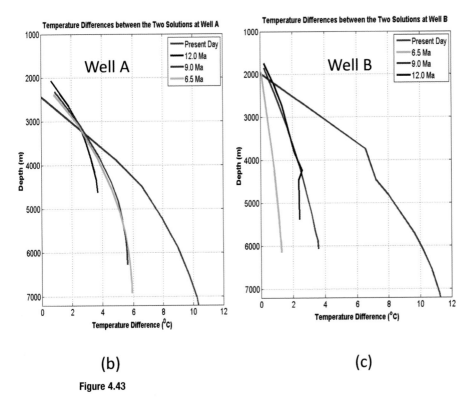

(b)

Figure 4.43

(c)

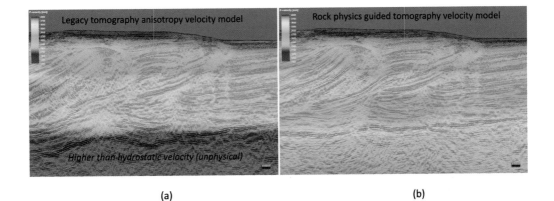

(a)

(b)

Figure 6.9

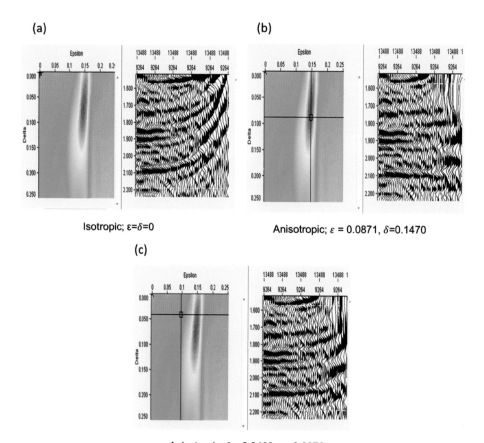

(a)

Isotropic; ε=δ=0

(b)

Anisotropic; ε = 0.0871, δ=0.1470

(c)

Anisotropic; δ = 0.0403, ε =0.0970

Figure 5.9

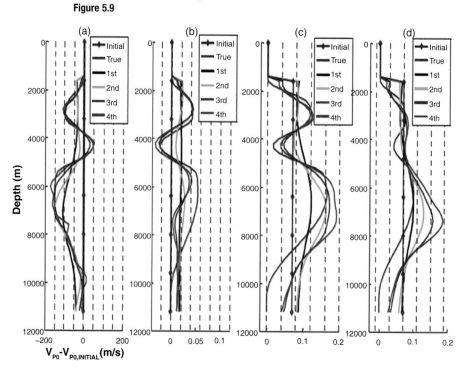

Figure 5.10

Figure 5.11

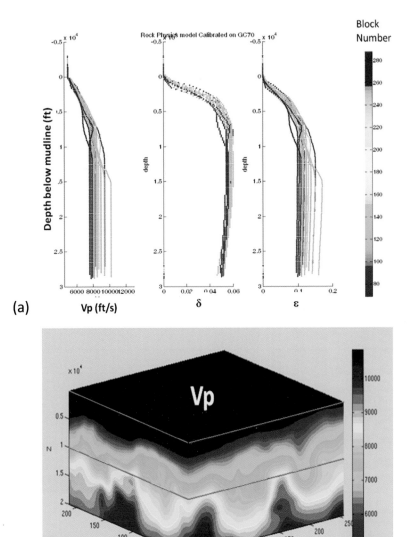

Block
Number

(a)

(b)

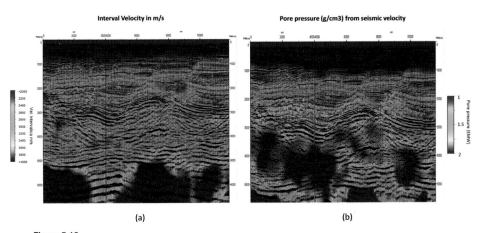

(a)   (b)

Figure 5.13

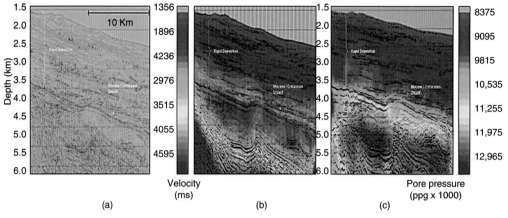

Velocity
(ms)

Pore pressure
(ppg x 1000)

(a)   (b)   (c)

Figure 5.14

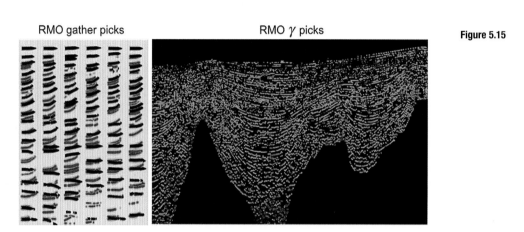

RMO gather picks   RMO γ picks

Figure 5.15

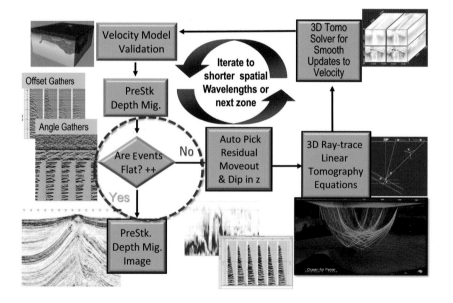

Figure 5.16

Figure 5.17

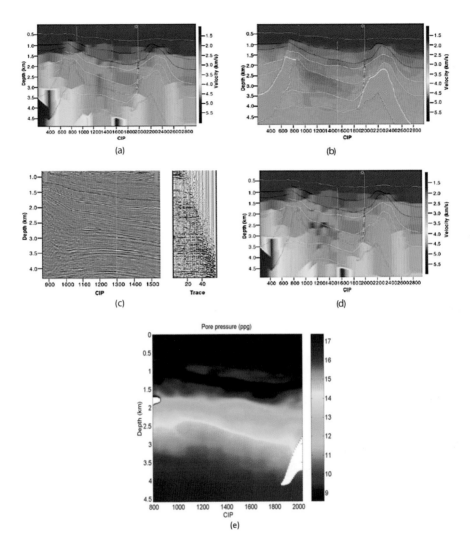

(a)

(b)

(c)

(d)

Pore pressure (ppg)

(e)

Figure 5.18

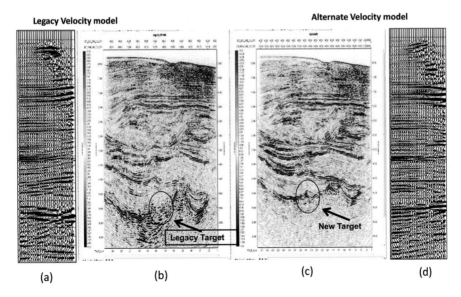

Legacy Velocity model

Alternate Velocity model

Legacy Target

New Target

(a)

(b)

(c)

(d)

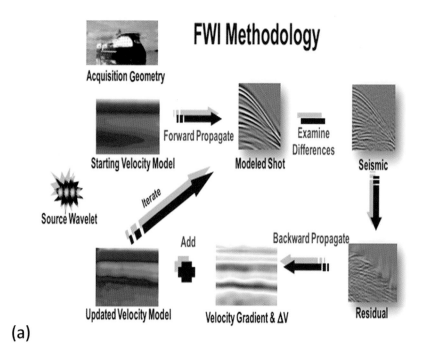

(a)

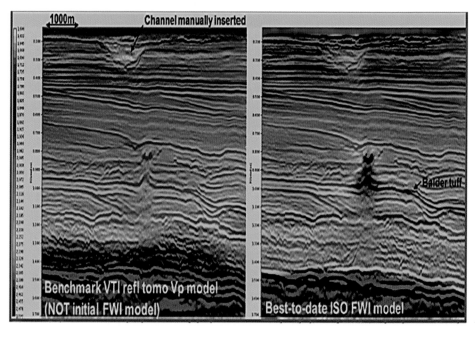

(b)

Figure 5.19

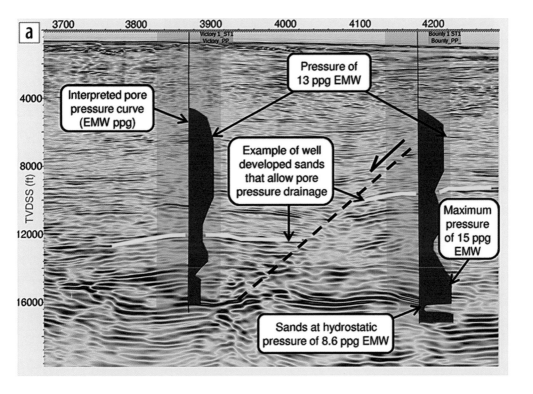

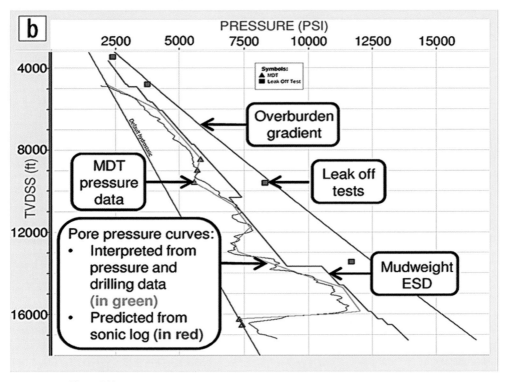

Figure 5.20

(a)

Figure 5.23

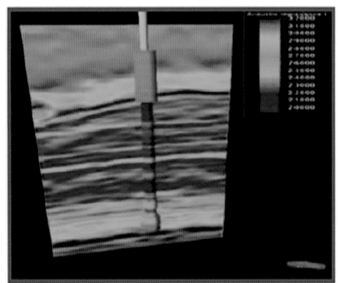

(b)

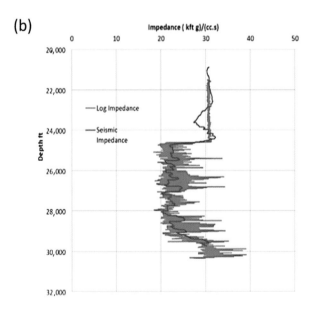

Pore pressure gradient (ppg)

(a)

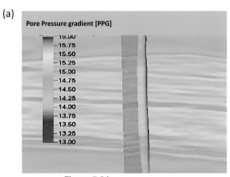

Figure 5.24

Pore pressure gradient (ppg)

(b)

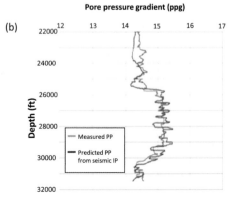

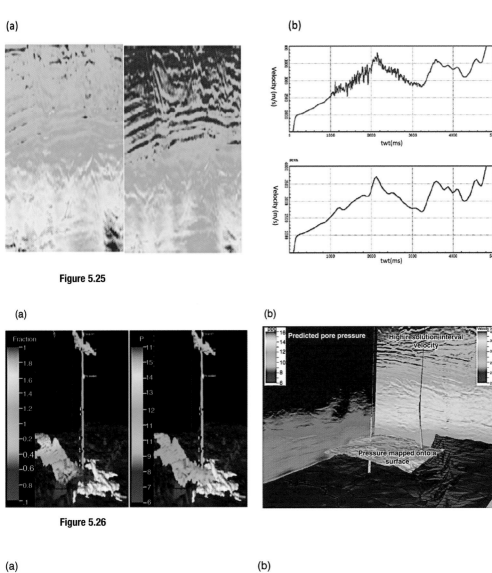

(a)

(b)

Figure 5.25

(a)

(b)

Predicted pore pressure
High resolution interval velocity

Pressure mapped onto a surface

Figure 5.26

(a)

Pore pressure profile for background shale

Sand geobody

Excess pressure in sand

Pore pressure profile corrected for sand pressure

(b)

Water

Oil

Gas

Figure 5.27

Figure 5.29

(a)

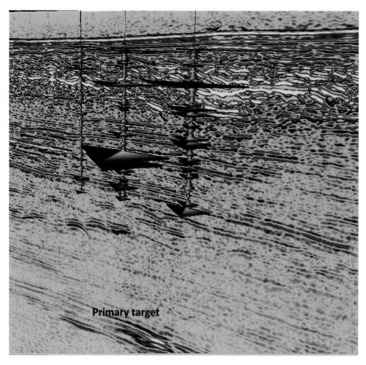

Primary target

(b)

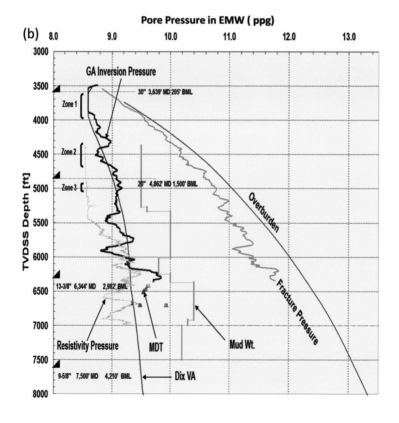

Figure 5.31

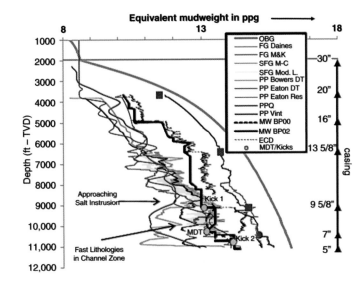

Figure 6.10

(a)

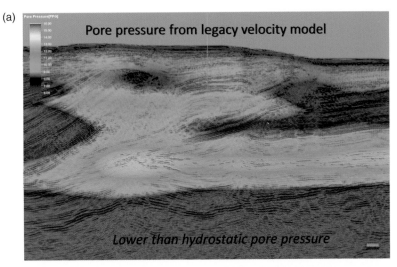

Pore pressure from legacy velocity model

Lower than hydrostatic pore pressure

(b)

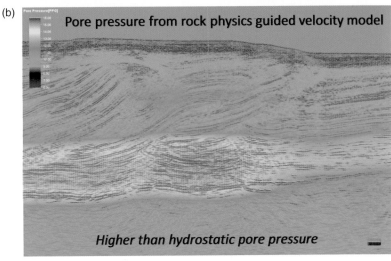

Pore pressure from rock physics guided velocity model

Higher than hydrostatic pore pressure

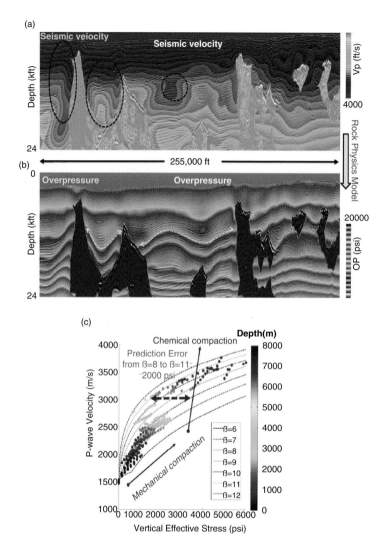

Figure 6.13

(a)

Seismic velocity

Seismic velocity

Depth (kft)

dV (s/ft)

4000

24

← 255,000 ft →

Rock Physics Model

(b)

0

Overpressure

Overpressure

Depth (kft)

20000

dP (psi)

24

(c)

Chemical compaction   **Depth(m)**

Prediction Error
from ß=8 to ß=11:
~2000 psi

8000

7000

6000

5000

4000

3000

2000

1000

0

4000

3500

3000

2500

2000

1500

1000

P-wave Velocity (m/s)

Mechanical compaction

------ ß=6
------ ß=7
------ ß=8
------ ß=9
------ ß=10
------ ß=11
------ ß=12

0 1000 2000 3000 4000 5000 6000
Vertical Effective Stress (psi)

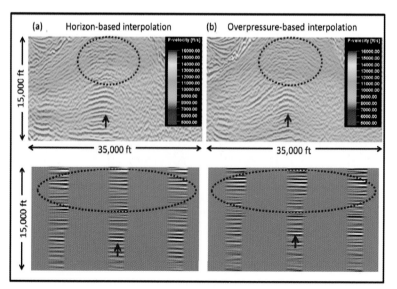

Figure 6.14

(a) Horizon-based interpolation   (b) Overpressure-based interpolation

15,000 ft

P-velocity [ft/s]

16000.00
15000.00
14000.00
13000.00
12000.00
11000.00
10000.00
9000.00
8000.00
7000.00
6000.00
5000.00

← 35,000 ft →

15,000 ft

P-velocity [ft/s]

16000.00
15000.00
14000.00
13000.00
12000.00
11000.00
10000.00
9000.00
8000.00
7000.00
6000.00
5000.00

← 35,000 ft →

a)

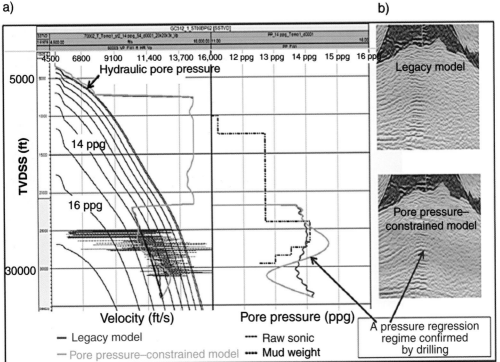

Figure 6.17

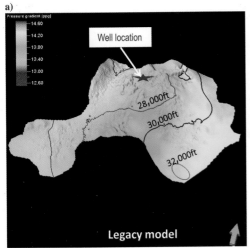

Figure 6.18

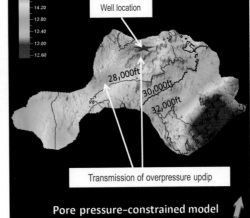

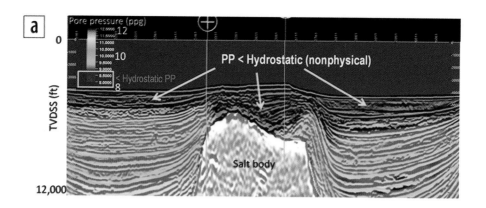

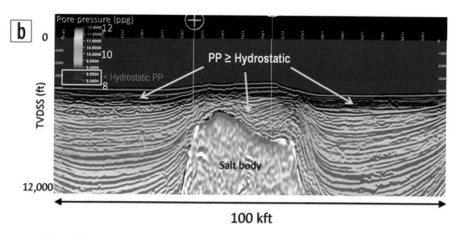

Figure 6.20

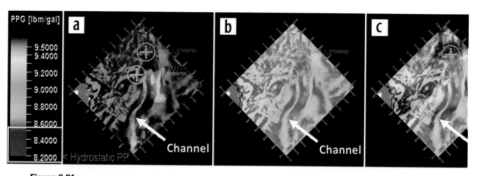

Figure 6.21

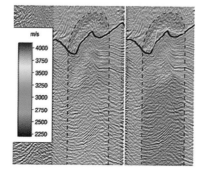

Figure 8.2

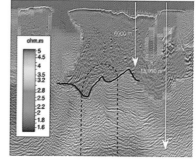

Figure 8.3

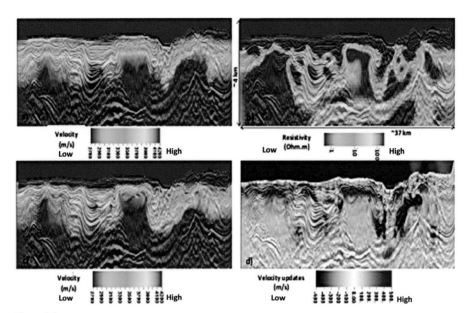

Figure 8.4

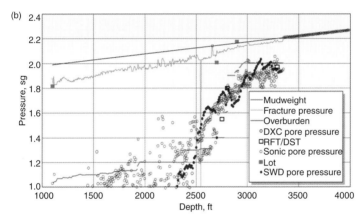

Figure 9.5

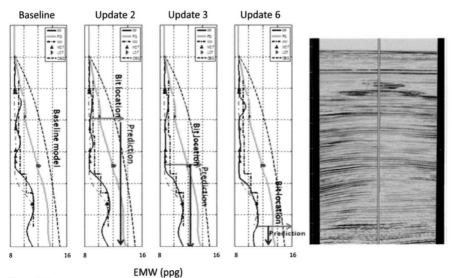

Figure 9.7

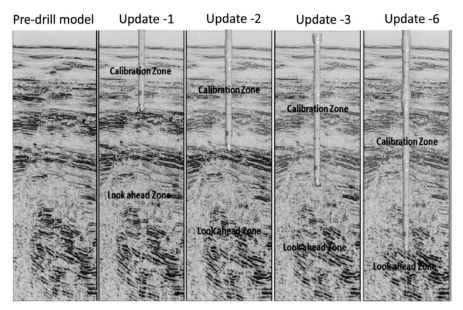

Figure 9.8

a

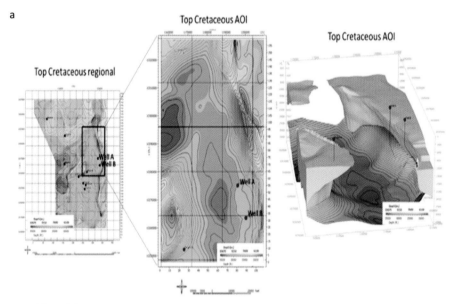

Figure 9.9

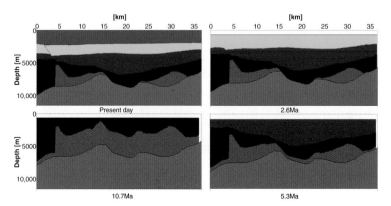

Figure 10.10

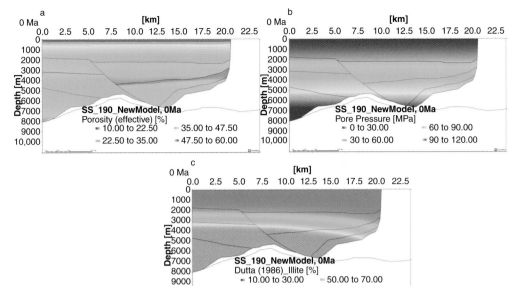

Figure 10.11

Figure 11.8

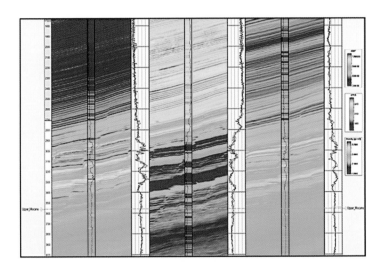

Figure 11.10

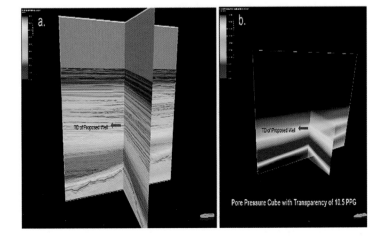

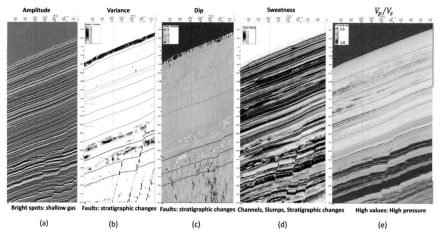

Amplitude

Variance

Dip

Sweetness

V_p/V_s

Figure 11.11

Bright spots: shallow gas   Faults: stratigraphic changes   Faults: stratigraphic changes   Channels, Slumps, Stratigraphic changes   High values: High pressure

(a)   (b)   (c)   (d)   (e)

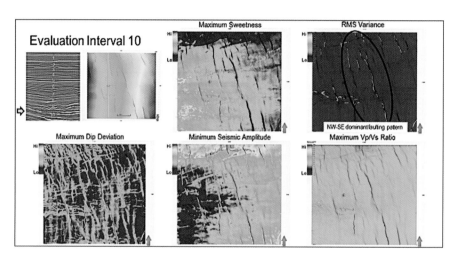

Evaluation Interval 10

Maximum Sweetness

RMS Variance

Figure 11.12

NW-SE dominant faulting pattern

Maximum Dip Deviation

Minimum Seismic Amplitude

Maximum Vp/Vs Ratio

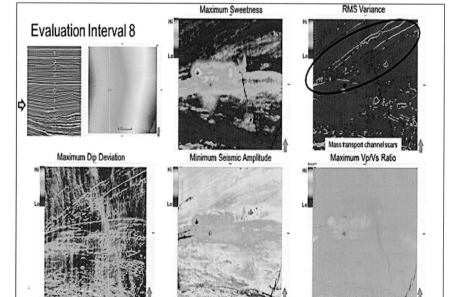

Evaluation Interval 8

Maximum Sweetness

RMS Variance

Figure 11.13

Mass transport channel scars

Maximum Dip Deviation

Minimum Seismic Amplitude

Maximum Vp/Vs Ratio

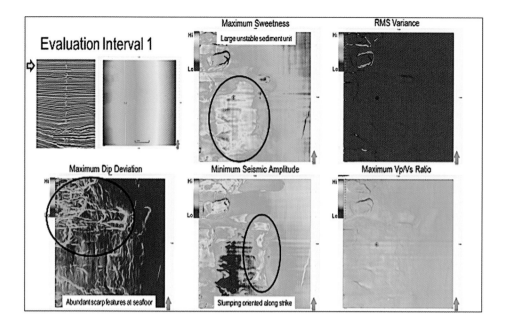

Figure 11.14

Figure 11.15

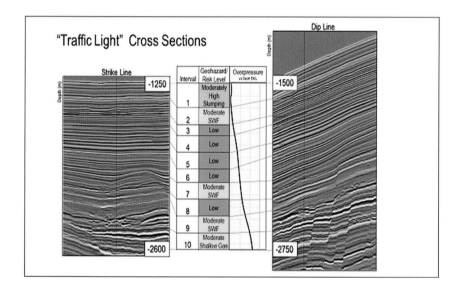

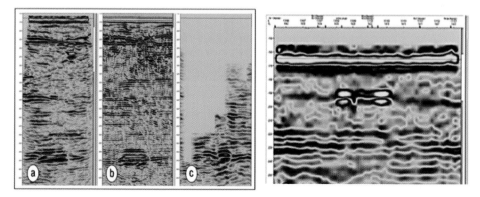

Figure 11.20

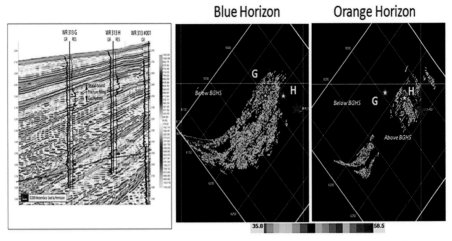

Blue Horizon   Orange Horizon

Boswell et al., 2009

Figure 11.23

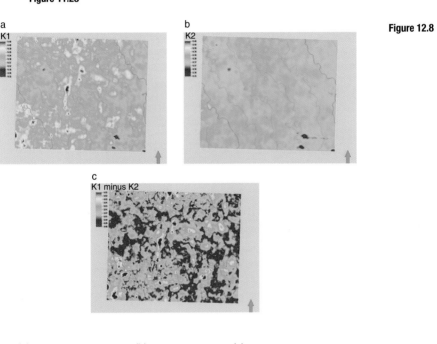

a
K1

b
K2

Figure 12.8

c
K1 minus K2

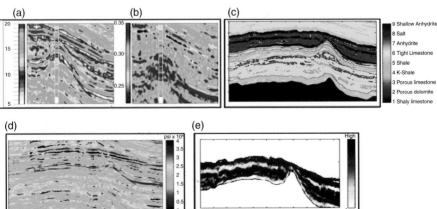

(a)   (b)   (c)

9 Shallow Anhydrite
8 Salt
7 Anhydrite
6 Tight Limestone
5 Shale
4 K-Shale
3 Porous limestone
2 Porous dolomite
1 Shaly limestone

(d)   (e)

psi x 10⁴

High

Low

Figure 12.9

Figure 12.11

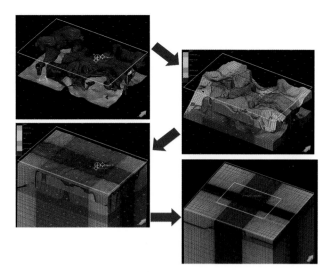

Figure 12.12

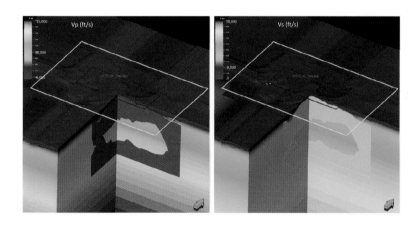

Figure 12.13

FITs from public data

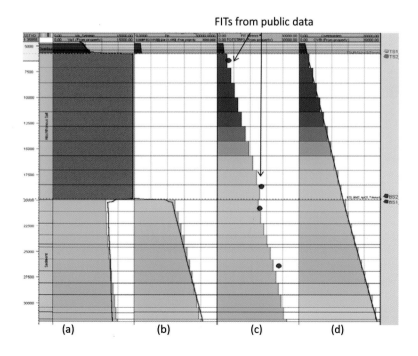

(a)   (b)   (c)   (d)

a

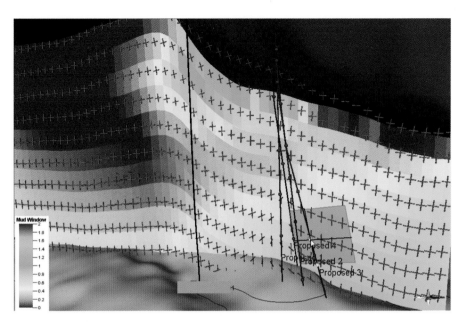

b

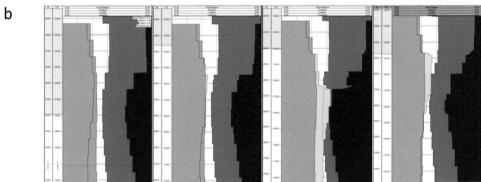

Figure 12.14

a

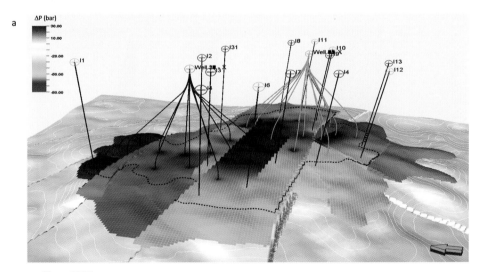

Figure 12.15

b

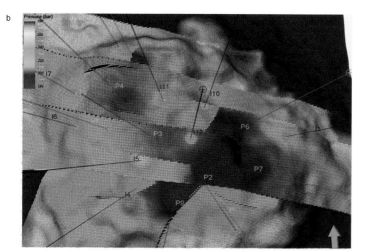

c

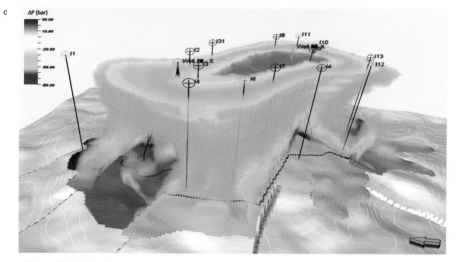

Figure 12.15 (cont.)

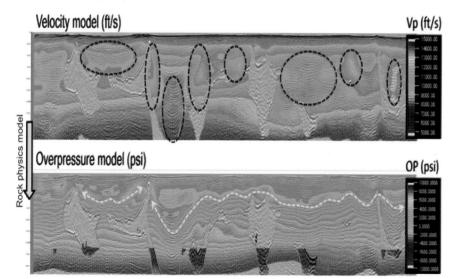

Figure 13.3

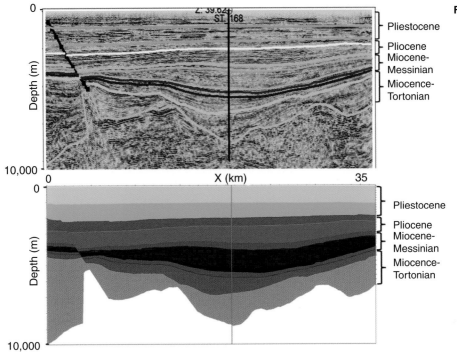

Figure 13.7

Pliestocene
Pliocene
Miocene-
Messinian
Miocence-
Tortonian

Pliestocene
Pliocene
Miocene-
Messinian
Miocence-
Tortonian

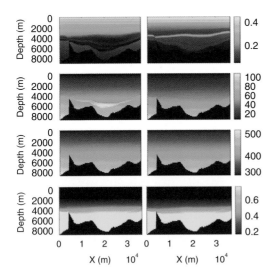

Figure 13.8

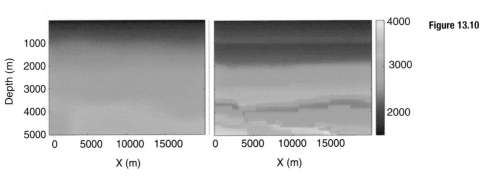

Figure 13.10

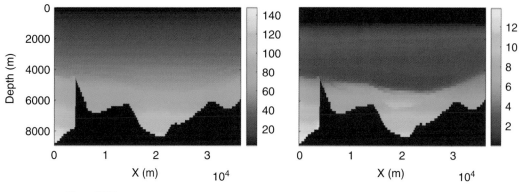

Figure 13.14

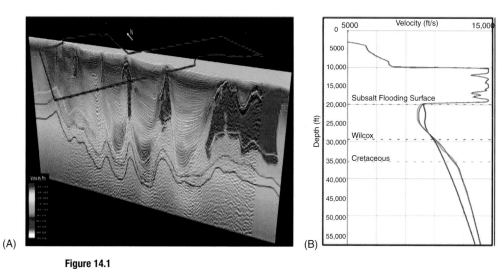

(A)

(B)

Figure 14.1

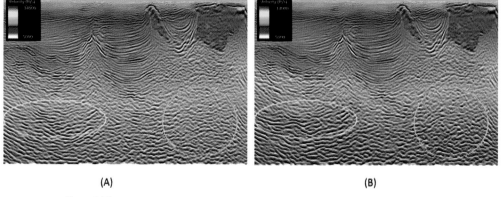

(A)

(B)

Figure 14.2

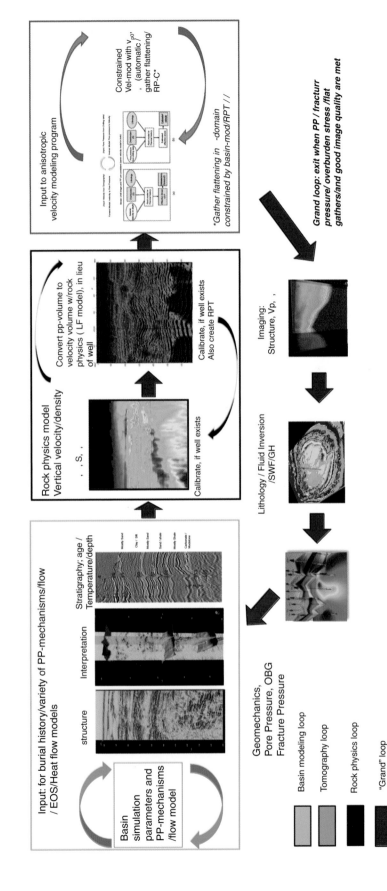

Input to anisotropic velocity modeling program

Constrained Vel-mod with v $_{po}$, (automatic / gather flattening/ RP-C*

*Gather flattening in -domain constrained by basin-mod/RPT //

Rock physics model Vertical velocity/density

Convert pp-volume to velocity volume w/rock physics (LF model), in lieu of well

Calibrate, if well exists Also create RPT

Calibrate, if well exists

, , S, ,

Input: for burial history/variety of PP-mechanisms/flow / EOS/Heat flow models

Stratigraphy; age / Temperature/depth

Mostly Sand
Clay / Silt
Mostly Sand
Sand / shale
Mostly Shale
Carbonate/ Mudstone

Interpretation

structure

Basin simulation parameters and PP-mechanisms /flow model

Geomechanics, Pore Pressure, OBG Fracture Pressure

Grand loop: exit when PP / fracturr pressure/ overburden stress /flat gathers/and good image quality are met

Imaging: Structure, Vp, ,

Lithology / Fluid Inversion /SWF/GH

Basin modeling loop

Tomography loop

Rock physics loop

"Grand" loop

Figure 14.4

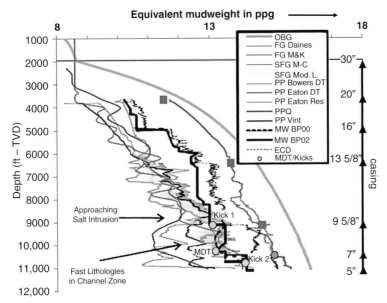

Figure 5.31 A comparison of the pore pressure prediction using seismic and sonic based velocity (blue) and Q-based methodology (green) as discussed in the text. There are also various other pore pressure prediction results from resistivity and Eaton method using seismic as indicated in the inset. Various fracture-pressure gradients, overburden gradient, and measurements such as LOT, MDT, and mudweight data are also shown and identified in the inset. Modified after Salehi and Mannon (2013). (A black and white version of this figure will appear in some formats. For the color version, please refer to the plate section.)

approach. First, amplitude decay as shown in equation (5.35) assumes that Q is constant and independent of frequency. This is clearly not the case as it would violate the causality principle (Brennan and Stacey, 1978). Second, it is known that Q is very sensitive to partial gas saturation (Dutta and Ode, 1979; Dvorkin and Mavko, 2006; Mavko et al., 2009). So, amplitude decay in pay zones (oil or gas) in overpressured interval must distinguish between the two sources of frequency loss – partial hydrocarbon saturation and abnormally high pore pressure. Third, Q also depends on lithology, porosity and fluid type (Valle et al., 2017; Toksoz et al., 1978) contrary to statements in the literature (e.g., Salehi and Mannon, 2013). In our opinion, the frequency or Q-based technique for pore pressure analysis should be considered an interesting technology – with some potential but fraught with many uncertainties in its current implementation.

5.5   Summary: Seismic Velocity Analysis and Guidelines for Applications to Pore Pressure

Among all geophysical methods seismic velocity-based methods are the most widely used and often the only ones available for *predrill* geopressure prediction. It is also the most misused. In our opinion this is mainly due to lack of interdisciplinary communication between the geophysical and the end-user communities such as drillers. The various seismic tools for obtaining velocity information that we discussed in this chapter have many limitations. Traditionally the seismic community (acquisition, processing, and interpretation) addresses its products with only one user application in mind – imaging. That is OK. However, for geopressure prediction we need to know whether the image is *geologically* consistent and the velocities are *physical*. Qualifying the depth accuracy and velocity uncertainty (e.g., Osypov et al., 2013) is an important step toward generating better velocities for pore pressure prediction. Modern processing technologies address the issue associated with geological consistency to some extent, and can integrate well data and checkshot information into the tomographic model building, but in locations where velocity is poorly resolved by the surface seismic reflection experiment, a major issue – whether the velocities are physical – is usually not addressed. This is a limiting factor. For pore pressure prediction we need fit-for-purpose tools. The velocities should be tested whether these are compliant with rock physics principles and yield velocities that are within expected bounds; such as the *highest* possible value for normally pressured rock and *lowest* possible value for fracturing the rock.

In Figure 5.32 we show a summary of various velocity tools that we have discussed in this chapter. Plotted is the complexity of the tool (horizontal axis) versus value added for pore pressure (vertical axis). On the lower end we have conventional velocity analysis with the requirement of dense picking. (Dense picking is often lacking.) It has lowest resolution; of the order of 200–400 ms vertically and ~ 1000 ft laterally. At the upper end we have acoustic FWI. It yields a better image and can handle lateral velocity heterogeneity very well as it is built on reflection (CIP) tomography as

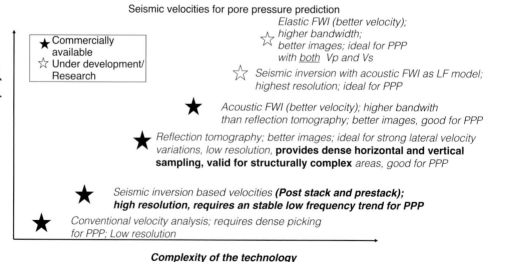

Figure 5.32 Seismic velocities for pore pressure prediction showing complexity of technology and value-added proposition.

a starting velocity model. Our limited experience with this technology suggests that it is very well suited for geopressure and geohazard detection and prediction. Seismic inversion based velocity models add a lot more to resolution – more than what is feasible from either reflection tomography or FWI alone. However, these need a good low-frequency (LF) model to build on. Both conventional velocity model with dense picking and models based on reflection tomography are found to be very useful in providing needed LF support. The missing bandwidth in the poststack or prestack inversion techniques is usually supplied by a priori models such as band-passed well log data. If we plan to predict pore pressure at a fine (reservoir) scale then we should use high-resolution seismic inversion based approaches as discussed here. These include P-impedance from poststack inversion, and P- and S-wave impedances from prestack inversion such as AVO. Both require a stable low-frequency background or a trend based on well(s). GA-technology is based on elastic FWI and it does not require a background low-frequency trend.

All velocity models for pore pressure analysis *must* include anisotropy corrections. Even conventional time domain velocities must account for this effect. Both VTI and TTI models are commonly used. Orthorhombic anisotropy models are beginning to get some traction and can provide information about the state of stress and fractures in the subsurface. This information is related to geomechanical application and effective stress as will be discussed in later chapters. We discussed several methods to estimate anisotropy parameters. The minimum number of parameters (for P-waves) is three (VTI) – vertical velocity and Thomsen's ε and δ parameters. We remarked that V_{nmo} is not the same as the vertical velocity. Using V_{nmo} instead of vertical velocity will cause errors and the resulting error would be distributed to ε and δ. Checkshot, zero-offset VSP or band-passed sonic velocities (checkshot corrected) are the best alternatives. A good estimate of "true" vertical velocity can be obtained using rock physics principles (as will be discussed in the next chapter).

CIP tomography and FWI (acoustic at this time) are very good sources of P-velocity for geopressure analysis. However, some of the inherent nonuniqueness in these technologies should be mitigated throughout the earth model building cycle.

Geopressure prediction in complex geologic areas such as subsalt remains a problem although several publications indicate that CIP tomography with rock physics guidance is an emerging area of optimism. The key to success is the ability to extend the low-frequency velocity trend from above the salt to a meaningful trend below the salt. However, the biggest breakthrough is the ultra large-offset seismic acquisition with continuous azimuth coverage. This provides some useful signal below salt and helps improve images. Subsalt geopressure estimation is an active area of research currently and it appears that integration of data and models from various diverse sources such as seismic and basin modeling may be a viable area of investigation. This will be discussed in Chapter 10.

Shear wave velocities, either from direct S-wave recording or from mode-converted events are great sources for pore pressure applications. However, there is a paucity of such data in the industry as well as publications that deal with pore pressure. The reason is mostly economic – high cost of data acquisition. Shear velocities from ocean-bottom node-acquisition and mode-converted shear waves are good sources for geopressure prediction in the offshore. Pseudo-shear velocity estimates from 1D elastic FWI method are useful substitutes for real shear log, if real shear logs are unavailable. However, such models need calibration.

Seismic attenuation or Q-based methods for pore pressure prediction are promising, and limited published examples do look interesting. We do need more case studies. However, there are many fundamental issues that remain to be addressed, some of which we discussed above. Additional research is needed before the technique can be used confidently.

6   Integrating Seismic Imaging, Rock Physics, and Geopressure

6.1   Introduction

In Chapter 5 we discussed various velocity analysis techniques for imaging and geopressure prediction. We noted that of all well-known seismic velocity analysis techniques, the ray-based or CIP tomography technique is well suited for pore pressure prediction. High resolution FWI techniques are also promising although far more expensive today. The strengths of CIP tomography technique is that it uses *all* the traveltime from seismic data to build an anisotropic velocity model, it accounts for lateral heterogeneities of the subsurface and provides a geologically consistent model. We also noted some limitations of this technique in Chapter 5, the main one being the nonuniqueness of the model. It can produce a velocity model that may not be physical because it relies solely on the criterion of delivering flat seismic gathers. Furthermore, anisotropy increases the number of parameters per grid and thus increases the degrees of freedom in the velocity model building.

Accounting for anisotropy is necessary for both imaging and pore pressure analysis. Both alignment of clay minerals in shales and the effect of layering in geology imply transverse isotropy rather than isotropy. Salt bodies as well as tectonic events can cause stress perturbations that further complicate velocity variations due to perturbations in anisotropy parameters. Thus building anisotropic velocity models for imaging is a challenge. Conventional velocity analysis and tomography of surface seismic might not provide a satisfactory answer because a number of models could equally well explain the observed data. Such is also the case with full waveform inversion (FWI).

We commented in Chapter 5 that a velocity model that is acceptable for imaging may *not* yield a good model for geopressure prediction. This means that there is a risk of having to deal with multiple realizations of the subsurface model – some for imaging and the others for pore pressure analysis. In other words, the velocity models underlying imaging can be different from that for pore pressure – separate calibrations would be needed for each. This also means the structural image derived by the iterative velocity model building and imaging workflow discussed in Chapter 5, will not be kinematically consistent with velocities that have been manipulated to fit well log data through various types of calibration process. In addition, any improvement in one, for example, image would not necessarily translate into an updated pore pressure image if the two were not linked. In this chapter, we propose a workflow that tries to mitigate the differences and we hope will let us move closer toward building a true *earth model*.

6.2   Rock Physics Guided Velocity Modeling (*RPGVM*) with Reflection (CIP) Tomography for Pore Pressure Analysis

In most cases, both tomography and FWI use linearized inversion schemes to fit the nonlinear forward model (Tarantola, 1987), be it raytracing (for CIP tomography) or finite difference seismogram (for FWI). Linearized inversion schemes offer a large reduction in computational cost. Their shortcoming is that the minimization of the objective function can converge into a local minimum. Furthermore, due to non-uniqueness of the solution more than a single model can fit the data. Therefore, both CIP tomography and FWI rely heavily on the assumption that the initial model is close to the true model. When this assumption fails, there is a high possibility of convergence into a velocity model that satisfies the imposed convergence criterion but may provide *geologically* and *physically* improbable models. To handle this, we require additional constraints on the model. The rock physics guided workflow that is suggested in this chapter provides some constraints on velocities that add to the data fit criteria a requirement for the model to be geologically *plausible* and physically *possible*.

Anisotropic velocity models can be built with forward modeling using rock physics principles, geomechanics, and basin modeling. The most commonly used models are still based on the assumption of *transverse isotropy*. A transversely isotropic material is one with physical properties which are symmetric about an axis that is normal to a plane of isotropy. This transverse plane has infinite planes of symmetry and thus, within this plane, the material properties are the same in all directions. There are two transversely isotropic cases that we need to consider: vertical transverse isotropy (VTI) and tilted transverse isotropy (TTI). An example of both are given in Figure 5.8 in cartoon form and a real field example in Figure 6.1. The number of parameters to describe P-wave anisotropy increases to *three* in vertical transversely isotropic (VTI) media (vertical velocity and Thomsen's ε and δ parameters) and *five* in tilted transversely isotropic (TTI) media (vertical velocity, Thomsen's ε and δ parameters and two tilt angles, Dip_1 and Dip_2). While VTI may be a simpler representation, TTI is probably geologically more plausible. It is not always possible to isolate the proper axis of symmetry (VTI or TTI) from data alone (Bakulin et al., 2009). In many cases, an assumption of structurally aligned TI symmetry or STI is used, where the axis of symmetry is assumed to be aligned normal to bedding. The bedding dip is calculated directly from the depth image.

As a result of the increased number of parameters, the uncertainty in the parameter space increases. Bachrach (2010) used differential effective medium (DEM) theory from rock physics combined with well logs and empirical models of shale diagenesis to build anisotropic velocity models. Petmecky et al. (2009) derived anisotropic velocities for imaging from a 3D basin model to capture the pressure, depositional, fluid flow, and salt movement histories of a basin. Matava et al. (2016) used finite elastic deformation theory to calculate the effect of stress anomalies caused by salt movements on velocity.

VTI

TTI

(a)

(b)

Figure 6.1 A field example of geologic rock formations. (a) VTI medium (Source: https://en .wikipedia.org/wiki/Transverse_isotropy) and (b) TTI medium (Source: www.geoexpro.com/ articles/2010/04/depth-imaging-seeing-the-invisible). VTI medium needs three parameters to describe P-anisotropy (vertical velocity and two Thomsen's parameters, ε and δ), whereas a TTI medium needs two *extra* parameters (two tilt angles) in addition to the three required for VTI medium.

The examples in Chapter 5 and the discussions therein point toward the need for constraints on tomography when well control (such as checkshot, VSP, or sonic logs) is either unavailable or if available, deemed unsuitable for calibrating seismic velocity. Even when using a well or a set of wells to calibrate the vertical velocity and attendant anisotropy parameters, we note that the traditional calibration approaches can only constrain anisotropy locally near the well. The model building, however, requires 3D extrapolation of well data away and beyond the well(s). Stochastic modeling is a well-known method for this purpose. But it has a limitation such as how reliable is the extrapolation of data (using, for example, geostatistical methods such as "kriging" or "co-kriging") away from well(s).

Recent developments in anisotropic velocity model building show that integrating additional data, such as rock physics and pore pressures, can constrain the velocity inversion process. Dutta et al. (2015a, 2015b) combined rock physics and pore pressure constrained velocity models to create velocity bounds for tomography. These constraints not only reduce uncertainty in the tomography process, but also produce a velocity model that is able to predict ranges of physical pore pressure. This extra constraint forces the vertical velocity to be within a physically expected range yielding a pore pressure that is bounded below by hydrostatic pore pressure and above by fracture pressure. These bounds can be further refined if actual pore pressure and fracture pressure data are available. The use of rock physics compliant velocity model enables us to estimate vertical velocity without having to rely on normal moveout analysis, which often produce poor estimates of velocity. Li et al. (2016) used stochastic rock physics modeling (Bachrach, 2010) to build model covariance matrices to constrain wave equation migration velocity analysis (WEMVA). Following Dutta et al. (2015a, 2015b), we discuss a workflow that *combines* rock physics, and pore pressure constraints to improve anisotropic velocity models for structural *and* pore pressure imaging using tomography. Le (2019) used a rock physics compliant pore pressure model to constraint anisotropic velocity fields using FWI approach. In Chapter 10 we will discuss how pore pressure from basin history modeling can provide additional constraints on velocity model building.

6.2.1   Rock Physics Guided Velocity Modeling (*RPGVM*) Workflow

Concept

The basic concept can be summarized as follows:

- Seismic velocity estimates in depth can be transformed into effective stress and pore pressure estimates. This would require an estimate of lithology and overburden.
- Similarly, knowledge of pore pressure from drilling data such as pore pressure measurements (RFT/DST) or estimates from mud weight data, together with estimates of lithology and overburden, can be transformed into estimates of velocity.
- Iteration between pore pressure and velocity estimates based on calibrated transform is the central concept. It can be used in different ways within the velocity model building workflow. The integration can be via simple QC of velocity estimates to ensure consistency with a reasonable and geologically plausible pore pressure regime. More complex applications can benefit seismic velocity model building by helping anisotropy estimates and even be used directly within the tomography operators. In this chapter, we describe a few examples applying this concept on real data.

Figure 6.2 describes the general concept behind this new workflow. The model(s) in Figure 6.2a, termed *forward model*, describes a general process that is well known – with some knowledge of lithology and thermal history and given an estimate of overburden stress one can obtain an estimate of pore pressure using seismic velocity as discussed in Chapter 3. The model in Figure 6.2b is the *inverse model*. In this process, an estimate of velocity is obtained from drilling data. Note that the effective stress estimate here is not derived from seismic data. Rather, the effective stress estimate is derived from sources *other* than the seismic data such as the pore pressure estimates based on drilling data (e.g., mud weights, penetration logs, or measured RFT or MDT and LOT data) and an estimate of overburden stress using the same rock model(s) as used in the forward model. It is clear that the two velocity estimates – one from the forward model (that contains kinematic information) and the other from the inverse model (i.e., derived from drilling data and/or mechanical earth modeling) would be different. In the *RPGVM* workflow, the aim is to *reconcile* the two velocities by using an iterative approach, for example, in the tomography loop for velocity model building. An alternate approach would be to use this constraint on the FWI process for model building (Le, 2019). Upon convergence, this process is expected to yield a velocity model that will not only flatten all common-image-point gathers but also ensure that the resulting velocities are *physically* possible (constrained by physical domain of pore pressure and fracture pressure). Geological plausibility is ensured via

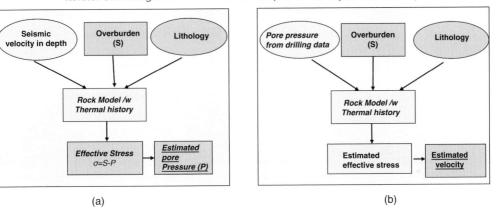

(Input: Velocity from Tomography)   *(Input: Pore Pressure from Drilling data)*

Forward Model: Velocity to Pore Pressure   Inverse Model: Pore pressure to Velocity

Iterate: Until image and PP are consistent (same velocity model in both)

(a)   (b)

Figure 6.2 Conceptual workflow for rock physics guided velocity modeling. (a) Forward modeling that uses seismic velocity to predict effective stress, pore pressure, and overburden stress using a given rock model. (b) Inverse model that uses drilling data to predict estimates of velocity using the same rock model that is used in (a). We try to reconcile the velocity in (a) to the estimated velocity from (b) as well as the estimated pore pressure in both.

dip and steering filters as discussed in Chapter 5. In the FWI process, this would be a part of the initial model building (Le, 2019). The details of this workflow are shown in Figure 6.3. The essential steps are: Gathering input parameters, creating a rock physics model, building an initial velocity model for input to tomography, and finally, performing tomography model updates for arriving at a possible solution. Some key elements are discussed in more detail below.

Input Parameters

The input parameters for the new workflow are similar to what are needed for building a plausible geologic model prior to seismic analysis. The goal is to establish a stratigraphic and structural framework needed to build an initial model. This is based on interpretation of stacked seismic data (after migration) from a legacy model in depth domain with a legacy "image." It is ensured that the standard seismic processing workflow has been applied to enhance the appropriate S/N ratio and that the ensuing prestack seismic gathers are "flat" in the angle (or offset) domain. De-multiple operation should have been performed on the data set. Typically, a legacy velocity model (typically after legacy tomography and Kirchoff PSDM) is used in preparing an "initial" velocity model for input to start the new tomography that will be constrained by a rock physics model.

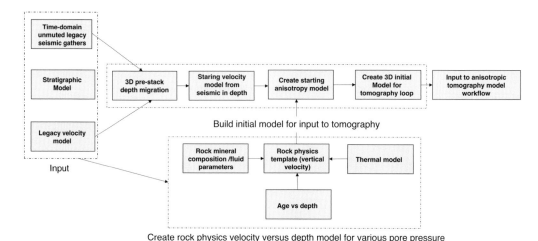

Figure 6.3 A schematic workflow to implement the conceptual model shown in Figure 6.2. The workflow uses the concept of Rock Physics Guided Velocity Modeling (*RPGVM*) as discussed in the text.

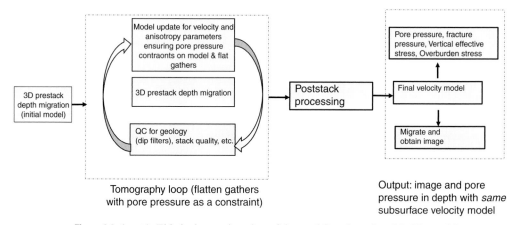

Figure 6.3 (cont.). This is the continuation of the workflow introduced in Figure 6.3.

Rock Physics Model

This is a very important step in the workflow. It allows the analysts to assess the range of velocities from available data and establish some bounds on the modeled velocities by constraining the parameters using the rock physics model. The method we propose to constrain velocity for tomography is to use rock physics to create an integrated earth model including rock properties such as effective stress and pore pressure and its relation to velocity. For shale dominated basins, the Dutta-model (Dutta, 2014, 2016) includes effective stress changes due both to mechanical compaction and chemical diagenesis such as S/I transformation in shale. Other diagenetic models can also be built such as those for sandstone diagenesis, provided appropriate activation energies are available. Alternate models such as the Eaton model and Bowers's model can also be used. However, since these latter models do not include chemical diagenesis effects in a predictive way, accounting of this must be made at some later stage by trend-shift, if using the Eaton model or including the onset of the unloading profile with well log, if using the Bowers's model. Otherwise, predicted pore pressures would be wrong.

The Dutta-model relating velocity (V) to vertical effective stress (σ) was discussed in Chapter 5 and will not be repeated here. An outcome of this model is the creation of a rock physics template (RPT) as discussed in Chapter 4 (see Figure 4.12). RPT provides an estimate of vertical velocity to start the inversion process for building a velocity model. It also guides the process so that the velocity is never allowed to wander outside of the bounds during the inversion process. These bounds are very useful to manage the nonuniqueness by restricting the parameter space of the tomography and FWI (Le et al., 2018; Le, 2019). In the tomography based inversion approach, we use RPT to define the bounds after calibrating the template with well data (measured pressure, RFT/DST/mud weights, and LOT and petrophysical well logs).

The RPT shown earlier is based on a thermal history construction using steady-state heat flow assumption. Szydlik et al. (2015) and Al-Kawai (2018) relaxed this restrictive assumption and applied basin modeling algorithms to data sets from the Gulf of Mexico. Their approach includes all thermal transients due to deposition, uplift and erosion of sediments. Thus, an improved RPT can be constructed with output from basin modeling. This is discussed in Chapter 10.

Anisotropy Parameter Estimation

The goal in this step is to build a starting or initial velocity model with anisotropy correction. In Chapter 5 we discussed three approaches to estimating anisotropy parameters. Any of these models can be used. A simplification can be achieved with the RPT that can be used in lieu of the vertical velocity. There still remain two other parameters for VTI. These are Thomsen's parameters ε and δ.

Implementation of *RPGVM* introduces a simple process and an assumption that focuses on estimating the δ parameter with the heuristic assumption that $\varepsilon \approx 2\delta$ for the initial model. This reduces to the determination of only one parameter, δ. The essential steps for this are shown in Figure 6.4. In Figure 6.4a, we show checkshot correction of sonic log used in Figure 6.4b. Figure 6.4b shows displays of seismic traveltime from legacy tomography velocity analysis along with RPT as well as a *new* initial model at the well location. Both models (legacy and the new initial model) flatten the gather at the well location. The new initial model is created by migrating the gather with the new velocity making sure that it is "low frequency" and that it is physical, namely, at every depth it is within the *bounds* of the RPT as opposed to the legacy model that makes excursions outside the RPT. This is clearly evident at deeper depths. At these depths the pore pressure gradient is *lower* than the hydrostatic pressure gradient and the velocities are *higher* than what is physical at these depths. The new velocity model is created by lowering the velocities at these depths as shown under the label "*new initial*" velocity (vertical). Obviously, this will not flatten the gather. The gather flattening is accomplished by adjusting Thomsen's δ parameter such that the gather

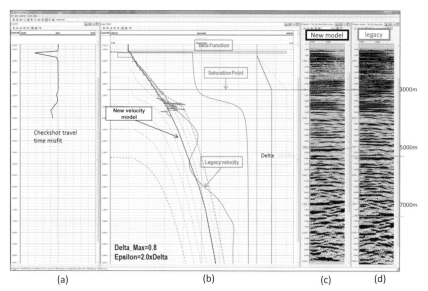

Figure 6.4 Anisotropy parameter estimation using rock physics template (RPT) and seismic gathers after depth migration with the vertical velocity from RPT. (a) Checkshot correction applied to the gather. (b) Seismic traveltime from legacy tomography velocity analysis along with the rock physics template (RPT) and a new initial model at the well location. Both models (legacy and the new initial model) flatten the gather at the well location, as shown in panels (c) and (d). Panel (b) also shows the diagenetic function at the gather location along with the anisotropy parameters ε and δ. (c) Gather flatness with the new vertical velocity model and the anisotropy parameters in (b). (d) Gather flatness with the legacy velocity model that was also anisotropic. The fact that we can flatten gathers with multiple velocities shows the nonunique-ness of the tomography. After S. Yang (personal communication, 2015).

at the largest angle (offset) was flat (panel b). Thus it is a two-layer model for δ: at seabed, it is zero and a linear extrapolation is used from seabed to a maximum value of $\delta(\delta_{\max})$. It is held constant to that value at all depths beyond that. As remarked earlier for the anisotropy parameter, ε, we assumed $\varepsilon \sim 2\delta$ as an estimate for the starting model. This newly created anisotropic velocity model that flattens the gather is shown in panel c and compared with the legacy model in panel d. The *new* (initial) model has *lower* vertical velocity than the legacy model (the marked events in panel c are at shallower depths than those in panel d). Furthermore, the vertical velocity is bounded by the physical bounds of the RPT. This new model is subsequently propagated below seabed in 3D by adjusting water depths and seabed temperature, thermal gradients, and burial rates of sediments, as appropriate. In panel b, we note that the β-function (chemical diagenesis parameter of the Dutta-model) correlates well with the depth where the anisotropy parameter δ reaches the maximum value. We have not found any rigorous reason for this except for the fact that the clay platelets (illite and smectite) are beginning to undergo an "ordering" process.

A rock physics basis for the observations just made is provided in Figure 6.5 where we show results from laboratory measurements of anisotropic P-wave velocities versus pressure (Scott and Abousleiman, 2004) on shallow marine sediments. The authors state

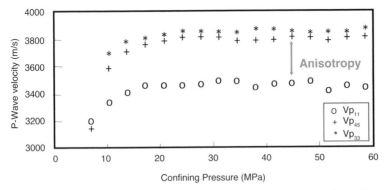

Figure 6.5 Anisotropy parameters as measured in the laboratory on marine sediments as shown in the inset. Pressure-induced anisotropy in porous sandstone is observed in laboratory tests to be as large as the inherent anisotropy of layered rocks. Anisotropy is induced early in loading and is effectively "locked in" (at ~12 MPa) with little change thereafter as pressure continues to increase.

that *the pressure-induced anisotropy in porous sandstone is observed in laboratory tests to be as large as the inherent anisotropy of layered rocks. Anisotropy is induced early in loading and is effectively "locked in" with little change thereafter as pressure continues to increase.* The "*locking in*" of anisotropy occurred at ~ 12 MPa (termed "saturation point" in Figure 6.4). This is consistent with our empirical observation of the occurrence of δ_{max} at shallow depths where the β-function indicates the beginning of the transition from smectite to illite layers based on thermal history estimates. This was pointed out by Dr. Sherman Yang of Schlumberger (personal communication, 2015).

In summary, for a VTI model the vertical velocity is created using a rock physics model that relates velocity to effective stress and is bounded by the RPT within acceptable pressure range to begin the iteration. The δ parameter is adjusted by flattening the gathers at the farthest offsets at shallow depth in a two-layer model (zero at seabed and increasing linearly with depth to a maximum value and holding it constant thereafter at deeper depths). In addition, we approximate $\varepsilon \approx 2\delta$ to build the anisotropy model for starting our initial model. The tomography does the rest. A typical value of the δ parameter is within 0.5 to 0.8. The two-layer model for δ as discussed here is consistent with similar observations made by Jones (2010).

Angle Gathers in Pore Pressure Gradient Domain and Final Tomography Model

The initial anisotropic velocity model thus created in the *RPGVM* workflow is used subsequently in the tomographic velocity model building. The starting model is considered to be physical as it honors the pore pressure bounds. During subsequent iterations, we also use the RPT to make sure that not only the vertical velocities remains bounded within the pore pressure bounds but the gathers also remain flat. This is accomplished by *creating angle gathers in the pore pressure domain directly.* This is similar to building a velocity model upon exiting salt, for example (Deo et al., 2014). An example is given in Figure 6.6. All gathers in the upper panel a show reverse time migrated (RTM) angle gathers that used

(a)

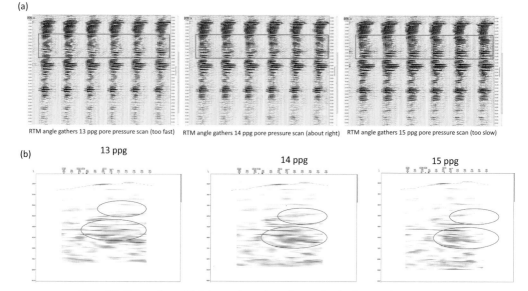

RTM angle gathers 13 ppg pore pressure scan (too fast)   RTM angle gathers 14 ppg pore pressure scan (about right)   RTM angle gathers 15 ppg pore pressure scan (too slow)

(b)

13 ppg   14 ppg   15 ppg

Figure 6.6 Examples of RTM – angle gathers for a seismic data set in a constant pore pressure gradient domain. All the gathers in panel (a) are created using velocities appropriate for indicated constant pressure gradients in pounds per gallon (pore pressures are not constant). The box around the events are intended to highlight the focus zone (in depth). This approach is used from top to bottom to flatten the gathers and create a pore pressure model without using well data. The final product would be a velocity model in depth that will be physical; the gathers would be flat and the image quality will be acceptable when depth migrated using the same velocity model. (b) Corresponding structural semblances of RTM stacks of panel (a) at indicated pore pressure gradients. This indicates lateral geologic control on the model. (14 ppg shows best structural continuity. Black indicates maximum semblance while white indicates minimum semblance.) Modified after Deo et al. (2014).

velocities from the RPT and are created to make sure that these are consistent with constant "pore pressure gradients" (in ppg) as indicated in panel a. These panels are very similar to *"velocity scans"* used commonly by the seismic data processing community, except that the range of velocities are now limited to yield a *"chosen constant pore pressure gradient in ppg"* and not constant velocity. In this example, the structural semblance plots shown in Figure 6.6b confirms better structural continuity at 14 ppg rather than 13 or 15 ppg. This observation is important to make so that the subsequent inversion process is limited to within these ranges.

There is another quality control criterion that one can use to assess the value of *pore pressure gradient scan*. This is displayed in Figure 6.7. In Figure 6.7a we show conventional semblance plots; each semblance is made of nine traces from 9 ppg to 15.5 ppg as noted. Figure 6.7b shows the angle gather stacked to get one trace per ppg pressure gradient scan. Figure 6.7c shows stack quality – the darker image represents higher stack power. Thus darker to the right means improvement in stack power with higher pore pressure gradient (lower interval velocity) and vice versa. In Figure 6.7d we show gathers stacked in *pore pressure gradient (ppg) domain*. This approach

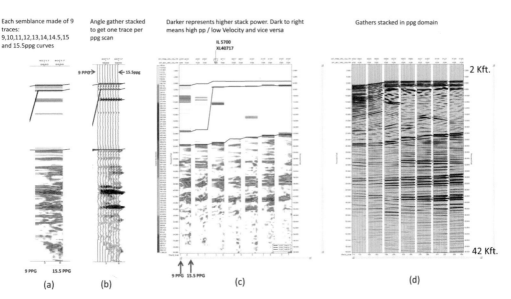

Figure 6.7 A quality control plot that shows the stack quality of the initial velocity model. See text for discussions. (a) Each semblance is made of nine traces: 9,10,11,12,13,14,14.5,15, and 15.5 ppg curves. (b) Angle gather stacked to get one trace per ppg scan. (c) darker image represents higher stack power. Darker to lighter means high pore pressure/low velocity, and vice versa. (d) Several gathers stacked in ppg domain. Modified after Deo et al. (2014) and B. Deo (personal communication, 2014).

conveys to seismic processors the value of stacking *directly* in *ppg domain* – increasing velocity with greater depths for better stack may not be advisable. From this example we see that the gather flatness criteria in the pore pressure gradient domain indicates the acceptable velocity ranges are closer to 14 ppg. The final velocity model is then created by allowing the tomography inversion process to continue but staying within the pressure gradient range of 14 ± 1 ppg. This guarantees the angle gathers would be flat and will be within the physical bounds of the pore pressure gradients limits as provided by the RPT. The image quality does not degrade during the process. Actually, as we shall see from the examples in this chapter the image quality is improved significantly when the same velocity model is used to depth migrate the data and create the new image. This process does not *eliminate* the inherent nonuniqueness associated with tomography; however, it does help *manage* the nonuniqueness better by restricting the range of parameters during inversion.

6.3   Example Applications of Rock Physics Guided Velocity Modeling for Geopressure and Imaging with CIP Tomography

6.3.1   Makassar Straits, Deep Water, East of Kutei Basin, Indonesia

In this example from offshore Indonesia, Makassar Straits, East of Kutai basin, the goal was to obtain a 3D model for pore pressure (Dutta et al., 2015b). According to

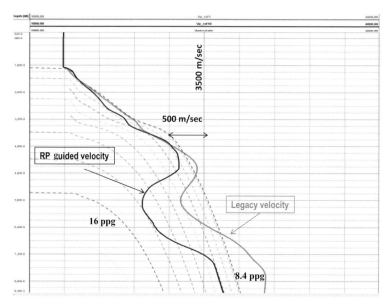

Figure 6.8 Vertical velocity field for the anisotropic (TTI) velocity model showing the RPT along with the final velocity model after tomography that is constrained by the RPT and the legacy velocity model. "Legacy" denotes the original anisotropy model generated with conventional tomography workflow, which was also based on TTI anisotropy parametrization.

Wikipedia, Kutai is the largest and deepest Tertiary age basin in Indonesia. Quote from Wikipedia: "Plate tectonic evolution in the Indonesian region of SE Asia has produced a diverse array of basins in the Cenozoic era. The Kutai is an extensional basin in a general foreland setting. Its geologic evolution begins in the mid-Eocene and involves phases of extension and rifting, thermal sag, and isostatic subsidence. Rapid sedimentation related to uplift and inversion began in the Early Miocene." There was a well drilled to a depth of ~4 km and the data from that well was used to calibrate the rock physics template. This well was also used to create a vertical velocity (V_Z) model and a new set of anisotropy parameters by flattening the gathers as discussed in section 6.2. The starting model was created to follow the velocity versus depth function for the 12 ppg iso-pore pressure gradient on the RPT. Gather flattening with anisotropy parameter adjustments as discussed in section 6.2 yielded new ε and δ parameters. The new velocity function after tomography inversion and guided by pore pressure is shown in Figure 6.8 at the well location and is compared with the legacy velocity model. We note that the legacy model – also an anisotropic velocity model, has unphysical velocities when compared with the calibrated RPT and the new rock physics guided velocity model. As noted in Figure 6.8 the legacy velocity model yielded lower than hydrostatic pore pressure at several depth intervals and violated rock physics principles. This suggests that the legacy velocity model is unphysical at certain depths.

The tomography was run in 3D with the RPT as a constraint on the inversion. The final 3D velocity fields and pore pressure images are shown in Figures 6.9 and 6.10, respectively. Velocity from the rock physics guided velocity model yields pore

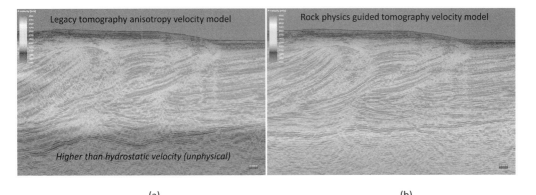

(a)   (b)

Figure 6.9 3D anisotropic (TTI) velocity model. (a) Legacy anisotropic velocity model. Note the unphysical velocity model at deeper depths. (b) Rock physics guided anisotropic (TTI) velocity model. Note velocity regression in the middle. In addition, there is no unphysical velocity anywhere. After Dutta et al. (2015b). Published with the permission of Indonesian Petroleum Association (IPA). (A black and white version of this figure will appear in some formats. For the color version, please refer to the plate section.)

pressure at deeper depths that are *physical* everywhere as opposed to the one from the legacy model that yielded pore pressures at certain depths that were below hydrostatic– a physically very unlikely realization! An example of the final stacked image after RTM migration is shown in Figure 6.11b. This is compared with the legacy image in Figure 6.11a – this image used the anisotropic tomography based velocity model that was derived using the *conventional anisotropic tomography* workflow. That the pore pressure is physical (Figure 6.10b) and the image is vastly improved (Figure 6.11b) suggests that the underlying velocity field is not only geologically plausible but also physical. That the tomography achieved converged results for velocity with both workflows suggests that the tomography based inversion methodology is inherently nonunique; we are just picking a model that we consider physical because it yields reasonable pore pressure. (There is no well control here to prove the numerical validity at deeper depths; however, it does not yield pressures less than hydrostatic anywhere which would have been unphysical based on our criterion.)

6.3.2   High-Pressure and High-Temperature Area of a Syn-Rift Basin, Offshore, India

This example is from a complex geologic area in a high-pressure and high-temperature (HPHT) environment in offshore India (Dutta, 2015; Dutta et al., 2014). The basin where the study was carried out came into existence following rifting along eastern continental margin in early Mesozoic. Increased gradients for the river systems and increased sediment load coupled with significant sea level fall during the Neogene and triggered sediment-induced tectonics in the shelf and slope parts of the basin created

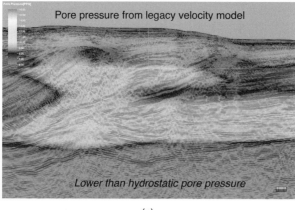

Pore pressure from legacy velocity model

Lower than hydrostatic pore pressure

(a)

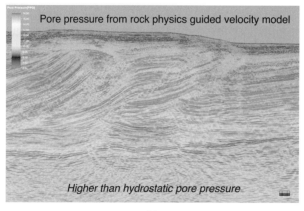

Pore pressure from rock physics guided velocity model

Higher than hydrostatic pore pressure

(b)

Figure 6.10 Pore pressure gradients (in ppg) in 3D. (a) Pressure gradients using the legacy velocity model. Zones of pressure gradient regime below hydrostatic gradient are highlighted in magenta color. (b) Predicted pore pressure gradients (in ppg) using the rock physics guided tomography workflow. The pore pressure gradients are physical everywhere. Color bars in both figures indicate pore pressure in ppg starting from 6 ppg to 16 ppg in step of 1 ppg. After Dutta et al. (2015b). Published with the permission of Indonesian Petroleum Association (IPA). (A black and white version of this figure will appear in some formats. For the color version, please refer to the plate section.)

desirable prospects for exploration. The exploration targets were based on a legacy model (VTI depth migrated velocity model.) Figure 6.12a is the legacy model (depth – imaged) where we also show the planned well trajectory. Shown also are two targets –shallow and deep. The final 3D model of the predicted velocity function and pore pressure (in the depth domain) was used to migrate and generate the image shown in Figure 6.12b. The associated seismic gathers at the well location are also shown in Figure 6.12. The legacy gather is on the left of Figure 6.12a and the gather that used the new velocity function is shown on the right of Figure 6.12b. The new image not only shows that the targets are at "shallower" depths but also the entire

Legacy tomography (conventional) based velocity model   Rock physics guided velocity modeling with Tomography

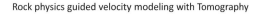

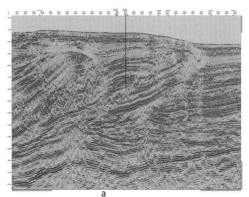
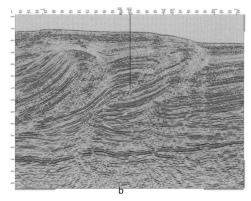

Figure 6.11 Stacked images after RTM migration. (a) Image after conventional tomography. (b) Improved image after rock physics guided tomography workflow. After Dutta et al. (2015b). Published with the permission of Indonesian Petroleum Association (IPA).

image is better focused as compared to the legacy image. Figure 6.12 continues to the next figure. The left figure shows velocity (m/s) versus depth (m) along with the RPT (from hydrostatic to 18 ppg models) used in the project. The right figure shows the predicted pore pressure gradients in ppg versus depth at the well location. We note that the velocity function used here for pore pressure gradient is *exactly* the same function as used for structural imaging (Figure 6.12b). This model of pore pressure gradient was used for the final well planning purpose. The existence of the narrow mud window (not shown here) at depths greater than 4 km as well as the predicted pressure gradients were verified by drilling.

6.3.3   Gulf of Mexico, USA (Miocene)

This example is from the Gulf of Mexico, USA (Liu et al., 2018). The area contained complex salt bodies. However, the study focused on the suprasalt areas. The purpose of the study was to show the value of *RPGVM workflow* to build a tomography velocity model (both VTI and TTI) and manage the inherent nonuniqueness of the tomography. Another premise of the work was to address the interesting geologic phenomenon associated with geologic structure and pore pressure as discussed earlier in Chapter 4 in the context of *lateral fluid transfer* as a mechanism of geopressure. Lopez et al. (2004) also reported on this phenomenon and showed that at the crest of an anticline structure, the P-wave velocity usually is the lowest compared to that at the base or the flank of the anticline structure. This is due to the *lateral fluid transfer* through *permeable and dipping* horizons, causing the vertical effective stress to be lower at the crest than the flanks or base of that anticlinal structure. Liu et al. (2018) used this phenomenon to build a large scale 3D velocity model using seismic data and used the RPT and *RPGVM* approach as discussed here to build a final velocity model for imaging and pore pressure. Figure 6.13a shows a large scale velocity model converted to the

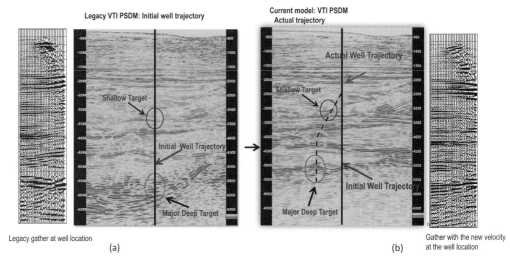

Legacy gather at well location

(a)

(b)

Gather with the new velocity
at the well location

Figure 6.12 (a) A legacy stacked image after VTI tomography and Kirchoff PSDM. Initial well trajectory along with two targets (one shallow and the other at deep) is indicated. The angle gather at the well location is shown on the left (legacy gather). (b) Final image after implementing the rock physics guided tomography workflow for velocity model building (*RPGVM*) and subsequent pore pressure bounded tomography (labeled as current model). The angle gather at the well location is displayed on the right of the figure. Note the shifts in the location of the targets. Not only have the targets moved shallower but they have migrated sideways and, consequently, the image is better focused than the legacy image in (a). During drilling, the well trajectory was modified as indicated; otherwise, the objective would have been missed. After Dutta et al. (2014).

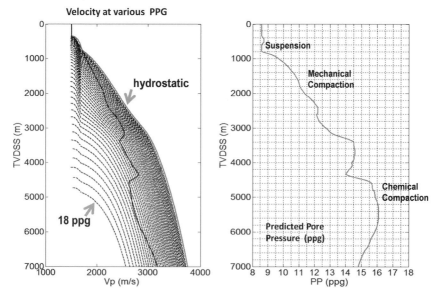

Figure 6.12 (cont.). The figure on the left shows the final velocity at the well location (heavy line) along with the RPT. The figure on the right shows the predicted pore pressure gradient (ppg). The well was drilled through a tight mud window as predicted but encountered no problem.

overpressure (OP) domain. Here OP is defined as the *difference* between the pore pressure and the hydrostatic pressure. We see a general conformation between the OP model (Figure 6.13b) and the geologic structure. This suggests that the derived velocity model is geologically plausible. This QC provides a validation to the velocity model derived by seismic data as indicated by circles in Figure 6.13a. Figure 6.13c shows a plot of vertical effective stress and P-wave velocity. Compiled data are from 11 wells in the northern Gulf of Mexico and is color-coded by the depth below mud line. The vertical effective stress estimates are based on pore pressure from field measurements. These data points suggest that as the rock is buried about 3000 m deeper below mud line, chemical diagenesis begins to contribute significantly to *lowering* of vertical effective stress (and increase in pore pressure). This is represented by an *increase* of β values as indicated. Lower β values (e.g., 6–8) indicate mechanical compaction only. Neglect of chemical diagenesis would have caused a *decrease* in overpressure by as much as 2000 psi for a velocity of 3 Km/s rather than an *increase*. This would have been unphysical. Validity of the model is provided in Figure 6.13d where we show pressure profiles versus depth. Shown are overburden gradient (OBG), fracture gradient (FG), driller's mud weight while drilling the particular well (MW), predicted pressure gradient from *RPGVM* workflow, and the postdrill best guess ppg as supplied by the driller after drilling.

Initial velocity model building is an important step in the tomography workflow. A typical initial velocity model building approach often involves interpolating velocity profiles from well locations along geologic horizons. This leads to a *structurally conforming* initial velocity model such as in Figure 6.14a. However, based on the principles of pore pressure and effective stress impact on velocity, as illustrated above, the velocity would *not* follow the dipping horizons for the geologic model discussed here. Instead, slower velocities will exist at the *crest* of an anticline structure and faster velocities exist at the base of a syncline structure (Figure 6.14b). This is an interesting finding as reported first by Petmecky et al. (2009) and confirmed here. It suggests that we should find a way to capture these important geologic features in the initial velocity model phase before executing the tomography workflow. This can help manage the nonuniqueness of the model better, especially for poorly illuminated areas, such as subsalt. The process led to a significant change in the anticline structure and CIP gathers after migration, as pointed out by the red arrows in gathers shown in Figure 6.14.

6.4   Subsalt Pore Pressure Applications

6.4.1   Introduction

With the recent advances in wide or continuous azimuth and ultralong-offset acquisition techniques coupled with routine use of Reverse Time Migration (RTM) and anisotropic tomography for velocity modeling, exploration for hydrocarbon in subsalt prospects have matured significantly due mainly to improved imaging below salt.

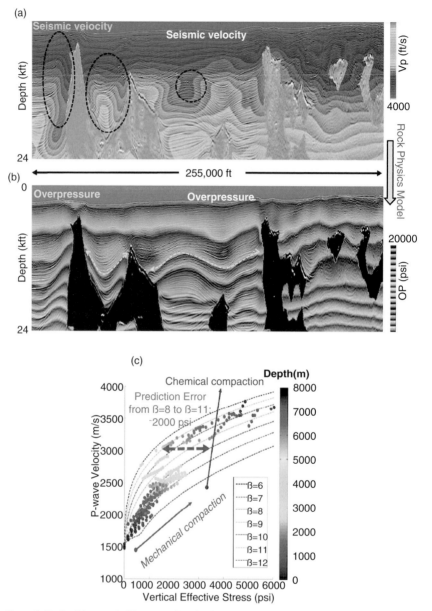

Figure 6.13 An (a) extended larger-scale seismic velocity model (b) converted to overpressure. The overpressure follows – geologic features and suggests that the slow and fast velocity trends derived by tomography are plausible features caused by fluid connection along geologic horizons. (c) Rock physics model for P-wave velocity versus vertical effective stress relation involving both mechanical compaction and chemical diagenesis as denoted by the parameter β. Data points color-coded by depth are from 11 wells in the Gulf of Mexico and show that β is increasing as a function of depth as described by Dutta (2016). After Liu et al. (2018). (d) Pressure profile versus depth. Shown are overburden gradient (OBG), fracture gradient (FG), driller's mud weight while drilling the particular well (MW), predicted pressure gradient predicted from the *RPGVM* workflow, and the postdrill best guess ppg as supplied by the drilling team after drilling. Chemical compaction refers to chemical diagenesis. After Liu et al. (2018). (A black and white version of this figure will appear in some formats. For the color version, please refer to the plate section.)

(d)

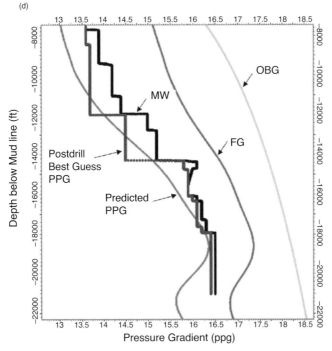

Figure 6.13 (cont.).

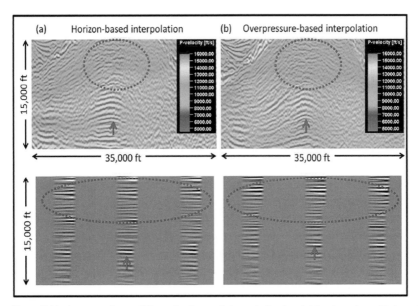

Figure 6.14 A comparison of the velocity model created using (a) horizon-based interpolation and (b) new workflow of overpressure-based interpolation. Notice at the crest of the anticline structure in (b) there is a velocity low area, caused by transmission of high overpressure up the dip leading to a lowering of the effective stress at the crest. This also led to a significant change in the anticline structure and CIP gathers after migration, as pointed out by the red arrow. After Liu et al. (2018). The color bars indicate velocity starting from 5000 ft/s to 16000ft/s in increment of 1000 ft/s. (A black and white version of this figure will appear in some formats. For the color version, please refer to the plate section.)

However, there still remain significant drilling challenges, such as reliability of underlying velocities to impact unknown regimes of subsalt pore pressure, presence of extensive rubble or shear zones and unexpected occurrence of unconsolidated rocks. The examples given below are intended to show that the *rock physics guided and pore pressure–constrained* velocity modeling approaches may provide help for subsalt pressure prediction. Rock property based velocity trend below salt helps fill in the null space often associated with subsalt tomography inversion.

6.4.2   Deepwater Green Canyon, Gulf of Mexico, USA

This study area was in the Green Canyon Area of deepwater, Gulf of Mexico (Dutta et al., 2015a). This is a geologically complex area and imaging is known to be affected by salt. The seismic data used was wide azimuth and the cable length was 8.3 Km (both in-line and cross-line). The goal was to predict pore pressure below salt and improve the image quality, as the legacy image was suboptimal. The study area included three wells, two of which were used as offset wells to parameterize and constrain the rock physics model and derive anisotropic parameters using the workflow described earlier. The third well (GC-943) was used as a blind test for pore pressure. The legacy velocity model was from anisotropic FWI (acoustic) results. Water depth ranged from ~ 3200 - 4400 ft. Checkshot, sonic, density, and mud logs were used to calibrate the overburden and velocity models for building the rock physics template.

RTM angle gathers have been shown to be useful for tomography and subsalt velocity updates (Zhou et al., 2011; Huang et al., 2011). An additional constraint for faster convergence is to have a good initial model. This was done using *RPGVM* workflow. The complete workflow for the project is given below:

a. Evaluate the existing "best" velocity model for consistency with rock physics and geological information as a starting point to build an *initial* model for tomography (legacy model).
b. Invert for Thomsen's parameters V_z, ε and δ using either the well velocities, if available, or RPT and the gather flattening criterion for epsilon and delta ($\varepsilon \approx 2\delta$) as discussed in this chapter.
c. Run tomographic updates to flatten the gathers above salt with the newly derived anisotropic parameters.
d. Model the salt-body.
e. Build the rock physics template subsalt using all available information from offset wells. If no wells are available, use RPT- based vertical velocity as a guide below salt.
f. Gather flattening in pore pressure domain – generate RTM angle gathers from the rock physics template for each chosen pore pressure gradient step (say, from hydrostatic to fracture pressure gradient limits in steps of 0.5 ppg).
g. Choose velocity bounds and starting model based on step (f) using the following as QC:

I. *Gather Flatness* – evaluate the gather flatness of RTM angle gathers for each pore pressure gradient.
II. *3D Structure-Oriented Semblance* – generate 3D structure-oriented semblance to determine the best image (image with most continuous structure). Structure-oriented semblances are calculated in curvilinear windows around the event which are derived from the structural tensor (Hale, 2011).
III. *Conventional Semblance* – generate 1D stack response of the angle gathers corresponding to each pore pressure gradient values.
h. Using the above criteria, choose the best pore pressure function as a starting model.
i. Run pore pressure–constrained tomography subsalt using velocity bounds derived from the previous steps. Use final updated velocity for pore pressure prediction and imaging.

While modeling the overburden, the effective stress due to salt buoyancy and the higher thermal conductivity of salt (compared to the sediments around the salt) were taken into account (Dutta and Khazanehdari, 2006). Generally, salt has a much higher thermal conductivity and a much lower density than the surrounding sediments. The density affects the overburden and hence, the effective stress, which in turn changes the rock velocity below salt. The thermal conductivity affects the thermal gradients of sediments around salt and therefore, the effective stress. Furthermore, onset of chemical diagenesis and its extent is altered by the presence of salt, which in turn affects the relationship between velocity and effective stress. As discussed earlier, flattening the RTM gathers directly in the pore pressure gradient domain process was used to create velocity models corresponding to various pore pressure gradients. The generated RTM angle gathers and structural semblances were performed to quality control the velocity model. This established the range of possible pore pressure gradients. Having established a range of plausible and physically reliable velocities, the velocity model corresponding to 14.5 ppg was chosen as an ideal candidate for the starting the further tomography inversion. Models corresponding to 13.5 and 15.5 ppg provided lower and upper bounds on inversion, respectively. Using this starting model, iterations of constrained tomography was conducted to update the velocity models. The updated and final velocity was then converted back to pore pressure, and this pore pressure prediction was then compared with the mud log-based results from the blind well, as shown in Figure 6.15. The model correctly predicted the increase in the pore pressure below salt. This had initially caused the well to be shut down and necessitated an increase in the mud weight from 14.2 ppg to 14.8 ppg. The model constrained the prediction of the subsalt pore pressure to within 1 ppg. The model also improved the RTM image below salt as shown in Figure 6.16. Figure 6.16a shows an image from an in-line, while the image in Figure 6.16b is from a cross-line. The areas of improvements are highlighted in these figures.

6.4.3 Stampede Field, Green Canyon, Deepwater Gulf of Mexico, USA

The target area within the Stampede Field, Green Canyon Area is inside a survey area covered with ultralong-offset (14.3 km) and full-azimuth seismic data (Vigh et al.,

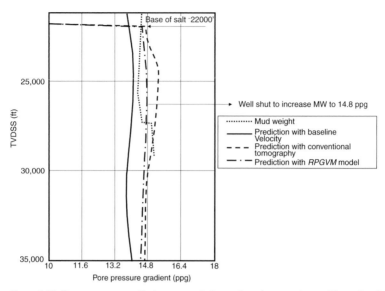

Figure 6.15 Pore pressure prediction curves below salt and comparison with mud weights at the blind well location GC943. The solid curve shows prediction with baseline velocity (legacy model). The short-dashed curve represents prediction with unbounded subsalt RTM angle gather tomography velocity (pore pressure–based starting model). The long-dashed curve is prediction with pore pressure–constrained tomography velocity using RTM angle gathers. The dotted curve shows mud weights from the blind well. TVDSS denotes true vertical depth subsea. After Dutta et al. (2015a).

2013). Eight offset wells outside the target area (all supra-salt) were used to parameterize and constrain the rock physics model and derive anisotropic parameters. The goal was to address the subsalt pore pressure estimation using the workflow described above (Dutta et al., 2015a). The legacy model was based on anisotropic (VTI) acoustic FWI. That model provided a good image above salt but was not optimal below the salt. A test using the legacy velocity model for pore pressure was not satisfactory. The velocity results from the *RPGVM* was used to predict a new model for subsalt pore pressure. Through this approach, it was possible to predict a *regression* in pore pressure subsalt, which had initially caused many sidetrack wells due to mud loss and necessitated a decrease of mud weight from 14.1 ppg to 12.5 ppg, as shown in Figure 6.17. This is in agreement with the new velocity and pore pressure model.

From an analysis of geologic structural contour maps and interpretation, the velocity reversal in Figure 6.17 is interpreted as fluid charging through permeable and dipping beds into the top of the anticlinal structure (*centroid or lateral pressure transfer effect*). This caused transmission of overpressure up dip and caused a higher pore pressure gradient at the crest of the structure. This is highlighted in Figure 6.18b and compared with the legacy model in Figure 6.18a. From pore pressure–constrained velocity model, one can generate images in depth domain using RTM migration for subsalt. The final image from the pore pressure–constrained model has better focusing and a more rounded anticline structure subsalt, leading to an improvement in imaging and facilitating the structural interpretation

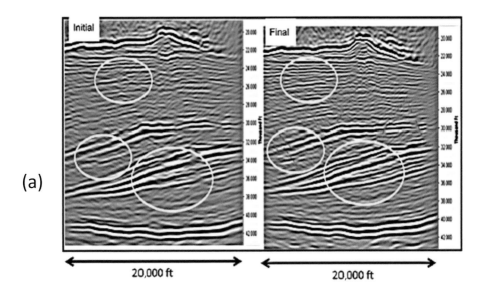

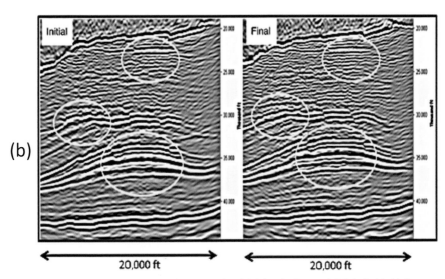

Figure 6.16 (a) RTM (in-line direction) image using initial velocity model (labeled initial) versus improved image from pore pressure–constrained tomography velocity model (labeled final). (b) RTM (cross-line direction) image using initial velocity results (labeled initial) versus improved image from pore pressure–constrained tomography velocity (labeled final). After Dutta et al. (2015a).

as shown in Figures 6.19a and 6.19b. The velocity anomaly obtained by tomography is interpreted to be a footprint of the lateral pressure transfer process. It is a geologic phenomenon. We compare these results with the legacy model that used the same seismic data as input and calibration with sonic logs and checkshots but

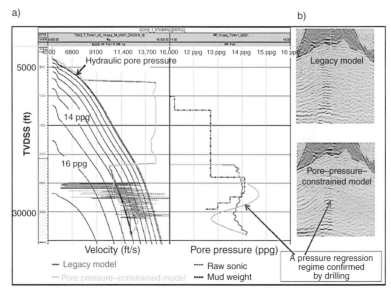

Figure 6.17 Comparison of velocity profiles at the blind test well. (a) The black curves in the background comprise the rock physics template, with pore pressure gradient ranging from hydrostatic to 17 ppg. (b) Predicted pore pressure gradients are compared to the equivalent mud weight at the same blind test well. Note the improved match to the mud weight at the well from the pore pressure–constrained tomography model versus that from the legacy model. The insets show the extent of the reversal in 3D model from the pore pressure–constrained tomography. The predicted reversal in the pore pressure gradient continues all the way below the salt. The legacy model did not yield a convincing case for the observed reversal. TVDSS denotes true vertical depth subsea. After Dutta et al. (2015a). (A black and white version of this figure will appear in some formats. For the color version, please refer to the plate section.)

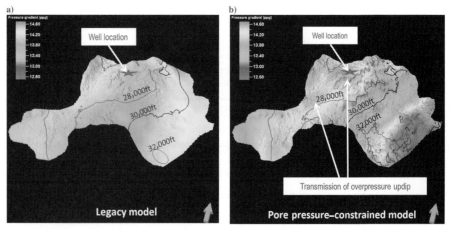

Figure 6.18 Overlay of pore pressure gradients (in ppg) on a subsalt surface for the legacy model in (a) and pore pressure–constrained tomography model (*RPGVM*) in (b). Note the high pore pressure gradient (ppg) on the crest of the pore pressure–constrained model. This is interpreted as due to transmission of overpressure up dip along permeable beds to the structure. The color bars show pore pressure gradient in ppg starting from 12.6 ppg to 14.6 ppg in step of 0.4 ppg. After Dutta et al. (2015a). (A black and white version of this figure will appear in some formats. For the color version, please refer to the plate section.)

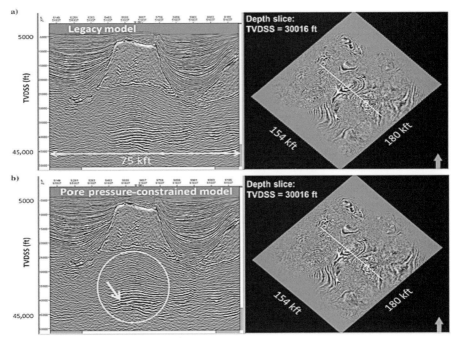

Figure 6.19 Comparison of cross section and depth slice images for (a) the legacy model and (b) the pore pressure–constrained model (*RPGVM*). Note the improvement in focusing subsalt and more rounded structure in the depth slices from the pore pressure–constrained model. True vertical depth subsea is denoted by TVDSS. After Dutta et al. (2015a).

a conventional tomography workflow. We find that the current model gives a better image and geologically plausible and physically possible velocity model.

There are three takeaways from the case studies for subsalt pore pressure prediction: (1) Fit-for-purpose seismic data, for example, ultra-long offset and full-azimuth seismic data were used in the examples; (2) accurate ray tracing from base of salt using RTM angle gathers yielded better starting model and pore pressure bounds; and (3) rock physics template and pore pressure constraints on velocity modeling helped stabilize the tomography inversion process and manage nonuniqueness. Ultra-long offsets provided higher fidelity seismic gathers and better illumination subsalt due to higher angles of incidence and diving waves. *If there is no illumination below salt, then predicting pore pressure from seismic data will be practically impossible.*

There are some lessons learned regarding the workflow for the subsalt modeling of pore pressure. First, we need to include all important overpressure generation mechanisms. In this case we needed to include *three* important pore pressure mechanisms: mechanical compaction, chemical diagenesis, and fluid transfer updip through permeable beds. While the mechanical compaction changes the porosity, and hence, effective stress and velocity, chemical diagenesis is practically independent of porosity. It is dependent on the activation energy of a modeled chemical process and is driven mostly by the thermal history of the basin (Dutta, 1986). Second, the effective stress alteration of the sediments due to the *thermal* and *buoyancy* fields associated with salt has significant effect on the effective stress regime of the sediments around the salt. The rock physics model used here includes these effects in an approximate fashion – more detailed studies would need to include complete geohistory simulation and to account for salt movement through geologic time using 3D basin modeling. Third, the use of a rock physics template is an embodiment of our knowledge of how sediments behave in a particular mini-basin as they are buried. This template is used to construct plausible bounds on the velocity below salt directly in the pore pressure domain prior to imaging. Fourth, when the standard tomography approach to subsalt velocity modeling is constrained by the bounds thus constructed, a more accurate velocity model may be produced *provided* there is some useful signal below the salt. The velocity model, when migrated with RTM, yields a better and more focused image below salt along with a more reliable pore pressure prognosis that can be used for drilling. In summary, the pore pressure and rock physics guided tomography approach extends the current operating envelope of seismic-based pore pressure prognosis to subsalt drilling applications such as casing and mud program design. More work needs to be done in the future using integration of better geologic models (such as realistic salt geometries constrained by realistic geohistory simulations that include salt movement over time) with high quality seismic data that will allow better estimates of anisotropy parameters.

6.5   Rock Physics Guided Velocity Modeling for Pore Pressure and Imaging with FWI

Full waveform inversion (FWI) is a very good technology for velocity model building for imaging (Vigh et al., 2009). However, not much application has been made to assess the usefulness of this technology for pore pressure estimation. Liu et al. (2016) made a useful contribution in this regard using the data set from the Stampede Field, Green Canyon Area of the Gulf of Mexico, USA. The application was made for suprasalt area using a two-step process. In step I they evaluated the legacy anisotropic tomography model and showed that although the seismic image was acceptable, it yielded unphysical pore pressure *above* salt. The pore pressure was below hydrostatic. This is highlighted in Figure 6.20a. In step II they corrected the model by using rock physics guided velocity modeling and pore pressure–constrained tomography. This yielded a slightly improved image and pore pressure that is equal to or higher than the hydrostatic pressure everywhere as shown in Figure 6.20b. This model was deemed useful as the *starting or initial* model for FWI iteration. To better guide and constrain the FWI process, Liu et al. (2016) used the RPT to guide the FWI process and limited the upper bound of velocity that is consistent with hydrostatic pore pressure during each stage of iteration. Figure 6.21c contains the final pore pressure model from FWI and it is compared with the legacy and the pore pressure–constrained tomography models in Figures 6.21a and 6.21b, respectively. The depth slices at 5344 ft showed

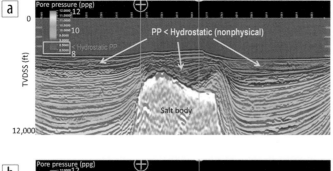

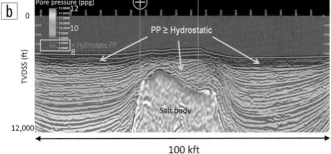

Figure 6.20 Comparison of (a) the legacy pore pressure model and (b) the new pore pressure model after rock physics–constrained tomography. The pressures in purple are lower than hydrostatic pressure and represent the unphysical nature of the underlying seismic velocity. After Liu et al. (2016). The color bars indicate pore pressure gradient in ppg starting from 8.0 ppg to 12.0 ppg in increment of 0.5 ppg. (A black and white version of this figure will appear in some formats. For the color version, please refer to the plate section.)

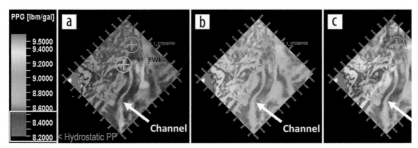

Figure 6.21 Comparison of pore pressure models in map view at 5344 ft TVDSS for (a) legacy model, (b) after pore pressure–constrained and rock physics–guided tomography, and (c) after FWI with rock physics and pore pressure constraints on the inversion. The pressures in purple are lower than hydrostatic (nonphysical pressure). While (a) has unphysical velocity (yielding lower than hydrostatic pore pressure), both (b) and (c) have no such deficiency. After Liu et al. (2016). The color bars indicate pore pressure gradient in ppg starting from 8.2 ppg to 9.5 ppg in increment of 0.2 ppg. (A black and white version of this figure will appear in some formats. For the color version, please refer to the plate section.)

that all the unphysical pore pressures in the legacy model (purple) are corrected in the new FWI model. A better-defined pore pressure, with strong contrast across a channel feature, can be observed in the new FWI model. Thus the image is also improved. Postdrill data at target well locations confirmed the increase in pore pressure above salt, which is also better predicted by the new FWI model.

This example above suggests that the FWI process benefits from a better starting model that uses rock physics guided velocity model workflow. These lead to stable and physically reliable velocity models from FWI. The lesson learned is this: as we expand the inversion space – conventional velocity analysis to tomography and then to FWI, we need to impose more and more constraints on *both* the forward and the inverse models. A totally mathematically based process is necessary but not sufficient. The constraints should be imposed through smart regularization or other innovative processes. Both acoustic and elastic FWI would impact subsalt targets and pore pressure. This area would benefit from further research. We encourage the seismic imaging community to evaluate goodness of velocity models by applying such models to quantify pore pressure and thus help bridge the gap between imaging and drilling communities.

6.6   Summary

In this chapter, we discussed technologies that yield *geologically plausible and physically possible* interval velocities from surface seismic data. The workflow is termed *RPGVM*. The approach can be used on *any* algorithm for interval velocity computations – be it conventional or based on inversions such as tomography and FWI. The goal is to define the parameter base associated with a particular inversion approach so that the inferred velocity model is *constrained by rock physics and bounds of pore pressure* (such as, for example, between hydrostatic and fracture pressure limits). Models based on other drilling parameters such as rate of penetration of drill-bits and D-exponents (see Chapters 7 and 8) that yield a plausible range of pore pressure can also be integrated with this approach as well. We used pore pressure as a criterion to define what a physical velocity model should be. There may be other indicators that we have not tested but consider to be appropriate. Some of these are lithology indicators calibrated with well logs, fluid type or fluid saturation, interval Q or attenuation (seismic, VSP, sonic logs), geologic markers, or well-known horizons, to name a few.

At the writing of this book, anisotropic tomography is the most reliable approach for geopressure prediction with seismic data. We showed the value of rock physics templates for deriving velocity models with transverse isotropy (VTI and TTI) symmetry. Orthorhombic tomography is becoming popular and the current approach could be extended to manage its inherent nonuniqueness.

Acoustic FWI approaches have gained traction for imaging but extension to pore pressure or geohazard applications are lacking. However, since these approaches use tomography as a starting model, it increases the size of the parameter base considerably and consequently, adds more nonuniqueness to the velocity models besides being handicapped for "cycle skipping." Nonetheless, the derived velocity models are of high-fidelity and our limited experience suggests that the models are very good for geopressure analysis. Elastic- FWI is in research stage at the writing of this book. We believe that when this technology is fully developed, it would be ideal for imaging of many subsurface properties such as pore pressure, fracture pressure, and lithology and fluid properties, beside conventional imaging products.

7   Methods for Pore Pressure Detection
Well Logging and Drilling Parameters

7.1   Introduction

Earlier we discussed how pore pressure measurements are done using direct methods such as using repeat formation testers (RFT), drill stem test (DST), or dynamic modular test (MDT) in boreholes. However, such methods cannot be used in low-permeability formations such as shales where overpressuring occurs. Indirect methods such as those based on shale compaction curves are the basis for obtaining pore pressure estimation from seismic and various wireline tools. Wireline log analysis is one of the major methods employed to obtain pore pressure estimates from analogue wells, if available, prior to developing a predrill pore pressure model for well planning purpose. Other methods such as seismic while drilling (SWD), logging or measurements while drilling (LWD or MWD) and vertical seismic profiling (VSP) are very useful technologies for good quality data on pore pressure estimates in shales and are routinely used by the petroleum industry. However, all these use indirect pressure measurements.

Although various well-logging methods used in the detection and evaluation of overpressure are postdrill techniques, these provide very useful information and help drillers and geoscience engineers construct area trends and help calibrate subsurface prediction models of pore pressure. A complete discussion of geophysical well-logging techniques for abnormal pore pressure is contained in the book by Fertl (1976) and by Bigelow (1994). An excellent review of principles of well logging and uncertainties are discussed in Mondol (2015), as well as in many excellent books on well logging such as Ellis and Singer (2008), Hearst et al. (2000), Liu (2017), and Serra (2008). Therefore, we shall not embark on a detailed description of each geophysical tool used for pore pressure measurements. Rather, we shall focus on methods that use data from some "key" and "commonly" used tools for pore pressure detection and measurement. The commonly used tools for pressure analysis are

- velocity measurements using tools such as wireline sonic (V_p, V_s)
- resistivity/conductivity measurement tools
- bulk density measurement tools

Some of the other tools that are not-so-commonly used for pore pressure analysis are: magnetic resonance logging tools, borehole gravity measuring tools, dielectric measuring tools, and pulsed neutron tools (employed in cased borehole). In principle, any wireline tool that uses porosity, either directly or indirectly as an attribute contains indirect information about pore pressure. Data from all tools require interpretation that extract *signal* (porosity, pore pressure, etc.) from *noise* (environmental corrections, for example). Lithology and fluid determination is essential prior to quantifying pore pressure from extracted signals. All log-derived pore pressure estimates require calibration such as those from RFT/MDT/DST tools.

7.2   Logging Tools

Well logs are obtained from tools that are lowered into borehole soon after drilling a segment of the borehole. Measurements of various formation properties are made at regular intervals, typically, between 5 and 20 cm. Data processing is done either on land or in rig platforms for offshore wells. The tools can be run either in open boreholes or in cased holes after casings are installed in the well. The *drilling mud* (drilling fluid) plays a significant role on the quality of the data. Drilling muds can be either water-based or oil-based and contain clays and other fine-grained natural or synthetic material. Because the pressure in the drilling mud usually exceeds the formation pressure, the mud begins to invade the virgin formation. This creates a *mudcake* which lines the borehole wall as shown in Figure 7.1. The *mud filtrate* (fluids in the drilling mud) also enters the formation, creating a *flushed* zone where all the primary formation pore fluids are replaced by the fluid from the drilling mud. Beyond this zone, there is another zone where the formation fluid is partially replaced by the invading mud filtrate. This zone is called the *transition zone.* The virgin formation fluids occupy the *uninvaded zone.* The extent of each zone depends upon the viscosity and density of the fluids and the permeability of the formation. Thus invasion of mud is farthest in porous sandstones and carbonates. It is less so in low permeability rocks such as shale and tight reservoir rocks. However, in overpressured shales, the extent of invasion could be large due to higher porosities in these formations. The properties of "mud cakes," "mud filtrate," and the extent of the "invaded zone" affect the data acquired by various wireline tools. Therefore, the acquired data must be corrected. This is known as "environmental corrections." There is some science – some theoretical, but mostly empirical, behind these corrections. While wireline logging companies provide correction charts for various tools and these are very helpful, the accuracy of correction diminishes as the diameter of the borehole, the thickness of mudcake and the depth of invasion, increases. The environmental corrections also include formation temperature and fluid chemistry (mud and formation). While some tools, such as resistivity are greatly impacted by both temperature and salinity of formation water and mud, others such as acoustic measurement (sonic) tools are not affected by these environmental factors. It is for this reason sonic tools are most commonly used for reliable pore pressure information.

Logging tools measure radioactivity (e.g., *gamma ray, neutron, and density* logs), electrical properties (e.g., *spontaneous potential or SP, induction,* and *resistivity* logs), acoustic properties (*sonic* logs of various types), and nuclear magnetic resonance (*NMR*

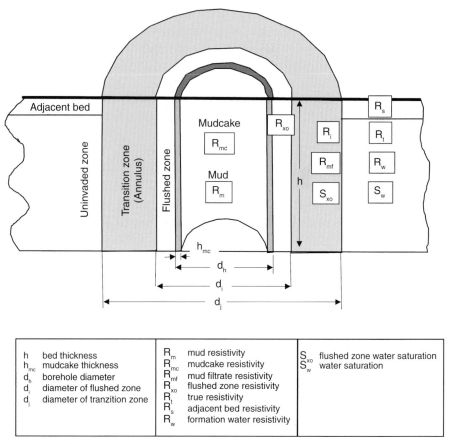

Figure 7.1 During drilling, mud and mud filtrate invade the virgin formation around the borehole since the hydrostatic pressure within the mud is higher than the formation pressure. This creates formation of mud cake, flushed zone, and the transition zone as one moves away from the center of borehole. The extent of the invasion is higher for porous formations and lowest for the tight formations such as shales, tight sandstones, and other low-porosity carbonates. However, for overpressured shales, mud and mud filtrate penetration could be large.

logs). Thus the measurements are *indirect* and rock and fluid properties and their environment such as pore pressure are inferred. Other logs that measure the properties of the borehole are *caliper, temperature, dipmeter,* and *image* logs. Most logging tools have limited range of depth of investigation. The tools that use radioactivity methods have the least range. Electrical tools have a wide range of depths of investigation – from few centimeters (for micro-tools) to several meters (for induction logs, for example). The sonic tools have a depth of investigation of ~60 cm, and they are not affected by the temperature of the environment as opposed to resistivity tools. Common sonic tools are borehole compensated sonic (BHC), long-spaced sonic (LSS), full waveform or array sonic, and dipole shear imager (DSI). We note that the sonic tools do not measure velocity directly but measure the reciprocal of the velocity or slowness.

Most logs can now be recorded while drilling. In this case there is no need to remove the drill string. These logs are called *MWD* (measurements while drilling) and *LWD* (logging while drilling). These tools provide real-time data. However, since the data are recorded while drilling is going on – a very noisy and destructive environment, the data quality could be compromised. They provide data that are often of poor quality, lower resolution, and less coverage. So most oil companies always record wireline logging data in addition to MWD/LWD logs to calibrate such while – drilling data. Another disadvantage is that the data transmission is via *mud pulses* that is restricted by the available bandwidth. (Mud pulses are created by closing a valve that generates pressure fluctuations and these fluctuations propagate through the mud and are deciphered at the surface by pressure sensors and computers.) Current mud-pulse telemetry technology offers a bandwidth of up to 40 bit/s (bits per second). The data rate drops with increasing length of the wellbore and could be as low as 0.5 bit/s to 3.0 bit/s at a depth of 35,000 - 40,000 ft (10,668 - 12,192 m). Alternatively, the data are recorded downhole and retrieved later when the tool is brought back to the surface. Most MWD tools can operate continuously at temperatures up to ~150°C, with some sensors available with ratings up to ~175°C. MWD tool temperatures may be 20°C lower than formation temperatures measured by wireline logs, owing to the cooling effect of mud circulation, so the highest temperatures encountered by MWD tools are those measured while running into a hole in which the drilling fluid volume has not been circulated for an extended period.

7.3 Pore Pressure from Well-Logging Methods

Hottman and Johnson (1965) laid the foundation of detecting and quantifying pore pressure from log-derived shale values – resistivity and velocity or transit time. Since then many other geophysical tools are used for pore pressure measurements. Invariably, most logging devices used to detect and measure pore pressure are essentially shale porosity measuring devices, and indicate overpressure because of abnormally high or low values of any of the following characteristics of over pressured shales: density, porosity, transit time, conductivity, resistivity, drilling rate, water salinity, and temperature gradient, as shown schematically in Figure 7.2. It shows a conceptual display of responses from various geophysical well logs in normal and abnormally high-pressure environment. Porosity is the common variable in the indicators of all of these tools.

7.3.1 Acoustic or Velocity Measurements

These tools provide the best data for inferring formation pressure because this type of log measurements is usually less affected by borehole size, formation temperature, or formation water salinity. Typically, a caliper log (a measurement of the borehole diameter) is used in conjunction with acoustic tools to quality control the sonic transit time data. The basic principle of acoustic (sonic) log device is the transmission of

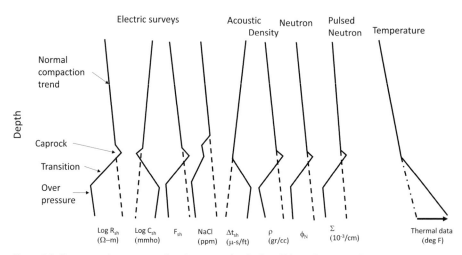

Figure 7.2 Conceptual response of various geophysical well logs for normally pressured and geopressured rocks. Modified after Fertl and Timko (1971).

a sound pulse that is recorded at receivers placed at fixed distances away from the transmitter(s). The sources can be either monopole or dipole and in some cases, quadrupole. In monopole excitation the transmitter emits energy uniformly around the tool. In consolidated formations, this energy excites three waves that travel down the borehole wall. These are: compressional, shear, and Stoneley waves. The compressional wave travels away from the transmitter with a velocity appropriate for the mud (V_f). When these waves reach the borehole wall, they are reflected, refracted, and converted according to Snell's law. For angles of incidence less than the compressional-wave critical angle, part of the energy is transmitted into the formation in the form of compressional wave, another part is converted as a refracted shear wave and the third part is reflected back into the mud as a compressional wave. The transmitted waves travel at speeds appropriate for compressional and shear waves for formation, close and parallel to the borehole wall, while continuously radiating energy back into the mud as converted compressional waves, at the same compressional wave critical angle at which it entered. It is this radiated energy that is detected by the receivers. If the formation shear wave velocity is less than the mud velocity, then no shear wave is refracted along the borehole wall and no shear wave is recorded by the receivers. The waves that propagate along the borehole walls are termed Stoneley waves and these are always present. These provide information about shear wave velocity through additional modeling. To avoid additional modeling requirements and additional parametrization, the industry always uses direct shear wave data from dipole tools. Additional details of these acoustic tools and the mathematics of elastic wave propagation in boreholes can be found in Cheng et al. (1992), Mondol et al. (2015), Paillet and Cheng (1991), Tang and Cheng (2004), and Timur (1987). It should be mentioned that the quality of the acoustic (sonic) log is affected by the mud cake that is always created during drilling. Furthermore, "cycle skipping" could pose a serious problem when the mud properties are affected by gas (it lowers the velocity of the fluid). (In sonic logging, if the first arrival, namely, first cycle, is not identified, due to weak signal, usually through bad borehole geometry causing washouts, later cycles will be picked, giving erroneous traveltimes.) For pore pressure measurements, the industry invariable uses the recorded compressional wave velocity logs. It is regrettable that the shear waves are rarely used for this purpose.

Since the pioneering work by Hottman and Johnson (1965), sonic logs are the primary source of estimating pore pressure as discussed in Chapter 4. This method uses shale transit time as an indicator of overpressure. Many practitioners use Eaton's method and equivalent depth method with data from acoustic tools to evaluate pore pressure. These methods were also discussed in Chapter 4. In fact, any of the velocity based methods that we discussed in Chapter 4 can be used with data from sonic logs. Since the pore pressure mechanisms are valid for shale, it is important that logs for shale indicator such as gamma ray log be used in conjunction with sonic logs. We should use transit time values appropriate for shale intervals only. Petrophysicists use V_{sh} as a shale volumetric fraction defined by

$$V_{sh} = \frac{\Gamma - \Gamma_{\min}}{\Gamma_{\max} - \Gamma_{\min}} \qquad (7.1)$$

where Γ is the measured gamma ray value from log at a depth Z and the suffixes min (sand line) and max (shale line) denote minimum and maximum gamma ray readings, respectively. The cutoff values define the shale picks. These cutoff values are very subjective. We also note that the sand line and/or shale line may be at one gamma ray value in one part of the well and at another gamma ray value at another level. Thus the model is impacted by the experience level of petrophysicists. Typically, computations are carried out by averages over large intervals such as 20-40 ft. Consequently, sonic transit time values are also averaged over the same interval. A point-by-point computations of V_{sh} and Δt_{sh} could be meaningless since the recorded sonic transit times are measured over an interval that is averaged over a distance dictated by the source-to-receiver distance.

As stated in Chapter 4, the Hottman-Johnson (HJ) method consists of measuring shale - interval transit time from sonic tools. A normal compaction trend is first established as shown in Figure 7.3a which is a plot of ln (Δt) – logarithm of interval transit time of shale versus burial depth (Z). We can write this as

$$\ln(\Delta t_n) = A + BZ \qquad (7.2)$$

where A and B are constants and the suffix n stands for normal pore pressure condition. In overpressure conditions, the data points for interval transit time would deviate from the normal trend, toward abnormally high transit times for a given burial depth, since the porosity is higher at that depth than if the shale were normally pressured. Figure 7.3b shows the nature of the departure from the normal trend. Hottmann and Johnson (1965) use two variables in their approach: Δt_{ob} for observed shale transit time at a given depth and Δt_n for the normal transit time at that depth. This is shown in Figure 7.3b. The deviation parameter, $\delta(\Delta t) = \Delta t_{ob} - \Delta t_n$ is the difference between the two. In the HJ-method, the $\delta(\Delta t)$ parameter is used for pressure calibration. Others, such as Eaton (1975), use the ratio of the two variables as indicator of pore pressure. Hottmann and Johnson (1965) did not specify the assumptions regarding either the compaction model or the attribute model. If we assume that the compaction model is described by an Athy-type model

$$\phi = \phi_0 e^{-K\sigma} \tag{7.3}$$

and the attribute model is described by a relation of the type

$$\Delta t = a\phi^b \tag{7.4}$$

Then we have

$$\ln\Delta t = \ln(a\phi_0^b) - Kb\sigma \tag{7.5}$$

In terms of overburden stress S and pore pressure P, using Terzaghi's law, equation (7.5) becomes

$$\ln \Delta t = \ln(a\phi_0^b) - Kb(S - P) \tag{7.6}$$

For normally pore pressured condition, we have similarly

$$\ln \Delta t_n = \ln(a\phi_0^b) - Kb(S - P_n) \tag{7.7}$$

Subtracting equation (7.7) from equation (7.6), the following relationship is established between transit time and pore pressure:

$$P = P_n + \ln\left(\frac{\Delta t}{\Delta t_n}\right)^{(1/bK)} \tag{7.8}$$

which is the fundamental representation of the calibration curve of the HJ-method. In terms of the average bulk density $\langle \rho_b \rangle$, we can approximate the following for the overburden stress, S, at a depth, Z, as follows:

$$S = \langle \rho_b \rangle gZ \tag{7.9}$$

where g is acceleration due to gravity. Using equation (7.9) in equation (7.7), the following relationship between sonic traveltime and depth for normally compacted shales is established:

$$\ln(\Delta t_n) = \ln(a\phi_0^b) - Kb(\langle \rho_b \rangle - \rho_w)gZ \tag{7.10}$$

which is the other fundamental curve used in the Hottman and Johnson approach (see Figure 7.3a) and it suggests that a plot of $\ln \Delta t_n$ versus depth is a straight line. We note that a plot of $P - P_n$ versus $\ln\left(\frac{\Delta t}{\Delta t_n}\right)$ also will be a straight line. However, Hottman

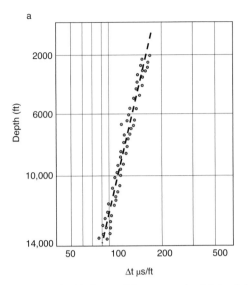
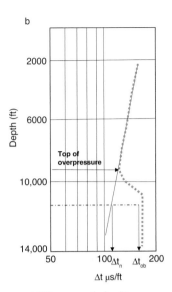

Figure 7.3 (a) Shale traveltime versus burial depth for Miocene and Oligocene shales Upper Texas and Southern Louisiana Gulf Coast. (b) Shale traveltime versus burial depth showing the departure of the observed shale traveltime Δt_{ob} from the normal compaction values Δt_n at the same depth. Modified after Hottmann and Johnson (1965).

and Johnson decided to use pressure gradient (pressure divided by depth) as a variable, instead of pressure, and plotted it against the difference $\delta(\Delta t)$, instead of $\ln\left(\dfrac{\Delta t}{\Delta t_n}\right)$. It is noted that the curvatures of Hottman and Johnson's calibration curves shown in Figure 4.5 are determined by two fundamental parameters, the exponent b shown in the attribute model of equation (7.4) and the shale compaction coefficient K in the equation (7.3).

Eaton's method for acoustic logs can also be derived in a simple form, which may lead to a clearer interpretation of the exponent in his method. Eaton's method implicitly uses the following equation for the compaction law in terms of the effective stress and porosity:

$$\sigma = c\phi^d \tag{7.11}$$

and

$$\Delta t = a'\phi^{b'} \tag{7.12}$$

for the attribute law in terms of shale traveltime and porosity. From equation (7.11) we have

$$\frac{\sigma}{\sigma_n} = \left(\frac{\phi}{\phi_n}\right)^d \tag{7.13}$$

And from equation (7.12), the following is also valid:

$$\frac{\Delta t_n}{\Delta t} = \left(\frac{\phi}{\phi_n}\right)^{b'} \tag{7.14}$$

Combining equations (7.13) and (7.14), we get

$$\frac{\sigma}{\sigma_n} = \left(\frac{\Delta t_n}{\Delta t}\right)^{\frac{d}{b'}} \tag{7.15}$$

This is Eaton's original equation, provided one recognizes that he sets the exponent in equation (7.15) to 3.0. In the published literature there is no explanation why the Eaton exponent must be equal to a fixed value of 3.0. It seems ad hoc. As the discussions above show, the coefficient is obviously data dependent. We propose that a better way to define Eaton's exponent is to carry out a two-step process: First, plot $\ln(\Delta t_n/\Delta t)$ versus $\ln \phi$ and second, plot $\ln(\sigma/\sigma_n)$ versus $\ln \phi$. The slopes of these two cross-plots, will automatically determine Eaton's exponent for a given data set, without having to use a fixed exponent as is commonly done. It also suggests that errors in the attribute law (its slope) can be compensated by the errors in the slope of the compaction law. Thus Eaton's *recommendation* of using a constant exponent of 3.0 and varying the slope of the compaction trend to match a given set of limited and discrete pressure data is an implicit admission that the exponent is *not* an independent parameter in the model. The exponent must necessarily be data dependent. Some authors (Sarker and Batzle, 2008) suggests that the Eaton exponent be depth dependent. Others such as Bowers (1995) suggest an exponent ~5. We tend to agree with these observations and stress that users must check the data quality and assess the validity of using an exponent of 3.0.

Chapman (1994a) proposed the following for the normal compaction curve in terms of shale sonic traveltime:

$$\Delta t_n = (\Delta t_0 - \Delta t_m)e^{-\frac{z}{\lambda}} + \Delta t_m \tag{7.16}$$

Here Δt_0 is the value of the shale traveltime at the *mudline*, Δt_m is the traveltime through the matrix material, and λ is the scale length in the Athy exponent (Athy, 1930a). Zhang (2011) suggested using above depth dependent normally pressured shale traveltime model in the Eaton model and thus remove the inherent uncertainty in adjusting the slope and intercept of the normal compaction trend of shale to fit a set of pore pressure data. This yields

$$P = S - (S - P_n)\left(\frac{(\Delta t_0 - \Delta t_m)e^{-\frac{z}{\lambda}} + \Delta t_m}{\Delta t}\right)^{3.0} \tag{7.17}$$

Here P is pore pressure, S is overburden stress, and P_n is the hydrostatic pressure. The range of values for Δt_0 is ~ 170 µs/ft or 558 µs/m to ~ 200 µs/ft or 656 µs/m. Since very rarely we have sonic log to mudline, this parameter needs to be guessed. Equation (7.16) states that logarithm of shale traveltime under hydrostatic condition is *not* a straight line as assumed by Eaton. (Eaton's approach to computing normal compaction trend will yield unphysical value for low porosity at deeper depths.)

An example of the application of sonic log-based pore pressure analysis is shown in Figure 7.4. The application is from a deep subsalt well. This bypass well was drilled to a depth of 34,189 ft (Zhang, 2011) and required setting casing in a 15,000 ft thick section of tectonically active salt. The authors used a depth-dependent model for pore pressure estimation using sonic transit times, combining Wyllie's model (Wyllie et al., 1956) for relating porosity to transit time, and Athy's model (Athy, 1930a) for depth-dependent porosity trend and derived the following (Zhang, 2011)

$$P_{ppg} = S_{ppg} - (S_{ppg} - P_{n,ppg}) \frac{\ln(\Delta t_{ml} - \Delta t_m) - \ln(\Delta t - \Delta t_m)}{cZ} \tag{7.18}$$

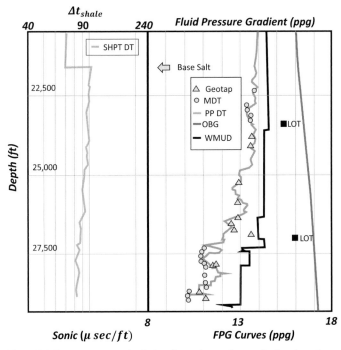

Figure 7.4 Pore pressure calculation from the sonic transit time in a deepwater subsalt formations of the Gulf of Mexico. In this figure, the gamma ray and shale base lines (not shown) are used to define a *trend* for shale sonic traveltimes labeled as SHPT DT (left track). Calculated pore pressure from the filtered shale transit time (PP DT) is shown in the right track with comparison to the measured formation pressures (MDT). OBG denotes overburden gradient, LOT denotes Leak-off test values and WMUD denoted mud weights. GEOTAP is Haliburton's Formation Pressure Tester in real-time that uses a slimhole LWD package. Modified after Zhang et al. (2011).

where P_{ppg} is pore pressure gradient in ppg, S_{ppg} is the overburden gradient in ppg, $P_{n,ppg}$ is hydrostatic or normal pore pressure gradient in ppg, Δt is sonic transit time in $\mu s/ft$; Δt_0 is sonic transit time at mudline, Δt_m is the sonic traveltime for mineral grain, Z is depth below mudline, and c is Athy's constant. Results in Figure 7.4 were obtained by using $\Delta t_{ml} = 120\ \mu s/ft$, $\Delta t_m = 73\ \mu s/ft$, $P_{n,ppg} = 8.75$ ppg, and $c = 0.00009$.

7.3.2   Electrical Resistivity/Conductivity Measurement Tools

Resistivity logging measures the subsurface electrical resistivity, which is the ability to impede the flow of electric current. This helps to differentiate between formations filled with salty waters (good conductors of electricity) and those filled with hydrocarbons (poor conductors of electricity). Resistivity and porosity measurements are typically used to calculate water saturation. Resistivity is expressed in ohms, and is frequently charted on a logarithm scale versus depth because of the large range of resistivity. The distance from the borehole penetrated by the electric current varies with the tool, from a few centimeters to several meters.

Because the resistivity logs contain information about porosity, such as through the Archie's law, it is quite extensively used for pore pressure quantification. The method relies on the observation that under normal compaction, the shale resistivity increases in value with increasing depth (due to porosity reduction). When overpressured formations are encountered, the resistivity curve shows a departure from the normal trend line (just as in the case of acoustic logs) and reads lower values of resistivity in shales. This departure is calibrated against measured pore pressure. An example of this is given in Figure 7.5. The technique was pioneered by Hottman and Johnson (1965) and Ham (1966). There are various types of resistivity tools depending upon the depth of penetration (into the formation and away from borehole). Best tools are those that yield measurements away from the "invaded zone."

Measuring pore pressure is essentially an attempt to make porosity determination in thick and uniform shale intervals throughout the logged interval. While acoustic tools are capable of providing accurate estimates of formation pressure, they are not run throughout the borehole due to borehole size limitations, especially at the shallow part of a borehole. That is not the case for either resistivity or conductivity tools – these tools are run on virtually every borehole and usually through the entire length of the borehole drilled below surface casing. The steps to estimate pore pressure from resistivity logs are summarized below (Hottmann and Johnson, 1965; Foster and Whalen, 1966):

1. Establish a "normal compaction trend (NCT)" of an area of interest by plotting the logarithm of shale resistivity (usually from measurements from several wells in the area of interest)
2. Make a similar plot for the well in question
3. Note the top of the overpressured zone (in depth) by noting the depth at which the plotted points diverge from the NCT

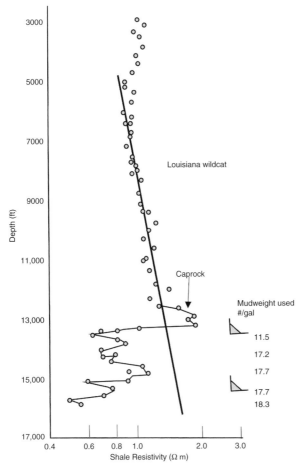

Figure 7.5 Shale resistivity plot versus depth from US Gulf Coast well (Louisiana). In normally compacted shale, resistivity increases with depth. In overpressured rocks, resistivity of rocks reverses. In this example, limey caprock caused slight resistivity increase just above the point where a 9 5/8 inch casing was set. Modified after Bigelow (1994).

4. The pressure gradient (in either psi/ft or pounds per gallon, ppg) of a reservoir at a given depth is estimated as follow:
   a. The ratio of the extrapolated NCT to the observed shale resistivity is determined
   b. The fluid pressure gradient corresponding to the calculated ratio is found from a "calibration" chart, an example of which is given in Figure 7.6
5. The reservoir pressure is obtained by multiplying the FPG value by the depth
6. Repeat above steps for all depths

An example of estimating pore pressure from resistivity log is shown in Figure 7.7.

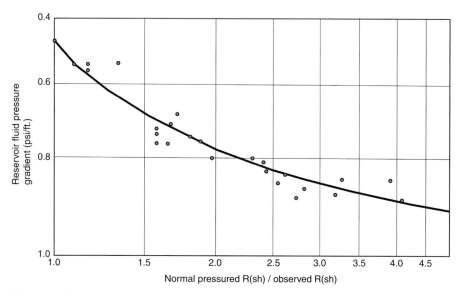

Figure 7.6 Relationship between shale resistivity ratios $(R_{sh,nor}/R_{sh,obs})$ and reservoir fluid pressure gradient (psi/ft) in overpressured Miocene/Oligocene formations of the US Gulf Coast. Modified after Hottman and Johnson (1965).

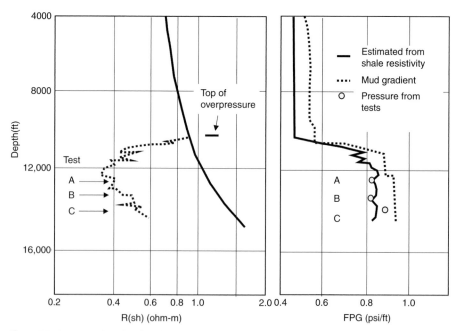

Figure 7.7 An example of estimating pore pressure in shale from resistivity log. Modified after Hottman and Johnson (1965).

Eaton's method (Eaton, 1976) is another popular method for quantifying pore pressure using electrical resistivity logs as pointed out in Chapter 4. Eaton's equation for resistivity log is given below:

$$P = S - (S - P_n)\left(\frac{R}{R_n}\right)^{1.2} \tag{7.19}$$

where P is pore pressure, S is overburden stress, and R is the resistivity of shale. The symbol, n, denotes the value of the variable under normal or hydrostatic pore pressure condition. The exponent in equation (7.19) is set at 1.2 as per Eaton (1976). Below we show the basis for analysis of pore pressure using resistivity logs as outlined above (equation (7.19)) in more detail.

Eaton's method requires establishing a normal compaction trend line versus depth for resistivity. Eaton (1976) did not provide a rock physics basis for resistivity of normally pressured shales. Following Chapman (1994a) who suggested a normal compaction trend line model for acoustic transit time of shales (equation (7.16)), we assume that the following describes the depth-dependent normal compaction values of resistivity of shales (R_n)

$$R_n = R_0 e^{bZ} \tag{7.20}$$

or in terms of logarithm variables

$$\ln R_n = \ln R_0 + bZ \tag{7.21}$$

The preceding equation suggests that the logarithm of R_n versus depth is a straight line as shown in Figure 7.5. With this assumption, Eaton's model (equation (7.19)) for pore pressure in terms of resistivity can be written in more general terms as

$$P = S - (S - P_n)\left(\frac{R}{R_0 e^{bZ}}\right)^{\zeta} \tag{7.22}$$

Here R_0 is the resistivity of clays at mudline, b is a constant, and ζ is Eaton's exponent (whose typical value as recommended by Eaton is 1.2). The advantage of this formulation is that once the slope of the normal compaction trend values for resistivity b is determined, we can concentrate on deriving a proper value of the exponent ζ, without having to adjust the slope of the NCT or engage in point-by-point calculations as done by Hottmann and Johnson (1965).

In the literature, shale Formation Factor, F, is used frequently to predict pore pressure from electrical logs (Foster and Whalen, 1966). Formation factor is defined as

$$F = R/R_w \tag{7.23}$$

where R_w is the resistivity of pore water of shale and R is the resistivity of shale. In the literature there is no recipe provided to measure resistivity of pore water in shale. We *assume* that it is equal to or related to the resistivity of pore water in *nearby sands (this is an assumption and may not be valid)*. Using Archie's law, we can write

$$F = a/\phi^m \tag{7.24}$$

where a and m (cementation exponent) are constants. Typically, $a = 1$ and $m = 2$ may be used. From equations (7.23) and (7.24), we can solve for porosity in terms of resistivity:

$$\phi = a^{\frac{1}{m}}(R_w/R)^{\frac{1}{m}} \tag{7.25}$$

For normally compacted shales, one can also write

$$\phi_n = a^{\frac{1}{m}}(R_w/R_n)^{\frac{1}{m}} \tag{7.26}$$

We can further simplify the above relation using Athy's Law (Athy, 1930a) for compaction in terms of the vertical effective stress, namely,

$$\phi = \phi_0 e^{-K\sigma} \tag{7.27}$$

and for normally compacted shales we have

$$\phi_n = \phi_0 e^{-K\sigma_n}$$

Thus, from equations (7.25) and (7.26),

$$\phi = \phi_0 e^{-K\sigma} = a^{\frac{1}{m}}(R_w/R)^{\frac{1}{m}} \tag{7.28}$$

and similarly,

$$\phi_n = \phi_0 e^{-K\sigma_n} = a^{\frac{1}{m}}(R_w/R_n)^{\frac{1}{m}} \tag{7.29}$$

Taking logarithm of the ratio of equations (7.28) and (7.29), we get

$$\sigma = \sigma_n - \frac{\ln(R_n/R)}{mK} \tag{7.30}$$

which upon using Terzaghi's law yields the following expression for the pore pressure P:

$$P = P_n + \ln(R_n/R)^{\frac{1}{mK}} \tag{7.31}$$

Or in terms of the pore pressure gradients,

$$P/Z = P_n/Z + \frac{\ln(R_n/R)\frac{1}{mK}}{Z} \qquad (7.32)$$

Thus, by relating the shale resistivity that exists under overpressured conditions, to that which would exist under normal conditions, formation pressure can be estimated provided we know m, K, and R_n. The cementation index m is the slope of the line of logarithm of formation factor F when plotted versus depth. K is obtained from Athy's compaction model (equation (7.27)). Resistivity of normally pressured shale R_n is obtained by plotting resistivity values of shale versus depth and deciding the normal compaction trend as shown in Figure 7.4 and equation (7.21). The basis for the plot shown in Figure 7.5 is given by equation (7.32).

Dutta's model for effective stress prediction from velocity as shown in Chapter 4 can be extended to resistivity logs, by combining his compaction model as given in equation (4.24), with Archie's laws shown in equations (7.23) and (7.24). From equation (4.23), one finds

$$\varepsilon = \frac{\ln(\sigma_0/\sigma)}{\beta} \qquad (7.33)$$

and therefore

$$\varepsilon_n = \frac{\ln(\sigma_0/\sigma_n)}{\beta} \qquad (7.34)$$

The void ratio ε is defined by

$$\varepsilon = \frac{\phi}{1 - \phi} \qquad (7.35)$$

Relating porosity and void ratio to resistivity via Archie's law, we get

$$\phi = \frac{\varepsilon}{1 + \varepsilon} = a^{\frac{1}{m}}(R_w/R)^{\frac{1}{m}} \qquad (7.36a)$$

and

$$\phi_n = \frac{\varepsilon_n}{1 + \varepsilon_n} = a^{\frac{1}{m}}(R_w/R_n)^{\frac{1}{m}} \qquad (7.36b)$$

From the above it is easy to show after some algebraic manipulations that

$$\sigma = \sigma_n^X \sigma_0^{1-X} \qquad (7.37)$$

where the coefficient X is related to the formation factor F by

$$X = \frac{(F_n/a)^{\frac{1}{m}} - 1}{(F/a)^{\frac{1}{m}} - 1} \qquad (7.38)$$

where F_n is the formation factor under normally pressured condition. Equation (7.37) is an extension of Dutta's velocity and effective stress technique to resistivity. In terms of prepressure, the above equation can be rewritten as

$$P = S - (S - P_n)^X \sigma_0^{1-X} \qquad (7.39)$$

Here σ_0 is that value of effective stress when porosity of the rock approaches zero. We reiterate that the formation factor F, denotes the formation resistivity of shale, and it requires a knowledge of the resistivity of pore water in the shale. This is difficult to measure directly. It is commonly *assumed* that it is equal or related to the nearby water-filled sand as derived from SP logs (Foster and Whalen, 1966). However, this is only an assumption and thus may not be valid for all geologic conditions. It is definitely not valid at the structural high of reservoirs embedded in overpressured shales or wells in the proximity of salt.

In summary, the resistivity approach to quantification of pore pressure is given in equation (7.31) and Figure 7.5. The behavior of resistivity logs versus depth, is a result of a two-step process. Step 1 involves recognizing the cause of overpressure and postulating an underlying "compaction model" in terms of fundamental variables such as porosity and vertical effective stress. These models must be depth-independent. Step 2 involves using a model to relate porosity to a geophysical log-based parameter. This may be termed as an "attribute model." For resistivity logs, we used the well-known Archie's relation that relates porosity to resistivity, equations (7.23) and (7.24). The final step is to eliminate porosity from the two governing models – the compaction model and the attribute model. This yields the desired results in equation (7.31) for pore pressure. These models are depth-independent and are transferable from basin-to-basin. However, in doing so care must be exercised to focus on obtaining estimates of parameters such as resistivity of water, and Archie's cementation exponents.

7.3.3   Density Measurement Tools

This is a commonly used geophysical well-logging tool to measure bulk density of rock formations in a continuous manner in a borehole. Since bulk density is a function of rock and fluid density and thus contains porosity as a parameter, it is quite useful for geopressure detection and measurements. The density tool first came into being in the 1950s and by 1960s it was adopted by the hydrocarbon industry. The tool uses a radioactive source and multiple detectors at fixed distances apart. (Typically Cs-137 is used as a radioactive source that is a medium-energy neutron emitter.) The scattered neutrons are captured by the detectors and are related to the formation's electron density and hence, its bulk density. Multiple detectors compensate for the borehole effects such as borehole damage and the effect of fluid density in much the same way as for sonic logs. Borehole density measuring tools provide reliable porosity data for sandstones and limestones because the grain densities of constituent minerals are well known. On the other hand, the density of clay minerals in mudstones is highly variable, depending on environment, type of clay mineral (smectite, illite, chlorite, etc.) and many other factors. So for shales, this tool is not always reliable for density estimates. Calibration of the tool in this type of environment is required.

As noted earlier, other wireline logs such as neutron and pulsed neutron measurements, nuclear magnetic resonance techniques, and downhole gravity data measurements can also detect geopressures. We did not discuss these as these are not commonly used for geopressure quantification. We refer the readers to Bigelow (1994) and Fertl (1976).

7.4 Recommendations on Use of Wireline Logs for Pore Pressure Analysis

Wireline log-based analysis and quantification of formation pressure has long been the "gold standard" for reliability. This is always used to calibrate predictions made with seismic velocity. This is a subject that merits some discussion. We should realize that just because a tool is in the borehole (acoustic and resistivity tools, for example), it does not measure pore pressure directly. Information regarding pore pressure is obtained by *indirect* means. In addition, all borehole logs require calibration – the data are affected by borehole dimensions, borehole conditions, temperature, and the type of fluid in the borehole, among other things. Although engineering developments of the downhole tools have been spectacular in the last several decades, many tools either malfunction or yield not so-reliable data in high-temperature and high-pressure (HPHT) environments. *User be aware* is the guidance here. A good practice is to have a team of multidisciplinary geoscientists access the data and provide guidance. This team should quality control the data while it is being acquired as well as during the postwell analysis. This implies the following: (1) log data acquisition is of acceptable quality; (2) environmental corrections such as borehole damage or washouts, bit size changes, or lithology variations are not present, and if present, have been corrected; (3) reasonably thick shale zones are available for pressure computations; (4) logs are properly calibrated for reliable pore pressure estimates. This improves the odds that well logs are run properly and measured shale parameters are analyzed properly. If the overall log quality is questionable, then pore pressure calculations from those log(s) should be avoided.

While judging the data quality of logs or evaluating the reliability of the predicted pore pressure values, we need to be aware of the local geology. Offset wells, just because they are close, may not provide a confidence on the log data or the derived values from the logs. If two adjacent wells are nearby but drilled into two different fault blocks, and the fault blocks are known to act either as a pressure barrier or pressure transmitter, the pressure profiles from the adjacent wells might be identical, similar, or totally different. We refer to Bigelow (1994) for a lucid discussion on the subject and many examples on the pitfalls of using well logs for pore pressure evaluation.

The acoustic or sonic log is the best log for quantitative pore pressure evaluation since the data from such tools are relatively unaffected by changes in hole size, temperature, and formation water salinity. A similar statement also applies to VSP tools. Since sonic log is a good porosity indicating tool, some authors use the Wyllie et al. (1956, 1958, 1963) relation to compute porosity from sonic logs (Zhang et al., 2011; Singha and Chatterjee, 2014) and use that porosity to estimate pore pressure. We advise against this practice. This relationship (also known as the Wyllie time-average equation) is only valid approximately for water saturated consolidated sandstones at very high effective pressure. In this model the following velocity–porosity transform is used:

$$\frac{1}{V_p} = \frac{\phi}{V_f} + \frac{1-\phi}{V_m} \tag{7.40}$$

where V_p, V_m, and V_f are the P-wave velocities of the saturated rock, of the mineral material making up the rocks (quartz, for example), and of the pore fluid, respectively. Mavko et al. (2009) noted correctly that the time-average equation is heuristic and cannot be justified theoretically. They noted that the argument that the total traveltime can be written as the sum of the traveltime in each of the phases is a seismic ray theory assumption and can be valid only if the (1) wavelength is small compared the typical pore and grain sizes and (2) the pores and grains are arranged as homogeneous layers perpendicular to the ray path. Because neither of these assumptions is valid in general, the agreement with observations is only fortuitous.

The resistivity or conductivity tools are very sensitive to temperature and pore water salinity. These must be accounted for and corrections *must* be made prior to using the data from these tools for pore pressure calculations. The formation water resistivity may be corrected from its value at laboratory temperature to formation temperature either by use of a chart found in most logging manuals or by Arp's empirical formula (Assad et al., 2004). In Fahrenheit, Arp's formula is

$$R_{w2} = R_{w1}\left(\frac{T_1 + 6.77}{T_2 + 6.77}\right) \tag{7.41}$$

and in centigrade, Arp's formula is

$$R_{w2} = R_{w1}\left(\frac{T_1 + 21.5}{T_2 + 21.5}\right) \tag{7.42}$$

where R_{w1} and R_{w2} are formation water resistivity at temperatures T_1 and T_2, respectively. In our experience, many practitioners of pore pressure calculations do not apply this correction and we consider it a bad practice.

Pore fluid properties can also have a significant effect on pore pressure predictions. This is because resistivity and velocity are both affected by the type and properties of the pore fluid. Changes in the salinity of brines will change resistivity, because pore fluid conductivity increases with salinity; thus a salinity increase (e.g., adjacent to or beneath a salt dome) could be misinterpreted as an increase in pore pressure. Fluid conductivity is also a function of temperature.

Substitution of hydrocarbons for brine will increase resistivity, because hydrocarbons do not conduct electricity; this can mask increases in pore pressure that often accompany the presence of hydrocarbons. Because hydrocarbons are more compliant and less dense than brines, compression-wave velocity will decrease, and shear wave velocity will increase as hydrocarbon saturation increases. High gas saturation or API index for oil will amplify this affect. Because a change from water to hydrocarbon affects resistivity and compressional velocity in opposite ways, simultaneous pore pressure analyses using both measurements can sometimes identify such zones. It is more difficult to identify and deal with changes in fluid salinity.

7.5   Drilling Parameters for Pore Pressure Analysis

7.5.1   Introduction

In Table 7.1, modified after Dutta (1987a), we show a summary of methods to predict and detect pore pressure. This table is a modified version of a table that appeared first in Fertl (1976). It is clear from the table that the industry uses a host of drilling methods to evaluate pore pressure – the methods use variables that are different from predrill techniques such as velocity and amplitude of seismic waves.

Below we discuss the drilling parameters for pore pressure estimation. These parameters could be classified in two groups: *drilling rate (penetration) parameters* and *drilling mud parameters.*

7.5.2   Drilling Rate Penetration Parameter as a Pore Pressure Indicator

Drilling rate is dependent on such things as the weight on bit, rotary speed, bit type, and its condition during drilling and its size, hydraulics, and formation properties such as lithology and its pore pressure state. Under ideal conditions such as constant weight on bit, rotary speed, bit type, and hydraulics, the drilling rate in shale *decreases* with depth due to consolidation of rocks as depth increases. However, in pressure transition zones (cap rocks, for example) and in overpressured zones the porosity is higher compared to the case if it was normally compacted at that depth and consequently, penetration rate *increases*. However, any other major lithologic changes would also cause changes in rate of penetration. Fluctuating rotary torque and the erratic action of the drill bit on the bottom of the borehole also affects penetration rate. So the rate of penetration as a pore pressure indicator is not unambiguous. Most mud-logging companies record penetration

Table 7.1 Techniques for predicting and detecting pore pressure

Time	Source of data	Pressure indicator
Before drilling	Surface geophysical detection	Seismic velocity
		Gravity
While drilling	Drilling parameters (A)	Penetration rate
		Measurement while drilling
	Drilling parameters (B)	Mud-gas cutting
		Pressure kicks
		Flowline temperature
		Pit-level and total pit volume
		Hole fill-up
		Mud flow rate
	Shale cutting parameters	Bulk density
		Shale formation factor volume, shape, size, sand/shale ratio
	Correlation between new and existing wells	Drilling data
After drilling	Subsurface geophysical detection	VSP
	Well logging	Sonic, resistivity, density, neutron, downhole gravity
When well is tested or completed	Monitoring pore pressure variations in short zones	Repeat formation tester
		Drill stem test
		Pressure bombs

logs and these logs must be interpreted and corrected prior to use for pore pressure analysis.

Of all the drilling parameters used to indicate pore pressure, the d-exponent method as developed by Jordan and Shirley (1966) is the most popular method. It is dependent of the rate-of-penetration parameter, R (expressed by drillers in foot per hour) and is widely used as an indicator of high pore pressure. This parameter indicates how fast each foot of a well is drilled. It is the first parameter received as the well is drilled. We noted earlier that it is affected by pore pressure as well as lithology changes. Thus deciphering the effect of pore pressure from this parameter requires a good knowledge of the lithology being drilled such as the difference between carbonates and clastics. Jordan and Shirley (1966) discussed a method, still widely used, to normalize rate-of-penetration data and change in rotary speed, bit weight, and bit diameter in order to detect overpressure. Jordan and Shirley (1966) and Bourgoyne et al. (1986) suggested the following equations for the d-exponent

$$R = K(RPM)^{E}\left(\frac{W}{D_B}\right)^{d} \tag{7.43}$$

where

R = rate of penetration in feet per hour
K = drillability constant
RPM = rotary speed in revolutions per minute
E = rotary speed exponent
W = weight on bit in 1000 pounds
D_B = bit diameter in inches
d = bit weight exponent or d-exponent

Jordan and Shirley (1966) assumed $K = 1$ and $E = 1$ and rearranged the above to express the d-exponent in the following form:

$$d = \frac{\log(R/60RPM)}{\log(12W/10^6 D_B)} \tag{7.44}$$

The ratio (R/60RPM) in equation (7.44) is always less than 1.0. The absolute value of $\log(R/60RPM)$ varies inversely with R. Therefore, the d-exponent varies inversely with the rate of penetration. Basically, plots of d-exponent versus depth show a decreasing trend with depth. In transition zones and overpressure environments, the calculated d-exponent values diverge from the normal trend to lower than the normal values. When lithology is constant, the d-exponent gives a good indication of the state of compaction (i.e., porosity) and pressure. Quantitative pressure evaluation can be made with the (1) Eaton method (Eaton, 1972), (2) Zamora method (Zamora, 1972), or (3) the equivalent depth method. The pertinent equations are

$$P = S - (S - P_n)\left(\frac{d}{d_n}\right)^{\alpha} \qquad \text{(Eaton method)} \tag{7.45}$$

$$P = P_n\left(\frac{d_n}{d}\right) \qquad \text{(Zamora method)} \tag{7.46}$$

where P is the pore pressure, P_n is the normal pressure, S is the overburden stress and the exponent, α, is the Eaton exponent (typically varying between 1.2 to 1.5), d is the d-exponent, and d_n is the normal trend of the d-exponent. Any decrease in the d-exponent when drilling an argillaceous sequence is a function of the degree of undercompaction and of the value of the associated overpressure. The equivalent depth method was discussed earlier in Chapter 4.

Since the d-exponent is influenced by mud weight variations, a modification was introduced by Harper (1969) and Rehm and Mcledon (1971) to include the effect of mud weight on the d-exponent in order to normalize the d-exponent for the effective mud weight:

$$d_c = d\left(\frac{M_n}{M}\right) \tag{7.47}$$

where d_c is the modified or corrected d-exponent, M_n is the mud weight under normally pressured condition, and M is the actual mud weight used. An example is given in

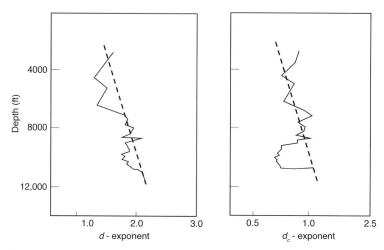

Figure 7.8 Comparison of d-exponent and d_c exponent versus depth in the same well. Note that d_c exponent defines overpressure zone more clearly. Modified after Fertl et al. (2002).

Figure 7.8 where we show a comparison of the d-exponent and the d_c-exponent in the same well. The protective casing in the well was set at 8700 ft (2652 m). We note that the d_c-exponent defines overpressure more clearly than the d-exponent. The procedure to calculate pore pressure from the d_c-exponent is given below:

1. Calculate d_c over 10- to 50-ft intervals.
2. Plot d_c versus depth (use only data from clean shale sections).
3. Determine the normal line for the d_c versus depth plot.
4. Establish where d_c deviates from the normal line to determine abnormal pressure zone.
5. Calculate pore pressure using one of the three methods suggested above (Eaton, Zamora, or equivalent depth).

An example of the pore pressure calculations using the d_c-exponent method is shown in Figure 7.9. Note that pore pressure gradient is expressed in units of specific gravity S_g. ($S_g = 1$ denotes the normal pore pressure gradient). Figure 7.9a shows the pressure gradient from Eaton's method with an exponent of 1.2 and Figure 7.9b shows the computed pressure gradient using Zamora's method. The rate-of-penetration logs in both cases are identical. We note that the normal trend line for the d_c-exponent was *predetermined* by analogue means. The results are compared with established pressure gradients in the area by Statoil as quoted by Stune (2012). The choice of a normal trend line for the d_c-exponent is not easy. This is because the rate of penetration is affected by many factors such as change in bit, bit conditions, mud conditions, and age and degree of consolidation of formations, to name a few. We refer to Fertl et al. (2002, Chapter 6) for a lucid discussions of these effects.

The d_c-exponent method is still widely used with considerable success in most basins in the world. However, this practice remains mostly in the drilling community and the

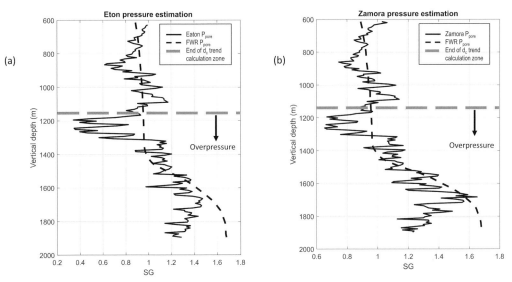

Figure 7.9 An example of pore pressure gradient calculations (in specific gravity units) using d_c-exponent. (a) Eaton's method. (b) Zamora's method. The normal trend line for d_c-exponent was set up to 1150 m. The locally calibrated pressure gradient is plotted for comparison purpose (denoted by FWR Ppore). It is supplied by former Statoil as reported by Stunes (2012). The top of overpressure is indicated in both figures. Note that pore pressure gradients are expressed in SG units in both figures. Modified after Stunes (2012).

geophysics community has shied away from this technology. The d_c-exponent methodology provides several operating advantages (Moutchet and Mitchell, 1989): it is a low cost methodology and thus has a minor financial impact on exploration; the method can be performed in real time during drilling; implementation and application is simple, and it does not require highly skilled personnel on the rig to acquire data. However, quality control of the data and the results require some expertise who are familiar with common drilling practices in the area under investigation and its imprint on the data. In particular, the use of PDC-type bits with a shearing cutting action (instead of the chipping action that Jorden and Shirley [1966] assumed in their model) will lead to d_c-exponent plots that differ from tri-cone or bi-cone bits in the same formations. While the method can be used successfully with care, one must always validate the information presented by d_c-exponent plots by examining multiple or other pore pressure indicators.

7.5.3   Drilling Mud Parameters as Pore Pressure Indicator

In Table 7.1, the drilling parameters listed under (A) are pressure indicators as a well is being drilled while those listed under group (B) are *delayed* indicators as these are obtained after the mud carries the information after some delay – the delay being caused due to the time required for mud to return to the surface.

Torque and drag parameters fall under group (A) and these are *indirectly* related to overpressure and directly to drill-pipes and wellbore. Drag is the excess hook load over free handling load. Torque measured at the surface is the sum of torque of the drill bit and torque resulting from the friction of the drilling column against the well walls. Because of increased wall-to-wall contact of drill pipe with wellbore, torque typically increases with depth. However, high-pressure zones can cause a reduction in torque and thus it could be indicative of overpressure. Torque is, nonetheless, not easy to interpret. Experienced drillers monitor this routinely while drilling a well. Drilling engineers note how these (torques) vary with depth and make a call whether the penetrated formation is overpressured by integrating other indicators from group (B) parameters. Torque and drag are not reliable parameters for pressure indicator in deviated holes or in holes with severe dog-legs. Occasionally, the drill bit may get stuck in overpressured zones due to high plasticity of clays contained in the over-pressured formation. When that happens, the torque is also reduced.

Delayed Indicators

Many parameters obtained by the mud-logging unit can also be used to detect formation pressure changes. For example, changes in drilling fluid such as total gas content, temperature, density, or salinity or changes in characteristics of the formation samples gathered in the shale shaker, such as density, shape, and amount, may relate to zones of overpressure (Moutchet and Mitchell, 1989). These are, however, delayed indicators. Again, the delay is due to the time required for mud to return to the surface.

All of these parameters provide *qualitative* pressure information and require local calibration for improving reliability. Considerable literature exists concerning the various engineering and related aspects of the use of drilling data for the detection and evaluation of formation pressure; however, most of the literature falls beyond the scope of this book. While reviewing the published literature on drilling performance data and many pressure manuals from oil companies, it became clear that the authors of these literatures are mindful of the effect of mud weight on drilling parameters. For example, most of the pressure indicators are affected by the type of drilling mud (oil, water, or synthetic), and the weight of the mud, i.e., whether the well is drilled underbalanced (not a good practice) or overbalanced (typical case).

In Figure 7.10 we show a schematic of various drilling indicators for pore pressure. Mud-gas logging is a popular technique for overpressure detection in real time while a well is being drilled. As early as 1945, the use of gas measurement was used as an indicator for drilling through overpressured zones (Pixler, 1945). A classic article by Rochon (1968) discusses the use of mud-gas anomalies as an aid in controlling drilling mud weights and hence, overpressure. These authors cited that in many cases, mud-gas shows are useful to indicate flow into the wellbore if formation permeabilities are favorable to allow it. However, this is merely a qualitative or semiqualitative technique for real-time pressure detection as there are many false indicators that affect the reliability of this indicator such as drilling through high gas readings while drilling through *pay zones,* existence of connection gas while raising the Kelly during drilling process and during downtime when mud pumps are shut down for bit trips and repair.

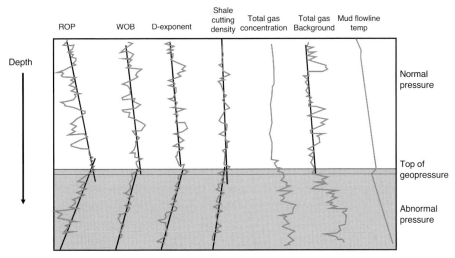

Figure 7.10 A schematic of various drilling indicators of geopressure. Source: http://petrowiki
.org/images/4/42/Vol5_Page_0374_Image_0001.png.

Mud-gas is often used an indicator of high pressure. Gas in the drilling mud often
increases because methane is often dissolved in the pore water of many overpressured
shales. As the cuttings and cavings come up the hole, the gas escapes and can be
detected in the mud. However, gas in the mud may also be caused by oil- or gas-bearing
formations and by organic-rich shales. Thus it may not be a reliable indicator in all
cases.

Flowline temperature is another qualitative source of pressure indication while
drilling a well. Thermal gradient change associated with excess trapped water in an
overpressured zone causes a dogleg anomaly in measured temperature gradient as
shown in Figure 7.10. This essentially suggests that with high pore pressures, porosity
is also higher accounting for higher fluid content and thus resulting in a thermal
gradient anomaly.

Boatman (1967) showed the use of shale bulk density measurements on cuttings as
an overpressure indicator. The key to using density as an indicator of overpressure is
the mineralogical composition of the rock (lithology). Since overpressured shales
retain their porosity at greater depth than do normally pressured shales at the same
depth, their bulk density is smaller than the density of normally compacted shales.
Thus, a departure from the normal compaction trend will indicate abnormally high-
pressure zones. Very rarely one uses density measurements on cuttings as high-
pressure indicators by itself due to inherent uncertainties – the chief ones being
depth uncertainty. Therefore, such measurements are usually combined with other
indicators such as reconstructed lithology logs from cuttings, pore pressure from
resistivity and conductivity logs and rate-of-penetration log. An example is shown
Figure 7.11. A density trend was established as shown in Figure 7.11a and its deviation
from the normal compaction trend was established. The rate of penetration is also noted
in the second panel. The lithology descriptions are given as annotations. Measured

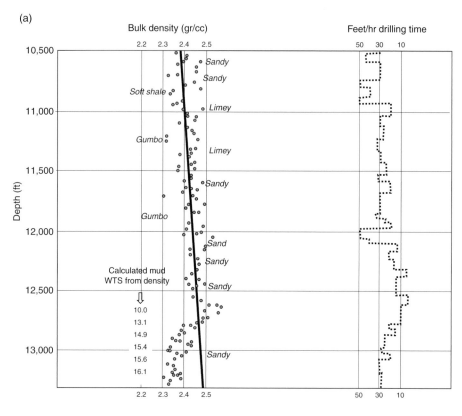

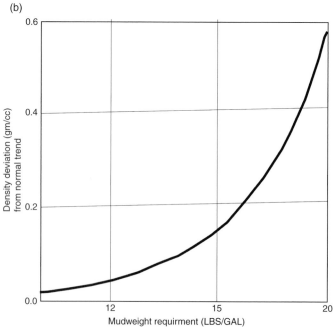

Figure 7.11 (a) Shale density measurements on cuttings at the rig site as pore pressure indicator in real time. Shown are density measurements and drilling rate versus depth along with predicted mud weights, actual mud weights used, lithology description and trip-gas amount. Modified after Boatman (1967). (b) Calibration chart that converts density deviation from the normal compaction trend to mud weights. Modified after Boatman (1967).

deviation in bulk density from the normal construction was transformed into pore pressure in ppg using the calibration chart shown in Figure 7.11b. The calculated mud weights are also shown in the leftmost panel in Figure 7.11a and compared with the actual mud weights used. The occurrence of high pressure is verified with the trip-gas measurements from within the reservoir unit as noted in Figure 7.11a. We note that the actual mud weights used are higher than those predicted using density measurements. This is the normal case. However, excessively high mud weights can cause severe drilling problems as noted earlier. In principle, Boatman's technique is similar to bulk density measurement from standard wireline logs. However, the cuttings method has the distinct advantage of a much shorter lag time, since the only delay is the sample lag time. Furthermore, lithologic descriptions of the shale cuttings help evaluate natural density fluctuations when seeking a normal compaction trend. Limitations of this method are: (1) all cuttings are not representative of the formation being drilled; (2) measurements are affected by the washing and drying procedure used; and (3) most shales are altered by a reaction involving salinity differences of the water in the shale and of the drilling mud. Experienced mud-logging engineers also infer the occurrence of overpressure by analyzing the shape of drill cuttings. The shapes of cuttings and cave-ins from under-compacted shales may be different than those from normally compacted shales.

Mud-logging engineers always monitor the level of the mud in the mud-tank. A rising level of mud in the tanks indicates that more mud is coming out of the hole than is going in. This is called a "*kick.*" This happens because formation fluids are entering the hole. It is due to the fact that the formation fluid pressure is higher than the mud weight in the borehole. This situation can lead to a well blowout. This is extremely serious, and proper steps must be taken to get the gas, oil, or water out of the hole. The most common method is to close the blowout preventers and stop the pumps. After a few minutes, the pressure at the top of the drill pipe will equal the pressure in the formation minus the weight of the column of mud. This is the excess pressure that must be balanced by *increasing* the mud weight. The pumps are then started to circulate the extraneous fluid out of the hole. The drill pipe pressure is carefully controlled with the choke. If the equilibrium drill pipe pressure is exceeded, the well may lose circulation, and if it is too low, the well may blow out.

To summarize, real-time indication of overpressure during drilling can be made using real-time drilling parameters. Some of these are discussed very briefly above. We note that these are either qualitative or semiqualitative indicators of geopressured zones. However, when combined with a viable predrill model, say, based on any of the geophysical methods as discussed earlier (velocity based techniques, for example, using either seismic data or well logs from analogue wells) these methods become very useful and provide reliable information of overpressure in real time. *We suggest that multiple indicators of pore pressure is always preferred over that from a single method.*

8   Gravity and EM Field Methods Aiding Pore Pressure Prediction

8.1   Introduction

We begin with an observation in this chapter. Not much published literature exist on applications of gravity and electromagnetic (EM) methods for pore pressure analysis. Nonetheless, it is our goal to highlight these potential field methods so that exploration geoscientists can include these in their radar for quantitative evaluation of geopressure in the future. That nonseismic methods such as gravity and EM methods are useful for exploration is beyond question. These methods are responsive to lateral variations in rock properties for structural definition of subsurface features such as salt domes, detecting steep discontinuities associated with faults, uplifts, and igneous intrusions and lateral and vertical variations of basement rocks. However, applications of these nonseismic measurements have been limited to structural analysis only (Blakely, 1995; Reynolds, 1997). We believe that these methods are also applicable to geopressure investigation. Given an understanding of how the tools behind the potential field technologies work in their current form, we believe that these methods will not replace seismic methods for geopressure analysis for years to come mainly due to low resolution. Rather, they will add to it. Despite being comparatively low resolution, they have some advantages. For example, at a comparatively low cost, airborne potential field surveys can provide coverage of large areas. For EM techniques, two methods are routinely used for oil exploration: marine seismic reflection and electromagnetic seabed logging (SBL). Marine magnetotellurics (MT) or marine Controlled Source Electro-Magnetics (CSEM) can provide pseudo-direct detection of hydrocarbons and other targets by detecting resistivity changes over geological traps (signaled by seismic survey). This helps in minimizing the number of wildcats (Sainson, 2017).

The links between electrical resistivity and pore pressure for EM measurements and a link between density and pore pressure for gravity measurements have been established in previous chapters from a borehole centric point of view. Surface electromagnetic and gravity surveying methods can provide information about the same properties in space at a larger scale. However, as will be discussed below, the estimation of bulk density from gravity surveying and electrical conductivity from EM surveying is an ill posed inverse problem where solutions are nonunique. Therefore, the quantitative use of density and electrical conductivity estimates using single domain surface measurements (i.e., only gravity and/or only EM from surface surveying) may not provide sufficient resolution needed for well planning. However, this by no means suggests that these methods are not useful in general. A practical approach therefore is to use joint inversion of different measurements to improve accuracy of the material properties of interest. To that end linking seismic velocity to density and electrical conductivity can improve both seismic and EM and gravity interpretations as well.

In this chapter, we briefly review some of the basics of gravity and EM methods and show how they can be used individually and together with seismic data to improve density, seismic velocity, and resistivity estimation. As noted earlier, we did not find any published example of the prediction and detections of geopressure using nonseismic techniques – just structural analysis of objects in complex geologic areas using joint inversion methodology.

8.2   Gravity Method

8.2.1   Theory

Newton's law of gravitation relates the force to mass of two bodies by

$$\mathbf{F} = -\frac{Gm_1 m_2}{r^2}\hat{\mathbf{r}} \tag{8.1}$$

where $G = 6.67 \times 10^{-11}$ [m^3/(Kg s^2)]. Combining this expression with Newton's second law of motion, $\mathbf{F} = m\mathbf{a}$, yields $\mathbf{a} = -\frac{GM}{R^2}\hat{\mathbf{r}} = \mathbf{g}_N$. Here $\mathbf{g}_N \sim 9.81$ m/s^2 is the acceleration due to gravity associated with spherical nonrotating homogeneous earth with mass $M = 5.977 \times 10^{24}$ kg and $R = 6371$ km. In the literature, the unit of gravity is gal $= 10e-2$ m/s^2. Often milli-gal (mgal) or micro-gal units are used where $\mathbf{g}_N \sim 98,100$ mgal.

The gravitational potential field U is defined such that its gradient gives the component of the gravity with respect to the gradient reference system, i.e.,

$$\nabla U = \mathbf{g}, \quad U = \int_{\infty}^{R} \mathbf{g} \cdot \mathbf{r} dr = \frac{Gm}{R} \tag{8.2}$$

Gravity Anomaly due to a Point Mass

The perturbation in gravity observed at point P(x, y, z) due to a point mass associated with an infinitesimal rectangular box at point P(x', y', z'), where $\delta\mu = \rho\delta\xi\,\delta\psi\,\delta\zeta$ is given by

$$\delta g_r = \frac{G\delta m}{r^2}, \quad \delta g_z = \frac{G\delta m}{r^2}\cos\theta = \frac{G\delta m(z'-z)}{r^3}$$
$$r^2 = (x-x')^2 + (y-y')^2 + (z-z')^2 \tag{8.3}$$

When the angle θ is small (Figure 8.1), we get $\delta g \approx \delta g_z$.

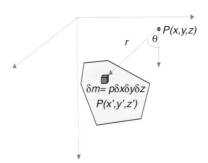

Figure 8.1 Concept of gravity and a perturbation observed at a point P due to a point mass in the box.

For generally irregular shaped geobodies the gravitational anomaly can be calculated using the superposition principle to become at the limit the integral

$$\Delta g = \iiint \frac{G\rho(z' - z)}{r^3} dx' dy' dz' \qquad (8.4)$$

Thus, the forward model of the gravity problem is well defined.

8.2.2 Measurement of Gravity and Gravity Surveying

There are two types of gravity measurements: direct methods which determine g directly and are good to an accuracy of about 1 mgal – such methods include large pendulum and falling body techniques and measurements by gravimeters. Direct methods are hardly used these days due to long measurements and low precision. Modern gravity surveying uses gravimeters which provide relative measurements of gravity. Their precision is from 0.01 to 0.001 mgal typically, and as the measurement time is very fast, they are suitable for large data collection.

The factors that influence gravity are latitude, elevation, topography of the surrounding terrains, and the density variation in the subsurface. It is important to note that while we are interested in the subsurface, the subsurface density variations are smaller than latitude or elevation effects. Therefore, gravity surveying require good location and high-precision accuracy in elevation. Observed data are not used directly. Rather there are many corrections needed to eliminate factors affecting the reading including drift and tidal correction, free air correction, and many other environmental corrections. The reader is referred to basic textbooks on gravity surveying, such as LaFehr and Nabighian (2012) and Hinze et al. (2013).

8.2.3 Inversion of Gravity Data

The inversion of gravity data is the process where a computer code is used to search for a suitable distribution of density in the subsurface (the model **m**) such that the misfit between observed and modeled data is minimized, i.e.,

$$\rho(x,y,z) = \arg\min\left\{\sum_{i=1}^{n}\left\|\Delta g_{obs\_i} - \Delta g_{calc\_i}(\rho)\right\|\right\} \qquad (8.5)$$

where n is the number of observations.

The gravity inversion problem is nonunique as there are many shapes and distributions of density in the subsurface that will have similar gravitational response. Therefore, the minimization of the data misfit is not sufficient to provide a solution and additional a priori information is used to regularize and stabilize the inversion process. Often the use of regularized inversion reduces the spatial resolution associated with gravity distribution in the earth. The use of seismically derived structures and multiple domain (multiphysics) where gravity is inverted together with seismic and/or EM data often provide superior solutions to single domain gravity inversion and helps to increase the resolution. We will discuss the integration of gravity with other geophysical methods later in this chapter. It is important to note that the gravity method is the only method that provides density distribution in the subsurface directly.

8.3   Electromagnetic Method

Electromagnetic (EM) methods try to resolve the subsurface electrical properties, mainly the subsurface conductivity using measurements from EM surveys. We will discuss here briefly two main methods used within the context of exploration and development. One is Controlled Source EM method where an *active* EM source is used to generate strong induction currents that are measured and inverted for material properties. The other is the magnetotellurics (MT) method that is a *passive* method where magnetic and electric natural sources associated with solar activity induce transient electrical fields that produce currents and are measured using passive sensors.

8.3.1   Theory

A detailed discussion on propagation of EM fields in conductive earth is beyond the scope of this chapter and we refer the reader to textbooks on EM such as Jackson (1999), Kaufman et al. (2014), and Zhdanov (2017). A very brief discussion here is aimed at providing a brief understanding of the governing physics associated with the method.

All EM fields in matter are governed by Maxwell's equations given by

$$\begin{aligned}
\nabla \cdot \mathbf{B} &= 0, \\
\nabla \cdot \mathbf{E} &= \rho/\varepsilon \\
\nabla \times \mathbf{E} &= -\frac{\partial \mathbf{B}}{\partial t}, \\
\nabla \times \mathbf{B} &= \mu\sigma\mathbf{E} + \mu\varepsilon\frac{\partial \mathbf{E}}{\partial t}
\end{aligned} \qquad (8.6)$$

where $\mathbf{E}$ and $\mathbf{B}$ are the electric and magnetic fields, respectively, μ is the magnetic permeability, ε is the electric permittivity, σ is the electrical conductivity, and ρ is the electrical charge density (Jackson, 1999).

Applying the curl operator to Maxwell's equations one gets the following wave equation in a lossy-medium:

$$\nabla^2 \mathbf{E} = \mu \left(\sigma \frac{\partial \mathbf{E}}{\partial t} + \varepsilon \frac{\partial^2 \mathbf{E}}{\partial t^2} \right)$$
$$\nabla^2 \mathbf{B} = \mu \left(\sigma \frac{\partial \mathbf{B}}{\partial t} + \varepsilon \frac{\partial^2 \mathbf{B}}{\partial t^2} \right) \qquad (8.7)$$

Assuming $\mathbf{E}$ and $\mathbf{B}$ can span in the frequency domain using nonvanishing oscillatory functions proportional to exp (iωt), one can derive the well-known Helmholtz equation in the frequency domain for both $\mathbf{E}$ and $\mathbf{B}$ fields:

$$(\nabla^2 - \gamma^2) \mathbf{E} = 0$$
$$(\nabla^2 - \gamma^2) \mathbf{B} = 0 \qquad (8.8)$$
$$\gamma^2 = i\omega\mu\sigma - \omega^2\mu\varepsilon$$

When $\sigma \gg \omega\varepsilon$, the wave is diffusive. The EM responses are governed by the skin depth associated with the decay of a diffusive EM wave in a conductive media, and the skin-depth is given by

$$\delta = \sqrt{\frac{2}{2\pi f \mu \sigma}} \qquad (8.9)$$

The EM method is aimed at resolving the subsurface conductivity associated with the propagation of diffusive waves. Frequency or time domain solutions to the EM wave equations are used to generate numerically synthetic phase and amplitude response of the EM field which are then compared to the measured observations. For deep sounding, the popular surface or seafloor methods for EM surveying are MT and CSEM methods. In both cases transient EM fields are measured and stored on the surface recorders.

8.3.2   Measurement and Inversion

Both EM and MT surveys are measured using antennas located at the seabed or on the surface of the earth. Data are recorded with periods of $0.04–10^4$ s (25–0.0001 Hz) for MT and CSEM utilizing frequencies up to ~10 Hz. The inversion of the EM data, much like gravity inversion is performed by minimizing an objective function associated with data misfit terms. The distribution of electrical resistivity in the earth is estimated by trying to minimize a data misfit term subject to a priori constrains, which are typically smoothness or minimum deviation from a prior model.

Much like the gravity problem, the solution to the EM problem from surface EM measurements generates typically a "blob," that is, a low-resolution resistivity image of the subsurface. It is not possible to assign directly specific boundaries or geometry associated with the EM image. The sensitivity to large resistivity contrasts makes EM inversion suitable for resistive body identification (either hydrocarbon reservoirs or salt bodies). However, as the solution to the EM method is nonunique the resistivity image obtained through regularized inversion may not be a true representation of the earth resistivity due to a heavy smoothing/damping used in the solution.

8.4   Joint Inversion

The term joint inversion refers to the use of different methods *together* through formal minimization of a misfit function. In general, data from two or more physical domains (e.g., traveltime and gravity, or traveltime, gravity and EM) are used to estimate the material properties associated with the earth. In many cases the assumption is a common structure (i.e., geometrical constrains on the distribution of the material properties in the earth) or petrophysical correlations, which are either empirically derived from well log data or from more fundamental calibrated rock physics relations. Joint inversion attempts to improve the inference of the earth material properties by adding observations of different domains which in principle will add resolution and improve predictions when compared to results from a single domain.

In general, there are two types of inversion: A cooperative inversion and a simultaneous joint inversion (Moorkamp et al., 2016). A cooperative inversion is a method where data is inverted as a single domain and the solver iterates between the two domains until a certain tolerance level is reached for both domains. The misfit function minimized is a single domain misfit function and a coupling term between the two is used. A simultaneous joint inversion is a method where the algorithm minimizes simultaneously the misfits associated with observation of fields associated with both domains. In this term an explicit coupling term is needed to enable the projection of the model update jointly on the material property grids used for modeling the data misfit.

8.4.1   Illustrative Examples: Simultaneous Joint Inversion of Seismic, Gravity, and MT Data

De Stefano et al. (2011) have demonstrated the application of simultaneous joint inversion using MT and gravity data to improve subsalt velocity estimation. The formulation is based on rock physics based links which relates the domain parameters (velocity and density for gravity and velocity and resistivity for MT). A typical choice for velocity-density coupling is Gardner's empirical equations which relate P-wave velocity to density using two empirical parameter a_1 and b_1, i.e., $\rho = a_1 V_P^{b_1}$. Resistivity r is related to seismic velocity empirically via log-linear relation of the type $\log r = a_2 V_P + b_2$.

Thus, if we consider the model parameter space for the joint velocity-density-resistivity inversion from seismic data, gravity, and MT we can denote V_P to be model parameter $\mathbf{m_1}$, density defined as $\mathbf{m_2}$ and resistivity as $\mathbf{m_3}$. Then we can define the joint objective function to be a sum of data misfit terms for each domain, and regularizations terms which are made not only by the distance from a priori model for each domain but from deviation from the cross-terms defined by the empirical rock physics relations between the different model parameters, and from a structural link. Thus the link functions are

$$
\begin{aligned}
\Psi(\mathbf{m}_1, \mathbf{m}_2) &= |a_1\mathbf{m}_1{}^{b_1} - \mathbf{m}_2|^2 \\
\Psi(\mathbf{m}_1, \mathbf{m}_3) &= |a_2\mathbf{m}_1 - \log \mathbf{m}_3 + b_2|^2
\end{aligned}
\tag{8.10}
$$

The structural link which relates the domain geometrical properties is given using the cross-gradient term between the model parameters. The cross-gradient is given by

$$
\Psi(\mathbf{m}_1, \mathbf{m}_2) = \|\nabla\mathbf{m}_1 \times \nabla\mathbf{m}_2\|
\tag{8.11}
$$

This function requires a parallelism between model gradients. It is well known that gradients are perpendicular to the edges of the objects. Therefore, this forces the output models to have the same shapes in the same spatial positions. Therefore, the simultaneous joint inversion objective function can be written as the weighted sum which adds single-domain and cross-domain cost functions to be minimized jointly. Defining the single-domain i^{th} cost function as the sum of data misfit and deviation from the prior model we have,

$$
\Phi_i(m_i(r)) = \left\| W_{d,i}(g_i(m_i) - d_i) \right\|^{p_i} + \int_{V_i} \Gamma_i(m_i - m_{prior,i})dv
\tag{8.12}
$$

where $W_{d,i}$ is a matrix of data weights, associated with the noise level of the ith domain data. The term Γ_i is a model weight matrix that penalizes deviation from the prior model. The model weight matrix Γ_i can also ensure that the solution has desirable features such as smoothness, and is also known as the regularization matrix (Parker, 1994). In the above equation Φ_i is the misfit norm. Then we can write the simultaneous joint inversion (SJI) objective function as a sum of the single domain and the link functions:

$$
\Phi_{SJI}(m) = \left(\sum_i \alpha_i \Phi_i(\mathbf{m}_i(r)) \right) + \sum_j \beta_j \int_{U_j \subseteq V_{int}} \Psi_j(\mathbf{m})dv
\tag{8.13}
$$

Here V_{int} is the volume of interest, U is the subvolume of V where links have been evaluated, $\Psi_j(\mathbf{m})$ is a generic cost function that links all the domains involved in the inversion, α_i stands for the weight of the ith domain, and β_j denotes the weight of the jth link function Ψj. Notice that r is a continuous position vector and that the continuous $\mathbf{m}_i$ functions depend on it. The SJI objective function is then minimized using numerical optimization algorithms such as nonlinear conjugate gradient solver to produce an estimate that minimizes the misfit to all observations and constraints simultaneously. De Stefano et al. (2011) presented a case study from the Gulf of Mexico (GOM) that compares the results of conventional common image point

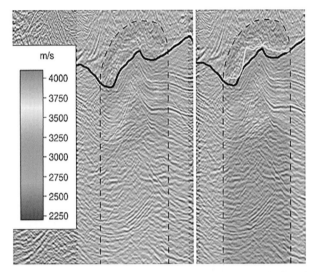

Figure 8.2 Final seismic velocity image comparing single domain (left) to simultaneous joint inversion results (right). (A black and white version of this figure will appear in some formats. For the color version, please refer to the plate section.)

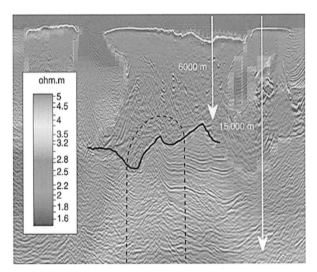

Figure 8.3 Final magnetotellurics (MT) resistivity section from simultaneous joint inversion. (A black and white version of this figure will appear in some formats. For the color version, please refer to the plate section.)

(CIP) tomography to SJI results where 1D MT solvers and 3D gravity solvers have been used (see Figure 8.2).

In Figure 8.2 a comparison between CIP tomography outputs using single domain (left) is compared to results of SJI over the inversion zone in dashed lines. Note that the joint inversion enabled better interpretation of salt bottom (yellow line) and showed the presence of autochthonous salt (cyan line). Also we observe a better definition of a low velocity zone at the salt boundary and it may be associated with high pore pressure effect which caused velocities to drop to much lower velocities than in the surrounding. The MT inversion results from SJI are presented in Figure 8.3. Note that here MT was used to invert the entire model and not just the dashed zone where CIP tomography was performed. Another example of joint inversion (CIP tomography based on seismic and CSEM data) for subsalt depth imaging application in Red Sea (Colombo et al., 2017) is presented in Figures 8.4 and 8.5. Figure 8.4 shows joint inversion results including (a) the starting velocity model from CMP traveltime-offset curves, (b) the resistivity model from 3D CSEM inversion used as a reference model through cross-regularization, (c) the final velocity model from joint inversion, and (d) velocity updates (c–a). Figure 8.5 shows depth imaging results from the resulting inversion. The ones from joint inversion have considerable improvement over the one from the legacy velocity model that used a traditional approach based on seismic data with no constraints. The authors needed to test the final velocity model for applicability to pore pressure. Until that is done we will never know the answer.

In summary, the approach based on joint inversion of multidisciplinary data for structural clarity of geology is demonstrated to be very useful. However, we need to take the next step and answer the question just how good is the resulting velocity field. One way to demonstrate this is to evaluate the quality of predicted pore pressure and other geomechanical properties such as overburden stress and fracture pressure using the velocity model.

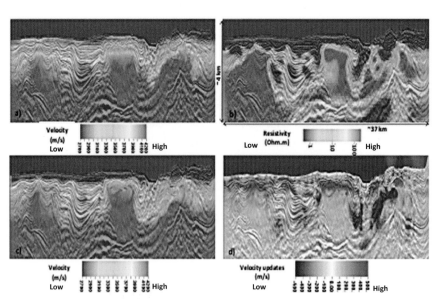

Figure 8.4 Joint inversion results including (a) the starting velocity model from CMP traveltime-offset curves starting from 2700 m/s to 4100 m/s in step of 200 m/s., (b) the resistivity model from 3D CSEM inversion used as a reference model through cross-regularization in units of ohm.m starting from 1 to 100 in step of 10., (c) the final velocity model from joint inversion starting from 2700 m/s to 4100 m/s in step of 200 m/s., and (d) velocity updates (c–a) in m/s starting from -500 to +500 in step of 100. After Colombo et al. (2017). (A black and white version of this figure will appear in some formats. For the color version, please refer to the plate section.)

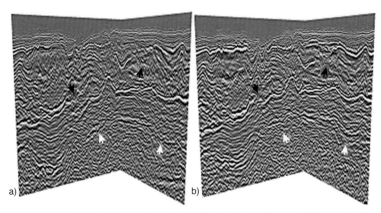

Figure 8.5 Depth imaging results using (a) the final velocity model from seismic data only and (b) the velocity model from joint seismic and CSEM inversion. Improved imaging of the salt is shown by black arrows, while white arrows indicate the basement. After Colombo et al. (2017).

8.5   Concluding Remarks

The ability to integrate different measurements (multiphysics) provides constraints on seismic velocity and resistivity subsurface properties estimation as it provides additional constraints to poorly posed geophysical inverse problems. It is important to remember that single domain inversion methods by itself may not uniquely resolve the subsurface parameter fields due to both lack of coverage and full subsurface illumination, and inherent nonuniqueness associated with the mathematical inversion procedures. Adding more information from different geophysical fields improves overall sensitivity and resolution. That the clarity of images in complex geologic settings such as subsalt or subbasalt are facilitated by joint inversion of seismic and nonseismic data sets the foundation for future work on geopressure. Our fervent hope is that motivated readers would add value to such projects by research in the near future.

9 Geopressure Detection and Prediction in Real Time

9.1 Introduction

In this chapter we shall discuss several techniques for detection and analysis of geopressure in real time. Undercompacted shales generally have a lower acoustic impedance (product of density and velocity) than those that follow a normal compaction trend. Departure from the normal compaction trend may indicate potential drilling hazards due to overpressure. Techniques that can monitor acoustic impedance can be used to indicate the existence of such potential hazards, and thereby, help in designing the casing and mud program. Prediction of pressure ahead of the bit starts with the best predrill model. In frontier wells, seismic data are the only data available. Seismic interval velocities from analysis of stacking velocities, tomography and impedances from reflection sequence analyses, in conjunction with a predrill rock model, can be used to develop a predrill pressure versus depth profile. This has been used with considerable success in the petroleum exploration and drilling industry. The limitations, however, are the lack of resolution in the reflection seismic data and uncalibrated velocity models. Thus, a strategy can be developed that can update this so-called static model in real time using additional data acquired *during* drilling. The strategy has several steps: (1) develop and implement the best predrill model that is based on available data prior to drilling a well such as seismic data, geologic data and a model for the expected lithology, faults and offset well information; (2) follow the model and engage in continuous calibration of the model (behind the bit) using all drilling data such as D-exponent logs and various mud log data as discussed earlier; and (3) update and project ahead-of-the-bit pore pressure estimation using data from various MWD techniques such as resistivity and sonic in real time.

9.2 Strategy for Real-Time Update and Prediction Ahead of the Bit

Before we proceed with the techniques for pore pressure prediction ahead of the bit in real time, we need to understand two things: *when is this needed and what is the meaning of "real-time" prediction in this context?* The answer to the first question depends upon the pore pressure versus depth variation. If the pressure profile changes gradually with depth, then real-time prediction may not be needed other than using the prediction while drilling as a "comfort factor." However, such is not the situation in many cases. For example, if the onset of high pore pressure is abrupt, one would definitely benefit by the prediction-while-drilling capability. Furthermore, if there is a pressure regression – a major cause of drilling problems and well safety – one would definitely require this capability. Unpredictable pore pressure regimes such as when a wellbore crosses a fault or a series of faults would also benefit from real-time prediction. Another cause of well problems is the existence and detection of thin permeable layers (subseismic scale). In this case real-time pore pressure prediction technology will definitely add value.

The next question is: How often do we need to predict pressure in real time? Do we need pressure prediction at every turn of the drill bit? Every time a connection is made? During every "bit trip"? How far ahead of the bit do we need to make such a prediction? Certainly, it is not necessary to predict pore pressure at every turn of the bit although technology may allow it. Just how can one implement this feature since mud program and other details of the drilling and the rig program are already in place and any change requires a warning well ahead of time! The key is to manage the downtime during drilling such as during "connection" and "bit trips" effectively. The strategy then is to calibrate behind the bit and predict ahead-of-the-bit during this downtime so that the drilling personnel can make informed decisions regarding, if and when, mud weight and well plan or trajectory need to be changed.

In Figure 9.1a we show a general schematic of how the process works. The process has three major phases: a predrill planning phase, a baseline modeling phase and a real-time updating phase. The predrill phase should begin as early as possible and definitely well ahead by several months, at least, of the date when the drilling is scheduled. During this phase, a static model is developed that accounts for all possible (and expected) geological and geophysical variations, quantifying and understanding all uncertainties. At this stage, analogues, or offset well data and seismic data play critical roles. The seismic model for predrill pore pressure is established at this stage with assigned uncertainty. The work in this phase should be done with close communication with the drilling community – an absolute must! Every operator maintains a "drilling design or well construction" team. It is imperative that the G&G staff be involved with predrill analyses and work in close contact with this sister team in the organization. This is mainly due to the fact that the drilling community is the end-user of the G&G team's predictions and therefore, both teams must understand the weaknesses of each team and try to eliminate them as much as possible. Some of these weaknesses – perceived or real, are partly due to communication gaps that exist between these two interdisciplinary teams.

The second phase deals with a small subset of the data volume used in phase 1. We call this the *volume of interest* (VOI) – it is a small volume of the data that is used during the first phase. The main reason is that the drilling activity happens in a small *area of interest* (AOI). However, another reason is that during phase 3, the drilling model must be updated quickly in real time – this is possible if the VOI just encompasses the AOI. Typically, the AOI is equivalent to one OCS block – *about* nine square miles (5000–5760 acres). Blocks in offshore Texas and deeper water areas are 5760 acres. Blocks in the shallow water off Louisiana are 5000 acres. (There are odd sized

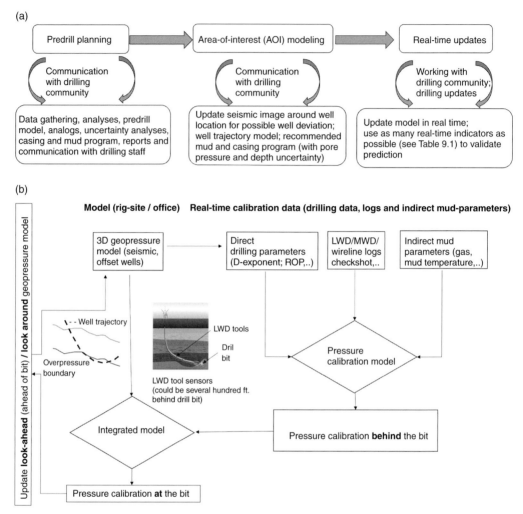

Figure 9.1 (a) A schematic of the overall process employed for geopressure prediction in real time. (b) The schematic of the use of real-time data to calibrate a geopressure model in real time behind, at, and ahead (and around) the bit.

blocks near protraction area boundaries.) The main emphasis of this phase of the work is to *increase the resolution* of the model prediction so that the drilling community can use it. Well trajectory and mud weights are predicted for multiple options and scenarios as necessitated by the geology – for example, will the well path cross fault(s) and if so, how well do we know the geometry and extent of the fault and its throw? How well can we predict pore pressure *across* the fault to be crossed? Is the fault a sealing fault or a nonsealing fault? These types of modeling and predictions are carried out in this phase and a report is produced that outlines multiple contingencies for drilling operations. The personnel carrying out the work during this phase must be an interdisciplinary team comprising personnel from both drilling and G&G communities.

Table 9.1 Real-time pore pressure indicators

Real-time indicators	Comments
Gas (total background gas; connection and trip-gas; mud-cut gas)	Underbalanced drilling increases total background gas while drilling through gas-bearing formations; while background gas increases with ROP, for a constant ROP, background gas will increase with higher pore pressure; connection and trip-gas occurring also indicate overpressure formations; mud-cut gas indicates abnormally high pore pressure and underbalanced drilling.
Rate-of-penetration (ROP); D-exponent	An increase in ROP may indicate high-pressured formations and underbalanced drilling; gradual increase in D-exponent or modified D-exponent with respect to NCT may indicate that the formation being drilled is overpressured.
LWD resistivity	Gradual decrease in resistivity (expressed in ohm-m) with respect to NCT may indicate drilling through overpressured formation.
LWD sonic	Gradual increase in interval transit time with respect to NCT may indicate drilling through high-pressured formations.
Mud volume	An increase in mud pit volume may indicate a "kick" – namely, formation pressure is greater than the mud weight.
Flowline temperature	Geothermal gradient is increased in high-pressured environment.
Bulk density of cuttings	Lower bulk density as compared to NCT values is an indicator of high-pressured formation.

Otherwise, there is a high probability that the results will not impact the third phase (drilling phase) to the extent that is desirable.

The third phase deals with real-time updates during drilling. This is an important phase during which real-time data (see Table 9.1) are acquired, processed and integrated with the model in phase 2 so that an updated look-ahead and look-around (drill bit) image of geopressure is made. This way, as drilling continues, the cone of uncertainty will reduce. This is also an intense phase of the activity. This is because many real-time indicators as outlined in Table 9.1 are acquired during this phase and these data must be analyzed, quality controlled and integrated with the drilling model to calibrate the model (behind and at the bit) and make predictions ahead of the bit. A schematic of how this is done is shown in Figure 9.1b. We point out that all real-time sensors such as the MWD/LWD tools are at least hundred or more feet behind the bit. Therefore, calibration must be done in three steps – **behind** the bit (using all processed logs including wireline), **at** the bit (mostly real-time logs), and **look ahead** of and **around** the bit (mostly calibrated seismic model). All activities must be made as a team.

The acceptance of the model prediction by the drilling personnel requires several items: how reliable are the predictions and the attendant uncertainties? Is their sufficient time available to *digest* the predictions and consequences by both the drilling supervisor and the company man and then implement the predictions? What is the role of all service contractors associated with drilling? These are not easy issues and currently communication and data access in a timely fashion remain important bottlenecks. A minimum of three to four hours may be required (at the time of writing this book) for this process to complete. Typically, wells are drilled by mud weights that are overbalanced – mud weight is higher than the formation pore pressure. If the predicted mud weight is within the "drilling mud window" dictated by the difference between the mud weight and the mud weight equivalent of the fracture pressure there is usually no need to change the existing drilling plan. The drilling can continue. However, if an impending high pressure (or a low pressure) zone is expected that was not a part of the prediction in phase 2, a contingency plan must be implemented before entering the formation. This may require increasing or decreasing mud weight, deviating the well trajectory or using a contingency casing. Updating and predicting a seismic image may also be needed at this time. This will require updating and calibrating the velocity model with borehole data such as sonic, checkshot and various drilling data related to pore pressure and then migrating the seismic data over the VOI with the new velocity (anisotropic) model to create a new image. We note that the ray path required to image the VOI is in general much larger than the VOI and thus the updated velocity model should carry local information from the well and expand it throughout the updated velocity model required for re-imaging. This can be done using steering filters among other techniques (e.g., Bakulin et al., 2010a). We note that the seismic data is the only data that is "truly" **ahead** of the bit. The rest of the data as discussed here and in Table 9.1 are from either "at the bit" or "behind the bit."

9.3   Pore Pressure Prediction Methods in Real-Time

Pore pressure prediction in real time must deal with continuously monitoring and measuring pressure in shale. This is because a petroleum basin contains predominantly shale lithology and the drilling activity happens mostly in that rock type. However, reservoir rocks such as sandstones create problems due to pressure transmission through permeable zones, causing higher pressure at structural updip locations. We noted earlier that shale is exceedingly variable in all of their properties, and various types of burial metamorphisms affect its mechanical, chemical, and electrical properties. This variability further complicates the definition of a well-defined shale normal compaction curve as shale compaction characteristics vary considerably. Direct measurements of pore pressure in shales using tools such as RFT/DST are currently impossible because shale has extremely low permeability. Hence indirect methods are the only alternative. Most of the indirect pressure evaluation methods are based on pressure estimates from seismic data, wireline and MWD/LWD logs and basin modeling. Use of shale compaction curves or normal compaction trend analysis is required in most of the methods. These curves are typically empirical, being based on regional experience or analogue data. Other methods such as seismic while drilling (SWD), logging while drilling (LWD), and vertical seismic profiling while drilling (VSP-WD) are being used by the industry to estimate pore pressure in real time. These methods are discussed in this chapter. Nonetheless, all these methods are based on *indirect* estimates of pressure in shales. Quite frequently an assumption is made that the pore pressure in shales is in equilibrium with adjacent permeable reservoir formations so

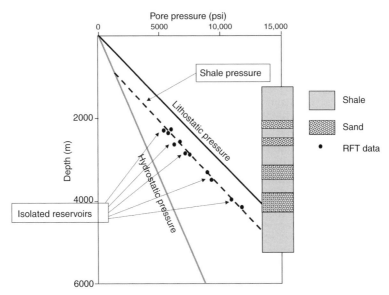

Figure 9.2 An example of fluid pressure in shale that is in communication with the adjacent Sands, Nile Delta, Egypt, as reported by O'Connor et al. (2008) and Mann and Mackenzie (1990).

that RFT/DST measurements can be integrated with indicators that provide indirect pore pressure estimates in shales. This assumption may or may not be true. In Figure 9.2 we show an example from the Nile Delta, Egypt, where the shales were noted to be in pressure communication with the adjacent reservoirs as verified by pressure measurements using RFT (O'Connor et al., 2008). This was attributed to Mann and Mackenzie (1990) and was quoted in O'Connor et al. (2008). This technique works in cases where shales are relatively thin and their pore pressure can bleed through to the reservoir sands. Otherwise, especially, in thick shales, pore pressure could be higher than that indicated by the RFT/DST tools.

In Chapters 4–7 we discussed various methods for pore pressure prediction. There are normal compaction trend (NCT) based methods such as the Eaton-method (Eaton, 1975) and effective stress based methods such as the Bowers (1994) and Dutta methods (Dutta, 1986, 2014). All of these methods can be applied in real time using velocities from seismic or borehole and resistivity/conductivity data. While currently it is not possible to reshoot seismic in real time, it is possible to update models based on seismic data by real-time data from borehole, such as checkshot and VSP, sonic, and resistivity logs. Add to this, the data based on drilling parameters such as ROP and D-exponents and many mud-based parameters such as gas and flowline temperature which provide additional constraints on the model, we have a powerful ahead-of-the-bit prediction technology. Here we will discuss a little more on the *D*-exponent technology as it is the only method that yields pore pressure at the drill bit location in the borehole assembly (BHA) as proposed by Jorden and Shirley (1966). Other LWD tools such as resistivity or sonic tools yield pore pressure at the location where these tools are attached in the BHA – typically quite far behind the bit ($\geq 100 ft$).

The corrected D-exponent D_{xc} was defined in the Chapter 7 as

$$D_{xc} = D_x(P_{hy}/P_{mw}) \tag{9.1}$$

where D_x is the D-exponent, P_{hy} is the hydrostatic pore pressure gradient in mud weight equivalent, and P_{mw} is the mud weight being used downhole during drilling. We note that Jorden and Shirley (1966) devised the D-exponent technology valid for *roller cone* bits. However, the drillers today use PDC bits quite often, especially when drilling through overpressured formations. The PDC bits crush rocks with a shear cutting action while the roller cone bits use chipping action. Therefore, the D-exponent results must be recalibrated again when the driller changes from a roller cone to a PDC bit. The method requires D-exponent values in the normally pressured section. This is important and requires establishing the *baseline* for the hydrostatic pore pressure gradient P_{hy}. Zhang and Yin (2017) recommended the following depth-dependent model for the normal D-exponent

$$D_n = D_0 + aZ \tag{9.2}$$

where D_0 is the shale D-exponent at the mudline, a is a calibration parameter, and Z is the depth below mudline. Quoted values for D_0, and a are 1.5 and 0.00015, respectively, when Z is expressed in feet. These authors noted that if the borehole size, weight on bit, and other parameters change, the normal compaction trend will have to be recalibrated. Thus the methodology should be used by experienced drillers in close collaboration with the G&G personnel in the team. An example of the application is given in Figure 9.3 for a shale-gas well. Upon establishing the normal pressure trend for D-exponent as shown in Figure 9.3a, the authors used the Eaton-method to calculate pore pressure as shown in Figure 9.3b. While this method yields pore pressure in real time, it has severe weaknesses. It assumes that ROP is controlled by pore pressure only – faster penetration in higher porosity and overpressured shale as compared to that for normally compacted shale. This may not be always valid as lithology may change with depth. In addition, the ROP also changes with weight on bit, type of bit used, and the borehole size. Therefore, we need to calibrate this method frequently with other indicators such as LWD data from resistivity or sonic. Real-time pore pressure prediction can be done either on the rig site or remotely in the office where real-time data is instantly accessible. The key steps in the workflow are given below (Fertl, 1976, Chapter 4; Dutta, 1987a, Chapter 5; Zhang and Yin, 2017):

1. Construct a predrill model for pore pressure using any of the techniques discussed earlier in this book (see Chapters 4 and 5) with uncertainty analyses and include potential risk factors. This model should be calibrated with offset well data, if available.
2. Establish a NCT if using a method that requires it.

3. Apply the model to the well being drilled in real time.
4. Make sure that the real-time drilling data are available to be used in the model. This will require special coding (WITS format). This will ensure that LWD and MWD data can be automatically loaded to the model established in steps 1–2.
5. Use the model to predict pore pressure in real time. This prediction must be compared with the downhole mud weight, for example, by comparing with the Equivalent Circulation Density (ECD). This is necessary because we need to know whether the mud weight is less than the pore pressure gradient (underbalanced drilling). Furthermore, comparison should be made with other real-time pore pressure indicators such as flowline temperature, well influx, mud pit volume (gains or loss), background gas, or kicks.
6. Establish a "confidence" limit on the prediction and then communicate the results to the drilling personnel. They need to know whether the well is being drilled "underbalanced" (potentially a dangerous situation) or slightly overbalanced (good practice) but not grossly overbalanced (not a good situation).
7. Ascertain with technical experts whether any unplanned drilling operation is needed such as adjusting ECD and mud weights.

In summary, the pore pressure detection and management of drilling operation while drilling through overpressured formations using LWD, MWD, and mud-

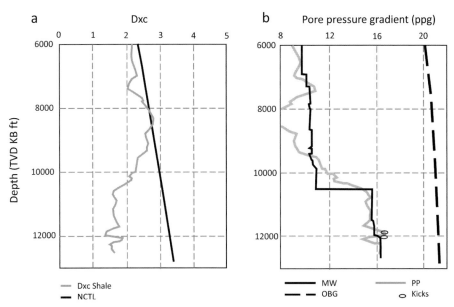

Figure 9.3 Pore pressure obtained from corrected D-exponent for a gas well using a linear depth-dependent trend for the normally pressured section. (a) Corrected D-exponent of shale versus depth (low frequency trend based on gamma ray log that is not shown) and the normal compaction trend line labeled as NCTL. (b) Computed overburden gradient in ppg, calculated pore pressure gradient in ppg, actual mud weights in ppg, and the pressure data inferred from "pressure kicks" in the well. Pore pressure gradient in ppg is based on the low frequency trend of modified D-exponent in (a). Modified after Zhang and Yin (2017).

logging data is very effective in real time as it allows multiple calibrations and updates of predrill models. In this context, one must distinguish the data from *at the bit* to that from *behind the bit*. We have seen many instances, where the drilling community extrapolates the data from behind the bit and uses them to estimate the pore pressure ahead of the bit. This is only possible when a formation follows a certain uniform compaction or undercompaction trend. Otherwise, this could yield erroneous estimate – the error will be larger if the estimate is made farther from the bit.

9.4   Seismic-While-Drilling Technology for Real-Time Pore Pressure Prediction

With the increasing use of 3D and 4D seismic data for pore pressure prediction, it is now possible to make pore pressure predications more reliably in real time and create three dimensional pressure profiles. Seismic while drilling (SWD) is the name given to those seismic workflows for predicting pore pressure *while the drilling operation is continuing*. SWD can be subdivided in two groups:

1. drill-bit seismic while drilling (drill-bit SWD or reverse-VSP with drill-bit as a seismic source)
2. vertical seismic profiling while drilling (VSP-WD)

In the past several decades, the drill- bit SWD technique utilized the acoustic energy radiated by a *tri-cone* bit to yield real-time information during drilling by providing time-to-depth and look-ahead information. Another technique is vertical seismic profile while drilling (VSP-WD), which consists in recording the seismic signal generated by a *surface* seismic source (e.g., an air gun on the rig) on seismic sensors integrated *inside* the downhole borehole assembly (BHA).

9.4.1   Drill-Bit SWD

We have learned earlier that seismic velocity can be related to pore pressure via some assumptions on the pore pressure mechanisms and a rock physics–constrained earth model that relates velocity to pore pressure or effective stress. However, seismic velocity must be calibrated properly to improve its reliability. Wireline VSP or checkshot data can be used for this purpose. However, acquiring such data requires drilling to be stopped and additional tools to be brought for acquiring data. This downtime could be costly. To circumvent this problem, drill-bit SWD technology can be used. It requires no downtime and it is totally noninvasive to the drilling process as no additional tool is lowered in the borehole (see, e.g., Aleotti et al., 1999; Miranda et al., 1996, 2000; Polleto and Miranda, 2004).

Drill-bit seismic uses the acoustic energy radiated by a working *rollercone* drill-bit to determine the seismic time-to-depth ratio as the well is being drilled. The energy required for drilling is supplied to the bit by rotation of the drillstring, causing the cones to roll over at the bottom of the hole. As the cones roll over, the teeth penetrate and gouge the formation, destroying the rock. Each tooth impact applies an axial force to the bottom of the hole and an equal and opposite force to the drillstring. The succession of axial impacts as the bit drills radiate compressional or P-waves into the formation and cause axial vibrations to travel up the drillstring. The bit acts as a dipole source for P-waves radiating energy upward toward the surface and downward ahead of the bit (Hardage, 1992). At the surface for land wells, and at the seafloor for offshore wells, geophones, hydrophones, or a combination of both are used to detect the P-waves. Accelerometers, placed near the top of the drillstring, detect the axial vibrations traveling up the drill pipe (see Meehan et al., 1998, for a more detailed description of this technique).

The bit-generated signal is continuous in nature, and timing information must be extracted. In Figure 9.4 we show tool configurations for two scenarios: case A for onshore and case B for offshore. The analysis techniques in both cases are based on the pioneering work by Rector (1990). The offshore geometry was designed and utilized by BP (where the senior author of this book was employed at that time and participated in the project) in cooperation with Schlumberger during the mid-1990s. For the offshore case, the sensors are lowered in the water from the rig. This configuration requires special handling of the tool, especially when the rig is a dynamically positioned ship in deepwater. Referring to Figure 9.4, correlating the drillstring sensor signal with the seismic sensor signals gives the traveltime difference between the formation path and the drillstring path. Once ΔT_{rel} is known, the time taken for the axial vibrations to travel along the drillstring, ΔT_{ds} can be determined, and the absolute traveltime from bit to surface, ΔT_f can be calculated. The

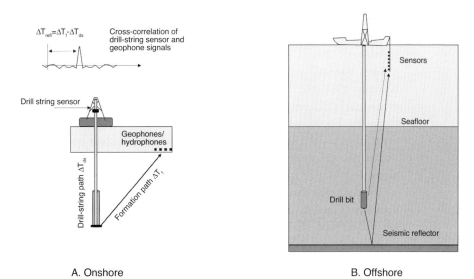

A. Onshore   B. Offshore

Figure 9.4 A schematic of drill bit SWD tool. (a) Onshore. The sensors are both at the top of rig and spread in a array on the surface. (b) Offshore. Sensors are lowered from the drillship into the ocean. Correlation of the rig sensor signal with the seismic sensors gives the relative traveltime difference between the formation path and the drillstring path. If the traveltime is known, the traveltime from the bit to the surface through the formation can be determined (Rector, 1990). Modified after Dutta et al. (2002b).

time-to-depth ratio is calculated using the direct radiation from the drill-bit to the surface. The energy that propagates downward ahead of the bit is reflected back to the surface by impedance changes in the formation. This energy can also be detected and processed to produce a seismic image of the formation ahead of the bit. When used in combination with the surface seismic, such look-ahead images monitor the critical horizons as drilling progresses.

Although simple in concept, significant technical hurdles must be overcome to produce a robust and reliable measurement technique. The most important information derived is the formation traveltime, and it is essential that all factors that affect the accuracy, both relative and absolute, of this measurement are understood. Quantifying the size of the possible timing errors gives confidence in the measurement and helps ensure the technique is correctly applied (Meehan et al., 1998). Particular attention must be paid to the drill string traveltime measurement and to the effects of processing on the phase of the signals. Working rigs create a great deal of noise, and sophisticated signal processing methods must be employed to extract the drill bit signal (Meehan et al., 1998). Differentiation of the time-to-depth measurement with respect to depth gives an estimate of the local formation velocity.

The drill-bit seismic technique can be used in conjunction with a conventional wireline VSP to enable real-time prediction of the depth to overpressured zones. Figure 9.5 shows an example (Dutta et al., 2002b). The top part of Figure 9.5a shows an acoustic impedance inversion of an intermediate wireline VSP acquired at a TD of 2000 m. The sudden drop in acoustic impedance just before 2.2 s two-way traveltime is interpreted as the onset of overpressure. The bottom part of the plot shows how the time-to-depth information from the drill-bit seismic is used to update the predicted depth to the top of the overpressured zone. Where the bit has reached a depth of 2200 m, the hazard depth is predicted to be at 2707 m (using a least-squares extrapolation). As drilling progresses, more time-to-depth information becomes available. Where the bit has reached 2400 m, the new depth-to-hazard prediction is 2753 m. The closer the bit approaches the hazard, the more accurate the prediction becomes. Figure 9.5b shows a composite of pore pressure in real time from the drill-bit seismic velocity data, mud weights and pore pressure estimated from post drill sonic log as well as D-exponent data (R. Meehan, personal communication, 1997).

This technique relies upon a successful inversion of the wireline VSP data and the updating of the current time-to-depth ratio using the drill-bit seismic data. If the drill-bit seismic look-ahead image could be inverted for acoustic impedance, it would be more convenient, eliminating the need for the intermediate wireline VSP. The poorer SNR of the drill-bit seismic data and the lack of control over the source signature, however, mean that the data are not suitable for inversion. As the methodologies for acquiring and processing drill-bit seismic data improve and evolve, this situation may change.

Using a rotary-cone bit as a seismic source has several proven applications such as (Hardage, 2009):

a

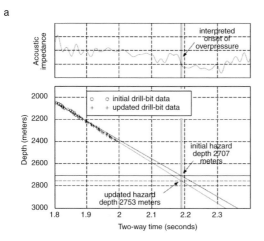

b

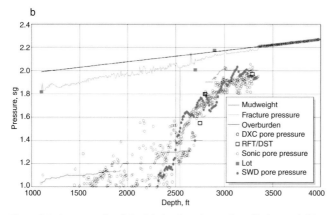

Figure 9.5 An example of the (a) depth-to-hazard prediction and (b) a composite of pore pressure ahead of the drill-bit using drill-bit SWD tool. Pore pressure is expressed in SG units (1.0 SG denotes hydrostatic pore pressure). After Dutta et al. (2002b). Reprinted by permission of the AAPG whose permission is required for further use. (A black and white version of this figure will appear in some formats. For the color version, please refer to the plate section.)

- Real-drill-time velocity checkshot information (locating the bit on the seismic)
- Overpressure detection in real time and ahead of the bit and predicting the depth to top of geohazard
- Guiding the bit to a target seen on surface-acquired seismic data
- Real-drill-time imaging ahead of the bit
- Real-drill-time depth-to-time conversion to know when the bit is reaching an important depth interval
- Positioning the bit at the top of an interval that needs to be cored
- Optimizing casing points

All of these applications, and others, were achieved with drill-bit seismic technology in the 1980s and 1990s. Even with these proven applications, drill-bit seismic technology is not as widely used today as it was 15-20 years ago. The principal reason has been the conversion to poly-diamond-composite (PDC) bits by drilling contractors. PDC bits cut by a scraping action – not by vertical impacts of chisel teeth as occurs with a rotary-cone bit. Effective seismic wave fields are difficult to achieve with PDC bits.

Technologies are now available that acquire drill-bit SWD data by embedding seismic sensors in the drill-string near the drill-bit (Hardage, 2009; Polleto and Miranda, 2004; Rao, 2002). With this technology, vertical seismic profile (VSP) data can be acquired while drilling with any kind of bit, including PDC bits, by using these downhole sensors together with a surface-based seismic source. At each depth where seismic information needs to be obtained, drilling action must cease for several minutes so that the downhole sensors are in a quieter environment when they record the seismic wave field produced by the source. The responses of the drill string sensors are stored in a downhole memory unit included in the drill string system. The data are retrieved when bit trips are made and the seismic sensor section is returned to the drill floor. This downhole seismic sensor technology allows numerous seismic applications to be implemented as a well is being drilled, the most important one being hazard prediction ahead of the bit. Furthermore, the drill-bit SWD technology does not interfere with drilling operations.

9.4.2   VSP-WD

Drill-bit SWD technique as discussed above uses the drill-bit as a source of seismic energy. Hence it is a *reverse-VSP* configuration. To overcome limitations associated with this technology, Schlumberger developed the VSP-while-drilling or VSP-WD technology. It is a transfer of wireline borehole seismic technology to drilling operations for real-time operations such as detecting geohazards ahead of the drill-bit. VSP-WD is almost identical to wireline service using the same surface source and downhole sensors. The main difference is that there is no direct cable connection between the tool and the surface (Greenburg, 2008). A schematic of the acquisition is shown in Figure 9.6. The tool, which contains a processor and memory, receives seismic energy from a conventional source such as an explosive device on land or an air gun suspended from a crane on the drillship in offshore. After acquisition, the seismic signals are stored and processed downhole, and checkshot data and quality indicators are transmitted up-hole in real time by connection with the MWD pulse telemetry system. Waveforms are recorded in the tool memory for further processing after a bit trip. The source is fired while making the drill string connection with drilling and mud circulation stopped, to prevent the effect of drilling noises on data acquisition process. A stand change takes some time (5–15 minutes) which is enough for three to five shots to be fired. The time/depth data are used to position the well on the seismic map at the wellsite or offsite. The technology uses two synchronized high-precision clocks (not shown on schematic in Figure 9.6) – one at the surface (clock starts when a shot is fired) and the other at the bit level (when the first arrival reaches the recorders). These clocks are accurate enough to measure time in milliseconds but also rugged enough to survive downhole environment. Some example applications are given below.

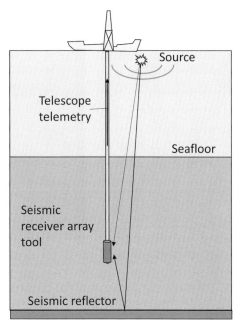

Figure 9.6 A schematic of a VSP-WD tool. Two clocks (not shown), one at the source level and the other at the bit level, record the timing of firing a shot and that when arrival is recorded at the bit level. We note that the downhole environment is very noisy and therefore the instrument is activated during the downtime of drilling such as the time interval during changing a stand of pipes.

9.4.3   Repeated Real-Time Updates of Seismic Velocity (Seismic-Guided Drilling with VSP-WD Tool)

This technique for pore pressure prediction ahead of the bit is based on the advanced velocity modeling methodology that we discussed in Chapter 6. However, that was a static model in which an improved seismic image is produced by imposing pore pressure as a constraint on the velocity modeling (tomography and possibly, FWI) and subsequent depth migration. In the context of pressure prediction in real time, one starts with the same static model in a small volume of interest (VOI) as discussed earlier in this chapter. This velocity model is then *updated multiple times* by using real-time checkshot data from VSP-WD tool (Greenberg, 2008) and other conventional LWD data. The calibrated velocity model is used to update the pore pressure model and the prediction is checked against pore pressure indicators from drilling parameters as discussed earlier. (This is behind-the-bit calibration but essential to establishing a reliable ahead-of-the-bit prediction.) The seismic volume over the VOI is also updated using the new velocity model by using appropriate depth imaging algorithm. (No new seismic data are acquired.) This results in look-ahead (and also look-around) images of pore pressure and structural image. A target such as the "top of over-pressure" and its "intensity" (pore pressure magnitude) as identified in the initial

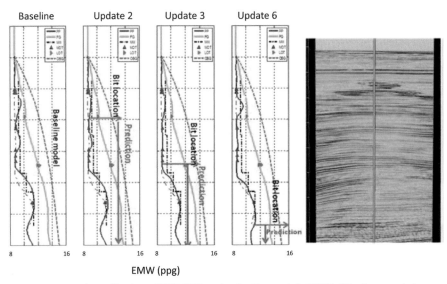

Figure 9.7 Example application of VSP-WD tool, after Dutta et al. (2012). The first panel shows a prediction of pore pressure in depth from the baseline velocity model, while panels 2–4 show various updates of look-ahead pore pressure. In each insert of these panels the horizontal axis denotes Equivalent Mud Weight in ppg starting from 8 ppg to 16 ppg in increment of 2 ppg. Blue: Pore Pressure Gradient; Green: Fracture Gradient; Magenta: Overburden Gradient; Dashed line: post-well mud weight (ppg); Blue triangle: MDT Pressure (ppg); Magenta triangle: LOT (ppg). The vertical red line in each panel indicates the bit location in depth and the predicted distance to the target zone. See text for updating procedure. Panel 5 shows the baseline seismic image. No new seismic data are acquired in this technique. In each figure, the vertical axis is depth increasing downwards. (A black and white version of this figure will appear in some formats. For the color version, please refer to the plate section.)

model (and prior to drilling) could be thus monitored and its location updated and placed on the seismic image (in depth) as the drilling continues. An example is shown in Figure 9.7 for a well in offshore, India (Dutta et al., 2012; Bhaduri et al., 2012). The predrill seismic image over the AOI is shown in panel 5 of the figure. As the drilling commenced, the real-time checkshot data from seismic-WD tool was used to update the velocity model and the image around the borehole in the VOI. Each update included the "top of overpressure" and its "intensity" (magnitude of pore pressure) well in advance of reaching the deep target. Figure 9.7 shows some examples of how these parameters varied as the drill bit continued its progress toward the target (Bhaduri et al., 2012; Dutta et al., 2012). For each update, recalibration, pressure prediction, and reimaging were done using the methodology discussed in Chapter 6. The first panel shows the static pore pressure (in ppg) versus depth model at the well location prior to drilling. The "blue" curve is the pore pressure (ppg), the "green" curve is the fracture pressure (ppg), and the "magenta" curve denotes the overburden stress (ppg). The postwell pressure, actual mud weights used (dashed blocky line) and fracture gradient data are also shown for the sake of reference. The second panel from the left is the result of update #2 and shows the same variables when the drill bit was located at a depth as indicated. The remaining two panels

Pre-drill model   Update -1   Update -2   Update -3   Update -6

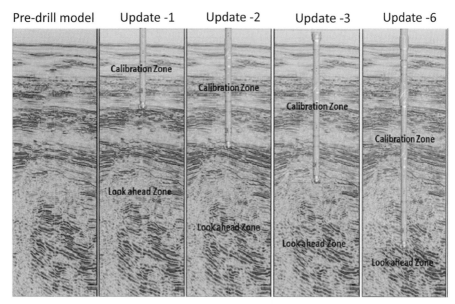

Figure 9.8 Imaging improvements with VSP-WD technology. The first panel shows Baseline image. The rest of the panels show images after successive updates as indicated. Each update is based on recalibration of velocity and remigration of seismic data. No new seismic data are acquired in this technique. Panels 2-5 show the location of the drill-bit, the images behind the bit level (Calibration Zone) and that of the zone ahead (Look-ahead Zone). Thus the data from the calibration zones are used to predict the look-ahead images as drilling continues. After Dutta et al. (2012). (A black and white version of this figure will appear in some formats. For the color version, please refer to the plate section.)

show the update #3 and the update #6 at indicated "bit-depths." In each case, the extent of the "prediction" zones *ahead* of the bit is also indicated. We note that during each update, both the "top" and the "intensity" of overpressure zone changed. The "top" of overpressure was predicted to within 5 m and the "intensity" of the pore pressure to within 0.1 ppg, when the drill bit was at the calibration zone, some 250 m away from the target. This example indicates that this type of predictions can be made well ahead of reaching a particular target so that the drilling decisions could be made well in advance, even in cases like the current well where drilling was fast – a total of ~90 days to reach TD. Figure 9.8 shows imaging improvements with seismic-guided drilling with VSP-WD technology during several updates. The first panel shows the baseline image and the rest of the panels show images after successive updates as indicated (ahead of the bit). Another interesting case study is presented by Rao and Chandrashekar (2013) in a HPHT area that used the same approach as we discussed here.

There are several clarifications that need to be made regarding this technology. First, the meaning of *real time* here is *different* than the one used earlier in this chapter when using real-time drilling parameters as pressure indicators. Since we are using seismic data as a main tool here, the real time means the *total duration of time* in which the well in question is being drilled (typically, ~90 days for the example presented here). Second, no new seismic data are acquired in this technology as opposed to drill-bit seismic-while-drilling technology. The same static seismic data is used here repeatedly; however, the anisotropic velocity model in each case and hence, the pore pressure, is updated numerous time and then the seismic volume is remigrated to create a new image (during each update). Third, the real-time checkshot data is actually behind the bit and only the interval transit times are sent up-hole – the waveforms are retrieved during bit change and so the quality of the velocity data cannot be judged easily. It is for this reason, wireline checkshot data are always acquired to quality control the data. Fourth, velocity and pressure predictions are made only during "*downtime*" of the drilling process since velocity modeling and reimaging requires extensive use of computers. The workflow is executed in office. However, the advantage is that the prediction is made many hundreds-to-thousands of feet ahead of the drill-bit as opposed to real-time updates using drilling indicators. In our opinion, the best practice would be to use *both* technologies in tandem to predict pore pressure ahead of the bit – establish confidence by calibrating behind the bit and then predict ahead of the bit (velocity, pore pressure, and image in 3D) using seismic (see workflow description given in Figure 9.1).

Brown et al. (2015) presented Saudi Aramco's version of real-time pore pressure prediction technology in their first deepwater exploration well in the Red Sea. The target was a subsalt Miocene syn-rift section located in over 2000 ft of water and beneath 9000 ft of halite and evaporite. Predrill pore pressure prediction relied on seismic velocities extracted from a wide azimuth 3D survey. This model had significant uncertainty. Hence, real-time pore pressure monitoring was used to manage the risk due to expected high overpressure. It was based on a comprehensive program that included logging while drilling (LWD), multiple look-ahead vertical seismic profiles (VSPs), velocity model updating and *rapid remigration* (prestack depth migration) around the wellbore to produce simultaneous improvements in imaging and depth estimates that were tied back to an evolving geological pore pressure model. It is not clear whether they used a velocity model that was physical or rock physics compliant. However, they needed to modify the seismic velocity model and calculated pore pressures in real time to provide accurate information for drilling operations. The project resulted in the successful drilling of a deepwater well in a high overpressure, low fracture gradient environment with minimal operational downtime.

It appears that there are close similarities between the technique developed by Saudi Aramco to that used by Dutta et al. (2012) and Esmersoy et al. (2013). In 2002, Halliburton discussed a new downhole tool that used an acoustic system to look ahead of the drill bit much like radar (Rao, 2002). This tool used reflected signals from formations ahead of the bit to relate to geohazards such as high-pressured rock formation. We are not sure about the status of this tool. If this tool is operational and commercial, then it can speed up the turnaround time of the existing seismic-guided drilling technology (Dutta et al., 2012; Rao and Chandrasekhar, 2013; Esmersoy et al., 2013) and make the existing technology obsolete because *new* seismic data would be acquired in real time with the new tool and pore pressure could be detected and predicted in real time. Furthermore, seismic images would benefit from new data and the model would be calibrated multiple times.

9.5   Geopressure Prediction in Real-Time Using Basin Modeling

Calibrated basin modeling technology is a complementary approach to pore pressure prediction using seismic techniques. In Chapter 10 we discuss the details of this technology. Recent advances in basin modeling, such as the implementation of advanced models for chemical diagenesis of rocks and coupling of 3D rock stress and rock failure to pore pressure has extended the operating envelope of pore pressure prediction using this technique (De Prisco et al., 2015; Hantschel and Kauerauf, 2009). There is a considerable amount of flexibility in the models – the chief ones being the ability to deal with the pressure *history*. The effect of palinspastic restoration of salt movement and its effect on pore pressure and the simulation of the effects of major faults and its effect on pressure relief points are other components that can be dealt with in basin modeling technology. The distinct advantage of basin modeling is the capability of pore pressure prediction in structurally complex areas such as subsalt formations or areas affected by salt withdrawals and subsequent formation of minibasins. In such areas, traditional seismic imaging techniques may not be useful due to *poor illumination* below or around salt bodies. However, joint usage of basin modeling with seismic-based techniques is emerging to be a viable approach. Some authors term this technology geophysical basin modeling (Brevik et al., 2011; De Prisco et al., 2015).

In a recent publication, Mosca et al. (2017) applied this approach to predict pore pressure while drilling a well. The case study was conducted in the Gulf of Mexico. The first step of the work flow is to create and calibrate a regional model based on a set of regional maps with the main goal being to provide the regional context. The second step is to create a smaller area of interest (AOI) model using high resolution structural and facies maps. This refined model is then used for pore pressure prediction at the prospect scale. This model is then *updated repeatedly* at the wellbore scale using calibration with real-time pressure data and then the prediction is made ahead of the bit. In this regard, the strategy is similar to that we described in Figure 9.1 for real-time updates with seismic and petrophysical well logs. It should be realized that the prediction is made during the drilling operations phase and not necessarily at every turn of the bit. The method described by Mosca et al. (2017) relies upon calibration of the model (pressure and temperature) at various stages – the regional, the AOI and the wellbore scale. The real-time input to the model for calibration comes from the same sources as discussed earlier in this chapter – MWD and LWD logs, drilling parameters, background gas, and data based on cuttings as well as any direct pressure measurements acquired during drilling. The regional model as discussed by Mosca et al. (2017) covered an area of 1400 sq mi (155 OCS Blocks; 3600 km^2) as shown in Figure 9.9a. The grid interval was 500 ft (150 m) and the model had a total of 24 million cells. The model was simulated in 47 hours on a cluster with four processors (simulation time can be reduced by increasing the number of processors). The AOI was 150 sq mi (17 OCS blocks; 400 km^2). The grid interval was kept the same as for the regional model (500 ft or 150 m) and this smaller model had a total of six million cells. This allowed the model updates to be carried out *overnight*. A critical element of the application was the description of faults along the

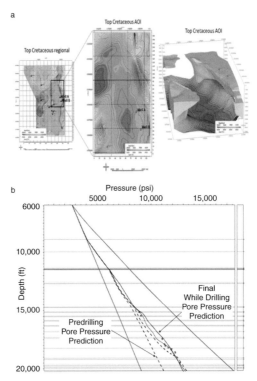

Figure 9.9 (a) An example of pore pressure prediction in real time using basin modeling. (left) Top Cretaceous structure map at the regional-scale (1400 sq mi (155 OCS Blocks; 3600 km^2)), (middle) AOI-scale for the area of interest (150 sq mi (17 OCS blocks; 400 km^2)), and salt including salt canopy (right, pink color) (150 sq mi (17 OCS blocks; 400 km^2)), with locations of the target well A and offset well B. (b) Pore pressure prediction calibration evolution while drilling. The pore pressure data (RFT) were measured after drilling the well. See text for details. From Mosca et al. (2017). (A black and white version of this figure will appear in some formats. For the color version, please refer to the plate section.)

proposed drilling path. The permeability values assigned to these fault planes were adjusted during the calibration phase to provide a better match to the pressure trends at the reference wells. Furthermore, the smaller AOI model could be run overnight so that pore pressure predictions would be done ahead of the bit *nearly* in real time.

The real-time prediction phase of the study began when the well reached a depth of approximately 9500 ft measured depth (4000 ft below the mud line). From this depth onward, a leak-off test (LOT) was performed at each casing point and that together with the real-time mud weight was used to calibrate the pressure profile and to predict the pore pressure ahead of the drill-bit until the next casing point. The evolution of the pressure profile while drilling is shown as a composite in Figure 9.9b. The discrete data points are acquired postdrill and are shown as reference. The key point to note is that the method predicted correctly the *reversal* of the pore pressure due to plumbing associated with a sand body. This allowed the drilling engineers to prevent lost circulation.

As far as we know, this is the first published example of real-time pore pressure prediction using basin modeling. This integrated approach uses a good 3D seismic image as a starting point and then adds value to this by providing pore pressure derived from basin modeling. This interdisciplinary approach makes the method very useful for studying plumbing of a complex basin such as those associated with salt and complex faults. The ability to play what-if scenarios is a particular strength of this technology. The case study shows that a regional framework followed by repeated calibration of pore pressure over a smaller AOI for the hydrodynamics of the basin is a very useful approach – it provides a broader geologic framework and the ability of multiple calibration and updates of pore pressure during the "while drilling" phase of a well. (Can we term this as *"drilling ahead-of-the-bit"* technology?) The attendant flexibility of testing various scenarios is an additional benefit. The calibration data used are the typical real-time data as discussed in this chapter – all real-time MWD/LWD logs, gas and ROP data, and MDT pressure. The turnaround time is dictated by the size of the AOI and available computer power. Currently, this is *overnight*. The turnaround time for the seismic technique is faster. This is because the size of AOI is smaller as compared to what Mosca et al. (2017) used. With the advent of faster computing technology in the future, we believe that the technology used by Mosca et al. (2017) would be very powerful.

9.6   Summary

In this chapter we discussed the rapid progression of technology, especially, downhole tools that impact real-time detection and prediction of geopressure. This technology benefited from even more rapid progress of computer technology. The upper level view of the real-time geopressure quantification workflow is outlined in Figure 9.1. This workflow allows drillers to drill wells safely by anticipating hazards ahead of drill bit and take necessary remedial actions. Of particular note is VSP while drilling (VSP-WD) tool and technique. It has all the advantages of wireline VSP – known locations of source and receiver and higher frequency than conventional seismic in its present form, but in real time. Its limitation is that of data transmission rate. It is worthwhile to remember that the biggest drawback of any downhole tool is the limitation imposed by mud-pulse telemetry. Perhaps wired-casing would be the answer to this. As far as the seismic-guided drilling technology is concerned, it is our assessment that it has the promise but lacks in turnaround time. Same comments apply to real-time pressure prediction using basin modeling technology.

Having said that, there is another challenge that we must face – ourselves. The data from the tools are voluminous and they arrive much more quickly than we – the humans can interpret. This is even more difficult when a multidisciplinary team is involved such as the ones we have been advocating in this book. In this regard, advances in machine-learning and artificial intelligence have great promise. It can overcome the chasm between the time when the data arrive and when it is interpreted and results transmitted for decision-making.

10 Geopressure Prediction Using Basin History Modeling

10.1 Introduction: Basin and Petroleum System Modeling

Basin and petroleum system modeling (BPSM) is a key technology in hydrocarbon exploration that reconstructs the dynamic burial history of sediments, including deposition, uplift, and erosion, and forward-simulates thermal history, diagenesis, and the associated generation, migration, and accumulation of petroleum (Peters, 2009). The modeling process can cover very large geological time ($\sim$ tens to hundreds millions of years) and spatial ($\sim$ tens to hundreds of km) scales. BPSM involves solving coupled nonlinear partial differential equations with moving boundaries. The equations govern deformation and fluid flow in porous media, together with heat transport and chemical reactions (e.g., kerogen maturation, and chemical diagenesis such as smectite-to-illite transformation). The equations arise from the usual principles of continuum mechanics as discussed in chapter 2, and include conservation of mass, momentum, and energy, with appropriate constitutive relations for the sediment and fluid properties, and kinetic reaction equations. Heat flow simulations along with transport of energy through conduction and convection, is a part of the basin modeling algorithm. The coupled system of partial differential equations usually have to be solved numerically on discretized time and spatial grids (1D, 2D, or 3D) with the integration of geological, geophysical, and geochemical inputs. This chapter will give a brief overview of basin modeling as it relates to pore pressure generation and develop a simple model to explore the interplay between deposition, compaction, diagenesis, fluid flow, and pore pressures. There are many excellent texts that describe in much detail the various aspects of basin and petroleum system modeling (Lerche, 1990; Makhous and Galushkin, 2004; Yang, 2006; Hantschel and Kauerauf, 2009; Wangen, 2010). Burnham (2017) gives details of modeling the chemical kinetics of kerogen maturation.

Typically, basin modeling is used to understand the evolution of the basin and the timing of hydrocarbon generation, migration, and trapping. These models help to assess basins for prospective oil accumulations. Outputs from numerical basin models include temperature and measures of maturity (e.g., vitrinite reflectance or hydrogen index) of the source rock. The basin models are calibrated by comparing these outputs with data on vitrinite reflectance and temperature from wells and core analyses. From the basin modeling, we can also get estimates of effective stress, porosity, and pore pressure, which when combined with the appropriate rock physics models can also give the trends of seismic velocity and density. Building the basin model requires as inputs structural horizons that describe the present day configurations of different formations. When there is not enough well control, these horizons are often interpreted from seismic data and have to be positioned in depth using an initial velocity model. However in situations such as subsalt or subbasalt where ray illumination is poor and velocity estimation by conventional methods is problematic, the estimate of the smooth background velocity obtained from basin modeling and rock physics can guide seismic imaging to constrain and update the initial approximate velocity model. Thus basin modeling can provide a very useful input for seismic imaging. On the other side, seismic attributes such as velocity cubes and amplitudes can provide very useful constraints on the basin model that can be used to calibrate the models. Seismic calibration of basin models is not limited to only well locations (as in traditional basin model calibration), but is available over a much larger spatial extent. Currently though seismic velocities are used for positioning of structural horizons in basin models, there is much more information in the seismic velocities and amplitudes that could potentially be used to improve basin models. Estimates of pore pressure from basin modeling are complementary with seismic methods of pore pressure estimation, and integrated together can yield superior results. It can also help improve seismic images through constraints on velocity model as discussed in this book.

10.2   Governing Equations for Mathematical Basin Modeling

The basic set of equations consists of the following.

Conservation of mass (or continuity equation):

$$\frac{\partial(f_i\rho_i)}{\partial t} + \nabla \cdot (f_i\rho_i v_i) = S_i(x,t) \tag{10.1}$$

where f_i, ρ_i, and v_i are the volume fractions, densities, and velocities, respectively, of the ith constituent in the porous medium, and S_i is the rate of mass production per unit volume per unit time, for example, from chemical reactions.

Conservation of momentum: For static equilibrium with gravitational body force this is written as

$$\nabla \cdot \boldsymbol{\sigma} + \rho \mathbf{g} = 0 \tag{10.2}$$

where $\boldsymbol{\sigma}$ is the stress tensor and $\mathbf{g}$ is the acceleration due to gravity.

Conservation of heat energy:

$$\frac{\partial(\rho CT)}{\partial t} + \nabla \cdot (\mathbf{q}_{heat}) = Q_{source} \tag{10.3}$$

where T is the temperature, C the bulk specific heat capacity of the rock, $\mathbf{q}_{heat}$ is the heat flux and the source term Q_{source} can include sources or sinks of thermal energy from chemical reactions as well as radiogenic heat production. Transport of heat and fluid are governed by Fourier's law and Darcy's law, respectively.

Heat transport (conduction, Fourier's law):

$$\mathbf{q}_{heat} = -\lambda \nabla T \tag{10.4}$$

where λ is the thermal conductivity tensor.

Heat transport (conduction and convection):

$$\mathbf{q}_{heat} = -\lambda \nabla T + v_{fluid}\left(\rho_{fluid}C_{fluid}T\right) \tag{10.5}$$

where v_{fluid}, ρ_{fluid}, and C_{fluid} are the fluid velocity, density, and specific heat capacity, respectively. Darcy's law governs fluid transport in porous media and the relation between the fluid flux $\mathbf{q}_{fluid}$ and the fluid pressure p_{fluid} is given by the following equation:

Fluid transport:

$$\mathbf{q}_{fluid} = -\frac{\kappa}{\mu}\left(\nabla p_{fluid} - \rho_{fluid}\mathbf{g}\right) \tag{10.6}$$

where μ is the fluid viscosity, κ denotes the rock permeability tensor, and $\mathbf{g}$ is the acceleration due to gravity. For multiphase flow (e.g., oil–water–gas) this equation generalizes to the multiphase Darcy's equation, involving individual phase fluxes and the relative permeabilities.

In addition to this basic set of governing equations, there are also the kinetic reaction equations for modeling various chemical transformations such as kerogen maturation or clay transformation (smectite-to-illite transformation). We also need constitutive relations that depend on the material properties, and the mineral and fluid constituents. These relations include compaction laws (i.e., porosity-effective stress relation, or void ratio-effective stress relation), porosity-permeability relations, constitutive relations describing the stress-strain behavior (elastic, poroelastic, poroplastic, etc.), and relations describing how thermal conductivity and heat capacity depend on the mineral and fluid constituents. Finally, of course, solving this system of coupled equations requires appropriate boundary and initial conditions. There are numerous commercial algorithms for solving the mathematical equations for predicting pore pressure using basin modeling.

As noted above, basin modeling requires input of various parameters in the model building stage such as porosity-effective stress-lithology, porosity-permeability-lithology and thermal-conductivity-lithology relations. The basin modeling based simulations depend upon the ability to create a hydrodynamic regime of a basin based upon the parametrizations just discussed. We note that the default options in many of the commercial packages may or may not be valid for a particular application. Hence, the decision process leading to acceptable choices of these parameters must include available calibration data such as RFT or MDT data for pore pressure and well logs and drilling data. There is another aspect that requires special attention. It has to do with building a realistic 3D geologic model (constrained by seismic data) that accounts for the effect of faults and geological facies distribution – both of these have significant control on plumbing, and hence, pore pressure distribution in the basin (Satti et al., 2015).

As far as carbonates are concerned, the current basin modeling algorithms may or may not model these rocks adequately. The hydrodynamic models in this case need "compaction curves." Such curves are usually not available for carbonates as the porosity in carbonate rocks evolve with processes that are mostly diagenetically controlled rather than by mechanical compaction. Diagenetically controlled processes require an understanding of chemical kinetics, rock fabric and texture and microscopic description of a variety of pore shapes and their aspect ratios. In conclusion, porosity loss in carbonate sediments is mostly due to chemical compaction and very little to mechanical compaction. Chemical compaction processes are pressure solution and pressure solution enhanced by subcritical crack growth. The predominance of one or the other mechanism is related to the fluid in presence (fluid chemistry) and to the nature of the grains. So it is not clear that common basin modeling algorithms can treat pore pressure evolution in basins dominated by carbonates. This is an ongoing area of research.

10.3   Basin Modeling: Compaction, Diagenesis, and Overpressure

In this section we will analyze some of the underlying physics related to compaction, diagenesis and overpressure as specifically applied in basin modeling in 1D and extend the concepts to 2D/3D basin modeling. This is in preparation for presenting an integrated approach in which we will *combine* the output of basin modeling to anisotropic velocity model building using rock physics models and show that it not only improves reliability of pressure prediction but also improves seismic imaging. In addition, the integrated approach leads to selection of better priors in model building that yields better overall uncertainty quantification (see Section 13.5).

10.3.1   Fluid Flow in Low-Permeability Rocks: A Simple Physical Model of Sediment Compaction with Diagenesis under Constant Gravitational Loading

We first consider a physical model for sediment compaction as shown in Figure 10.1 due to gravitational compaction of sediments under continuous loading of sediments. This model is very relevant for shallow hazard in deepwater where geopressures are known to occur a few hundred feet below seabed. It also illustrates how compaction disequilibrium can cause geopressure in low-permeability sediments. This is well documented in the Gulf of Mexico, USA (Fertl, 1976). In this model water-saturated low-permeability sediments (shale) are deposited on an impermeable basement – this is an assumption to restrict the fluid flow in upward direction only for ease of discussion and mathematical modeling. A nice exposition of this model is given by

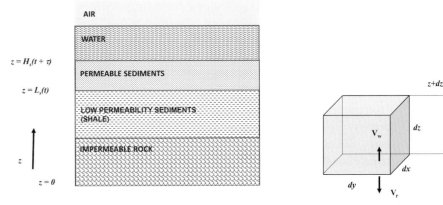

Figure 10.1 Gravitational loading and compaction of low-permeability sediments (shale) deposited on an impermeable basement followed by loading of permeable sediments (sand).

Smith (1971) and Dutta (1986, 1987b). An extension to 2D and 3D exist in commercial basin modeling algorithms (see below).

Shale-on-Shale Compaction (1D Model)

In Figure 10.1 we describe a compacting (subsiding) – accreting geological system composed of fluid-saturated sediments such as mudrock (shale) which is a low-permeability sediment and sand (high-permeability sediment) being deposited on an *impermeable* basement at specified rates. This assumption is made so that fluid flow is restricted upward only for the sake of simplicity. In real geologic situation such as those involving repeated sequences of sand and shale, flow will be in both upward and downward directions, even for the 1D case. Furthermore, we assume that there are *no source terms* that generate fluids at depth such as diagenesis (S/I transformation, for example). We will introduce the effect of diagenesis through "the backdoor" – via a constitutive relation that relates porosity to effective stress, and a thermal history dependent activation process for diagenesis as discussed in Chapter 3. In reality, fluid transport process in this simplified model should be coupled to heat transfer processes as described in Section 10.1. In the current simplified model, we are decoupling the two processes and assuming that diagenesis can be handled through time-temperature dependent porosity reduction process such as the equation (3.10). We also assume that the water depth does not change with time.

    There are two components to this idealized (but physical) model. The first component deals with *"shale-on-shale"* compaction. The second one deals with deposition of additional high-permeability sediments (sand) on top of the low-permeable (shale) sediment after shale deposition ceases. We term this *"sand-on-shale"* deposition. Here "sand" denotes any sediment that has much higher permeability at a given porosity than the shale at that porosity – sand permeabilities are of the order of 0.1 Darcy as opposed to shale which has permeability of the order of micro-Darcy to nano-Darcy; this means that the rate of fluid outflow from sand is more or less instantaneous compared to the outflow from the shale. We assume that the horizontal dimensions of the model in Figure 10.1 are much larger than the vertical dimension. This implies that fluid flow will be in the upward direction only. There are two sources of porosity loss with continued burial. At shallower depths (temperature below 60°C or so) gravity is the only source that causes compaction and fluid flow is dictated by Darcy flow (equation (10.6)). At deeper depths, diagenesis (smectite-to-illite transformation) kicks in which contributes further to change in porosity with effective stress (equation (3.10)). Since the permeability of the shale sediment is low, the rate of expulsion of water decreases with the increasing load, decreasing porosity, and overpressure begins at some depth. The magnitude of overpressure increases further when chemical diagenesis begins – this causes *additional* water release (bound water from clay platelets) in to the host shale that causes load transfer from solid to fluid (decrease in effective stress). The problem we propose to solve is to assess the development of geopressure, effective stress, and bulk density profiles versus depth for this *physical* model.

The coordinate system is shown in Figure 10.1. The base of the coordinate system is *fixed* at the sediment – impermeable rock boundary ($z = 0$). Both shale-sand and sand-water interfaces are *moving* boundaries – and, so are the layers of sediments due to loading. The height of the sediment column is denoted by z; the shale-sand interface is denoted by $z = L_s(t + \tau)$. The shale deposition ceases at $\tau = 0$ at which time the sand deposition starts. The compaction (reduction of porosity ϕ with increasing effective stress σ) is described by a constitutive relation of the type given in equation (3.10) for *both* sand and shale. However, as noted earlier, their permeabilities are vastly different. In the following equations rock grain properties are denoted by suffix r and the fluid properties are denoted by suffix w.

We first address the continuity equations for the solid and the fluid in a typical volume element as shown in Figure 10.1. This suggests that the time rate of change of fluid (solid) mass in an elementary cube must equal the net sum of rate of fluid (solid) mass accumulation in that cube and the rate of fluid (solid) mass that leaves the cube. The continuity equation was stated in equation (10.1). For the 1D model we have the following for the continuity equations for the solid and the pore water

Solid:

$$\frac{\partial[v_r \rho_r (1 - \phi)]}{\partial z} = -\frac{\partial[\rho_r (1 - \phi)]}{\partial t} \tag{10.7}$$

Water:

$$\frac{\partial[v_w \rho_w \phi)]}{\partial z} = -\frac{\partial[\rho_w \phi)]}{\partial t} \tag{10.8}$$

where v_w and v_r are the velocity of water and rock, respectively, with respect to the *fixed* coordinate system. Integrating the above continuity equations and assuming that neither ρ_r (grain density) nor ρ_w (brine density) change with z and t (rock grains and brine are incompressible), we obtain

$$v_r = \frac{[v_r(1-\phi)]_0}{[1-\phi]_z} + \frac{1}{[1-\phi]_z}\int_0^z \frac{\partial\phi}{\partial t}dz \qquad (10.9)$$

$$v_w = \frac{[v_w\phi]_0}{[\phi]_z} - \frac{1}{[\phi]_z}\int_0^z \frac{\partial\phi}{\partial t}dz \qquad (10.10)$$

We introduce a variable, Q, to denote the average volume rate of flow of water per unit area *relative* to the matrix. Thus,

$$Q = (v_w - v_r)\phi \qquad (10.11)$$

Using equations (10.9) and (10.10) in equation (10.11), we get

$$Q = [v_w\phi]_0 - \frac{\phi}{(1-\phi)}[v_r(1-\phi)]_0 - \frac{1}{(1-\phi)}\int_0^z \frac{\partial\phi}{\partial t}dz \qquad (10.12)$$

Differentiation of equation (10.12) with respect to z leads to the continuity equation in terms of the variable Q:

$$\frac{\partial[Q(1-\phi)]}{\partial z} + [v_w\phi + v_r(1-\phi)]_0\frac{\partial\phi}{\partial z} = -\frac{\partial\phi}{\partial t} \qquad (10.13)$$

Since the shale units under consideration lie on an *impermeable* bottom (no flow boundary), we have

$$[v_w]_0 = [v_r]_0 = 0 \qquad (10.14)$$

and consequently, the continuity equation (10.13) in terms of Q further simplifies to:

$$\frac{\partial[Q(1-\phi)]}{\partial z} = -\frac{\partial\phi}{\partial t} \qquad (10.15)$$

Next, we use Darcy's law to relate Q to the negative gradient of the fluid potential, Ψ as:

$$Q = -\frac{\kappa}{\mu}\frac{\partial\psi}{\partial z} \qquad (10.16)$$

where the proportionality constant, κ is the permeability and μ is the viscosity of water. The fluid potential gradient, $\frac{\partial\psi}{\partial z}$ can be expressed in terms of the fluid pressure P, and gravity (acting in downward direction, i.e., negative z-direction) as

$$\frac{\partial \psi}{\partial z} = \frac{\partial P}{\partial z} + g\rho_w \tag{10.17}$$

and therefore,

$$Q = -\frac{\kappa}{\mu}\left(\frac{\partial P}{\partial z} + g\rho_w\right) \tag{10.18}$$

Next we use Terzaghi's law (equation (3.15)) to relate effective stress σ to pore pressure P and overburden stress S as $\sigma = S - P$. The overburden stress S is given for the sediment and the water column as

$$S = g \int_z^{L_s(t)} \rho_b dz + P_0 + P_a = P + \sigma \tag{10.19}$$

where $L_s(t)$ is the coordinate of the sediment – water interface (see Figure 10.1), P_0 is the pressure due to the water column (ocean), P_a is the atmospheric pressure, and ρ_b is the bulk density of the sediments. From equation (10.19) we obtain

$$\frac{\partial P}{\partial z} + g\rho_w = -\left[g(\rho_r - \rho_w)(1 - \phi) + \frac{\partial \sigma}{\partial z}\right] \tag{10.20}$$

Then, re-writing equation (10.15) the basic compaction equation becomes

$$\frac{\partial \phi}{\partial t} = -(1 - \phi)\frac{\partial Q}{\partial z} + Q\frac{\partial \phi}{\partial z} \tag{10.21}$$

where the relative fluid flow (from equations (10.18) and (10.20)) is

$$Q = \frac{\kappa}{\mu}[g(\rho_r - \rho_w)(1 - \phi) + \frac{\partial \sigma}{\partial z}]. \tag{10.22}$$

These two coupled equations (equations (10.21) and (10.22)) are the final compaction equations that needs to be solved for specified initial and boundary conditions. In addition, we also need a constitutive relation (e.g., of the type in equation (3.10)) relating effective stress to porosity (or void ratio). The output field variable from solving the coupled system of equations is porosity (or void ratio). Once this is computed, we can predict other quantities such as pore pressure, pressure gradient, sediment column thickness, effective stress, and bulk density. If we invoke a relation between shale elastic wave velocity to porosity and effective stress (from a rock physics based model, for example), we could also predict a depth dependent velocity profile of the accreting-subsiding basin over time. This provides a link between pore pressures from basin modeling to velocity. In passing we note that the velocity thus estimated is "rock velocity" and it could be very different from the various seismic imaging velocities that we discussed in Chapters 5 and 6. Nonetheless, this is the basis for linking seismic velocity, imaging and pore pressure (Petmecky et al., 2009).

Boundary and Initial Conditions (Shale-on-Shale Compaction)

We assume that the initial depositional porosity at the sediment–water interface during shale deposition is a known constant, and at all times prior to the deposition of succeeding units of shale, the effective stress at the sediment–water interface is zero. Thus

1. $\phi(z,t) = \phi_s$ constant; $\ t > 0, \ z(t) = L_s(t)$ (10.23)

$L_s(t)$ is the coordinate of the top of the moving shale-water interface at time t and it is given by

$$\int_0^{L_s(t)} [1 - \phi(z,t)] = H(t) \tag{10.24}$$

Here $H(t)$ is a specified loading function and it is the volume of "dry" sediment per unit area at time t.

2. The second boundary condition is only applicable at the bottom of the shale column. We have assumed that at all times there is no flow of fluid either in or out of the base. Thus,

$$Q(z,t) = 0 \ ; \ t \geq 0, z = 0 \tag{10.25}$$

3. We also assume that the initial porosity profile is known, that is,

$\phi(z,t) = \phi_0(z); \text{t} = 0,$ $0 \leq z \leq L_s(t)$ (10.26)

where

$$\int_0^{L_s(0)} [1 - \phi(z)0] = H(0) \tag{10.27}$$

Since the top of the shale at $t = 0$ is located at $z = L_s(0)$, we have $H(0)$ as the initial volume of the "dry" sediment per unit area.

Solution of the Shale-on-Shale Compaction Equation

The solution of the compaction equation under a specified loading condition requires solving equations (10.21) and (10.22) with *moving boundary conditions* since new deposition continues to happen at the sediment–water interface as compaction continues due to gravity. To handle the moving boundaries efficiently, we introduce the *Lagrangian coordinates,* $[Z'(z,t), t']$ (see, e.g., Wangen, 2010) defined as follows:

$$Z'(z,t) = \int_0^z [1 - \phi(z,t)]dz \tag{10.28}$$

$$t' = t \tag{10.29}$$

Here $dZ' = (1 - \phi)dz = dz/(1 + \varepsilon)$ denotes the net (porosity-free) column of "dry" sediment. Here ε denotes the void ratio. Using the new coordinates the chain rule of differentiation for any function $f(z,t)$ gives:

$$\frac{\partial f}{\partial z} = \frac{\partial f}{\partial Z'}\frac{\partial Z'}{\partial z} = (1 - \phi)\frac{\partial f}{\partial Z'} = (1 + \varepsilon)^{-1}\frac{\partial f}{\partial Z'}, \tag{10.30}$$

$$\frac{\partial f}{\partial t} = \frac{\partial f}{\partial t'} + \frac{\partial f}{\partial Z'}\frac{\partial Z'}{\partial t} = \frac{\partial f}{\partial t'} + \frac{\partial f}{\partial Z'}Q(1 - \phi), \tag{10.31}$$

In equation (10.31) we also use equations (10.28) and (10.15) to get:

$$\frac{\partial Z'}{\partial t} = \int\limits_0^z -\frac{\partial \phi}{\partial t}dz = Q(1 - \phi) = Q(1 + \epsilon)^{-1} \tag{10.32}$$

The compaction equation (equation (10.21)) then can be written in the new coordinate system as

$$\frac{\partial \phi}{\partial t} + (1 - \phi)^2\frac{\partial Q}{\partial Z'} = 0 \tag{10.33}$$

where we have dropped the prime superscript on the new time coordinate. In the new coordinates

$$Q = \frac{\kappa}{\mu}(1 - \phi)\left[g(\rho_r - \rho_w) + \frac{\partial \sigma}{\partial Z'}\right] \tag{10.34}$$

Since we have the following that relates porosity to void ration ε

$$\varepsilon = \frac{\phi}{1 - \phi} \tag{10.35}$$

this implies

$$(1 - \phi)^2\frac{\partial \varepsilon}{\partial t} = \frac{\partial \phi}{\partial t} \tag{10.36}$$

and therefore, in terms of void ratio ε we have the following field equation for compaction:

$$\frac{\partial \varepsilon}{\partial t} + \frac{\partial Q}{\partial Z'} = 0 \tag{10.37}$$

together with,

$$Q = \frac{\kappa}{\mu(1+\varepsilon)} \left[g(\rho_r - \rho_w) + \frac{\partial \sigma}{\partial Z'} \right] \tag{10.38}$$

The above two equations are subject to the following boundary conditions in the new coordinates:

$$\varepsilon(Z',t) = \varepsilon_s; \, t > 0, \, Z' = H(t) \tag{10.39}$$

$$Q(Z',t) = 0; \, t \geq 0, \, Z' = 0 \tag{10.40}$$

where ε_s is the depositional void ratio at the sediment–water interface. The initial condition can be rewritten in the new coordinate system as

$$\varepsilon(Z',t) = \varepsilon_0(Z'); \quad t = 0, \quad 0 \leq Z' \leq H(0) \tag{10.41}$$

where ε_0 is the basement void ratio. The loading function $H(t)$ is related to $L_s(t)$ and ϕ through the following transformation,

$$H(t) = \int_0^{L_s(t)} [1 - \phi(z,t)]dz = \int_0^{L_s(t)} \frac{dz}{1 + \varepsilon(z,t)} \tag{10.42}$$

The top of the subsiding-accumulating column is given by,

$$L_s(t) = H(t) + \int_0^{H(t)} \varepsilon(Z',t)dZ' \tag{10.43}$$

since

$$z(Z',t) = \int_0^{Z'} [1 + \varepsilon(Z',t)]dZ' \tag{10.44}$$

The first term of equation (10.43) is simply the "dry" volume of sediment per unit area at time t and the second term is the volume of water per unit area at time t remaining in the shale. The loading function, $H(t)$ can be arbitrarily chosen. For the sake of simplicity, we can use a simple linear relation with time for a constant deposition rate Γ,

$$H(t) = H(0) + \Gamma t \tag{10.45}$$

When modeling smectite-illite diagenesis along with compaction, equations (10.37) and (10.38) are augmented by two other partial differential equations, one for the temperature as a function of depth, and another for the smectite fraction as a function of temperature (see Appendix C for details). These equations can be solved using finite-difference or finite element techniques provided we know the three constitutive relations, namely, (1) the relation between ε and σ and a model for the β-function in equation (3.10); (2) a model for the permeability κ as a function of porosity; and (3) a model for the water viscosity μ in Darcy's Law as a function of temperature. Solution of the coupled compaction and pore pressure equations with a moving boundary condition is challenging. In addition to Dutta (1986, 1987b), other early works addressing solution of these equations include Glezen and Lerche (1985), Lerche (1990), Audet and Fowler (1992), and Wangen (1992, 1993). As described in Appendix C, one approach is via another coordinate transformation from the moving Lagrangian Z' depth coordinate to a fixed dimensionless fractional coordinate. This approach is also described in Wangen (2010, Chapter 12) as well as in Lerche (1990, vol. 1), who extends it to two dimensions.

Sand-on-Shale Compaction (1D Model)

This is handled similar to what we have discussed for shale-on-shale compaction. The physical model is again shown in Figure 10.1. Water-saturated shale is continuously accumulating at a given loading rate on an impermeable base. The z-coordinate is fixed at the base. At the end of time t, water-saturated sand is deposited at a given loading rate that could be different from that of the shale deposition. As sand is deposited on shale, the total "dry" volume per unit area of the underlying shale must remain constant. This will be the value equal to what it had during the cessation of shale deposition. We note that gravity will continue to compact the shale, irrespective of whether any new deposition of sand occurs or not. The shale column, L, will decrease in thickness as water continues to be expelled. The compaction equation (10.37) will remain the same. The boundary conditions at the base of the shale column will also remain the same (no flow boundary), namely,

$$Q = \frac{\kappa}{\mu(1+\varepsilon)}\left[g(\rho_r - \rho_w) + \frac{\partial\sigma}{\partial Z'}\right] = 0; \quad t > 0, Z' = 0 \tag{10.46}$$

However, the other boundary condition will be modified. Since we have assumed that the permeability of sand is much higher than that of the shale, the sands loaded on the shale can be assumed to be *normally pressured* at all times after sand sedimentation starts ($\tau > 0$). This means that there is no excess fluid pressure at the sand-shale interface. The load at that interface varies with time with the rate given by the sand deposition rate. If γ is the rate of sedimentation of "dry" sand, then the effective stress at the sand-shale interface at time t will be,

$$\sigma(Z', \tau) = g(\rho_r - \rho_w)\gamma\tau; \quad \tau > 0, Z' = H(\tau) \tag{10.47}$$

Here we note that $\tau = 0$ signifies the moment the shale deposition ceases, at which time the volume of the "dry" shale per unit area is H (τ).

The initial condition for the shale-on-shale compaction given earlier will be modified by

$$\phi(Z',0) = \phi_g(Z',0); \quad \tau = 0, 0 \leq Z' \leq L_s, \tag{10.48}$$

with the assumption that we start counting time with the deposition of sand. In this case $\phi_g(Z',0)$ will be the porosity profile of the shale column at the *end of the shale deposition*. This completes the mathematical description of sand-on-shale deposition and compaction. The numerical algorithm to solve this will be identical to that of the shale – on-shale compaction as long as the "counting of time" during sand deposition is adhered to carefully.

Constitutive Relations for Sediments

To solve the compaction equations – both shale-on-shale and sand-on-shale – under gravitational loading of sediments, we need to specify several constitutive relations involving the rocks and the fluid. These are

1. fluid (water) density as a function of pressure, temperature, and salinity;
2. fluid (water) viscosity as a function of temperature;
3. shale and sand permeability as a function of porosity or void ratio;
4. shale and sand void ratio as a function of effective stress that includes diagenesis via the β function, as discussed in Chapter 3.

Water density as a function of pressure, temperature and salinity can be taken from the regression given in Appendix A (equation (A.1)). Viscosity of water as a function of temperature is known, for example, from the laboratory data of Korosi and Fabuss (1968). The data can be represented by a simple relation as given below for viscosity of water μ (T), at a temperature T with respect to the standard viscosity of water at 20°C (1.002 centipoise or cp):

$$\ln \mu_{20} - \ln \mu(T) = \frac{AT' + BT'^2}{T + C} \tag{10.49}$$

This relation is valid in the temperature range $20\,°C \leq T \leq 150\,°C$. Here A, B, and C are constants, T is temperature in °C, and $T' = T - 20$. Recommended values for the constants are:

A = 1.37023
B = 0.000836
C = 109.

Permeability of mudrock or shale depends greatly on the shale mineralogical composition and it is probably the most important quantity that controls shale compaction and pore pressure buildup. It is also fraught with uncertainty, as permeability data from actual measurements on shales are very few. There are limited data on laboratory measurements of high-porosity clays from shallow depths of the Gulf of Mexico. These data can be used as a model constraint at higher permeability than what is expected of shales from deeper depths. A literature search on this subject showed that most of the algorithms on basin history simulation *infer* permeability estimates for shales based on measurements of pore pressure in sands (as direct pore pressure measurements in shales are not possible), and assuming that the pore pressure in shale can be *estimated* from that in sands from geometrical and geological considerations. In most cases, it is not that simple and the results and ensuing models are neither reliable nor transferable from one mini-basin to another or in the different part of the same mini-basin. This is indeed a complex subject and worth further research. Dutta (1987b) provided a model for shale permeability that may be used for modeling purposes. This is shown in Figure 10.2 and is modified after Dutta (1987b, Figure 7). The permeability of shale (in the vertical direction; shale permeability is known to be highly anisotropic) is represented by the following empirical equation:

$$\kappa = m\varepsilon^n \tag{10.50}$$

where κ is the shale permeability in millidarcys (mD), ε is void ratio, and m and n are constants with values 0.0034 and 5.0, respectively. In 3D basin modeling, we require models for shale-permeability anisotropy as permeability in horizontal directions could be significantly different from that in the vertical direction. Various commercially available software have built-in assumptions for this (permeability anisotropy); users and practitioners of basin modeling discipline must be aware of these assumptions and their consequences. The permeability relations such as the one proposed by Dutta (1987b) and in equation (10.50) above suggest clearly that it is dependent upon stress (through void ratio which depends on the porosity) – any reduction in porosity through increasing vertical effective stress during sediment burial and compression is exaggerated into permeability reduction by orders of magnitude. Thus decrease of permeability with burial is directly related to pore pressure profile. The 1D model that we presented above can be used to understand the sensitivity of the predicted results on the porosity (or void ratio) and stress dependency of mudrock (shale) permeability.

A model for mudrock (shale) compaction (with chemical diagenesis) is needed to carry out the solution of the shale compaction under gravitational loading. We can use any of the mechanical compaction models as discussed in Chapter 3 had this been the only source of compaction. This was done by several authors (Bredehoeft and Hanshaw, 1968; Smith, 1970) both in the context of shale compaction as well as gypsum dehydration (Hanshaw and Bredehoeft, 1968). However, in the current model we must include chemical diagenesis. This was discussed in detail in Chapter 3. The compaction model below includes *both* mechanical compaction and chemical diagenesis (Dutta, 1987b; Dutta et al., 2014) that relates effective stress to void ratio and temperature (dependent on geologic time):

$$\sigma = \sigma_0 e^{-\beta(T)\varepsilon} \tag{10.51}$$

In equation (10.51), β is a function of temperature T and (geologic) time t as explained in Chapter 3 where we showed the explicit time dependency of temperature under steady state heat flow condition (an assumption). Thus β depends on *both* T and t. Its

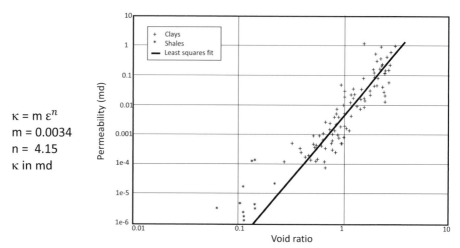

$\kappa = m\,\varepsilon^{n}$
$m = 0.0034$
$n = 4.15$
κ in md

Figure 10.2 Permeability versus void ratio of Gulf Coast clays and shales. Modified after Dutta (1987b).

explicit form under steady state assumption was given in equation (3.30). Alkawai (2018) tested this assumption of steady state heat flow and compared the results with a time-dependent solution for an area in deepwater Gulf of Mexico (Thunder Horse Field of BP). The steady state assumption was found to be inadequate for this area. Time-dependent heat flow conditions can be accounted for when using any commercial basin modeling software.

Numerical Results on Pore Pressure Prediction with Combined Effect of Mechanical Compaction and Chemical Diagenesis

In this section we provide results of numerical modeling of compaction of shale as discussed above. As a reminder, results include *both* mechanical compaction and chemical diagenesis. These computations are aimed toward understanding the variations in the magnitude and profiles of bulk density, fluid pressure gradient, and effective stress as a function of sediment deposition sequence and thickness, sedimentation rate, geothermal gradient, and permeability of shales. More details are given by Dutta (1986, 1987b). A typical example of sand-on-shale compaction is shown in (Dutta, 1987b; Figure 10). Here we start with approximately 5000 ft of normally pressured shale (saturated with brine) of porosity 57 percent being deposited on an impermeable basement at a rate of 1 ft of solid (zero porosity shale) shale per thousand years. The geothermal gradient is assumed to be 1.3°F per 100 ft. The sand deposition rate is also assumed to be the same as that of the shale. The resulting time histories of bulk density in g/cm^{3}, fluid pressure gradient (FPG) in psi/ft and (vertical) effective stress in psi are shown in Figure 10.3. These figures show the development of a *transition zone* in the geopressured shale below the hydrostatically pressured sand. This transition zone is characterized by a reduction in bulk density with depth, a rapid change in FPG, a sharp drop in effective stress, and development of a permeability

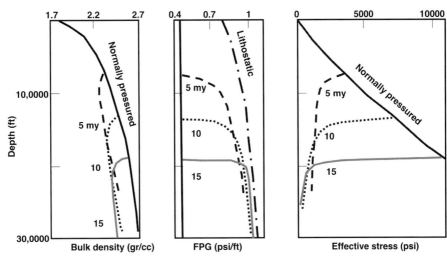

Figure 10.3 Results of 1D-basin modeling results as discussed in the text for sand-on-shale burial. Shown are the (left) bulk density, (middle) fluid pressure gradient, FPG in psi/ft, and (right) effective stress variations with depth and geologic time (sand deposition for 5 million, 10 million, and 15 million years after cessation of shale deposition).

barrier – all due to rapid burial of shales and lithology change (shale to sand) at the transition zone. The width of the transition zone is dependent upon factors such as the permeability of the sediments, geothermal gradient, and the past pressure history of the buried shale. It is possible to obtain more than one transition zone in a clastic sequence. It can be simulated by modeling variable sedimentation rates and permeabilities.

Miller et al. (2002) obtained similar results as discussed above. Their results on vertical and horizontal effective stresses are shown in Figures 10.4a and 10.4b, respectively, for pure mechanical compaction. The sediments are predominantly shale at the lower half and predominantly sand at the upper half. Both sediments were buried at a constant burial rate of 1.5 km/Ma. Figure 10.4a is a plot of the current and maximum total stresses (overburden), maximum horizontal stress, and pore pressure. Figure 10.4b shows the effective vertical and horizontal stresses. The authors noted that stress conditions shown in these figures are commonly *mistaken* for unloading. Rather these plots show that the current and maximum vertical stresses are the same, a result that indicates unloading has not occurred despite the effective stresses decreasing with depth. An example of the unloading feature is presented in Figure 10.5 (after Miller et al., 2002) and it includes *fluid expansion or chemical diagenesis effect* (termed as *stress unloading* in Miller et al., 2002). In cases such as this, the horizontal total stresses exceed the overburden, and pore pressures are very high.

Figures 10.6a and 10.6b show the effect of variation in sedimentation rate on vertical effective stress, σ, and FPG (psi/ft). The sediment–water interface temperature was taken to be 68°F and a constant geothermal gradient of 1.40°F/100 ft was used. The unloading feature is clearly visible in Figure 10.6a for vertical effective stress σ

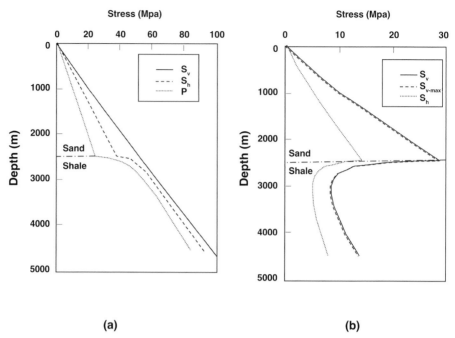

(a)   (b)

Figure 10.4 Predicted stress profiles for sand-on-shale compaction (mechanical) as predicted by Miller et al. (2002). (a) Total vertical stress (solid line), horizontal stress (long dashed line), and pore pressure (short dashed line) versus depth. (b) Variation of the effective stress with depth.

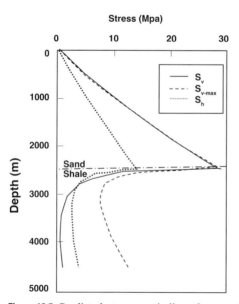

Figure 10.5 Predicted stresses as indicated versus depth for the sand-on-shale compaction that includes mechanical compaction as well as a fluid source at depth to simulate the effect of chemical diagenesis termed as unloading effect by the author (Miller et al., 2002).

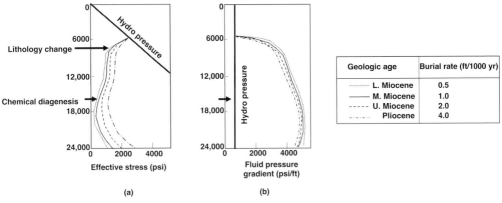

Figure 10.6 Predicted (a) effective stress and (b) fluid pressure gradient, FPG in psi/ft, for several burial rates typical of the Tertiary Clastic of the Gulf of Mexico. The model includes both mechanical compaction and chemical diagenesis. Note the second inflection point that indicates further lowering of effective stress and increased pore pressure due to chemical diagenesis.

and Figure 10.6b for fluid pressure gradient. In these figures the first inflection point (at about 9000 ft) is due to the onset of S/I diagenesis, and the second inflection point (at about 15,000 ft) is due to the termination of shale diagenesis. The onset of shale diagenesis is consistent with release of "bound" water as modeled by S/I kinetic transformation and reordering of smectite and illite layers (R2 order). As a result of this and stratification of the clay platelets, the effective stress drops further by as much as 1000 psi and fluid pressure gradient increases substantially.

Both Dutta (1987b) and Miller et al. (2002) demonstrated that predicted effective stresses as low as several hundred psi can easily occur in sand-shale sequences such as in the Tertiary Clastics Province of the Gulf of Mexico, USA, for *reasonable rates* of sedimentation and thermal gradients that are commonly encountered in hydrocarbon exploration. These low effective stresses, of the order of a few hundred psi, raise the distinct possibility of fracturing of *caprock* due to any lateral stress imbalance (horizontal stress) and thus could provide paths for secondary migration of hydrocarbons.

The geothermal gradient plays an important role in the pore pressure and effective stress evolution of clastics. In Figures 10.7a–10.7c we show the effect of varying geothermal gradient, from 1.3°F per 100 ft to 1.9°F per 100 ft keeping other parameters the same. It is clear that at a given depth, vertical effective stress (and hence the horizontal effective stress) can drop by a large amount causing a large increase in FPG. Pressures as high as the ones simulated by these model calculations will definitely compromise the integrity of *shale caprock* at greater depths.

Shale permeability is a critical parameter for geopressure development. While the industry has a very good understanding of the range of permeabilities of reservoir rocks (sandstone), very few definitive values are available for shales. Laboratory measurements are very difficult because the low-permeability of shales do not allow fluid flow through core plugs of reasonable dimensions. (We discussed this in great detail in Section 4.5.3.) For this reason, shale permeabilities cannot be measured in the

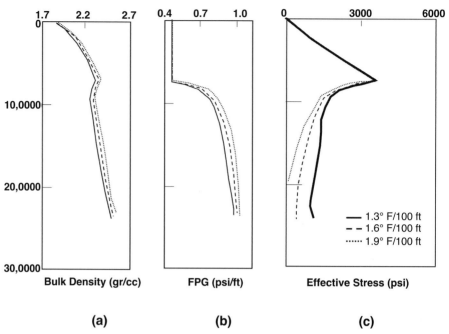

Figure 10.7 Effect of geothermal gradient on the predicted compaction profiles for sand-on-shale compaction that includes both mechanical compaction and burial diagenesis. Shown are (a) bulk density, (b) fluid pressure gradient in psi/ft, and (c) vertical effective stress versus depth for indicated geothermal gradients.

borehole by conventional pore pressure measurement techniques such as the ones used for pressure measurements on reservoir rocks. Yet, this is probably the most important parameter for geopressure buildup in shales.

An example of the porosity reduction in this model is shown in Figure 10.8a for shale-on-shale column that was deposited at a rate of 1000 ft per million years. Figure 10.8b shows the smectite reduction profile. The porosity profile of Figure 10.8a dictates the velocity versus depth profile shown in Figure 10.8c. It is the basis of the rock physics template discussed in Chapter 6. The velocity profile in (c) was derived using the Issler model relating slowness to porosity. Petmecky et al. (2009) followed a similar approach but used 3D basin modeling to derive a velocity model.

10.3.2   An Example of Pressure Prediction with Diagenesis Using Basin Modeling in 2D (E-Dragon Data Set)

The example models an E-W 2D cross section of a basin in the northern Gulf of Mexico. The basin lies in the Ship Shoal and South Timbalier areas as shown in Figure 10.9a. The corresponding poststack seismic image is shown in Figure 10.9b. With the aid of biostratigraphic data, the seismic section was interpreted for top of Pliocene, top of Upper Miocene (Messenian), top of Upper Miocene (Tortonian), salt top, and salt base.

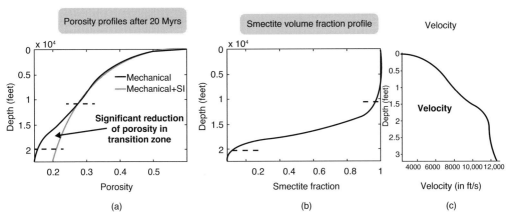

Figure 10.8 (a) Porosity reduction in the 1D basin model of Figure 10.1 when the chemical diagenesis is turned off and on. (b) The smectite metamorphism. (c) The corresponding velocity profile.

As can be seen, the section contains several salt bodies. The presence of salt bodies necessitates palinspastic restoration of the section to rigorously account for the effect of salt tectonics on pore pressure evolution. The restoration was accomplished by sequentially and kinematically restoring the compaction, Airy isostasy, movement on fault, and deformation due to salt movement for each layer. The deformation caused by salt movement was estimated by employing the concept of a regional datum (Rowan, 1993). Unfolding the layers above salt to a regional datum creates extra space in the unfolded section, which is assumed to be indicative of the deformation caused by salt movement. For restoring each layer, the water depth was used in an area of the section that seemed to be least affected by salt movement as a consistent estimate for the regional datum. Figure 10.10 depicts the restored sections at present day, 2.6 Ma, 5.3 Ma, and 10.7 Ma, respectively. The salt systems in the current geographical area have been commonly characterized as Roho salt system (Schuster, 1995). Roho systems are generally characterized by counterregional (dipping northward) salt feeders. Subsequent burial of sediments cause evacuation of salt and development of listric faults. The fault that we observe in the eastern part of the section is part of a larger listric fault with a counterregional feeder. The restoration at 10.7 Ma suggests that there was significant salt movement into the section, engendered by the counterregional feeder. Subsequent burial caused lateral evacuation of the salt out of the section, generating a listric growth fault, which soled into the resulting salt stock. Of particular interest is the central portion of the section, where rapid sedimentation in the Miocene caused rapid evacuation of salt, leading to the development of a salt weld.

The sequentially restored sections were used to model the temperature and pressure evolution of the section using a commercial basin modeling software. The number of layers used was kept the same as the present-day section used in sequential restoration. Lithologies were defined and assigned based on the gross shale content for each layer obtained from gamma-ray logs. The paleo-water depth was derived from the depth of sediment surfaces on each restored section. Various parameters of the assigned

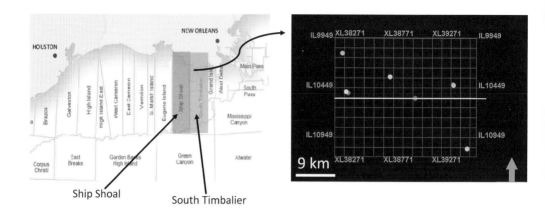

(a)

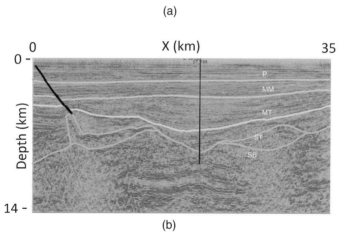

(b)

Figure 10.9 (a) The geographic location of the study area, E-Dragon II (left) and the base map showing available wells and the chosen 2D cross section (right). (b) The PSDM seismic image for the 2D cross section chosen for the study. Horizon interpretations for top of Pliocene (P), top of Miocene–Messianian (MM), top of Miocene–Tortonian (MT), and salt top and salt base (ST, SB) are overlain.

lithologies were calibrated using porosity, mud weight (upper bound of pore pressure) and bottom-hole temperature at two wells on the section. Figure 10.11 shows computed porosity, pore pressure, and illite fractions. The pore pressure plots at these wells show increase in pore pressure in the Miocene as expected, also reflected in the overlain mud weight data. It is important to note that mud weight data provides an upper bound on the actual pore pressure of the subsurface. The temperature calibration was achieved by tuning the parameters related to heat flow and matching the measured bottom-hole temperature in several wells.

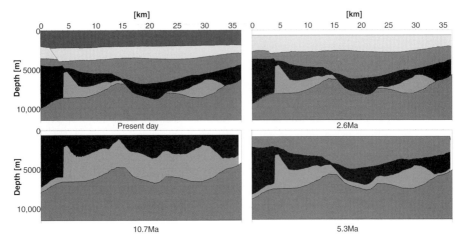

Figure 10.10 Palinspastic restoration of the section. The ages of the sections are present day, 2.6 Ma, 5.3 Ma, and 10.7 Ma clockwise from top left. (A black and white version of this figure will appear in some formats. For the color version, please refer to the plate section.)

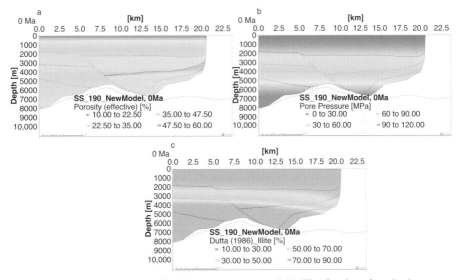

Figure 10.11 (a) Computed porosity, (b) pore pressure, and (c) illite fractions from basin modeling in E-Dragon area (seismic data of Figure 10.9b is used to construct present-day horizons as inputs for the basin model). (A black and white version of this figure will appear in some formats. For the color version, please refer to the plate section.)

10.4   Basin Modeling in 3D

The previous examples of basin modeling in 1D and 2D show the power of basin models to predict pore pressure. With the advent of high speed computers, this technique has evolved dramatically to the extent that applications are being conducted in large scales – more than 1400 sq mi (155 OCS Blocks; 3600 km^2) in 3D and in real-time drilling application (Mosca et al., 2017). While seismic models for pore pressure prediction as discussed in earlier chapters predict pore pressure – given *present day* geology, with considerable accuracy and success (as compared with drilling results), basin modeling provides an alternate technique that describes the pore pressure distribution using transient models that depend on fluid and heat flow during geologic times along with other phenomena such as changes in facies and geologic structures in paleo-times. So this approach tells us not only what the pore pressure is today but allows us to investigate the effect of uncertainties in geology (various parameter uncertainties) in the past and its imprint on the present day pressure distribution. An illustrative example can be found in Mosca et al. (2017). These authors noted quite correctly that the basin model building requires a lot of input parameters that need to be correctly defined to get good pore pressure prediction results. Essential steps are: structural maps with depth converted horizons and fault maps and surfaces; facies maps with details of lithology and environment of deposition; mathematical description of various mechanisms of pore pressure and various boundary conditions related to fluid and heat flow to simulate the physics of the model. Naturally, the list of parameters is very large and most, if not all, have to be guessed, especially, those dealing with the past. Thus calibration is a very important step in the process. This is done with well logs from boreholes and pressure measurements.

There are many commercial packages available for basin modeling. All use well-known physical principles and constrained models. The depth converted horizons and fault surfaces are typically obtained from 3D seismic data and interpretation. Most utilize depth-migrated 3D seismic data along with wireline logs and checkshot data for this purpose. This enables the calibration of the ensuing velocity model. Wireline logs such as gamma-ray and density logs are used to calibrate the lithology model and build confidence on the depositional model. Bottom-hole temperature data are used to calibrate the heat flow model. Typically, pore pressure is not the main focus of a 3D basin modeling package – maturation of kerogen, expulsion of hydrocarbon, and fluid migration are the main drivers. Older software packages did not have modules dealing with various processes that deal with shale metamorphism and sophisticated models of chemical kinetics dealing with maturation. Most modern basin modeling packages have no such limitation.

Our personal experience with 3D modeling software showed that we should be wary of some of the default models and parameters. Some authors (Mosca et al., 2017) account for additional effects such as the smectite-illite transition, cementation models, aquathermal pressuring (Luo and Vasseur, 1992), and the effects of low- and high-permeability faults. Quite often the lithologies with default parameters, fail to reproduce the overpressures demanded by the measured calibration data (RFT). The calibration of porosity-effective stress, porosity-permeability and properties of fault (sealing versus non-sealing) are essential steps in pressure prediction. Generally, the main process for the formation of pore pressure is the transfer of overburden load to low-permeability rocks (background pressure) and the equilibration of the overpressure in the high-permeability rocks. This requires realistic structure maps and correct juxtaposition of various fault types and lithologies. For geomechanical applications, one should use appropriate 3D stress models as discussed in Chapter 12. In this context Biot formulation for various material parameters are highly recommended. It is clear that the entire process of model building and calibration is time consuming and many simulations are needed to match observed measurements in boreholes. An example of pressure prediction in 3D can be found in Satti et al. (2015).

Mosca et al. (2017) showed the value of 3D basin modeling for predicting pore pressure in basins containing salt. Because of the importance of salt, a 3D structural reconstruction of salt geometries through time is needed to be performed. Salt thickness maps at different time steps for both autochthonous and allochthonous salt needed to be imported from the software package. The authors showed that 3D salt restoration in the last 1–2 million years was a key component for the present day pressure and stress distribution in their study area (Mississippi Canyon area of the Gulf of Mexico, USA), and noted that the proper salt deformation can only be achieved by detailed mapping of the syndepositional sediments, using correctly imaged prestack depth-migrated 3D seismic volumes. In this basin, they also used basin modeling for pore pressure prediction in *almost* real-time (overnight) which was hitherto considered to be impossible. (We discussed this in Chapter 9.) When this approach is combined with other techniques that we discussed in Chapter 9, especially the ones that use seismic data (surface and borehole), we do have a recipe for reliably monitoring and predicting geopressure ahead of the bit.

11 Geohazard Prediction and Detection

11.1 Introduction: What Is Geohazard?

According to Komac and Zorn (2013) geohazard is a relatively new scientific term related to *natural hazard* studies. It indicates geomorphological, geological, or environmental processes, phenomena, and conditions that are potentially dangerous or pose a level of threat to human life, health, and property, or to the environment. It includes subaerial and submarine processes, such as *earthquakes*, volcanic eruptions, floods, erosion, debris flows, rock falls, and other types of *landslides* and *tsunamis*. Human-induced processes such as buried submarines, explosives, and shipwrecks may also be considered as geohazards. In the petroleum industry geohazard refers to *all* hazards that are either at the surface (e.g., at the sea bed – either man-made or naturally occurring) or in the subsurface at shallow depths (typically, less than 2 km for deep-water sediments) that can adversely affect the environment when disturbed by drilling activities and impact subsequent drilling performance. Sometimes the drilling community uses the term *geohazard* loosely to denote any hazard that can be encountered while drilling a well – this includes both shallow hazard and deep hazard such as those associated with high overpressured formations. The International Association of Drilling Contractors defines shallow hazards as *adverse drilling subsurface conditions that may be encountered prior to the setting of the first pressure containment string and the emplacement of the blowup preventer (BOP) upon the well* (see http://drillingmat ters.iadc.org/glossary/shallow-hazards/). (Blowout Preventer or BOP is a large valve at the top of a well that may be closed if the drilling crew loses control.) Prior to this there is no way to shut-in the well. "Shallow" is commonly understood by the industry to be the interval above the setting depth of the first pressure containment string.

We shall refer to all *shallow* hazards as geohazards. If these hazards are not mitigated properly, it can result in a disaster causing loss of life and property and damage to the environment. This is why a site survey is required before spudding any well. This is especially important in deepwater, where geohazards are known to cause serious problems such as the blowouts and loss of the production template at the Mars/ Ursa site (Ostermeier et al., 2002). The guidelines for operations associated with geohazards are stipulated by the BOEM (Bureau of Ocean Energy Management, formerly the Minerals Management Service [MMS]) of the USA Department of the Interior. While these guidelines and enforcement policies in the USA are implemented after careful studies and are adhered to by the drilling community, such may not be the case outside of the Federal Waters of the USA. The policies in many countries are either lacking or incomplete and rules of enforcement are often inconsistent. In fact during our research, we could not find any *documented* case of shallow hazards *occurrence* (such as well control issues, shallow waterflow associated with drilling disturbances) outside of the USA and Europe; whatever we found were mostly of anecdotal nature.

Geohazards have been classified into three categories by the US Department of Interior (www.boem.gov/):

1. seafloor geologic hazards: fault scarps, gas vents, hydrate mounds, unstable slopes, slumping, active mud gullies, crown cracks, collapsed depressions, furrows, sinkholes, mass sediment movement, surface channels, pinnacles, and reefs.
2. subsurface geologic hazards: faults, gas-charged sediments, abnormal pressure zones, gas hydrates, shallow-water flow sands, and buried channels.
3. man-made hazards: pipelines, wellheads, shipwrecks, ordnance, communication cables, and debris from oil and gas operations including the effects of hurricanes.

In this chapter we shall address issues associated with subsurface geologic hazards. High aquifer pore pressure and frequent occurrence of gas that contributes further to higher pore pressure due to buoyancy pressure are major contributors to subsurface geologic hazard.

11.2   Shallow-Waterflow-Sands (SWF)

Of all subsurface geologic hazards, shallow-water-flow (SWF) sands pose the most significant drilling challenge. Shallow-water flows are flows from *overpressured sands* encountered at *shallow depth* below the mud line in *deepwater* regions of the world. We note that shallow-water-flow sands are encountered in deepwater areas *only* and these are caused by overpressured sands (Furlow, 1998). Thus the word "shallow" refers to *shallow part of the stratigraphy* as these sands are known to occur in water depths of 450 m or more and typically, between 300 m and 600 m *below* the mud line. A SWF (hazardous) sand forms when sand-rich deposits from turbidity flows, landslides, or a drop in sea level are sealed with a condensed zone (clay/mudstone and/or gas hydrates). When the seal is breached – either by drilling activity or by natural causes such as landslide or earthquakes, sand flows with the water. Flow rates as high as 25,000 bbls/day (~730 gal/min) have been reported. A video presentation at the "Shallow-Water Flow" Forum (organized by MMS at Woodlands, Texas, in June 1998) showed a SWF producing plumes of sand and debris that boiled up 60 ft from the seafloor. As far as we know, SWF sands have been encountered not only in in the Gulf of Mexico, but also in the West of the Shetlands, the Norwegian Sea, Southern Caspian Sea, and the North Sea. Although SWF sands probably occur in other geographic areas, we cannot document their occurrences.

Figure 11.1 depicts the environment where SWF sands occur. The sands are usually unconsolidated or poorly consolidated. The seal is either clay/mudstone or gas hydrate. The pressure builds in the trapped water–sand suspension due to the weight of sea water and continued sediment deposition, resulting in the buildup of high pore fluid pressure (overpressured aquifer). Once breached, the flow rate can be very high because of thick, high-permeability nature of these sands, low water viscosity, and sufficient pressure differential. Once the flow begins it is very difficult to stop. This makes it difficult, and

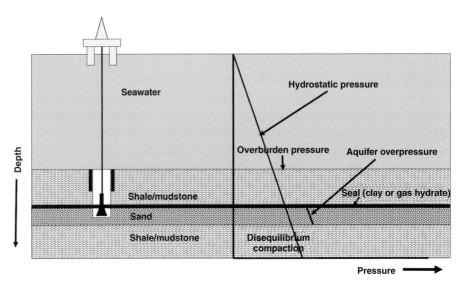

Figure 11.1 A schematic showing the environment where SWF occurs. It is known to occur only in deepwater (water depth greater than 450 m) and in the shallow part of the stratigraphy below the mudline (typically 300 m below mudline) in areas associated with a high rate of sedimentation. This causes overpressure when additional sands and clays are deposited that encase the sand, causing its pore pressure to increase. Tilt of the sand has a significant effect on the pressure gradient due to "centroid effect."

sometimes impossible, to get a good *cement job* around the casing. The presence of natural pressure release as well as ocean-bottom features (e.g., sea floor cracks, craters, fissures, mounds, and mud volcanoes) can indicate the presence of such pressurized zones that can be identified on high-resolution seismic images.

Drilling programs should either avoid potential shallow hazards or have contingency plans to control such events. However, well control with variable mud weight can be challenging, especially in deepwater due to a narrow drilling window, as the gap between the pore pressure and the fracture pressure (drilling mud window) is small in shallow sediments. When drilling into these poorly consolidated overpressured sands one can face severe drilling problems that include sediment pile-up at the well head, cratering, possibility of blowout, and loss of slot or entire template. Failure to control the flow of pressurized sand slurry can cause serious well-control issues and lead to expensive non-productive time. Since sediment–water flow and subsequent pile-up of sediments at the well head increase with time, it requires quick action by the drilling team. For drilling in water depths of less than 650 m, the industry practice is to run an extra casing and BOP for preventing flow. For depths greater that 650 m, the drilling is, typically, *riserless* and since pore pressure is already very close to fracture pressure before spudding the well, the fracture gradient may be too low for mud in riser. This will require "dynamic kill" to control the well – a very expensive process to *kill* a blowout well by pumping viscous fluid into the wellbore at high rates to generate enough friction to suppress the flow.

Billions of dollars have been lost to date for remediation and prevention of shallow-water flow problems in the Gulf of Mexico alone where the presence of SWF layers is common. This is because the Gulf of Mexico is a salt withdrawal basin and high sedimentation rate is of common occurrence. Lately, this problem has also been a concern to drillers in many other deepwater clastic basins of the world (anecdotal and no reference is available), such as Mediterranean, Angola, Nigeria, India, and Brazil. Therefore, identification of SWF sands *prior* to drilling is very important in reducing drilling risks. Unfortunately, the industry's track record is not very good in this regard; it spends far more for *mitigation* of SWF hazards and less on *identification* and *avoidance*. We shall discuss the important role of surface seismic data for SWF identification prior to drilling (see below).

In the case of the Shell's Ursa (MC 810) disaster, the original tension leg platform site had to be abandoned as 10 of the 21 drill slots became unusable due to buckling of the casing strings believed to have resulted from the evacuation/wash-out of the sediment supporting the casing (Eaton, 1999). There were reports of drill strings being blown up the hole with uncontrolled pressures in excess of 10.65 ppg. The massive flow as observed by a remotely operated vehicle resulted in a huge buildup of sediment on the ocean floor leading to burial of the template and several subsea well heads. Ostermeier et al. (2002) reported that resistivity measurements indicated that four separate zones in the MC 810 wells were washed out to a diameter of more than 8 feet resulting in the buckling of the casing. The site is characterized by high over-pressures close to the seafloor, resulting in very narrow drilling margins (pore pressure gradient close to within 70 percent of the fracture gradient). These high near-surface pressures are attributed primarily to geologically recent (last 100,000 years) rapid sedimentation on the continental slope by the Late Pleistocene Mississippi River.

11.2.1   Identification of SWF Sands

The objective of shallow hazard analysis is to *identify, map, and quantify* the risks posed by shallow hazards that will impact exploration and development drilling operations. This is usually done through producing a "traffic light" map over the prospect to be drilled; the map is used mainly for avoidance purpose. Guidelines for shallow hazard analysis are mandated by the BOEM for US-based drilling operations. A listing of the BOEM Notices to the Leaseholders (NTLs) may be accessed through www.mms.gov/, or www.gomr.m ms.gov/ for the GOM, or for a summary listing of NTLs the OCS BBS at www.ocsbbs. com/ntls.asp. NTLs pertinent to shallow hazard analysis include the following:

1. NTL No.2008-G05 – Shallow Hazards Program/Requirements
2. NTL No. 2005-G07 – Archaeological Resource Surveys and Reports and 2008-G20 – Revisions to the List of OCS Lease Blocks Requiring Archaeological Resource Surveys and Reports
3. NTL No. 2000-G20 – Deepwater Chemosynthetic Communities
4. NTL No. 2008-G04 – Information Requirements for Exploration Plans and Development Operations Coordination Documents

The reprocessing of conventional recent-vintage 3D seismic is normally sufficient to meet the requirements specified by the BOEM. For further discussion we refer to the literature (Mallick and Dutta, 2002). Below we address data processing practices for conventional seismic data for geohazards use. Magnetometer surveys and side scan sonar are *not* a requirement in water depths greater than 200 meters *unless in areas of high archeological potential* (see NTL No. 2005-G07). Archeological surveys using these survey instruments is required for exploration and development activity over OCS Blocks identified by the MMS to have shipwrecks and other artifacts of archeological significance (NTL No. 2008-G20).

The identification and mapping of chemosynthetic communities is addressed in NTL No. 2000-G20. Again high-fidelity 3D seismic is also permitted to map sea floor chemosynthetic communities and shallow geologic features in the vicinity of wells, platforms, anchors, and anchor chains which include hydrocarbon charged sediments associated with surface faulting, acoustic void zones, mounds, or knolls and gas or oil seeps. We have found that reprocessed conventional 3D seismic data can also be used for gas hydrate identification and risk quantification (Dutta and Dai, 2009). These occur in similar geologic areas where SWF sands tend to occur and the associated hazard is mainly due to release of gas when hydrates are disturbed either due to human intervention or due to nature.

11.2.2   Geological Environment of the Occurrence of SWF Sands

Shallow-water flow deposits in the USA are mostly found within the primary paleo-river depocenters of the Late Pleistocene low-stands (Ostermeier et al., 2002). Briefly, there are several geological environments for SWF sands. These are

1. channel sands: deposits of clean channel sands laid down in a deltaic environment shift over time so that new sand deposits create a superimposed structure;
2. turbidite sands: sands deposited in the Gulf of Mexico on the continental slope, which runs at a steep angle between the continental shelf and the deepwater basin; in strong turbidity currents, these sands slide down the slope at high speeds, to form turbidity sands on the sea floor, and this results in rapidly formed, coarsely graded turbidity deposits;
3. rotated slump block: in this condition, the face of the continental shelf slides down like an undersea avalanche of sand, to deposit rapidly onto the deep basin.

These three forms of rapid deposition lay the foundation, literally, for SWFs. The next ingredient is a sealing layer of shale (or in proper temperature and pressure conditions – gas hydrates) which deposits above the sands. This low-permeability material encloses the sands, as well as the water and gas pressure they contain. On top of the shale are further rapid deposits of sand and shales which build to increase the pressure within the sands. These sand deposits grow in thickness at the rate of 500–7000 ft per million years, creating shallow pressure zones near the surface (Furlow, 1998). The pore pressure buildup in the sand is a result of mechanical compaction disequilibrium as shown in Figure 11.2 (upper left). In the upper right of Figure 11.2, we show the

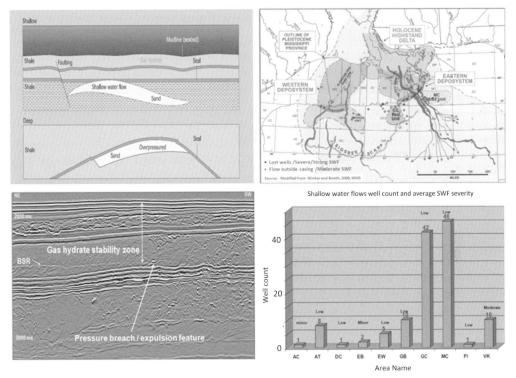

Figure 11.2 (upper left) Formation of shallow-water-flow sands. In the deepwater environment, these pressurized sands form when porous aquifer sands are trapped by a seal (either clay or gas hydrates or both) preventing dewatering and compaction. Pressure disequilibrium builds over time as the burial depth of the formation increases. (upper right) Map of Late Pleistocene depocenters in the northern Gulf of Mexico (MMS; adapted from Winker and Booth, 2000). Heavy dots indicate lost wells due to SWF, and lighter dots indicate flow outside casing. (lower left) Seismic evidence of natural pressure release features (breach event) associated with gas hydrates and SWF from the Mississippi Canyon area, GOM (interpretation). (lower right) A bar chart showing the occurrences of SWF incidents while drilling through the shallow sediments (courtesy BOEM).

occurrences of SWF incidents in a map of Late Pleistocene depocenters in the northern Gulf of Mexico. With sufficient burial, the pressurized water–sand slurry will eventually breach the seal resulting in natural expulsion and flow. Expressions of these natural pressure release features such as subsurface breach events, and ocean-bottom features such as seafloor cracks, craters, fissures mounds and mud volcanoes are evident in high-fidelity deepwater seismic data (Figure 11.2, lower left). Bar charts showing the distribution of these events since 1985 through 2005 are displayed in the lower right of Figure 11.2. (We contacted the US Department of Interior in June 2015 regarding an update on recorded incidents of SWF sands since 2005. We were informed that the United States Government was in the process of updating their data base for release to the public.)

11.2.3   Rock Properties of SWF Sediments

In situ measurements of elastic and other rock properties of SWF sediments have so far been very limited, because the SWF layers are associated with very low velocities. Measurement of such low velocities is difficult in a cased-hole environment (tool limitations), and open-hole logging under SWF conditions is hazardous. We therefore rely on the scarce information that have been published and also on the elastic property trends of regular sands and shales from laboratory measurements to determine the rock properties of SWF sands. In recent articles, Huffman and Castagna (2001) and Zimmer et al. (2002) reviewed the rock properties of SWF sediments. These sediments lie in the transition zone between suspended materials in fluid and rocks around critical porosity. These nearly unconsolidated sediments exhibit *low* bulk densities and *anomalously low* P- and S-wave velocities (V_P and V_S). With increasing fluid pressure, they lose cohesion, causing V_S to drop faster than V_P. This causes the P- to S-wave velocity ratio V_P/V_S to increase and the Poisson's ratio to increase. A V_P/V_S ratio of the order of 10 or higher, that is, a Poisson's ratio of 0.49 or higher is typical of these overpressured SWF sediments. Figure 11.3 summarizes some of the rock properties of SWF formations. Note that the high V_P/V_S and Poisson's ratio is a direct result of poor grain contact in these overpressured SWF sediments.

11.2.4   Geophysical Methods for SWF Detection

The most cost-effective SWF mitigation approach is to *avoid or minimize SWF risks* in the selection of the drill site. This can be achieved through multidisciplinary geotechnical analysis employing quantitative geophysical methods. In Figure 11.4, we present a summary of various geophysical techniques to identify and quantify risks associated with SWF sands. Strengths and weaknesses of each method are also listed (column two and three, respectively). Although the use of short cable

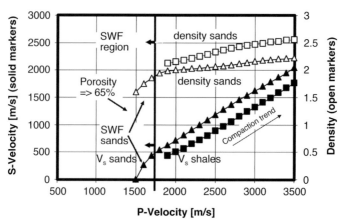

Figure 11.3 Elastic properties of sands and shales. Note that the SWF sands are characterized by low V_P, V_S, and density. These trends are based on limited log and laboratory data and interpretation.

Technique	Strengths	Weaknesses
Short cable site survey (Geohazard data)	• Highest temporal and spatial resolution • Very useful for stratigraphy	• Require extra acquisition • Short cable. Poor velocity resolution and no pre-stack inversion • No shear wave information
Direct shear wave acquisition	• Can get direct information on Poisson's ratio • SH data	• Require extra acquisition • Relatively expensive • Technically chalanging
Multi-component ocean bottom acquisition (P-source 4C seafloor sensors)	• Can get direct information on Poisson's ratio • High S/N • Large bandwidth, better azimuthal coverage	• Require extra acquisition • More expensive than towed streamer and dedicated site survey
Conventional 3D seismic processed at 2ms	• Data may be already available in multi-client libraries, no extra acquisition is needed • Large offset allows quantitative velocity analysis and pre-stack inversion • Can get Vp/Vs ratio • Cost effective	• Require some reprocessing • Signal to noise ratio may not be optimal for quantitative work • Survey may not be designed for optimal coverage at shallow depth • Lower bandwidth than dedicated site survey

Figure 11.4 Shallow-water flow identification – strengths and weaknesses of various geophysical methods used to identify SWF risk areas.

geohazard data is the most common, we note that such data cannot characterize the SWF sands in terms of reliable V_p/V_s ratio – and is used only for geologic interpretation and in some cases near offset poststack inversion. This is because these data are acquired with a short cable (~ 25 m to 100 m) and do not contain enough offset to utilize the properties of SWF sands fully to extract its inherent properties through quantitative methods such as detailed velocity analyses or pre-stack inversion. This is not the case when seismic inversion is used to extract SWF sand properties with conventional 3D seismic data that benefit from larger cable lengths and hence, more offset.

Since V_P/V_S ratios are key indicators of SWF sands, one can extract these indicators from seismic inversion on 3D data with large offsets. This has been demonstrated to add value to SWF risk assessment (Mallick and Dutta, 2002). The entire workflow is discussed in a US Patent (Dutta and Mallick, 2010). The quantitative seismic methods not only provide a means to identify SWF drilling risks but also allow us to quantify the *degree of risk* including effective stress and pore pressure estimation (see below where we present results from several case studies).

Integrated Approach for SWF Seismic Identification Using a Five-Step Workflow and an Example (Case Study 1)

The details of this workflow are addressed in Dutta and Mallick (2010). An overview of the steps in the workflow are shown schematically in Figure 11.5. The quantitative process uses stratigraphic analysis and physical properties of SWF sands to locate and minimize SWF risks in selecting the drill site. The process uses conventional 3D seismic data to identify, analyze, and quantify subsurface occurrence of these events. Using interpretation in conjunction with seismic attributes as stratigraphic indicators, V_P/V_S ratio, and pore pressure evaluation, the method provides a quantitative traffic light map for SWF risks quantification. The steps are elucidated below with a real example from an offshore, deepwater basin. The goal of this approach is to show the value of conventional 3D seismic data for SWF risk quantization that does not require any new (and perhaps, more costly) acquisition.

Step 1   *Seismic Modeling in 3D*

The first step in the workflow is *reprocessing* conventional 3D seismic data with a 2 ms (or even lower, if possible) sampling rate and 12.5 × 12.5 m (or lower) bin migration. This improves image resolution, especially in shallow parts of the stratigraphy, by preserving the high-frequency content of data, which is critical for interpreting shallow hazards. Though high-resolution processing preserves the frequency content of the data, an accurate *anisotropic* earth model is essential for good image and pore pressure prediction. We recommend using multiple iterations of rock physics guided common image point (CIP) tomography (see Chapter 6 for details) to estimate an interval velocity field that flattens seismic events in CIP gathers as well as leads to better target placements (in depth) and focusing of events. Rock physics guided velocity modeling incorporates the lithology-dependent relationship between porosity and slowness in conjunction with the thermal history model, porosity, and effective stress to generate a template of velocity curves corresponding to iso-pressure gradients calibrated to appropriate information from "offset wells" (nearby wells) (Deo et al., 2014). These templates are used to guide tomography inversion for velocity modeling. The results of

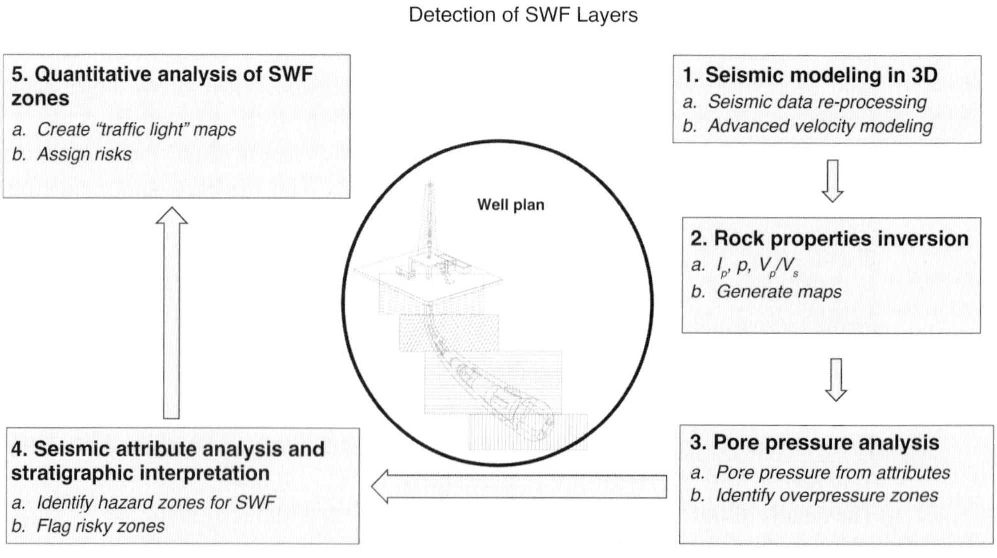

Detection of SWF Layers

5. Quantitative analysis of SWF zones
a. Create "traffic light" maps
b. Assign risks

Well plan

1. Seismic modeling in 3D
a. Seismic data re-processing
b. Advanced velocity modeling

2. Rock properties inversion
a. I_p, ρ, V_p/V_s
b. Generate maps

4. Seismic attribute analysis and stratigraphic interpretation
a. Identify hazard zones for SWF
b. Flag risky zones

3. Pore pressure analysis
a. Pore pressure from attributes
b. Identify overpressure zones

Figure 11.5 Five-step process for shallow-water flow-identification. After Dutta and Mallick (2010).

this high-resolution imaging and rock physics guided tomography show a significant improvement in both the resolution and the focusing of structures, especially the target faults as shown in Figures 11.6 and 11.7.

Step 2   *Rock Properties Inversion ($I_P, \rho, V_P/V_S$ Ratios)*
The second step in the workflow is simultaneous prestack seismic inversion to compute elastic properties, namely, acoustic impedance, V_p/V_s ratios, and density attributes. A well within the current seismic survey with a complete set of open-hole well logs with corresponding drilling and geological data was used for calibration during inversion. SWF sands are undercompacted and consequently, they have poor grain-to-grain contact resulting in low effective stress and high porosity. Effective stress is directly related to the seismic velocities; however, S-wave velocities are sensitive to small changes in effective stress near critical porosity; therefore, SWF sands have lower V_s. This leads to high V_p/V_s ratios for shallow-water flow features and is anomalous with respect to overlying and encasing sediment; typically the V_p/V_s range is 5–20 or more (Dutta and Mallick, 2010). Prestack inversion (AVO) is performed on the 3D seismic angle stacks data using the estimated wavelets and

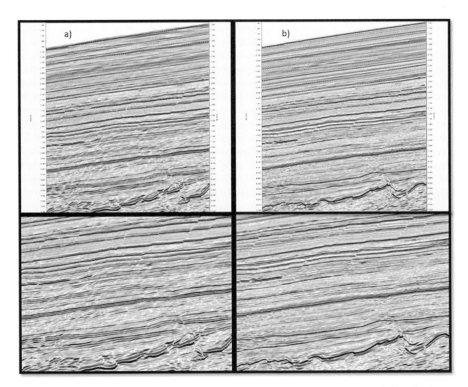

Figure 11.6 Images from an in-line cross section. (a) Legacy seismic image and (b) seismic image resulting from rock physics guided anisotropic velocity modeling and CIP tomography as discussed in Chapter 6. Note the image quality enhancement in (b). After Deo et al. (2016).

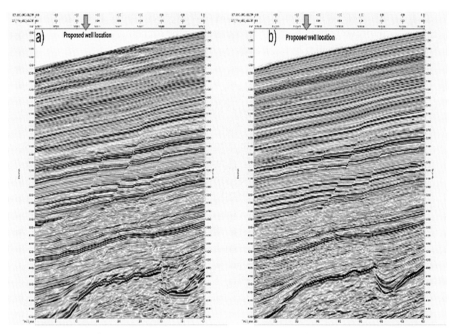

Figure 11.7 Images from a cross-line cross section. (a) Legacy seismic image and (b) seismic image resulting from rock physics guided anisotropic velocity modeling and CIP tomography as discussed in Chapter 6. Note the image quality enhancement in (b). Arrows indicate well location. After Deo et al. (2016).

low-frequency models from the rock physics guided CIP tomography from the earlier step. The primary outputs from the prestack simultaneous inversion are the elastic attributes, namely, P-impedance, V_p/V_s, and density (Figure 11.8).

Step 3   *Pore Pressure Analysis*

A fundamental step in the workflow includes creating a rock physics template that captures the relationship between rock velocities and effective stress. The template incorporates various rock physics components: (1) overburden stress versus depth, (2) porosity-velocity relationship, and (3) thermal history and chemical diagenesis of shale related to burial. The compaction model includes both a mechanical and a chemical diagenetic component as discussed in Chapter 4. Given a rock physics template – an outcome of the model that provides a relationship between rock velocity versus depth for various pore pressure gradients in ppg, the effective stress regime for normal compaction (lower bound of pore pressure) and the fracture pressure (highest possible pore pressure prior to hydraulic fracturing of rocks) can be predicted (Dutta et al., 2011). Figure 11.9a shows the template of iso-pressure gradient velocity curves and the final velocity from CIP tomography extracted at the proposed well location. The iso-pressure gradient curves range from hydrostatic to 14 ppg. Figure 11.9b shows the pore pressure, fracture pressure, and overburden stress gradients predicted at the

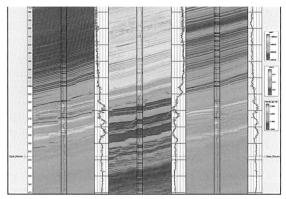

Figure 11.8 Simultaneous prestack inversion results (AVO) for (left) P-impedance in units of g/cc . m/s starting from 3000 to 12000 in step of 3000, (center) V_p/V_s starting from 1.8 to 2.4 in step of 0.3, and (right) density in g/cc starting from 1.8 to 2.7 in step of 0.3 at the well location. The red curve is the well log, the blue curve is the inverted trace, and the green curve is the low-frequency model from rock physics guided CIP tomography. From Deo et al. (2016). (A black and white version of this figure will appear in some formats. For the color version, please refer to the plate section.)

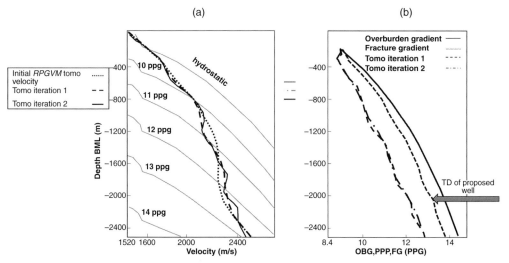

Figure 11.9 (a) Rock physics template with depth below mudline versus velocity. This figure shows iso-pressure gradient velocity curves (ranging from hydrostatic to 14.0 ppg) plotted with initial modeled velocity from rock physics guided CIP tomography (*RPGVM* as discussed in chapter 6) and two tomographic velocity iterations. Each iteration is guided by the rock physics template. (b) Pore pressure, fracture pressure, and overburden stress gradients (in ppg) predictions at the proposed well.

proposed well location. Figure 11.10a shows the predicted 3D pore pressure volume overlaid on the image. Figure 11.10b shows pore pressure transparency cube which shows only regions where pore pressure gradient is greater than 10.5 ppg. This cube highlights the areas with abnormal pore pressure and pressurized zones.

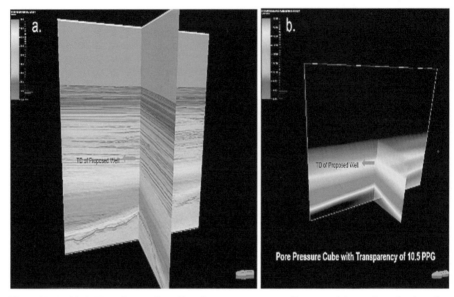

Figure 11.10 (a) A 3D volume of predicted pore pressure gradient overlain on the seismic cube. (b) Pore pressure transparency cube with pore pressure gradient greater than 10.5 ppg. This cube highlights only the areas with high overpressured zones (pressure gradients greater than 10.5 ppg). Here pore pressure is in ppg starting from 10.5 to 12.5 in step of 0.5. From Deo et al. (2016). (A black and white version of this figure will appear in some formats. For the color version, please refer to the plate section.)

Step 4   *Seismic Attribute and Stratigraphic Interpretation*

Once a high-resolution seismic section in depth is obtained, the next step is to carry out a detailed geologic interpretation of the data to identify zone(s) that may have geologic features associated with geohazards. For interpretation, 11 identifiable horizons were picked and used as boundaries for 10 zones of interest for potential geohazards. Seismic attributes were used for their efficacy in identifying geologic features and they were extracted over each study interval. Five attributes to aid in identification of potential geohazard are shown in Figure 11.11. From left to right, these are (a) *seismic amplitude* – useful for highlighting bright spots, especially shallow gas; (b) *variance* – an edge attribute useful for finding boundaries and discontinuities such as scarps and faults; (c) *dip deviation* – an attribute that highlights lateral changes in dip, especially faults and scarps; (d) *sweetness* – an attribute for visualizing stratigraphic units and their boundaries such as channels and slumps; and (e) V_p/V_s *ratio* – a high value on this attribute would be indicative of a SWF unit. Comparing the variation of these attributes in the initial strike and dip oriented cross sections, it was possible to identify various geologic features and evaluate the potential of associated drilling hazards.

Step 5   *Quantitative Analysis of SWF Zones and Traffic Light Maps*

The final step of the workflow is interpretation of seismic attributes and geophysical properties such as pore pressure and fracture pressure to quantify geohazard and generate a *"traffic light"* map. The study conducted the interpretation chronologically, beginning

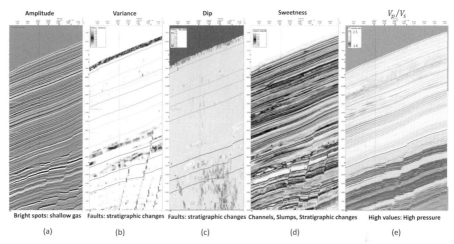

Amplitude	Variance	Dip	Sweetness	V_p/V_s
Bright spots: shallow gas	Faults: stratigraphic changes	Faults: stratigraphic changes	Channels, Slumps, Stratigraphic changes	High values: High pressure
(a)	(b)	(c)	(d)	(e)

Figure 11.11 Several attributes extracted from the inverted seismic cubes for SWF interpretation. (a) *Seismic amplitude* – useful for highlighting bright spots, especially shallow gas; (b) *variance* – an edge attribute useful for finding boundaries and discontinuities such as scarps and faults; (c) *dip deviation* – an attribute that highlights lateral changes in dip, especially faults and scarps; (d) *sweetness* – an attribute for visualizing stratigraphic units and their boundaries, such as channels and slumps; and (e) V_p/V_s *ratio* – a high value on this attribute would be an indicator for a SWF unit. From Deo et al. (2016). (A black and white version of this figure will appear in some formats. For the color version, please refer to the plate section.)

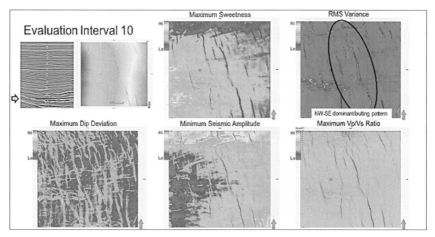

Figure 11.12 Seismic attributes extracted over Interval 10; note the fault pattern and fan system. These attributes are in arbitrary units, the cool and warm colors represent low and high values, respectively of attributes as noted. From Deo et al. (2016). (A black and white version of this figure will appear in some formats. For the color version, please refer to the plate section.)

with the bottom-most intervals and moving upward through the section. The lower-most interval examined in this study was Interval 10. This interval is characterized by a dominant faulting pattern, circled on the variance display as shown in Figure 11.12. Another important feature of this interval is the submarine fan, seen on sweetness and amplitude displays, and its feeder channels seen on variance attributes. Moving up the

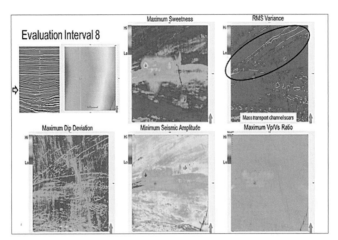

Figure 11.13 Seismic attributes extracted over Interval 8; note the sediment transport corridor and linear mass-transport scars. These attributes are in arbitrary units, the cool and warm colors represent low and high values, respectively of attributes as noted. From Deo et al. (2016). (A black and white version of this figure will appear in some formats. For the color version, please refer to the plate section.)

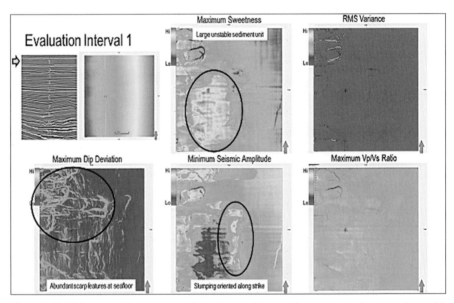

Figure 11.14 Seismic attributes extracted over Interval 1; note the numerous slumps and unstable slope features. From Deo et al. (2016). (A black and white version of this figure will appear in some formats. For the color version, please refer to the plate section.)

section to Interval 8 (Figure 11.13), the dominant feature is a large central sediment transport corridor, shown as warm colors on the sweetness attribute. Additional features are seen on variance including long, straight mass-transport scars (circled) and smaller round features, which may be the result of sediment dewatering. The shallowest interval extends from the water bottom to the top-most interval of sediment, shown in Figure 11.14.

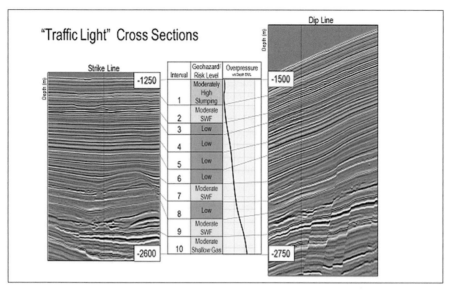

Figure 11.15 Traffic light map displaying the relative geohazard risk level for each interval examined in the study. Shown are the strike line, stratigraphic interval in the studied area (middle panel), geohazard risk levels for each interval, overpressure distributions, and the dip line. This is the final product that is helpful for locating the well head. The middle panel shows stratigraphic intervals (1-10 units), geohazard risk level for each unit (low to moderate) and overpressure distribution (low at shallow to high at deep) within each unit. From Deo et al. (2016). (A black and white version of this figure will appear in some formats. For the color version, please refer to the plate section.)

Many slump and scarp features dominate this interval, some nearly 1 km across, which are characteristic of an unstable slope. The dip deviation display highlights the many slump scarps, which are highly prevalent, and the sweetness map shows a large central block of sediment (circled) that appears to have shifted as a unit. Circled on the amplitude map is a strike-oriented area of unstable sediment, which has spawned a series of slumps.

The results of the interpretation in conjunction with geophysical properties are then used to create a "traffic light" map (hazard risk prioritization), as shown in Figure 11.15 for risk assessment and quantification – a final product. The risk potential of each interval is classified as low (green), moderate (yellow) or high (red). Thus, interval 2 has a moderate risk for SWF sands, due to the presence of semichaotic seismic character at the proposed well location (black line), but it was not upgraded to a more severe risk level due to the lack of anomalously elevated V_p/V_s ratios or predicted overpressure. By far, the most significant drilling hazard was determined to be areas of unstable, failure-prone sediments in interval 1 near the seafloor. These were listed as a moderately high geohazard risk. This unstable area covered a significant portion of the survey, and further examination concluded that sediment instability within these areas could pose a danger to permanent structures, such as a well heads.

Discussion and Conclusion

This case study for SWF hazard analysis shows that conventional 3D seismic data when processed in high-resolution mode significantly facilitates seismic attribute analysis and quantitative geologic interpretation of geohazards. Rock physics guided CIP tomography is found to be essential to provide a stable low-frequency velocity model for pore pressure calculation and V_p/V_s inversion using prestack seismic data. The workflow employed an integrated approach to provide a quantitative (high to low) matrix in terms of a traffic light map to reliably delineate geohazard intervals. To summarize, the key steps are: high-resolution seismic data reprocessing, low-frequency anisotropic velocity modeling using CIP tomography that is guided by rock physics templates, prestack simultaneous inversion for AVO attributes, pore pressure analysis, and detailed geologic interpretation that utilizes several attributes such as amplitudes, variances and sweetness, and V_p/V_s ratios in conjunction with pore pressure analysis.

Application of Full Elastic Waveform Inversion (1D) to Detection of SWF Sands with an Example (Case Study 2)

In the earlier example (case study #1), a simultaneous AVO-inversion method such as the ones described by Ostrander (1984), Rutherford and Williams (1989), and Connolly (1999) among others, was used to estimate the V_p/V_s ratio and the pore pressure. These AVO methodologies attempt to fit an approximate form of the P-wave reflection coefficient as a function of angle of incidence to the observed P-wave reflection amplitudes. An implicit assumption in the AVO theory is that the reflection events on prestack P-wave data are *primary* reflections only and are *not* contaminated by interference effects. Using synthetic data, Mallick (2001) has demonstrated that such an assumption is valid for relatively small (usually less than 25 to 30 degrees) incidence angles. At higher angles, other wave modes such as mode-converted reflections and interbed multiple reflections interfere severely with P-wave primary reflections. For an accurate estimation of V_p/V_s and Poisson's ratio from P-wave seismic data and relating these to geohazard risk assessment, AVO information from angles larger than 25 to 30 degrees is necessary. Therefore, conventional AVO limited to small angles of incidence may not be useful in all cases for SWF detection. A prestack waveform inversion scheme based on the *genetic algorithm* (GA) (Mallick, 1995, 1999) has been shown to provide an alternative as discussed below in case study #2 (Dutta and Mallick, 2010).

This case study employed the same five-step workflow as depicted in Figure 11.5. However, the inversion methodology used in step 2 utilizes an alternate approach based on *full (prestack) elastic waveform inversion* – FWI (Sen and Stoffa, 1991, 1992; Stoffa and Sen, 1991; Mallick, 1995, 1999). It offers a better solution because unlike traditional AVO, it uses an accurate forward modeling methodology based on a *full elastic* earth model for inverting prestack seismic data. The optimization to minimize the misfit between observed data and forward modeled waveforms uses the genetic algorithm (GA) as discussed in Section 5.4. Thus this approach accounts for interference of P-wave reflections with other wave modes. Consequently, these techniques can handle data to large angles of incidence, provide reliable estimates of V_P and V_S, and therefore V_P/V_S and Poisson's ratio. Because

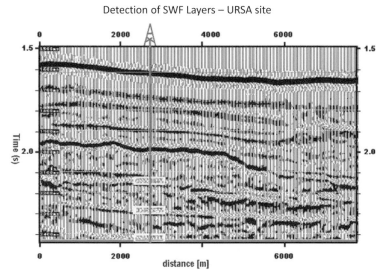

Figure 11.16 Stacked seismic data around the Ursa well site, Mississippi Canyon area of the Gulf of Mexico, USA. This is where the SWF sands caused explosions as reported by Ostermeier et al. (2002).

GA uses a statistical optimization process in searching for the best-fit elastic model, it also generates a measure of the error bounds or uncertainties in the estimation of the earth model. Furthermore, the methodology uses geologic model building aimed at understanding the stratigraphic and structural control on SWF formations. Some of these aspects have been discussed by de Kok et al. (2001). Full elastic 3D FWI is under research at various academic institutions and industry at the time of writing this book. The GA process for geohazard detection is described by an application in the following.

The example application is from a 3D seismic data volume from the Gulf of Mexico (Mississippi Canyon area, offshore Louisiana). In this area, SWF conditions are known to occur as remarked earlier in this chapter. The seismic volume encompassed the Ursa development site (Blocks 809, 810, 853, and 854) around well MC-854#2. The seismic data were reprocessed using 2 ms sampling interval for high-resolution trace inversion. A stacked section, reprocessed at 2 ms is shown in Figure 11.16. The MC 854#2 well is shown on this stacked section with markers at the three known SWF levels. Notice that on this stacked section the reflection strength of the SWF sands is weak, often making them less visible than other sands. This is due to the fact that the SWF reflections on the prestack data have a polarity reversal around 30 degrees, and summing of the negative and positive reflections tends to weaken the reflection strength on the stacked data. The results of prestack inversion are shown in Figures 11.17a and 11.17b. Figure 11.17a shows a comparison of observed prestack data with synthetic data computed using the inverted elastic earth model. Figure 11.17b shows the P- to S-wave velocity ratio (V_p/V_s), estimated from inversion. Note that the high values of V_p/V_s obtained from the prestack inversion matches with the zones where SWF layers were experienced during drilling.

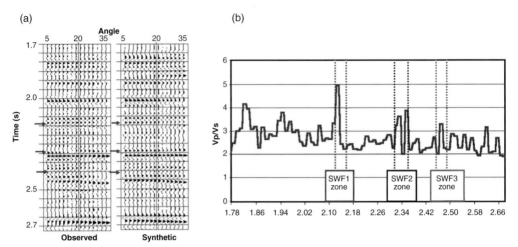

Figure 11.17 Results of 1D full-waveform elastic inversion (FWI) at the Ursa well site that experienced severe SWF explosions in the MC-854 #2 well site (Ostermeier et al., 2002). (a) Comparison of observed prestack data with synthetic data computed using the prestack inverted elastic model at the MC-854#2 well location (FWI). Reflections from the known SWF layers are marked with arrows. (b) V_p/V_s ratio obtained from prestack full elastic waveform inversion at the MC-854#2 location. The SWF zones correspond to high V_p/V_s values, as indicated. After Mallick and Dutta (2002).

Discussion and Conclusion

High values of V_p/V_s ratio is used as an indicator of SWF sands. Hence, multicomponent seismic data can also be used for this purpose (Huffman and Castagna, 2001). However, this requires additional data acquisition in deepwater that may be expensive. The examples discussed above suggest that 1D elastic-FWI of conventional 3D seismic data with long offsets can yield reliable quantitative information about the rock parameters associated with SWF formations. These formations often have a distinct seismic signature – a polarity reversals at large offsets (see Figure 11.17). Of course the success not only depends on the data quality and the choice of the discriminator, but also, and possibly more importantly, on the inversion method being used. In contrast to traditional inversion algorithms, a full elastic waveform prestack inversion is designed to obtain *both* low and high-frequency velocity variations through the *simultaneous* modeling of moveout and elastic reflection amplitudes. In addition, such an inversion correctly handles the interference of primary reflections with other wave modes, leading to an accurate elastic earth model to a resolution of 1/6th to 1/4th wavelength of the dominant seismic frequency (Mallick et al., 2000). Application of elastic 1D-FWI is therefore the key to the success of SWF detection at well locations. A word of caution – a typical reflection behavior (polarity reversal) depends not only upon the elastic properties of the SWF layers, but also on those of the shale/mudstone formations that surround the SWF sands. Thus synthetic modeling studies are highly recommended prior to using full elastic waveform inversion on real seismic angle gathers.

In addition, although identification of SWF is sensitive to V_p/V_s ratio as shown, reflection amplitudes on seismic data as functions of offset/angle of incidence are more sensitive to Poisson's ratio rather than the V_p/V_s ratio (Koefoed, 1955, 1962; Bortfeld,

1961; Shuey, 1985). Poisson's ratio and V_p/V_s are related to one another, and one can be derived from the other. For a typical shale/gas sand reflection, the Poisson's ratio of shale is of the order of 0.35 while that of gas sand is of the order of 0.1 (Mavko et al., 1998). This corresponds to V_p/V_s of shale of about 2 and that for gas sand of about 1.5. For SWF reflections on the other hand, SWF sands can have Poisson's ratio between 0.46 and 0.49. This corresponds to a V_p/V_s between 3.5 and 10. Notice that SWF reflections are characterized by a high contrast in V_p/V_s and relatively low contrast in Poisson's ratio. Gas sand reflections on the other hand are characterized by low contrast in V_p/V_s and high contrast in Poisson's ratio (see Figure 11.18).

11.3   Shallow Gas as Geohazard

Shallow gas accumulations have historically caused severe accidents in many areas where drilling for oil and gas has taken place, for example, shallow gas has been reported in approximately 27 percent of all wildcat and appraisal wells drilled on the Gulf of Mexico and Norwegian continental shelf. Several smaller kicks, and many blowouts have occurred as a result from drilling through shallow gas-prone areas. Occurrence of gas is common in the shallow part of the stratigraphy; however, the danger is associated with the fact that sediments containing gas are invariably overpressured due to buoyancy effect. The magnitude of overpressure is dependent on the vertical thickness of the gas column. The gas accumulation thickness is dependent on factors like: the gas trap height; the caprock permeability and lithology; the amount of gas supplied from the source and the caprock fracture gradient and how close this is to the pore pressure gradient.

In an area with near-horizontal bedding and small structural closure, the gas height, and corresponding over pressure is expected to be low. However, where there is significant structural relief, small stringers, too thin to show up on shallow seismic data, can contain large overpressure. Shallow high amplitude reflections and pull down effects can be seen frequently over the crest of fault blocks. In Viking Graben area, problematic shallow gas accumulations represent significant drilling challenges.

Detection of gas using surface seismic data is usually carried out by the well-known AVO techniques (Rutherford and Williams, 1989; Aki and Richards, 1980). Since we are interested in the shallow part of the stratigraphy – a section where we may also have the occurrence of SWF sands, it would be interesting to compare and contrast the two AVO profiles, namely, one from the shallow gas sand and the other for SWF sand. Figure 11.18 shows the P-wave reflection coefficient as a function of angle of incidence for a typical gas sand and two SWF reflections (SWF 1 and SWF 2). All these reflections coefficients were computed using the exact reflection coefficient formula, given by Aki and Richards (1980). For the gas sand reflection, the P-wave velocity, density, and Poisson's ratio for the overlying shale were assumed to be 2300 m/s, 2.1 g/cm^3, and 0.35 respectively, and those for gas sand were assumed to be 2250 m/s, 2.05 g/cm^3, and 0.10, respectively. For the SWF reflection, a P-wave velocity of 1550 m/s, density of 1.85 g/cm^3, and Poisson's ratio of 0.48 were used. In SWF 1, the surrounding shale/mudstone was assumed to have a P-wave

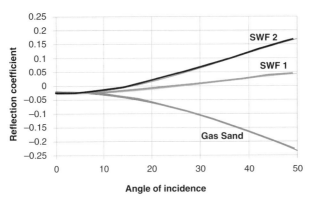

Figure 11.18 P-wave reflection coefficient as a function of angle of incidence for a typical gas sand and two SWF reflections (SWF 1 and SWF 2).

velocity of 1600 m/s, density 1.9 g/cm³, and Poisson's ratio of 0.44. In SWF 2 on the other hand, the P-wave velocity and density were kept the same while the Poisson's ratio was changed to 0.38. The AVO behaviors for gas sand and SWF sands are different – SWF sands have a reversal of polarity.

For geohazard detection, Mallick and Dutta (2002) argued that SWF reflections are characterized by a *high* contrast in V_p/V_s and relatively *low* contrast in Poisson's ratio. Gas sand reflections on the other hand are characterized by *low* contrast in V_p/V_s and *high* contrast in Poisson's ratio (see also Koefoed, 1955, 1962; Bortfeld, 1961; Shuey, 1985). The parametrization in Figure 11.18 suggests that for the gas sand reflection, the Poisson's ratio contrast is 0.25 while the contrast in V_p/V_s is 0.5. For the SWF 1 reflection, the Poisson's ratio contrast is 0.04 and the V_p/V_s contrast is 4. Finally, for the SWF 2 reflection, the Poisson's ratio and V_p/V_s contrasts are 0.1 and 4.5, respectively. For the gas sand reflection, although the V_p/V_s contrast is low, the Poisson's ratio contrast is high. This high contrast in Poisson's ratio causes the difference in P-wave reflection coefficient between small and large angles of incidence to be so big that it is clearly visible on real seismic data. This is the primary reason for the success of amplitude variation with offset (AVO) and other prestack methods such as prestack elastic full-waveform inversion in detecting gas sands from seismic data. For the SWF reflections on the other hand, the V_p/V_s contrast is large, but the Poisson's ratio contrast is not so large. Considering the range of angles available in prestack data to lie between 0° and 40°, gas sand reflection amplitudes change from –0.03 to –0.16, SWF 1 amplitudes change from –0.03 to 0.04, and SWF 2 amplitudes change from –0.03 to 0.13. A reflection behavior such as SWF 1 is likely to be hidden below the noise level, and therefore can be quite difficult to detect from prestack seismic data. In order to detect SWF reflections, AVO behavior must at least be similar to the SWF 2 reflection. Such differences allow the usage of seismic data for quantifying the hazardous effects of gas in shallow reservoirs (shallow gas) and those from aquifer pressured sands (SWF sands).

In both cases of geohazards, it is advisable to move the drilling rig away from these hazardous zones. However, this is not always possible, especially, if one is dealing with

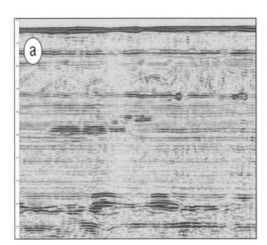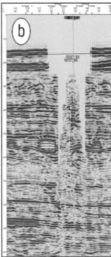

Figure 11.19 (a) Seismic data from a 2D high-resolution site survey. (b) 3D undershoot data taken from a 4D monitoring streamer survey using dual-vessel undershoot with a separate source vessel. Drilling hazards are identified in both data sets as bright spots. However, the streamer data cannot image the potential drilling hazards under the platforms. The 2D data are 100 m away from the platform, and the absence of data directly underneath the platform is clearly visible in the 3D undershoot shown in (b). After Ronen et al. (2012).

production platforms. This situation has been addressed by Ronen et al. (2012) in a production area. The authors noted that shallow gas has been a major drilling hazard in Forties Field in UK. A fire caused by shallow gas severely damaged the Delta platform in 1985 (see Ronen et al., 2012). The gas is easily identified as bright spots on seismic data, and there is an abundance of such data in the Forties Field (Figure 11.19, taken from Ronen et al., 2012). But all of the seismic data were acquired with towed streamers after the platforms were put in place. Hence, there was no usable seismic image of the shallow subsurface under the platforms. The authors carried out an acquisition of high-resolution seismic data using ocean-bottom nodes (OBN) with the objective of defining the exact depth and extent of the shallow gas hazards under the producing oil field platform. The OBN data confirmed the existence of shallow gas and validated the result from the conventional 3D seismic survey. Figure 11.20 shows that to verify the node data (a), the team cross-referenced and compared the interpreted results utilizing a common subsurface line 100 m away from the platform. The data from the OBN acquisition were compared with legacy 2D seismic data in (b) and the 3D undershoot data in (c). The figure shows downgoing mirror-imaged node data that provided 3D image of hazards below the platform as shallow as 20 m below seabed. Most importantly, these data can also identify hazards where streamer data are unavailable. Such an approach suggests that detection of shallow gas hazards below man-made obstacles such as drilling platforms is possible using new seismic data acquisition techniques. Once such data are available, one could evaluate the magnitude of overpressure by standard velocity and amplitude analysis methods as discussed in this chapter.

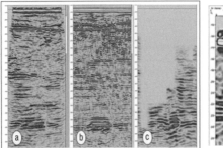 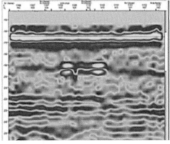

Figure 11.20 To verify the node data (a), the team cross-referenced and compared the interpreted results utilizing a common subsurface line 100 m away from the platform. The data from the OBN acquisition were compared with legacy 2D seismic data in (b) and the 3D undershoot data in (c). The node data clearly identified the same hazards where streamer data were available. Most importantly, these data also can identify hazards where streamer data are unavailable. The figure on the extreme right shows downgoing mirror-imaged node data that provided a 3D image of hazards below the platform as shallow as 20 m below the seabed. After Ronen et al. (2012). (A black and white version of this figure will appear in some formats. For the color version, please refer to the plate section.)

11.4   Gas Hydrate as Geohazard

Gas hydrates are naturally occurring, crystalline, ice-like substance composed of gas molecules (ethane, methane, propane, etc.) and water. In the earlier discussion on SWF sands, we noted (see Figure 11.1) that gas hydrate can form effective seals for high pressure aquifer sands. Although gas hydrates may not contribute directly to overpressured sediments, they could be related to these geologic events. Furthermore, the habitat of gas hydrates coincides with that of SWF sands and they are known to contribute to geohazards. Figure 11.21 shows a seismic traverse through a deepwater GOM area in which a bottom simulating reflector (BSR) is identified. BSR is indicative of gas hydrates. (The inset shows the stability zone of methane hydrates for an assumed thermal gradient of 2.5°C/100 m.) This indicates the *possible* presence of gas hydrates or more precisely, it identifies a *boundary* of impedance contrasts between free gas accumulations underneath an impermeable boundary. The BSR cuts across the stratigraphy and follows the sea bed. The stability of gas hydrates is dictated by whether the BSR is within the Gas Hydrate Stability Zone. This is also calculated using temperature gradient, pressure, and water salinity data as identified in Figure 11.21. In this remarkable seismic image, not only do we see a BSR but also *several possible zones of pressure breach and expulsion features*. This is possibly associated with naturally disturbed SWF sands. Thus a short description of how to locate this hazard (gas hydrate) and *avoid* drilling through the sediments containing gas hydrate deserves some discussion in this chapter.

Gas hydrates are found worldwide in both deep marine and permafrost areas. In the GOM alone, over 100 locations have been identified in the deep marine areas with known gas hydrate deposits (Figure 11.22a). Because each volume of solid methane hydrate contains as much as 164 standard volumes of methane, hydrates can be viewed as a concentrated form of natural gas equivalent to compressed gas but less concentrated

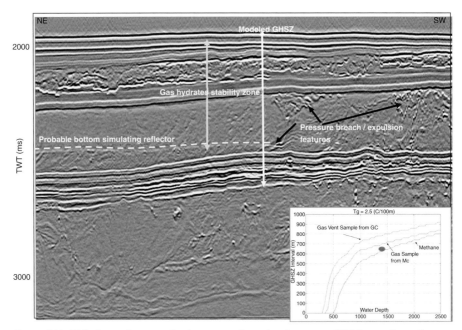

Figure 11.21 This figure shows a seismic traverse through a deepwater GOM area where a bottom simulating reflector (BSR) is seen. This could be associated with gas hydrates. We also note the presence of pressure/fluid expulsion features, possibly associated with breach of SWF sands. The habitats for these two hazards (gas hydrates and SWF sands) overlap in the GOM.

than liquefied natural gas. However, because gas hydrates are stable only under specific temperature, pressure, and salinity conditions, they can also pose a hazard to environment by becoming destabilized, especially, when drilling through formations containing gas hydrates in deepwater areas. There are several legitimate environmental issues: slope stability, the hydrate gun hypothesis (Kennet et al., 2003), and borehole stability (Birchwood et al., 2007; Reagan and Moridis, 2010). Wells penetrating a gas hydrate accumulation would alter the pressure/temperatures of nearby sediments, potentially destabilizing the hydrates and causing its dissociation into gas and water. This may result in a large change of volume and increase in pore pressure possibly causing the sediments to become fluidized. Sand sediments would, in general, contain a larger volume of gas hydrates than clay sediments due to higher effective porosity.

Figure 11.22b shows a plausible model for slope failure and release of copious amount of methane in the environment. Sea level changes (fall) can cause the base of hydrates to shift and trigger slumping on weak gas-charged zones and thus cause release of methane, for example, in the atmosphere, which in turn can contribute to the green-house effect. As far as the other concern – "calthrate gun hypothesis" – most scientists now believe that it is not serious. However, it is fair to say that the methane contribution from hydrate decomposition in marine waters is an important unknown in the global methane budget.

Typical ranges for physical properties of gas hydrates are given below (Gabitto and Tsouris, 2010):

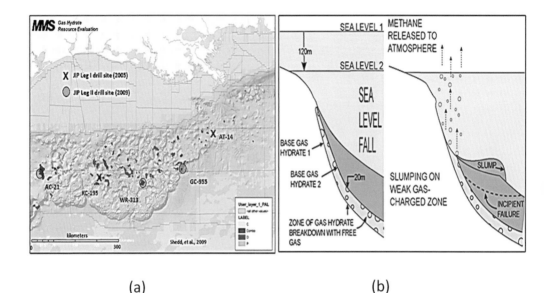

(a)   (b)

Figure 11.22 (a) Locations in the Gulf of Mexico, USA, where gas hydrates have been identified. (b) Possible model for slope instability and failure that can be caused by release of gas due to breakdown of gas hydrates.

Density: 0.912–0.940 gm/cc
P-wave velocity: 3778–3822 m/s
S-wave velocity: 1960–2000 m/s
V_p/V_s ratio: 1.92–1.91

No matter whether gas hydrates are viewed as hazard or resource (possibly both), we need to find it and characterize it using seismic data as a remote sensing tool. Dutta et al. (2009), and Shelander et al. (2010a), discuss the use of surface seismic data (3D) to map locations and quantify the amount of gas hydrates in unconsolidated sediments like those in the GOM. The differentiating physical properties of gas hydrates as shown above, allow the successful prediction of gas hydrate accumulation (Dutta and Dai, 2009; Dai et al., 2008a, 2008b; Shelander et al., 2010b) employing the same quantitative inversion process as applied for shallow-water flow analysis. The technology development in this application benefited by a major sponsor: the Department of Energy (DOE) of the USA through a joint industry project (JIP). The JIP was composed of over a dozen oil and gas and service companies. The objectives of the JIP were to ground truth seismic and gas hydrate modeling and develop a methodology for gas hydrate characterization. These objectives were achieved in the GOM and the major accomplishments are summarized by Boswell et al. (2009). Since then more studies were conducted as shown in the table below and the gas hydrate accumulation and detection models were verified by further drilling during the GOM JIP Leg II campaign during 2009.

In Figure 11.23 we show an example from the Walker Ridge, block 313 area of the Gulf of Mexico, USA, where extensive gas hydrate modeling was done prior to the

GOM JIP Leg II well sites (courtesy of Ray Boswell, Gas Hydrate Program Manager, Department of the Energy, USA).

Hole	API number	Latitude	Longitude	Water depth (ft)	Water depth RKB (ft)	Total depth of hole RKB (ft)	Total depth of hole below seafloor (ft)
AC 21 A	608054007000	26 55' 25.0550"	94 54' 00.8545"	4889	4940	6700	1760
AC 21 B	608054007100	26 56' 40.3922"	94 53' 36.4053"	4883	4934	6050	1116
GC 955 H	608114053700	27 00' 03.2836"	90 25' 35.4475"	6670	6721	8654	1933
GC 955 I	608114054400	27 01' 00.7414"	90 25' 17.2257"	6770	6822	9027	2205
GC 955 Q	608114054300	27 00' 08.5611"	90 26' 12.0500"	6516	6567	8078	1511
WR 313 G	608124003900	26 39' 48.7355"	91 41' 02.3996"	6562	6614	10200	3586
WR 313 H	608124004000	26 39' 46.0997"	91 40' 43.2051"	6450	6501	9770	3269

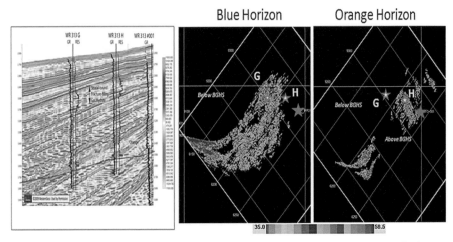

Figure 11.23 Walker Ridge. (left) Block 313 seismic and drilled wells (courtesy of Ray Boswell of USA – DOE) and (right) gas hydrate saturation starting from 35 percent to 58.5 percent for the two wells in block 313 (marked G and H). WesternGeco provided the conventional 3D seismic data for the study to the GOM JIP. These data were used for analytical work in the project. (A black and white version of this figure will appear in some formats. For the color version, please refer to the plate section.)

drilling campaign of Leg II of the GOM JIP during April 16–May 5, 2009. On the left side we show the seismic traverse through the drilled prospects and initial estimates of gas hydrate volume from the inversion of seismic data using the workflow discussed in Dutta et al. (2009) and Dai et al. (2008a, 2008b). The right side of Figure 11.23 shows gas hydrate saturation predicted by seismic modeling and these were verified by drilling. So, from the hazard *avoidance* view point, it is safe to conclude that seismic techniques can be used to locate gas hydrates and thus avoid drilling through them.

11.5   Geohazard Mitigation (Dynamic Kill Drill or DKD Procedure)

In spite of the success of geohazard detection techniques as discussed here, geologic complexities prevent 100 percent success rate using any predrill technique. In that case, the drilling community uses a mitigation technique known as the Dynamic Kill Drilling (DKD) technique (Vieira et al., 2014). As deepwater exploration and development move toward deeper water depths and reservoirs, conventional *riserless* drilling techniques, such as seawater sweeps, no longer provide the required equivalent density from the formation integrity test (FIT) to set the surface casing at the required depth. In that case DKD technique is commonly used. Vieira et al. (2014) report the following case study from offshore Angola where DKD technique was used. DKD is also widely used in the Gulf of Mexico and recently its use has been extended to China and Mexico (Vieira et al., 2014). The authors noted that the most significant challenge for its first application in deepwater Angola was the DKD fluid volume availability, as well as the logistics involved. Drilling practices and parameters that posed a challenge with regards to control included rate of penetration (ROP), flow rates, rotation (rev/min), and reaming.

The following is taken *verbatim* from Vieira et al. (2014) to illustrate how the procedure works:

The DKD 16.3-lb/gal fluid was mixed in the Luanda liquid mud plant (LMP) 2. Approximately 10,000 bbl. were mixed and then taken to the rig for storage until required for the 26-in. borehole section. The process continued until 30,000 bbls were created and sent offshore. The 26-in. interval was successfully drilled in 30 hours. Mud density was 10.7 lb/gal, which was achieved by combining a total of 5,544 bbls of 16.3-lb./gal DKD fluid mixed with 14,335 bbls of seawater. The mix-on-the-fly (MOTF) unit was rigged up on top of the pits. A line was attached from the pits holding the 16.3-lb./gal DKD fluid and another line was attached to the pit with the seawater. The two liquid streams were blended to achieve the required mud weight. The 16.3-lb/gal DKD fluid was reduced to form pad mud with a density of 11.5 lb/gal, which was pumped and placed in the open-hole section until the 20-in. casing was run and cemented. A total of 19,879 bbls of fluid was pumped downhole without any drilling or formation issues to conclude the first DKD application in Angola's deepwater.

11.6   Recommendations for Detection of Geohazards

1. Geohazard assessment should begin long before any drilling activity. It should begin with an assessment of sea floor conditions. Salient features of this study should comply with the requirements set forth in the BOEM recommendations to the leaseholders as discussed at the beginning of the chapter. Modern vintage 3D-seismic data is very useful for this purpose. If the generated maps lack resolution, 2D seismic data must be acquired, processed, and maps generated.

2. Sea floor maps using 3D-exploration seismic maps should be generated and hazard assessment or risk profile made by a multidisciplinary team (exploration geoscientists, drilling engineers, geotechnical engineers, and site investigation specialists). Some of the analysis techniques discussed in this book may be used. By displaying amplitude variations of seafloor reflections on seafloor renderings, rocky areas, slumps, buried channels, and shallow failure zones, hydrate mounds, and seafloor vents can often be detected and mapped.

3. The next step is to focus on *subsurface* geohazards such as *SWF sands, gas hydrates, and gas-charged shallow reservoirs*. Again, modern vintage 3D-seismic data should be used following a procedure similar to the 5-step workflow that we have discussed in this chapter. It is a common but inadequate wisdom that a "traffic light map" based on geologic interpretation of stacked data alone will suffice. It will not. It is just the beginning. We need to have a quantitative measure of *"how red is red and how yellow is yellow?"* It has been our experience that 2D-high-resolution seismic often yield traffic light maps that contain *"too many reds"!* Such data being acquired with shorter-cable *cannot* quantify the risk. Only modern vintage 3D-seismic and/or multicomponent seismic and possibly, node data can accomplish the risk quantification workflow as discussed here.

11.7   Concluding Remarks

Successful drilling in deepwater to find and lift hydrocarbons requires an assessment of drilling hazards that may be encountered in the shallow section of the geologic column below mudline. Shallow overpressured zones, and the potential for shallow-water flows (SWF) during drilling, are perhaps the most intractable geohazard encountered in that region. In fact, some operators feel that SWF is the single biggest risk and the only potential "showstopper" to deepwater development. During the 1997 joint industry SWF forum, it was reported that approximately 80 percent of deepwater wells (in the Gulf of Mexico) experienced SWF to some degree, and about 25 percent did not reach their drilling objectives because of SWF problems. The forum also lamented on the fact that the industry spends considerably more resources on *remedial* efforts such as DKD due to SWF problems rather than on *identification and avoidance*. We think that this situation needs to change.

We think that the use of 3D seismic data could be and should be a standard practice using the inversion technologies discussed here and in many other prior publications. The operators are now paying more attention to the use of 3D-seismic data (Alberty et al., 1997; Byrd et al., 1996; Campbell, 1997; Cowland, 1996; Hill, 1996). We showed that shallow-water flow hazards can be characterized through the use of indicators such as V_p/V_s, Poisson's ratios and density and also pore pressure that can be estimated quantitatively for the near seafloor sediments where conventional velocity analysis is not effective. This requires reprocessing and quantitative interpretation of derived attributes. Multicomponent seismic data can yield reliable V_p/V_s ratios *directly*, however, it requires ocean bottom cable or ocean buttomm nodes as receivers, which are typically more costly than towed streamer conventional aquisition. Deriving V_p/V_s through inversion of 3D surface seismic is more cost-effective and is applicable in all marine settings. The 3D-seismic-based processes recommended here provide valuable information not only with regard to resource exploration (e.g., gas hydrates) but also for drilling risk prevention, site selection, and predrill planning.

12 Petroleum Geomechanics and the Role of Geopressure

12.1 Introduction

Geomechanics is the science of how rocks deform, sometimes to failure, due to change in stress, temperature, and other environmental parameters. It includes disturbances caused by natural processes such as folds, faults, fractures, and other geological processes such as mountain building. Pore pressure is also very important as it plays a significant role in these processes. We should also note that the term *rock mechanics* is often used synonymously with geomechanics. These are not the same. *Rock mechanics is a branch of geomechanics* – it deals with the principles of continuum mechanics and geology to study the response of rocks to environmental forces caused by *man-made* factors which alter the original in situ or ambient conditions. The word *petroleum geomechanics* is used to describe the mechanical properties and behavior of geological formations which influence the exploration, development, and production of *oil and gas*. This not only includes drilling of wells that alter the stresses around boreholes (both vertical and deviated) thus contributing to borehole instability but also reservoir scale phenomena, such as faulting and fracture development over geological time, compaction, and subsidence caused by production and hence, changes in pressure, and induced faulting and seismicity. It also includes wellbore scale phenomena, such as wellbore stability during drilling, hydraulic fracturing, formation/casing interaction during production, sand production, and waste injection.

Geopressure plays an important role in *all* aspects of geomechanics. Changes in pore pressure profile alter the state of effective stress acting on the subsurface formations. Increase in effective stress can cause failure effects such as compaction (or yield) and shearing. Therefore, the ability to model the stresses and their responses to different pore pressure scenarios is of importance for engineering projects associated with any subsurface infrastructure including borehole design, hydrocarbon development, and production planning in the oil and gas industry. In civil and environmental engineering, projects associated with structure/foundation design, slope stability, and landscape management among others benefit from geomechanical modeling of poroelastic systems.

In this chapter, we will review the *basic principles of petroleum geomechanics* and discuss the use of mechanical earth modeling techniques to model stresses, strains, and stability criteria. Such studies can help (1) reduce drilling cost by lowering nonproductive time (NPT), (2) improve drilling safety and reduce exploration risk,

(3) increase reservoir performance, (4) predict wellbore stability prior to drilling and reduce occurrences of stuck pipe, lost circulation, sidetracks and formation collapse, and (5) decide whether safe drilling can continue with processes such as under-balanced or highly overbalanced drilling. As the cost of drilling wells have gone up in the deepwater (water depth > 3000 ft), nonproductive time or NPT due to unexpected issues including lost returns, differential sticking, and narrow pore pressure/fracture gradients has also gone up. Figure 12.1a shows average NPT % to *total drill time* (excluding weather-related NPT) for 66 nonsubsalt wells drilled September 2004 through December 2008 in water depth greater than 3000 ft in the Gulf of Mexico, USA. Similar statistics are shown in Figure 12.1b for subsalt wells. These statistics are discussed by James K Dodson Company (York et al., 2009; Pritchard and Lacy, 2011). Using data such as the ones published by York et al. (2009) and similar other data, Pritchard and Lacy (2011) concluded that "in some categories of complex wells, wellbore stability events are as high as 10 percent of the total deepwater well time and well control incidents over four times those of 'normal' wells." As it can be seen from Figure 12.1a, 5.6% of total well cost for these wells is attributed to wellbore instability. For the average cost/ft of $2281 that equals 127.73 $/ft. This would mean $2,550,000 for a 20,000 ft well. Costs for subsalt wells are even higher as shown in Figure 12.1b. Schlumberger states that geomechanical problems are associated with 40 percent of the drilling related NPT in deepwater and other challenging environments (Schlumberger, 2016; Plumb et al., 2000). Total cost of geomechanics related issues is estimated to be many billions of dollars (Knoll, 2016).

There are many text books, and technical papers on geomechanics related topics, such as Atkinson (1987), Bourgoyne et al. (1991), Hudson and Harrison (1997), Marsden (1999), Zoback (2007), and Fjær et al. (2008). However, the literature that deal with actual case histories with proper technical details in areas that cover safe and reliable operation of drilling (borehole instability) and its prediction (prior to drilling), and the design of wells, is fairly sparse. There are three main reasons: (1) petroleum geomechanics is mired with empirical formulations of models that require specific knowledge from a particular petroleum basin or a well in that basin; (2) oil and gas companies have a culture of maintaining secrecy and thus keeping (occasionally, perceived) competitive edge that results in a lack of knowledge-sharing; and (3) all geomechanical models need calibration with core data that is always sparse (and restricted only to reservoir intervals) that leaves the overburden model completely uncalibrated where most of the problems begin to occur. In addition, the sparsity of details of geomechanics models (e.g., calibration of overburden) led to many bad practices – the chief one being transporting a model from a particular well (e.g., in the Gulf of Mexico that is dominated by clastics) to a well in a totally different geologic environment (as for example, in the Middle East dominated by carbonates). We want to raise an awareness of the issue and hope that the users will reach a balanced approach – protect the integrity and the competitive edge of the company and yet, do not compromise on either the safety or the protection of the environment.

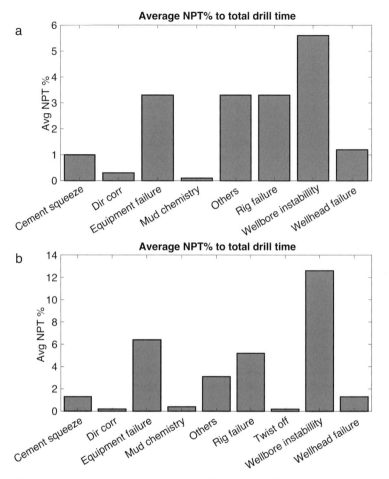

Figure 12.1 (a) Average *NPT % to total drill time* (excluding weather related NPT) for 66 nonsubsalt wells drilled September 2004 through December 2008 in water depth greater than 3000 ft. (b) Average *NPT % to total drill time* (excluding weather related NPT) for 38 subsalt wells drilled September 2004 through December 2008 in water depth greater than 3000 ft. After York et al. (2009).

12.2   Borehole Stability and Pore Pressure

The term stability implies a balance between strength and stress. When stress is higher than the strength of the material the material is labeled as unstable. Typical stability analysis consists of comparing the distribution of effective stress and material strength. Pore pressure affects both the strength of the material and the state of stresses. For example, the Mohr-Coloumb failure criteria states that the shear strength is a function of effective normal stress. Thus, increase in pore pressure will result in a decrease in material resistance to shear failure as effective stress is reduced. The consequences of borehole instability are many: mechanical failure, tight hole, stuck pipe, insufficient borehole cleaning, and many problems associated with wellbore trajectory. In general, borehole instabilities create and intensify problems. Failure envelope and shear and yield failure were discussed in Chapter 2. Below, we demonstrate the effect of pore pressure on borehole stability.

Basic Borehole Stability

Drilling into the earth often involves the replacement of rock mass by drilling fluids that exerts pressure on the borehole wall. In general, that pressure differs from the preexisting conditions. The effective stress that was present in the rock near a borehole before the well was drilled, is redistributed when the borehole is drilled. That redistribution can lead to yield or failure of the rock close to the borehole, which results in decreased stress near the borehole wall. If the strength of the rock will not sustain the new state of stresses associated with the pore pressure/rock volume, the borehole will fail. The modes of failure and the type of failure are related to the details of the pressures and strengths and are typically a part of the mechanical earth model used to evaluate borehole stability.

Stress Distribution around Borehole

Assuming plane strain, Kirsch (1898) derived a solution to a boundary value problem for the stress distribution around a borehole of radius a with borehole pressure P_b subject to a far-field stress with vertical stress σ_V, minimum horizontal stress σ_h and maximum horizontal stress σ_H. The solution is given in cylindrical coordinate (r, ϕ, and z) and is known as the Kirsch solution (Kirsch, 1898; Jaeger et al., 1979) given by

$$\sigma_{rr} = \frac{1}{2}(\sigma_H + \sigma_h)\left(1 + \frac{a^2}{r^2}\right) + \frac{1}{2}(\sigma_H - \sigma_h)\left(1 - \frac{4a^2}{r^2} + \frac{3a^4}{r^4}\right)\cos2\phi + \frac{a^2}{r^2}P_b$$

$$\sigma_{\phi\phi} = \frac{1}{2}(\sigma_H + \sigma_h)\left(1 + \frac{a^2}{r^2}\right) - \frac{1}{2}(\sigma_H - \sigma_h)\left(1 + \frac{3a^4}{r^4}\right)\cos2\phi - \frac{a^2}{r^2}P_b$$

$$\sigma_{zz} = \sigma_V - 2v(\sigma_H - \sigma_h)\frac{a^2}{r^2}\cos2\phi$$

$$\sigma_{r\phi} = \frac{1}{2}(\sigma_H - \sigma_h)\left(1 + \frac{2a^2}{r^2} - \frac{3a^4}{r^4}\right)\sin2\phi \tag{12.1}$$

The effective stress is given by $\boldsymbol{\sigma}' = \boldsymbol{\sigma} - P_P\mathbf{I}$. In Figure 12.2 we present the effective stress for the example following Sayers (2010). In this example $\sigma_V = 100$ MPa; $\sigma_H = 90$ MPa; $\sigma_h = 80$ MPa and $P_P = 60$ MPa. We change the borehole pressure from balanced ($P_P = P_b = 60$ MPa, Figure 12.2a), to underbalanced ($P_P > P_b = 40$ MPa, Figure 12.2b) and to overbalanced ($P_P < P_b = 100$ MPa, Figure 12.2c). Examination of the stress shows that the radial stresses change from positive to negative for the under-balanced case and the hoop stresses become negative for the overbalanced case. As the convention used is that positive stresses are compressive and tensile stresses are negative, this simple example shows that for different pore pressure

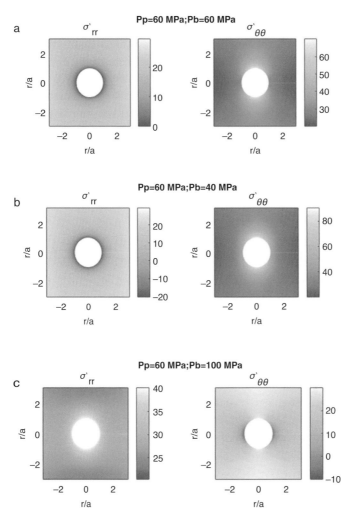

Figure 12.2 Plots of the Kirsch solution for different bore hole and pore pressure scenarios for horizontal far-field stress σ_H. Radial (left column) and hoop (right column) effective stresses associated with (a) balanced scenario where $P_P = P_b$, (b) underbalanced scenario where $P_P > P_b$, and (c) overbalanced scenario where $P_P < P_b$. Note that for the underbalanced case, the radial stress near the borehole is tensile, while for the overbalanced case, the hoop stresses along the horizontal plane are negative (tensile).

and mud weight scenarios the effective stresses vary from compressive to tensile. Thus, the borehole tensile strength is as important as is compressive and shear strengths. The far-field stresses and the pore pressure and borehole mud pressure govern the stability of the borehole. Typical 3D mechanical modeling will generate estimates of the far-field stresses while local analysis can be performed to evaluate the borehole stability conditions. For anisotropic material, nonlinear elastic constitutive relations, and complex borehole trajectories, numerical simulation of the stresses near the borehole are often used to evaluate its stability.

Formation Stresses and Failure Mode near a Borehole

Stresses applied to a formation are controlled by two sources that are different in origin. Radial stresses are controlled by the pressure of the drilling mud which are essentially controlled by the driller and as discussed earlier are dictated by the specific needs of operation. This pressure can be planned for and measured in real time by utilizing while-drilling sensors employed to monitor annular pressures during drilling. Axial and tangential stresses are controlled by the far-field stresses. Different modes of failure can be associated with distinct mechanisms. Bratton et al. (1999) classified common borehole failure modes into six modes of shear failure and three modes of tensile failure for a vertical wellbore based on the morphology and origin of the failure modes. These are described below and graphically shown in Figure 12.3.

Shear failure modes are as follows:

1. shear failure wide breakout (S_{wbo}): This mode (Figure 12.3a) occurs when the tangential stress is the maximum stress and the radial stress is the minimum stress. These two stresses act in the horizontal plane to cause the failure, which is centered at the azimuth of the minimum horizontal stress. This is the failure mode most commonly known as "wellbore breakout." It has the additional descriptor of wide because the failure tends to cover a large arc, from 30° to 90°.
2. shear failure shallow knockout (S_{sko}): This mode (Figure 12.3b) occurs when the axial stress is the maximum stress and the radial stress is the minimum stress. These two stresses act in a vertical plane, which is aligned with the minimum horizontal stress direction. The circumferential coverage is small and this mode could easily be confused with a vertical fracture.
3. shear failure high-angle echelon (S_{hae}): The S_{hae} mode (Figure 12.3c) occurs when the axial stress is the maximum stress and the tangential stress is the minimum stress. These two stresses act in the plane of the mud formation interface and center at the azimuth of the maximum horizontal stress. This failure mode makes high-angle fractures that cover up to a quarter of the borehole circumference.
4. shear failure narrow breakout (S_{nbo}): This mode (Figure 12.3d) occurs when the radial stress is the maximum stress and the tangential stress is the minimum stress. These two stresses act in the horizontal plane and center about the maximum horizontal stress direction. The angular coverage on the borehole is less than 30°.
5. shear failure low-angle echelon (S_{lae}): The S_{lae} mode (Figure 12.3e) occurs when the tangential stress is the maximum stress and the axial stress is the minimum stress. The failure is similar to the high-angle echelon, but it makes a low-angle fracture instead. Unlike the high-angle echelon, this mode is centered about the minimum horizontal stress.
6. shear failure deep knockout (S_{dko}): This mode (Figure 12.3f) occurs when the radial stress is the maximum stress and the axial stress is the minimum stress. Similar to

the shallow knockout, this failure is in the vertical plane but centered at the azimuth of the maximum horizontal stress.

The three tensile failure modes are as follows:

1. tensile failure cylindrical (T_{cyl}): The T_{cyl} mode (Figure 12.3g) occurs when the radial stress exceeds the tensile strength of the formation. The failure is concentric with the borehole and not visible on a wellbore image.
2. tensile failure horizontal (T_{hor}): This mode (Figure 12.3h) occurs when the axial stress exceeds the tensile strength of the formation. The failure creates horizontal fractures.
3. tensile failure vertical (T_{ver}): The T_{ver} mode (Figure 12.3i) occurs when the tangential stress exceeds the tensile strength of the formation. This failure creates a vertical fracture aligned with the maximum horizontal stress direction. This mode is exploited in hydraulic fracturing to enhance fluid flow.

Figure 12.3 shows a graphic representation of the different failure modes associated with different borehole conditions. For example, when low mud density is used in the borehole large tangential stresses and low radial stresses cause large differential stresses which may fail the rock. Failure occurs at a small angle to the maximum and tangential stresses, and is indicated by gray crosses shown in Figure 12.4. This figure shows that wide breakouts occur in the direction of the minimum horizontal stress. Shallow knockouts also occur in this direction. Failure modes due to a high equivalent circulation density occur in the direction of the maximum horizontal stress.

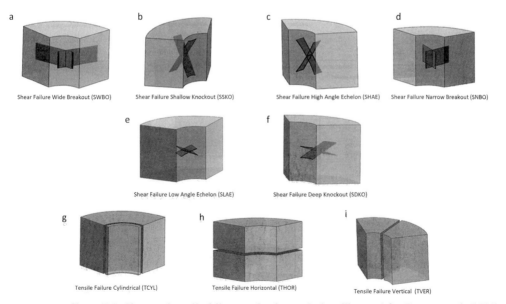

Figure 12.3 Shear and tensile failure modes for vertical wellbores. After Bratton et al. (1999).

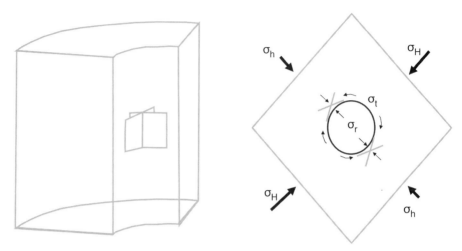

Figure 12.4 Example of wide breakout shear failure.

12.3   Petroleum Geomechanics Modeling

As discussed in the previous section, formation failure is related to both mud pressure during drilling and in situ state of stress. To mitigate the risk of borehole failure, a common approach is to use geomechanics modeling to improve well design and to analyze the state of in situ stresses and use them to derive a *safe* mud weight that will likely not cause instability issues. A typical plot depicting various components behind building a mechanical earth model or MEM is shown in Figure 12.5a, taken from Berard and Prioul (2016). The essential elements used in geomechanics modeling are mechanical properties, pore pressure estimation, and assessing magnitude and orientation of various stresses. We recall from Chapter 2 that we need to be concerned about three principal stresses – the vertical and two horizontal ones. Drilling must proceed in the direction of minimum horizontal stress. In Chapters 1 and 2 we discussed the concept of safe mud window. A schematic description of the concept of safe mud window in the context of wellbore stability is provided in Figure 12.5b. As is seen from this figure a low mudweight that is below the pore pressure gradient will result in a kick. If the mudweight is less than the breakout pressure gradient, shear failure will occur with rock fragments falling into the wellbore. On the other side, higher mudweight – higher than the magnitude of minimum stress will lead into invasion of the mud into the formation. This will result in mud loss. Increasing the mud weight further above the fracture pressure gradient will cause an induced fracture to be initiated in the wellbore wall.

Plots such as the ones in Figure 12.5a are the desired end products from a petroleum geomechanics modeling to prevent borehole instability related issues. In addition, the MEM also consists of depth profiles of various elastic and/or elastic-plastic parameters and rock strengths. To calculate the stresses and set safety margins, the industry uses a two-step approach. First, one builds a 1D-MEM and calibrates the parameters and the

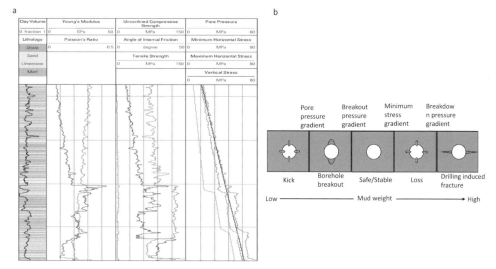

Figure 12.5 (a) A vertical log display of 1D-MEM commonly used in the industry. This log displays lithology including volume of clay (track 1), elastic properties (track 2), strength properties (track 3), and state of the stress (track 4) characterized by stress directions and magnitude (not shown). These calculations are used for proper mudweight design. (b) Concept of safe mud weight window for drilling. On one end are the "kicks" due to mud weight being less than the "safe" value, and on the other end is the "drilling-induced fracture" when the mud weight is higher than the "safe" value. Modified after Le and Rasouli (2012).

model with available borehole and stress data such as various leak-off tests and mini-frac data. Second, a 3D model is created for predictive purposes. Below we discuss how to build a mechanical earth model (MEM) with special attention to data sources. An important use of the MEM is to derive the safe mud weight window in vertical and deviated boreholes (Plumb et al., 2013). Currently, most MEM models are 1D in nature (Knoll, 2016; Plumb et al., 2013), although 3D modeling is getting traction due to the increased use of seismic data (Sengupta et al., 2011) and finite-element or finite-difference modeling algorithms designed to do 3D stress modeling (De Gennaro, 2012; Koutsabeloulis and Zhang, 2009).

12.3.1 Mechanical Earth Model (MEM) in 1D

A mechanical earth model quantifies the in situ stresses, elastic and poroelasic properties, and rock strength in the earth. The following are the major components of a MEM (see Figure 12.5):

1. structural model containing surfaces and faults from seismic data and guided by log data, and the litho-stratigraphic model that is guided by geology;
2. lithology analysis including mineral fractions and porosity;
3. elastic and poroelastic properties such as Young's modulus, Poisson's ratio, bulk modulus, shear modulus, and Biot's consolidation coefficients;

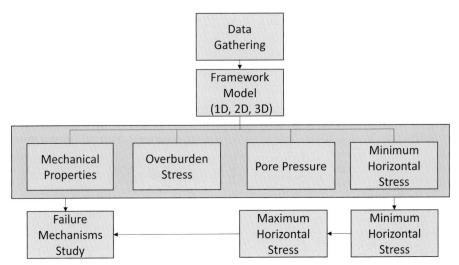

Figure 12.6 A simplified workflow to build a 1D-MEM. Modified after Ahmed et al. (2014).

4. rock strength such as unconfined compressive strength (UCS) and tensile strengths, and failure properties such as friction angle; and
5. in situ stresses such as pore pressure and fracture pressure and minimum horizontal stress (σ_h), maximum horizontal stress (σ_H), vertical stress (σ_V or S), and the direction of horizontal stress axes.

Figure 12.6 shows a typical workflow to build a 1D-MEM. Table 12.1 shows typical sources of information used to construct a MEM. The main source is the data from boreholes such as wireline logs, fracture and pore pressure data, as well as information from cuttings, cavings, and multiarm caliper to estimate stresses. Lately, model builders have begun using seismic data to some extent. Plumb et al. (2013) noted that the MEM is a living entity. In life-of-field-cycle analyses, the MEM evolves from exploration to development and then to abandonment. In its most complete form, the MEM consists of a full 3D description of pore pressure, stress, and mechanical properties. In practice, the complexity of the model evolves in step with the acquisition of new information. From exploration to development, the model evolves from a sparse set of 1D profiles to a full 3D description of rock properties and earth stresses. The degree of detail captured by the model will vary from field to field depending on the perceived operational risks (Plumb et al., 2013). A life-of-the-field cycle for MEM to minimize risks may look as follows:

1. Build a MEM. It represents all geological and rock mechanics information that currently exists in the field.
2. Use MEM to forecast wellbore stability along the planned well path.
3. Monitor the data while drilling to discover anomalies. They indicate flaws in the data or the MEM.

Table 12.1 Sources of data to build 1D-MEM as noted in Figure 12.6

Property needed for geomechanical model	Source logs	Other sources
Mechanical stratigraphy	*Conventional logs: resistivity, gamma, density, sonic*	*Seq. stratigraphy from seismic, cuttings, cavings, breakouts*
Pore pressure (P)	*sonic, checkshot, VSP, resistivity, conductivity*	*Int. vel. from seismic, MDT, RFT, DST, estimates based on mud weights*
Overburden stress (S)	*Any log that yields porosity estimates (density, resistivity)*	*Seismic (velocity–density–porosity transforms), cuttings*
Stress directions	*Multiarm caliper logs, borehole image log*	*Inferred velocity anisotropy (seismic), structural maps*
Minimum horizontal stress σ_h	*V_P, V_S logs, wireline stress tools*	*Existing data base, modeling*
Maximum horizontal stress σ_H	*Borehole image log*	*Wellbore stress models, database*
Various elastic parameters (Young's modulus-E, shear modulus-μ, Poisson's ratio-v)	*V_P, V_S logs and static-dynamic correlations*	*Seismic inversion (and static to dynamic correlations), tests on cores, database*
UCS, friction angle	*V_P, V_S logs, mechanical stratigraphy*	*Database, laboratory core tests on samples*
Failure mechanisms	*Borehole image log, oriented multiarm caliper log*	*Empirical observations, borehole modeling*

Source: Modified after Ali et al. (2003)

4. Analyze the anomalies to determine the sources of error. Immediate action on the rig can be initiated if required.
5. Correct the MEM as needed (e.g., abnormally low or high pore pressure).

Calibration of parameters and hence, the MEM, is an essential step in the workflow. Below we provide a narrative on how to obtain the parameters for 1D-MEM construction. This will be followed by discussions on how to build a 3D-MEM.

Elastic Properties of Rocks

Elastic properties that are needed for MEM construction are Young's modulus (E), Shear modulus (μ), and Poisson's ratio (v). There are two sources of information for this – each with its own strengths and weaknesses. For MEM construction we need data from *static* measurements such as those from cores as opposed to *dynamic* measurement such as those from petrophysical analysis of data from well logs or seismic data. Static measurements are, however, spotty and valid only in the area where cores were taken. For cost management, cores are taken only from a limited stratigraphic interval such as the ones where main reservoir units are located. This data, no matter how appropriate or valid do not provide enough information to generate a continuous MEM model. Furthermore, wellbore stability occurs mostly in the overburden such as in the over-pressured shale formations and very rarely shale core samples are taken – and definitely, not in the overpressured formations. Elastic properties from the dynamic measurements do not suffer from these deficiencies. It provides continuous coverage in depth including overburden and overpressured formations. However, the parameters need corrections as the MEM needs data from static measurements. The difference is mainly due to different domains of strain amplitudes employed in these measurements. The dynamic measurements are done using a very small energy pulse causing deformation which is reversible and so the dynamic moduli are obtained within a perfectly elastic regime. For core measurements, however, large strains have to be applied during loading process, some of which are irreversible. The measured moduli are therefore *not* purely elastic but introduce additional irreversible deformation caused by grain-scale friction and plastic deformation. This means the static strains are usually *larger* than the dynamic strains so the static elastic moduli are always smaller than the dynamic elastic moduli (Adisornsuapwat et al., 2013).

The dynamic elastic properties can be obtained from petrophysical analysis of interval transit times from sonic and dipole logs as well as seismic data. Recall that the relation between transit time (which is measured typically in μs/ft) to velocity (in m/s) is given by

$$V_p = \frac{0.3048 \times 10^6}{\Delta t_c}, \quad V_s = \frac{0.3048 \times 10^6}{\Delta t_s}$$

where V_P and V_S are the compressional and shear wave velocities while Δt_s is the shear wave slowness in μs/ft and Δt_c is the compressional wave slowness in μs/ft. Then, rearrangement of equations (2.42) with respect to velocity will give

$$\mu_d = \rho_b V_s^2 \tag{12.2a}$$

$$K_d = \rho_b \left(V_p^{\,2} - \frac{4}{3} V_s^{\,2} \right) \tag{12.2b}$$

$$E_d = \frac{9\mu_d K_d}{\mu_d + 3K_d} \tag{12.3}$$

$$v_d = \frac{3K_d - 2\mu_d}{6K_d + 2\mu_d} \tag{12.4}$$

where

ρ_b is the bulk density
μ_d is the dynamic shear modulus
K_d is the dynamic bulk modulus
E_d is the dynamic Young's modulus
v_d is the dynamic Poisson's ratio

In order to obtain static elastic properties from the dynamic properties, empirical correlations have to be used. The common practice is to cross-plot elastic properties from dynamic measurements to those from core measurements and devise a correlation factor. The literature is spotty; nonetheless, it provides many suggested correlations between static and dynamic elastic parameters over narrow reservoir intervals where the cores are taken

Table 12.2 Some empirical relations between UCS and static and dynamic Young's modulus from measurements on dynamic data

Lithology	Empirical relation	Reference
Igneous and metamorphic	$E_s = 1.263E_d - 29.5$	King (1983)
Igneous and metamorphic	$UCS = 4.31\left(\dfrac{E_d}{10}\right)^{1.705}$	King (1983)
Sedimentary rock	$E_s = 0.74E_d - 0.82$	Eissa and Kazi (1988)
Sedimentary rock	$\log E_s = 0.02 + 0.7\log(\rho E_d)$	Eissa and Kazi (1988)
Sedimentary rock	$E_s = 0.018E_d^2 + 0.422E_d$	Lacy (1997)
Sedimentary rock	$UCS = 0.278E_s^2 + 2.458E_s$	Lacy (1997)
Soft rocks	$UCS = 2.28 + 4.0189E_s$	Bradford et al. (1988)
Hard rocks ($E_s > 15$Gpa)	$E_s = 1.153E_d - 15.2$	Nur and Wang (1989)
Shale	$UCS = 0.77V_p^{2.93}$	Horsrud (2001)
Shale	$E_s = 0.076V_p^{3.23}$	Horsrud (2001)
Shale	$E_s = 0.0158E_d^{2.74}$	Ohen (2003)
Mudstone	$E_s = 0.103UCS^{1.086}$	Lashkaripour (2002)
Limestone	$E_s = 0.541E_d + 12.852$	Ameen (2009)
Limestone	$UCS = 2.94\left(\dfrac{E_s^{0.83}}{\phi^{0.088}}\right)$	Asef and Farrokhrouz (2010)

Note: V_P is in km/s and UCS is in MPa; E_S and E_d are in GPa.
Source: After Knoll (2016)

(Eissa and Kazi, 1988; Jizba, 1991; Wang and Nur, 2000; Haidary et al., 2015; Najibi et al., 2015; Knoll, 2016; Mavko et al., 2020). Table 12.2 (taken from Knoll, 2016) is a potpourri of sampling of some correlations. Suffixes s and d denote, respectively, static and dynamic values for the property under consideration. Before applications, it is always wise to check whether these correlations are valid in the particular area of application, especially if it is in a geologic area that is different from where the cores were taken.

Rock Strength Parameters (UCS and Friction Angle)

The unconfined compressive strength (UCS) and angle of internal friction (φ) of sedimentary rocks are key parameters for a MEM. Laboratory-based UCS and φ are typically determined through tri-axial tests on cylindrical core samples. Again, there is a paucity of such measurements and even if such measurements are made, such data are held as proprietary data. As a practical solution to this, a number of empirical relations have been proposed that relate rock strength to parameters measurable from geophysical well logs. Using such relations is often the only way to estimate strength parameter (UCS) in many situations. Carmichael (1988) classified the rock strength into five different groups based on UCS magnitude. A rock with UCS higher than 220 MPa is classified as a very high-strength rock and a rock with UCS less than 28 MPa is classified as a very low-strength rock. In Table 12.3, we give a few correlations for UCS for shales and sandstones (taken from Knoll, 2016, Table 4, p. 18) based on measurements using geophysical well logs (velocity and density) and derived Young's modulus, E from these logs.

Table 12.3 UCS for some typical lithologies

Lithology	UCS
Shale, North Sea	$0.77 (304.8/\Delta t)^{2.93}$
Shale, Gulf of Mexico	$0.43 (304.8/\Delta t)^{3.2}$
Shale, worldwide	$1.35 (304.8/\Delta t)^{2.6}$
Shale, Gulf of Mexico	$0.5 (304.8/\Delta t)^{3}$
Shale, North Sea	$7.97 E^{0.91}$
Shale	$7.22 E^{0.712}$
Sandstone, Gulf Coast	$1.4138 \times 107\, \Delta t^{-3}$
Sandstone, Australia	$42.1\exp(1.9 \times 10^{-11}\rho v_p^2)$
Sandstone, Gulf of Mexico	$3.87\exp(1.14 \times 10^{-11}\rho v_p^2)$
Sandstone, worldwide	$2.28 + 4.1089 E$
Limestone and dolomite, Middle East	$143.8 \exp(-6.95\Phi)$; Φ is porosity

Source: After Knoll (2016)

The angle of internal friction φ is a measure of the ability of a rock to withstand shear stress. It is the angle between the normal force and resultant force during failure due to a shearing stress. The tangent (shear/normal) is the coefficient of sliding friction. These parameters can be determined with laboratory tests on core data. A similar situation exists for friction angle as was the case with UCS – limited availability of static measurements. As with the UCS, it is often estimated using correlations to well log data. Below we show two published equations for the internal friction angle (Chang et al., 2006) that are commonly used in terms of V_p and porosity Φ:

Shale (Lal, 1999)	$\sin^{-1}\left[\dfrac{V_p - 1000}{V_p + 1000}\right]$
Sandstone (Weingarten and Perkins, 1995)	$57.8 - 105\Phi$

Chang et al. (2006) contains a good review of various empirical relationships for UCS and friction angle for various sedimentary rocks (sandstone, shale, limestone, and dolomite) to physical properties (such as velocity, modulus, and porosity). These equations can be used to estimate rock strength from parameters measurable with geophysical well logs. However, the usual caution applies – calibration of an MEM requires that parameters such as UCS and friction angle be calibrated as well for applicability of relations such as the ones presented above to a particular case.

Geopressure Data for MEM

The pore pressure is a very important component in a Mechanical Earth Model building process and is also critical to the calculation of horizontal stresses, wellbore stability analysis and other petroleum geomechanics applications. For reservoir intervals, in situ measurements such as RFT/MDT/DST are the most reliable ones. However, such data are usually not available for the overburden and in low-permeability sedimentary rocks such as shales. In that case we recommend using the methods discussed earlier in Chapters 4–7 that use geophysical well logs (sonic, resistivity, density), seismic, and drilling data. This is a viable approach – use seismic data to populate pore pressure in 3D and calibrate with well log data and RFT/MDT/DST measurements whenever possible. This is the only way to obtain *continuous* MEM data. In Chapter 9 we discussed pore pressure detection in real time. Techniques such as those are applicable to update pore pressure and hence, MEM, in real time. The required calibration points can be measured data as noted above or based on instability events encountered while drilling. Such events include the occurrence of kicks (usually because the mud weight is lower than the equivalent density of the pore pressure), loss of circulation (usually because of natural fractures or drilling-induced fractures due to too high a mudweight) or observations of instabilities in shales (shale breakouts) mainly due to wellbore shear failure or tensile failure.

Horizontal Stresses for MEM: The Least Horizontal Stress

Forces in the earth are quantified using stress tensors. We discussed this in Chapter 2. We also introduced the concept of principal stress. In most parts of the world at depths relevant for drilling, the total vertical stress or overburden stress is a principal stress. This means that the other principal stresses must act in horizontal directions. The larger of them is called greatest or maximum horizontal stress S_H (or $S_{H\max}$) and the smaller one minimum or least horizontal stress S_h (or $S_{h\min}$). The stress orientation around the world can be seen on the World Stress Map (Heidbach et al., 2016) where a lot of data has been compiled. Local perturbations occur and have to be considered for local geomechanical analysis. In situ stresses are related to each other. Just as vertical or overburden loading causes the rocks to be squeezed in that direction, it also pushes the rocks horizontally. This affects the horizontal stresses as the rocks are constrained by surrounding rocks. Thus the amount of horizontal stress depends on the Poisson's ratio (static values). Rocks with lower Poisson's ratio will have lower horizontal stress. The opposite will be true for rocks with higher Poisson's ratio. This will be in addition to whatever ambient stresses exist due to the tectonic components. The tectonic stress will always be a component in any stress modeling exercise in earth. It is also difficult to estimate.

The minimum horizontal stress can be directly measured using hydraulic fracturing or leak-off tests (LOT) or mini-frac tests. The minimum horizontal stress is taken to be the pressure at which the transition in flow regime occurs, namely, fracture closure pressure is achieved. However, such data are not routinely acquired. Even when these are acquired, the data points are usually confined to the reservoir interval and the measurements provide limited data points for the overburden.

A continuous estimation of the minimum horizontal stress thus requires using models. The following is a popular model for estimating the minimum principal horizontal stress, S_h from geophysical well logs (Avasthi et al., 2000)

$$S_h = \frac{v}{1-v}(S - \alpha P) + \alpha P + \sigma_{tect} \tag{12.5}$$

where v is Poisson's ratio, P is pore pressure, S is the total vertical stress (overburden stress), α is Biot's consolidation coefficient, and σ_{tect} is the tectonic stress. Both rock lithology (Poisson's ratio) and pore fluid pressure strongly affect horizontal stresses. Biot's consolidation coefficient α is a factor that helps to account for the deformation of a poroelastic material as the pore pressure changes. It illustrates how compressible the dry matrix frame is with respect to the solid material composing the matrix of the rock. The poroelastic constant also describes how effectively the fluid pressure counteracts the total applied stress. By definition, it varies between 0 and 1, but is typically between 0.7 and 1.0 for most petroleum reservoirs. In Chapter 2 we showed that

$$\alpha = 1 - \frac{K_d}{K_0} \tag{12.6}$$

where K_d is the bulk modulus of the dry rock framework and K_0 is the bulk modulus of the rock grains. Thus α describes the change of the bulk volume due to a pore pressure change while the stress is constant. Estimated values of α are from $\sim (0.7 - 0.8)$ to 1.0. In terms of Young's modulus E and Poisson's ratio v we have

$$K_d = \frac{E}{3(1 - 2v)} \tag{12.7}$$

Thus the methods that we discussed for obtaining E and v from geophysical well logs can be used for computing K_d using Gassmann's relation as discussed in Chapter 2. The modulus of mineral grains K_0 can be obtained from literature. We provide in Table 12.4 some commonly used values of bulk modulus of commonly occurring grain minerals (Song, 2012).

It is relatively easier to obtain the minimum horizontal stress than the maximum horizontal stress. In practice, the minimum horizontal stress can be estimated by acoustic log data analysis or measured through diagnostic fracture testing. It is worth

Table 12.4 Grain mineral bulk modulus of some common minerals

Grain mineral	Bulk modulus (Gpa)
Quartz	38
K-feldspar	37.5
Plagioclase	75.6
Calcite	70
Ankerite dolomite	80
Dolomite	80
Pyrite	143
Fluorapatite	86.5
Illite – smectite	23
Illite – mica	23
Kaoline	1.5
Chlorite	1.5
Clay	21.2
Illite/smectite (60/40 ordered)	36.7
Illite	60.1

Source: After Song (2012)

noting that stress estimation from acoustic log data must be validated and calibrated using stress measurements from diagnostic fracture testing. Having Biot's constant thus computed, the minimum horizontal stress (equation (12.5)) can be calculated by setting the tectonic stress as a calibration factor with data from measurements or instability events. In our opinion, this factor is often misused and has become the repository of various "unknowns" in the stress estimation workflows.

Horizontal Stresses for MEM: The Maximum Horizontal Stress

The maximum horizontal stress is widely considered as the most difficult component of the stress tensor. It requires knowledge of pore pressure, calibrated rock strength, vertical stress, and minimum horizontal stress data. The maximum horizontal stress can be estimated from image logs, frictional limit to stress and drilling-induced fractures data, and micro frac and caliper data. Similar to the minimum horizontal stress, we could use the poroelastic model of Chapter 2 and estimate maximum horizontal stress using equation (12.5). However, the magnitude of the tectonic stress term would be greatly different (higher) than the minimum horizontal stress. For maximum horizontal stress calculation, we need to use "breakdown pressure" from *micro* frac or XLOT data as defined in Figure 4.33 in Chapter 4 and adjust it for breakdowns seen in caliper and image logs. This is discussed in detail in Haidary et al. (2015). We note that image logs are not run frequently. This makes estimation of maximum horizontal stress difficult. Furthermore, in the literature the use of micro frac data for calibration is challenged by some authors (Lake and Fanchi, 2007). In summary, image logs provide the best estimate of maximum horizontal stress.

Stress Orientations

Analysis of borehole breakout data from image logs and four arm caliper data provide a direct evaluation of the stress orientations. By definition, minimum and maximum horizontal stresses are mutually orthogonal as these are principal stresses. Sometimes these can also be estimated from analyses of fault pattern and directions. It is known that breakouts are ellipsoidal wellbore enlargements caused by stress-induced failure (Aadnoy and Looyah, 2010). The breakouts occur whenever the circumferential stress exceeds the compressive rock strength. In a vertical well the spalling direction of the breakout parallels the minimum horizontal stress. In a homogeneous stress field, the direction of breakouts is usually consistent from the top to the bottom of the well. Analysis of caliper data reveals the breakout direction which is the direction of the minimum horizontal stress. Obviously, the maximum horizontal stress direction is perpendicular to that. The higher the difference between maximum and minimum horizontal stresses, the higher is the change in the breakout directions. For deviated wells, analysis of stress orientations are more complicated and we refer to Aadnoy and Looyeh (2010) for a more detailed discussion.

Determination of Safe Mud Weight Windows and Drilling
Rates for Drilling Deviated Wellbores

Exploration wells are typically vertical. However, appraisal and production wells are not. So are the wells for unconventional resources – most of the wells exploring from unconventional resources such as shale oil and gas are highly deviated. In many cases these are horizontal. Thus the drilling community must deal with preventing rock failure while drilling in the same geologic formation but with different deviations and at different directions. The problem of designing well trajectory and safe mud window in such a case is complicated due to the rotation of the induced stresses around the wellbore along its trajectory. In Chapter 2 we discussed principal stress estimation for vertical wells. Evaluation of principal stresses (magnitude and direction) becomes more complex for deviated wells as the stresses become azimuth and inclination dependent. Furthermore, in deviated wells, one may have to deal with changing stress regimes as well, for example, going from dominantly a normal stress regime ($\sigma_V > \sigma_{H\,max} > \sigma_{H\,min}$) to dominantly a strike-slip regime where $\sigma_{H\,max} > \sigma_V > \sigma_{H\,min}$. This change in stress regime will change the extent of the safe mud window and the optimum drilling direction in case of deviated wellbores. A good discussion of the problems and a solution is provided by Le and Rasouli (2012) and Fjaer et al. (2008). Briefly, the authors state that such computations are performed at successive steps: (1) in situ stresses must first be transferred to a Cartesian coordinate system comprising normal stresses and shear stresses in the direction of the deviated wellbore. (2) Once the normal and shear stresses have been determined then they must be converted into cylindrical coordinates which take into account their respective positions around the wellbore (i.e., orientation angle) at that specific azimuth and inclination. (3) Ascertain whether the stresses in the new system are principal stresses. Most likely, they are not because one of the shear stresses is likely to be nonzero. (4) Therefore, in order to do calculations of safe mud window for *both* breakouts (i.e., shear failure) and breakdown (i.e., tensile failure) these stresses should again be transferred into principal stresses so that Mohr–Coulomb principles may apply. These computations need to be done in real time and fed back to the drilling operations for safe drilling along a deviated wellbore path. The readers are referred to a case study where the above steps were applied for a highly overpressured shale-gas well in Australia (Le and Rasouli, 2012). An example is shown in Figure 12.7. Shown are the safe mud weights (in SG-units and also in ppg) for a horizontal well in different directions at a depth of 2800 m in a particular shale formation in a well in NW Australia (after Le and Rasouli, 2012).

There is another factor that contributes to the drilling challenges for deviated wells besides overbalanced mud weights. This has to do with the drilling rate, especially under the condition when breakouts have happened. Thus in deviated and near-horizontal wells, one not only needs to manage the mud weights properly as shown in Figure 12.7 but also reduce the drilling rate.

Anisotropic Poroelasticity and Stress Estimation in Unconventional Formations

For anisotropic formations with VTI symmetry the minimum and maximum horizontal stresses are related to the pore pressure and total vertical stresses. Equation (2.5) can be written in terms of the elastic stiffness tensor, C_{ij} the horizontal tectonic stresses and the anisotropic Biot's coefficient $\boldsymbol{\alpha} = [\alpha_V, \alpha_H, \alpha_h]$ (Sayers, 2010) as

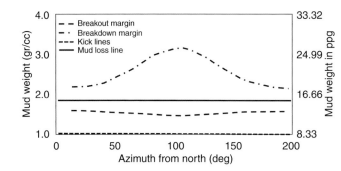

Figure 12.7 This example shows safe mud weight window for a horizontal well at different directions. The depth corresponds to 2800 m for a particular formation in a shale well in NW Australia. Modified after Le and Rasouli (2012).

$$S_h = \alpha_h p_p + \frac{C_{13}}{C_{33}}(S_V - \alpha_V p_p) + \left(C_{11} - \frac{C_{13}^2}{C_{33}}\right)\varepsilon_h + \left(C_{12} - \frac{C_{13}^2}{C_{33}}\right)\varepsilon_H$$

$$S_H = \alpha_H p_p + \frac{C_{13}}{C_{33}}(S_V - \alpha_V p_p) + \left(C_{12} - \frac{C_{13}^2}{C_{33}}\right)\varepsilon_h + \left(C_{11} - \frac{C_{13}^2}{C_{33}}\right)\varepsilon_H \qquad (12.8)$$

Note that for an anisotropic formation Biot's parameter in the principal coordinate system consists of three independent parameters for the vertical and two horizontal effective stresses such that the effective stress is given in terms of total stress and pore pressure as

$$S_V' = S_V - \alpha_V p_p, S_h' = S_h - \alpha_h p_p, S_H' = S_H - \alpha_H p_p \qquad (12.9)$$

The same equation can be written in terms of the strain ε and the anisotropic Young's modulus and Poisson's ratio as

$$S_h' = \frac{E v'}{E'(1-v)}S_V' + \frac{E}{(1-v^2)}(\varepsilon_h + v\varepsilon_H)$$

$$S_H' = \frac{E v'}{E'(1-v)}S_V' + \frac{E}{(1-v^2)}(\varepsilon_H + v\varepsilon_h) \qquad (12.10)$$

Under the assumption that the tectonic strains vanish one will get that for VTI media the poroelastic relation between vertical and horizontal stresses is given by $S_h' = S_H' = K_0 S_V'$ where $K_0 = \frac{C_{13}}{C_{33}} = \frac{E v'}{E'(1-v)}$. This is the VTI equivalent to equation (12.5) in anisotropic VTI media.

For material with orthorhombic symmetry the relation between vertical and horizontal stresses is given by

$$S_h = \alpha_h p_p + \frac{C_{13}}{C_{33}}(S_V - \alpha_V p_p) + \left(C_{11} - \frac{C_{13}^2}{C_{33}}\right)\varepsilon_h + \left(C_{12} - \frac{C_{23}C_{13}}{C_{33}}\right)\varepsilon_H$$

$$S_H = \alpha_H p_p + \frac{C_{23}}{C_{33}}(S_V - \alpha_V p_p) + \left(C_{12} - \frac{C_{23}C_{13}}{C_{33}}\right)\varepsilon_h + \left(C_{11} - \frac{C_{23}^2}{C_{33}}\right)\varepsilon_H \qquad (12.11)$$

And the relation between vertical and horizontal effective stresses are now given by

$$S'_h = K_1 S'_V, S'_H = K_2 S'_V \text{ where } K_1 = \frac{C_{13}}{C_{33}}, \ K_2 = \frac{C_{23}}{C_{33}}.$$

It is interesting to note that while orthorhombic analysis of anisotropic formation under stress is not often performed, wide-azimuth seismic data can be used to estimate the spatial distribution of the elastic parameters C_{33}, C_{13}, and C_{23} using orthorhombic AVA inversion technology (Bachrach, 2015). Therefore, K_1 and K_2 can be mapped. Figure 12.8 adopted from Bachrach et al. (2013) shows an example of strong variation in stress ratios within the middle Bakken formation.

In general, the elastic equation above assumed that the material is purely elastic. As noted by Sayers (2010) and Katahara (2009), the assumption of zero lateral strains can be interpreted as the steady state solution of system where all plastic strains are equal and is of opposite sign to the tectonic strain in the system. In other words, the assumption of zero lateral strain is in principle accounting for some plastic deformation which generates zero lateral total strains (tectonic elastic strain + plastic strain associated with irreversible deformation of the rock).

The assumption of elasticity to estimate horizontal stresses from vertical stress may not be valid at all times as some rocks, specifically high TOC rocks may behave in a viscoelastic manner (Sone and Zoback, 2014). Therefore, time-dependent deformation may play a role typically when using elastic properties estimated in recently drilled

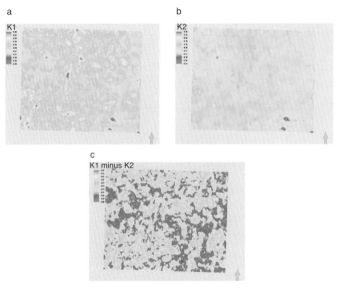

Figure 12.8 Example of spatial distribution of stress ratio coefficients in the middle Bakken formation within a few square kilometers: (a) K_1; (b) K_2; and (c) $K_1 - K_2$. The color bar ranges from 0.35 to 0.4 in (a) and (b) and from 0.0 to 0.02 in (c). This map is generated using azimuthal analysis of AVA data. Note the geomechanical hotspots, which show a large difference between K_1 and K_2. From Sengupta et al. (2011). (A black and white version of this figure will appear in some formats. For the color version, please refer to the plate section.)

boreholes where viscoelastic effects such as creep and relaxation still occur. However, practice shows that most effects will take place within the first 48 hours after drilling (Mavko and Saxena, 2016).

12.3.2   Mechanical Earth Model Building in 3D

The previous discussion was limited to 1D-MEM building for geomechanics. The key tools were geophysical well log–based continuous parameter description, calibrations based on laboratory tests or empirical models for converting dynamic model parameters to static model parameters, and field tests with hydraulic fracturing such as LOT or XLOT. There are two elements that can help improve the approach. First, the MEM building should be done in 3D. Second, wellbore stability analysis should include quantitative risk assessment through uncertainty analysis. We shall address the first issue here.

Three-dimensional mechanical earth modeling (MEM) is an established methodology concerned with the estimation of geomechanical properties in a heterogeneous 3D earth and is used extensively in the oil and gas industry for well planning and drilling and different scenarios of monitoring the reservoirs. Other geotechnical and civil applications consist of aquifer subsidence modeling and monitoring, slopes stability estimation and more.

The 3D MEM consists of the distribution of mechanical properties in 3D in accordance with all information available. Most of the time this information will come from well log data when available and seismic data which will be used to map the properties in 3D.

The 3D model of the subsurface mechanical properties is calibrated to well log data and is used to predict the state of stress in the earth and the expected deformation. In many cases the response of the subsurface to production, drilling, and stimulations such as hydraulic fracturing can be modeled for drill planning, well placement, and risk management.

Mechanical Properties in 3D from Seismic Inversion

To build a 3D MEM model one needs to assign geometry, material properties, and material constitutive equations. As well data are often sparse, a consistent approach to building a 3D-MEM would require extensive use of 3D seismic data in all aspects of a workflow. Because seismic wave propagation is related to the elastic properties of the subsurface a close link between seismic signatures and geomechanics exist. Thus, seismic-based subsurface parameter estimation (or seismic inversion) is a powerful technique which enables one to populate mechanical properties in 3D. The most popular methods to do this are based on poststack and/or AVA seismic inversion technology.

Poststack or near-offset seismic inversion techniques provide estimates of the acoustic impedance (or P-impedance often denoted as I_P which is the product of seismic velocity and density) from the reflection amplitude, and assumes seismic plane waves are being reflected from subsurface boundaries at normal incidence. Inversion of reflection amplitude into P-impedance enables one to map the heterogeneity in I_P in 3D and after proper dynamic to static transformation provide a constraint over one of the elastic parameters of the subsurface. Moreover, the P-impedance can be used to estimate pore pressure using many of the techniques as discussed in Chapter 5 by generating P-impedance and converting it to effective stress. Density from inversion can be used to estimate overburden stress and hence, pore pressure, by subtracting effective stress from the overburden stress by using Terzaghi law.

AVA prestack seismic inversion technology provides estimates of acoustic impedance, as well as V_P/V_S ratio and can also be used to map lithology from prestack seismic angle gathers (Avseth et al., 2005). Using different angle gathers that illuminate the subsurface at different directions, prestack inversion can be used to invert jointly for the layer elastic parameters (I_P and V_P/V_S ratios) as well as lithology. As in many cases geomechanical strength parameters and transformations are lithology-dependent, prestack inversion helps to provide much more information about many of the parameters needed for 3D MEM.

An example of a 3D-MEM using 3D seismic inversion methodology is shown in Figure 12.9 from a Middle Eastern deep carbonate reservoir (Sengupta et al., 2011). The field consists of tight, fractured carbonates (dolomite, calcite, and anhydrite), salt, and shales. In addition, high pressure and high temperature conditions, presence of H_2S and CO_2, narrow pore pressure/fracture pressure window and requirement of high pore pressure (~ 20 ppg mud weight) approaching overburden pressure made drilling a challenging operation in this area (Al-Saeedi et al., 2003). Figures 12.9a and 12.9b show acoustic impedance and Poisson's ratio from prestack seismic inversion. We also show the corresponding well curves. The tie is good. Figure 12.9c shows lithologies based on a probabilistic (Bayesian) classification estimated using seismic inversion and well logs. The color scale on the right of this figure indicates various lithologies.

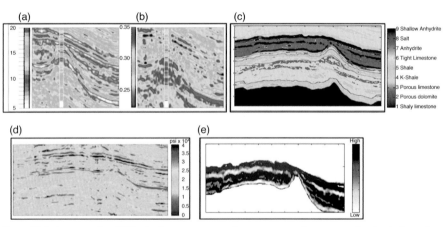

Figure 12.9 An example of 3D-elastic parameters and pore pressure from prestack seismic inversion of seismic data to build a seismic-guided 3D-MEM. (a) Acoustic impedance in (gr/cc x Km/s). (b) Poisson's ratio. (c) Lithologies based on a probabilistic (Bayesian) classification estimated using seismic inversion and well logs. The color scale on the right of this figure indicates various lithologies. (d) UCS estimates in (psi) and (e) a cross section of pore pressure derived from P-impedance. From Sengupta et al. (2011). (A black and white version of this figure will appear in some formats. For the color version, please refer to the plate section.)

This is a tedious but requisite process to obtain some clarity to up to nine lithofacies encountered here. The process used all well logs and benefited from a special well log processing – ELAN technique that generates rock models based on mineralogical composition, porosity, and water saturation. (The ELAN model is a popular probabilistic model that generates rock models given individual mineral compositions. Some details are contained in Mitchell and Nelson [1988].) Lithofacies analysis based on seismic inversion used here is based on the approach discussed by Avseth et al. (2005). It is known as Quantitative Seismic Interpretation or QSI. This allowed a separation of nine lithofacies. This is shown in the color scale on the right of Figure 12.9c. Multiple zones of porous dolomite (blue) and porous limestone (light blue) are revealed in the deep reservoir zone. These reservoirs are overlain with tight limestones (yellow) and shale (green). On top of these we see thick layers of salt (red) and anhydrites and dirty anhydrites (dark red). The lithologies indicate considerable lateral and vertical heterogeneities. The detailed analysis of lithology allows one to apply appropriate transforms to yield static moduli from dynamic moduli that are obtained from seismic inversion. Figure 12.9d shows a cross section of spatially varying UCS (in psi) from the 3D volume. Figure 12.9e shows a cross section of seismically derived pore pressure in the overburden zone containing the salt-anhydrite sequences that poses significant drilling hazards in this area. These models were tied to the 1D-MEM constructed using well logs and calibrated with core and other drilling data. The case study provided here suggests that seismic-driven geomechanical model building should be used whenever appropriate data are available. Using rock physics, geomechanics, and seismic and well logs we can estimate rock mechanical properties in a field in a physically meaningful way. Such approaches are predictive in nature for MEM in 3D and provide help in capturing the 3D heterogeneities of earth.

Stress Estimation from 3D MEM and Numerical Simulators

Three-dimensional mechanical earth modeling (MEM) involves solving the partial differential equation (2.29) in three dimensions for total stresses. The main benefit of 3D stress modeling is the ability to model or estimate the stresses associated with 3D heterogeneous earth where material properties can vary in space. To do that one needs to assign geometry, boundary conditions, and material properties with the material constitutive equations. The driving forces are usually gravity and far-field tectonic stresses (or strains). The localized stresses are distributed in 3D numerically using different numerical approximations to the boundary value problem. Most common solvers use the finite-element method to approximate numerically the solution in 3D. In geomechanics the constitutive equation is typically poroelastic (equations (2.77) and (2.78)) for a permeable reservoir while shale may be modeled using other constitutive elastoplastic models such as the Cam-clay model (Drucker and Prager, 1958). Strength and failure criteria (often parameterized using UCS, tensile strength and friction angle) are also provided for each element using the methods discussed in the previous section. Note that pore pressure in most static stress simulations is derived either from large-scale flow simulations (such as basin models) or from seismic

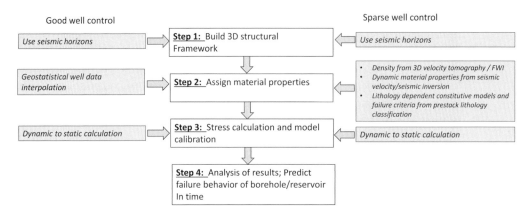

Figure 12.10 Schematic of a workflow to build 3D-MEM using seismic velocity analysis and prestack inversion.

velocity and/or regional well data information. The 3D MEM workflow typically includes the following four major steps as described in Figure 12.10:

1. Construct a 3D structural framework.
2. Assign material properties and strength to each element.
3. Calculate stresses and calibrate to well data.
4. Predict and analyze stresses, fracture gradients, and various drilling scenarios.

We will illustrate the workflow above and its benefits using the example given by Adachi et al. (2012) where a 3D MEM is used for fast well planning in the Gulf of Mexico.

The first step in the 3D MEM workflow is to construct a finite-element grid. For simple subsurface geometries one can use a stratigraphic framework based on interpreted horizons to define surfaces. In complex subsurface geometries such as salt-dominated provinces more complex surfaces may be used. Figure 12.11 modified from Adachi et al. (2012) shows an example of finite-element grid definition in a complex salt-dominated GOM region. In this example salt surfaces are first interpreted from seismic data together with key horizons. Then structured numerical grids that conform to surfaces are defined, layers that represent the overburden and intrasalt and underburden are added and finally the whole model is "encased" within sediment.

In the second step, after the finite-element grid is defined, material properties are assigned to the elements. Salt in this example is modeled as a linear elastic perfectly plastic Von Mises material (Fossum and Fredrich, 2002). Typical values for salt are density of 2.165 gm/cc, Young's modulus of 31 GPa and Poisson's ration of 0.25. Rock properties are assigned based on temperature profiles from regional studies accommodating the Smectite-to-Illite transition. In this example as no seismic inversion data are available, the material properties (Young's modulus, Poisson's ratio, UCS, tensile strength, and friction angle) are estimated from velocity based correlations. Shear properties are estimated using V_P/V_S regression such as the mudrock trend (Castagna et al., 1985). The elastic P- and shear wave velocities used are presented in Figure 12.12.

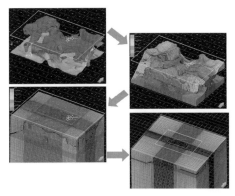

Figure 12.11 Various stages in constructing a finite-element grid in a salt-dominated region in the GOM (adopted from Adachi et al., 2012). Salt surfaces are defined (top left). Structured numerical grid conforming to surfaces is defined (top right). Layers that represent the overburden and intrasalt and underburden are added (bottom left). Finally, the whole model is "encased" within sediment (bottom right). (A black and white version of this figure will appear in some formats. For the color version, please refer to the plate section.)

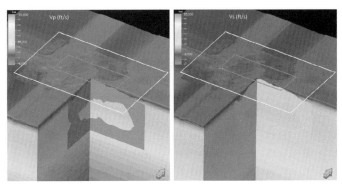

Figure 12.12 Material properties estimated from 3D seismic velocity cube. Modified from Adachi et al. (2012). (A black and white version of this figure will appear in some formats. For the color version, please refer to the plate section.)

In the third step – after the numerical grid is build and rock properties and pore pressure are populated, and the boundary conditions assigned, the stresses are calculated using any 3D finite-element solver to calculate the six Cartesian components of the stress tensor for each cell in the model. From these, the magnitude and orientation of the stress tensor is derived.

Model calibration is done by changing the boundary conditions to closely match borehole observation. Specifically, minimum principal stress (or fracture gradient) is compared to available LOT and FIT measurements as shown in Figure 12.13.

The results of the modeling can be used for well placement design where different well trajectories are considered with respect to the mud window. The three principal stresses are not always aligned with the vertical and horizontal directions as can be

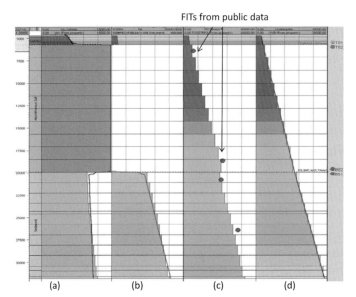

Figure 12.13 Comparison of minor principal stress computed using a finite-element model to FIT measurements. (a) Seismic velocity (the high-velocity layer is salt). (b) Pore pressure. (c) Minor principal stress. (d) Vertical stress. From Adachi et al. (2012). Vertical scale is from 5,00 to 32,500 ft . Horizontal scale on first track is from 0 to 15,000 ft/s and on second, third and fourth tracks is from 0 to 30,000 psi. (A black and white version of this figure will appear in some formats. For the color version, please refer to the plate section.)

clearly seen in Figure 12.14. Plotting the mudweight window attribute, pore pressure and shear and tensile failure together with the minimum horizontal stress shows clearly that some well trajectories will be subjected to large shear stresses and are not stable. This is a great help in well planning.

12.4   4D Geomechanics and 4D Earth Model Building

The term 4D geomechanics is borrowed from time lapse seismic and refers to time lapse geomechanical modeling of subsurface changes often due to production of hydrocarbon from reservoirs. However, this terminology also is valid to any other monitoring of various geomechanical effects that vary with time including subsidence, volcano monitoring, and any other time-dependent observations of deformation and stresses associated with natural or man-made activities. These stress changes can cause shear stresses along existing fault planes, eventually creating slip conditions and fault movement during depletion.

The need for better subsurface asset management has generated workflows where time lapse measurements of pore pressure and seismic response are compared with geomechanics simulations which map changes in pore pressure into stress and strains. The term 4D geomechanics refers to the modeling of pore pressure, stress changes, compaction, and subsidence with time. The equations used are typically the poroelastic equations for the

a

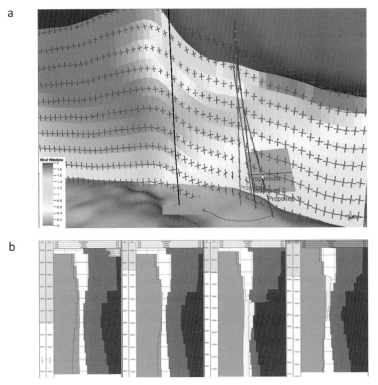

b

Figure 12.14 (a) Different well designs (Proposed 1–4) to reach subsalt target within their geomechanical context. The mud window attribute indicates the effect of the base of salt geometry on the "drillability" of a particular trajectory. Crosses represent the orientation of the minor (approximately horizontal) and major (approximately vertical) principal stresses. Notice the rotation of the principal stresses that follow the salt–sediment interface. (b) Pore pressure (gray), shear failure (yellow), tensile failure (dark blue), and minimum horizontal stress (light blue) along the different well trajectories 1–4. Note that trajectories 3 and 4 are predicted to have shear failure and are nondrillable based on this analysis. From Adachi et al. (2012). Units on all sections are (ppg). (A black and white version of this figure will appear in some formats. For the color version, please refer to the plate section.)

porous, permeable components of the sedimentary column while impermeable layers are often modeled using elastic or elastoplastic constitutive equations. The time-dependent changes are often related to human activities such as production and depletion or injection of enhanced oil recovery agents (CO_2 or brine) into the reservoirs.

Reservoir monitoring and surveillance technology have shown clearly that production-induced pore pressure changes affect the reservoir and the surrounding rocks (Addis, 1997; Hettema et al., 2000; Hatchell et al., 2003). The change in stress can trigger seismicity (Seagall, 1989) and can be imaged using seismic time-shift analysis. Many seismic attributes such as AVA response, shear wave splitting (Crampin, 1981) and anisotropic traveltime differences (Olofsson et al., 2003) can also be used. These effects can be analyzed using rock physics and calibrated to seismic observations. Time shifts which are the time difference between the traveltime of an image ray between repeated seismic surveys, say, baseline and monitor survey, have been of particular interest from the geomechanics point of view. This is because they can be directly related to strain changes in the reservoir as will be discussed below. As seismic velocities are directly related to the mechanical properties of the rocks, seismic data provides an opportunity to measure indirectly stress and strain through changes in elastic properties and with it the ability to improve the understanding of the reservoir behavior. Applications include production optimization, caprock mechanical integrity monitoring, and drill planning. As always pore pressure plays an important role. In the section below we will review some of the rock physics models used and show a case study where geomechanical simulations are used together with time lapse monitoring to better delineate reservoir pore pressure and compartmentalization.

12.4.1   Effective Stress Changes and Seismic Velocities

Linear elasticity assumes that the elastic stiffness is not dependent on stress or strain. Laboratory experiments on many rocks shows strong dependencies between the elastic stiffness $\mathbf{C}$ and the state of stress or strain of the sample. This nonlinear relation can be denoted as $\mathbf{C} = \mathbf{C}(\varepsilon)$ or $\mathbf{C} = \mathbf{C}(\sigma)$. We will assume here that nonlinear elastic material stiffnesses can be expressed as explicit functions of stress or strain and it is assumed that $\mathbf{C}(\varepsilon)$ can be expressed as $\mathbf{C}(\varepsilon(\sigma))$. When production-induced pressure changes are small in comparison to the state of stress in the reservoir, one can use a Taylor series expansion to express the relation between stiffness and strain as:

$$\mathbf{C}(\varepsilon' + \Delta\varepsilon) \approx \mathbf{C}(\varepsilon') + \left.\frac{\partial \mathbf{C}}{\partial \varepsilon}\right|_{\varepsilon'} \Delta\varepsilon = \mathbf{C}(\varepsilon') + \mathbf{c}\Delta\varepsilon \tag{12.12}$$

The gradient of the fourth-rank elastic stiffness tensor with respect to the second-rank strain tensor is a sixth-rank tensor $\mathbf{c} = \partial\mathbf{C}/\partial\varepsilon$ known as the third-order elasticity (TOE) tensor (Thurston, 1964; Prioul et al., 2004). In general $\mathbf{c}$ is not constant over the whole stress range that may be associated with production, but can be measured in the laboratory. For a VTI media the independent TOE coefficients are c_{111}, c_{112} and c_{123}. The stiffness is expressed as a linear function of strain as (Prioul et al., 2004):

$$C_{11} \cong C_{11}^0 + c_{111}\varepsilon_{11} + c_{112}(\varepsilon_{22} + \varepsilon_{33}), \quad C_{22} \cong C_{22}^0 + c_{111}\varepsilon_{22} + c_{112}(\varepsilon_{11} + \varepsilon_{33})$$
$$C_{33} \cong C_{33}^0 + c_{111}\varepsilon_{33} + c_{112}(\varepsilon_{22} + \varepsilon_{11}), \quad C_{12} \cong C_{12}^0 + c_{123}\varepsilon_{33} + c_{112}(\varepsilon_{22} + \varepsilon_{11})$$
$$C_{13} \cong C_{13}^0 + c_{123}\varepsilon_{22} + c_{112}(\varepsilon_{11} + \varepsilon_{33}), \quad C_{23} \cong C_{23}^0 + c_{123}\varepsilon_{11} + c_{112}(\varepsilon_{22} + \varepsilon_{33})$$
$$C_{66} \cong C_{66}^0 + c_{144}\varepsilon_{33} + c_{155}(\varepsilon_{11} + \varepsilon_{22}), \quad C_{55} \cong C_{44}^0 + c_{144}\varepsilon_{22} + c_{155}(\varepsilon_{11} + \varepsilon_{33})$$
$$C_{44} \cong C_{44}^0 + c_{144}\varepsilon_{11} + c_{155}(\varepsilon_{22} + \varepsilon_{33}) \tag{12.13}$$

where $c_{144} = (c_{112} - c_{113})/2$ and $c_{155} = (c_{111} - c_{112})/4$ and C_{ij}^0 is the unstrained reference VTI stiffness tensor.

Landro and Stammeijer (2004) showed that because time, depth, and velocity are related ($t = z/V$), the traveltime derivative $\Delta t/t$ can be expressed in terms of changes in layer depth (reservoir compaction $\frac{\Delta z}{z} = \varepsilon_{33}$) and the changes in seismic velocity,

that is, $\frac{\Delta t}{t} = \frac{\Delta z}{z} - \frac{\Delta V_P}{V_P} = \varepsilon_{33} - \frac{\Delta V_P}{V_P}$. Assuming that the main compaction is vertical, the traveltime can be directly related to the vertical strain and the TOE parameter which is often referred to as R factor (Hatchel et al., 2003; Fuck et al., 2009):

$$\frac{\Delta t}{t} \approx \varepsilon_{33} - \frac{\Delta V_P}{V_{P,base}} \approx \varepsilon_{33} - \frac{1}{2}\frac{\Delta C_{33}}{C_{33,base}} = \varepsilon_{33} - \frac{1}{2}\frac{c_{111}\varepsilon_{33}}{C_{33,base}} = (1 + R)\varepsilon_{33} \qquad (12.14)$$

where the R factor is expressed in terms of the TOE parameter as $R = -\frac{1}{2}\frac{c_{111}}{C_{33,base}}$. It is important to note that equation (12.13) is more general then the vertical approximation of equation (12.14) which uses only the vertical strains. Equation (12.13) maps the change in the three principal strain directions into the changes in elastic stiffness and can accommodate anisotropy and lateral strain or time-shift observations.

12.4.2 Using 4D Geomechanics to Model Production-Induced Pore Pressure, Strains, and Seismic Response

In 4D geomechanics we are interested in modeling the temporal changes in pore pressure, effective stress, and strains. Changes in stress in the earth will occur when aquifer and reservoirs are produced. The change in pore pressure over time will cause redistribution of the effective stresses and strains in the subsurface. The changes in stresses may cause failure in some prats of the subsurface and therefore modeling the effect of the production enables better reservoir management.

An example of 4D geomechanics modeling is presented in Figure 12.15 where a production scenario is being simulated using a coupled fluid-flow-geomechanical numerical simulator. A coupled fluid flow-geomechanical simulator solves for fluid flow in the porous media (usually using multiphase Darcy's law) and calculates the pore pressure distribution due to a certain production scenario. This method is extensively used also for CO_2 sequestration among other time varying geomechanical scenarios (Tillner et al., 2014; Cappa and Rutqvist, 2012). This pore pressure information is then coupled into a geomechanical simulation which solves for stress and strain distributions. There are different ways of coupling fluid-flow simulation and geomechanics. Biot's equation for fluid and solid displacement (Chapter 2) defines a fully coupled system. In many cases using the pore pressure from flow simulation as an input to the geomechanical simulation (known as one-way coupling) provides sufficient accuracy at a lower computational cost. In the example here a primary production scenario of five producers produce a heterogonous reservoir for 5 years, followed by a secondary scenario where five injectors are used to enhance hydrocarbon recovery (Hill et al., 2017; Bachrach and Padayesh, 2017). Figure 12.15a shows the geometry of the producer and injector wells together with the reservoir model. Figure 12.15b shows the pore pressure changes within the reservoir after 10 years of production. Figure

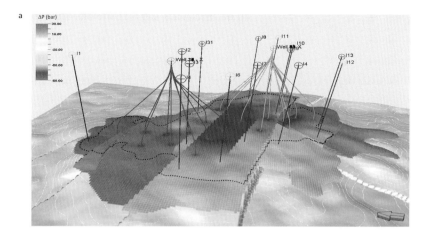

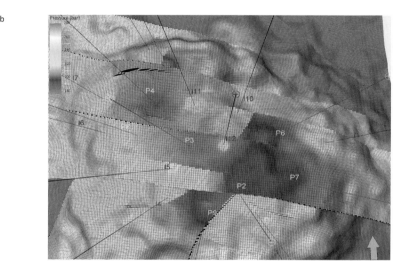

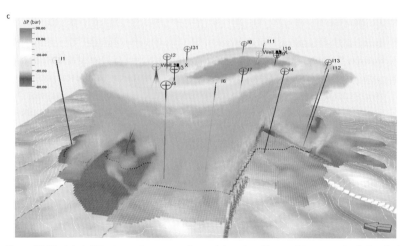

Figure 12.15 (a) A 3D view of a reservoir model and wells used in numerical simulation where producers and injectors are displayed. (b) Aerial view of producers (marked with "P") and injectors (marked with "i") and pore pressure distribution in the reservoir zone after modeling production for 10 years as described in the text. (c) Vertical strain estimates in 3D from coupled flow-geomechanical modeling showing 10 years of production. All pressure map units are in (bar). (A black and white version of this figure will appear in some formats. For the color version, please refer to the plate section.)

12.15c shows the modeled vertical strain associated with 10 years of production. The vertical strain can be mapped using equation (12.14) into seismic wave traveltime difference. This type of modeling provides a quantitative foundation for relating seismic observations to reservoir architecture and active pore pressure changes. Such forward modeling has been used by many authors to relate seismic observations to reservoir pressure changes and provides a basic framework for reservoir management (Barkved, 2012; Garcia and MacBeth, 2013).

12.5   Summary

Geomechanics and pore pressure are intimately related as both pore pressure and total stress are related to the effective stress. Geomechanics addresses pore pressure and stresses jointly. Strength and failure are important components of geomechanical analysis together with the elastic parameters. Analyses of the in situ stress can be done from a borehole-centric approach using 1D mechanical earth models but also can be done within the 3D aspect of the subsurface. We have demonstrated that in many cases where subsurface complexity is large, 1D MEM should be complimented with 3D stress modeling. Specifically, complex geometries and large contrast in material properties can rotate the stress tensor and affect borehole stability calculations. The elastic properties in 3D together with strength parameters can be estimated using seismic techniques such as tomography and seismic inversion. Elastic parameters used in geomechanics should be the static elastic parameters while strength parameters are typically estimated from laboratory derived correlations. In unconventional resource exploration and exploitation, formation anisotropy plays an important role and should be considered. Pore pressure changes due to production can cause an increase in effective stress when pore pressure is reduced due to reservoir depletion or a decrease in the effective stress due to pore pressure increase in an injection scenario or when wells are being shut down and stop pumping. The ability to look at the pore pressure and stresses jointly through numerical simulation can enable us to address different failure scenarios. The joint modeling and observation of stress, pressure and strain changes in the subsurface provide the basic building blocks for reservoir monitoring systems that can feed back observations into reservoir models and help in risk mitigation and cost optimization for subsurface resource exploitation. Lastly, geomechanics could be and should be used as a predictive tool rather than depending heavily on the offset-well concept. In this regard, prudent use of seismic data is highly recommended at all phases of petroleum geomechanics modeling.

13 Guidelines for Best Practices
Geopressure Prediction and Analysis

13.1 Introduction

Quantitative analysis of geopressure comprises a range of disciplines – seismic, well logs, drilling, mudlogs, rock physics, geomechanics, and basin modeling. Although there are some overlaps, it is a challenging task to write a best-practice document on the vast subject. So our attempt should be viewed just as stated – *an attempt*. This is not an apology but an admission on our part that although we have expertise in many aspects of geophysics, our views are certainly colored by what we have learned from the literature and from our work, and have not experienced the trials and tribulations faced by many of our colleagues in the sister disciplines such as drilling and mudlogging.

Geopressure in sedimentary rocks is a phenomenon involving complex interactions between rocks and fluids in a geologic setting that involves wide ranges of geologic time, temperature and pressure conditions. Furthermore, often the tectonic imprint of rock deformation and the structural evolution of the basin leave an indelible imprint on the magnitude and distribution of pore pressure that makes it harder to quantify in the predrill sense using currently known remote sensing tools such as seismic and commonly known borehole tools and technologies such as well logs. Even for sedimentary rocks, our knowledge of geopressure has been limited to a great extent to that of pore pressure in shales. Observation of geopressure in rocks other than shales are many (Bjørlykke, 2015). For sands, we often use a shortcut – the so-called centroid effect. What about carbonates? Although some efforts have been made in the literature – we discussed some of these efforts in the book (Chapter 4), most conclude that prediction methods for shales *do not* work well in carbonates, or for that matter, in any hard rock. Therefore, "best practice" statements in the context of quantitative prediction and analysis is limited mostly by our experience with workflows for pressure prediction in shales and extrapolation of that knowledge to non-clastics. There remains much to be investigated here.

Historically, early pioneers (Hottman and Johnson, 1965; Eaton, 1975) focused on shales not only because of its preponderance in sedimentary basins – more than 70 percent of any such basin is indeed composed of shales, but also because deposition of shales tends to be more massive and relatively homogeneous, thus making pressure prediction within shale units more manageable. Sands are, however, entirely different – pressure prediction within these units are difficult because of large variations of its rock properties, especially, composition, diagenetic alterations, porosity, permeability, and velocity. These variations plus its structural configurations cause the pore pressure within sands to be greater, equal or less than the pore pressure within the adjacent shale units. We discussed this in Chapter 3 and showed how one can relate the pressure in sands through some simple structural approximations – the so-called centroid effect (Bruce and Bowers, 2002; Swarbrick, 2002; Mouchet and Mitchell, 1989). As shown in Figure 3.19, the centroid effect occurs when an initially flat sand unit (reservoir) surrounded by and in equilibrium with geopressured shale is loaded asymmetrically and tilted during subsequent burial. The pore pressure gradient within the sand unit remains hydrostatic. At the depth of the centroid (usually heuristically taken to be the mean elevation of the sand), the shale and sand pressures are equal. In the up-dip direction of the sand unit, the pore pressure is higher than the shale while the reverse is the case in the down-dip direction. Because of this, pressure prediction in sands were considered unreliable using standard pore pressure prediction methodologies. Basin modeling approach to pore pressure prediction does not suffer from this limitation. The procedure includes the continuous interaction between sand and shale pressures during the entire structural evolution of the basin (Hantschel and Kauerauf, 2009).

In addition to sands, pore pressure prediction in rocks other than clastic, for example, carbonates and igneous rocks, is still a major challenge as noted earlier. In spite of this limitation, we can still present a few *guidelines* for pressure prediction. These guidelines for best practices can be divided into *three* broad categories: Subsurface geological *habita*t for pore pressure *(geology)*, *physics* of pore pressure generation (*models*), and *technology* for subsurface prediction (*tools*). We are going to leave best-practice discussions on subjects such as drilling and various other pressure detection tools such as MWD/LWD/Mudlogging to others to comment.

13.2   Subsurface Geological Habitat for Geopressure (Geology)

A study of processes related to geopressure development, prediction, and detection using state-of-the-art tools and technologies are extremely important from both energy resource estimation and production and environmental points of view. To meet the growing demands from operators in the oil and gas industry as well as regulatory agencies, many new businesses have sprung over many decades around pore pressure prediction. In this regard, engineering approaches based on some simple models have become the norm. The focus became the development of so-called user-friendly tools (software) rather than the "physics" behind pore pressure models. Although these simple models are easy to implement many are not fundamentally sound. Furthermore, all "unknowns" of these so-called simple models have been shoved under "calibration." While calibration is important and often required, it is not a substitute for failure to try to understand the fundamentals of this complex phenomenon and the limitations of each model. This is especialy important if we are to extrapolate these models to depths beyond the range of the calibration regime. As a result, many practitioners in the commercial domain often resort to empiricism and unduly heavy dependence on analogues that may not be correct analogues resulting in many failures and blowouts.

Therefore, before embarking on a pore pressure prediction project, the best practice is to first gather all necessary data and *understand the subsurface geological habitat of pore pressure* for a basin or a study area within a basin. The prediction at a well is the final outcome of this endeavor; however, we must remember that the rock and fluid distribution around and away from a well and its evolution through geologic time determine the pore pressure at that particular location. This understanding is critical to successfully predict subsurface pressure. Dickinson (1953) noted that there are two major components in understanding the distribution of geopressure – stratigraphy and structure of the basin or the mini-basin. This leads to an understanding of the rock types (distribution of shales and sands, for example) and their stress regime. If a basin is shale dominated and the overburden stress is mostly vertical, then the engineering approach based on compaction trend analyses may yield reasonable results for pore pressure (with calibration). Dickinson (1953) noted correctly that a reservoir containing high pore pressure must be effectively isolated from any other porous formation that contains normal hydrostatic pressure. Otherwise, the excess pressure would be dissipated. Thus, faulting and the stratigraphy that the fault(s) cut through will have an overriding effect on the pore pressure magnitude and distribution, irrespective of the cause of abnormally high pore pressure. This understanding is very important for a successful project. Otherwise, any "calibration" will not be sufficient to produce reliable results. An example is shown in Figure 13.1 (taken from Dickinson, 1953, Figure 10). These diagrams indicate that geopressure can occur near the top of the shale series in a mini-basin only if the porous bed (sand) is isolated by pinch-out or is faulted down against the shale series. In the absence of pinch-out of the reservoir abnormal pressures can only be preserved in upthrown blocks at depths below the top of the shale series which are greater in amount than the throw of the fault. Models and understanding of the pore pressure habitats such as the ones shown in Figure 13.1 are extremely important to analyze and understand prior to undertaking any quantitative geopressure analysis project. Otherwise results may be a futile exercise in a prediction project with very little in the way of further predictive capability.

In summary, an understanding of the overall geologic structures and stratigraphy along with fault distribution and their role in the "plumbing" of the basin are extremely important for geopressure prediction. Single-well prediction without this understanding can lead to failure.

It is very important to *understand the stress regimes* of a basin prior to embarking on a pressure prediction project. Stress orientation, for example, the orientation of three principal stresses as discussed in Chapter 2, would dictate not only whether a particular methodology for pressure prediction would work (say, undercompaction) but also affect pressure dissipation through fractures and its orientations. In Figure 13.2 we show schematically the orientation of three principal stresses in three dominant types of basins – extensional, compressional and pull-apart or strike-slip. About 80 percent of the sedimentary basins on earth have formed by *extension* of the plates (often termed lithospheric extension). Most of the remaining 20 percent of basins were formed by

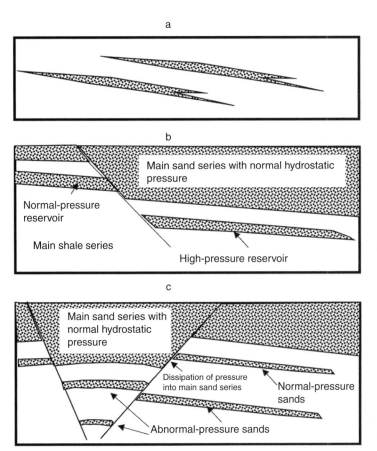

Figure 13.1 Types of reservoir seals necessary to preserve high pore pressure and its stratigraphic content. (a) Small reservoir sealed by pinchout. (b) Large reservoir sealed updip by faulting down against thick shale series, sealed downdip by regional facies changes. (c) Relative position of fault seals in upthrown and downthrown blocks. Modified after Dickinson (1953).

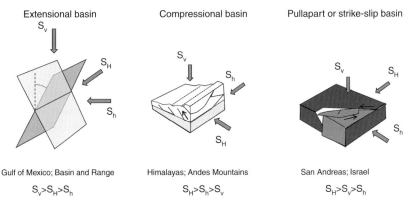

Figure 13.2 Stress regimes in sedimentary basins showing the three principal stress orientations. These have significant effect on velocity anisotropy. Images modified after http://plate-tectonic. narod.ru/tectonics5photoalbum.html.

flexure of the plates beneath with various forms of loading. Pull-apart or *strike-slip* basins are relatively small and form in association with bends in strike-slip faults, such as the San Andreas Fault. Stress orientations in a basin dictates the fracture orientation. Opening mode fractures are typically created perpendicular to minimum principal stress. Fractures control not only pressure dissipation but also create stress-induced anisotropy in the medium that affects overall velocity anisotropy. This is in addition to inherent intrinsic anisotropy of rocks (shales, for example) and layer-induced anisotropy such as the one discussed earlier in the context of transverse isotropy (VTI or TTI). For extensional basins, the minimum principal stress is horizontal and the maximum principal stress is vertical. In this case, the stress-induced maximum velocity will be vertical and the stress-induced minimum velocity will be horizontal. This is not the case for compressional basins. Here the maximum principal stress is horizontal and the minimum principal stress is vertical. Consequently, the stress-induced maximum velocity will be horizontal and the stress-induced minimum velocity will be vertical. Fracture pressure in this type of basins could be greater than the overburden pressure calculated from integrating bulk density logs, and if not accounted for correctly, it could lead to suboptimal well design. Furthermore, pore pressure estimate would be incorrect due to incorrect use of stresses in the Terzaghi model. In strike-slip basins, both maximum and minimum principal stresses are horizontal and the intermediate principal stress is vertical. This indicates that estimates of stress-induced velocity anisotropy would require data that correlates velocity variations with azimuth. Conventional seismic processing and analysis techniques to estimate anisotropy based on conventional seismic data and hence, pore pressure, will not work unless the data can provide information on azimuthal anisotropy (Bachrach and Sengupta, 2008). The simple well calibration method will not yield reliable pore pressure and fracture pressure estimates in this case.

The literature (and this book as well) has considerable discussions on the phenomenon of the "centroid effect," which is associated with the structural configuration of permeable and nonpermeable beds and the flow of fluids (oil, water, and gas) in those rocks (Stump et al., 1998; Heppard and Traugott, 1998). We discussed this in Chapters 3 and 4. In Figure 13.3, taken from Liu et al. (2018), we show an imprint of centroid effect on 3D anisotropic seismic velocity from tomography and the corresponding overpressure section. The effect of structure on fluid transmission up-dip causing overpressure is fairly convincing. What is not shown is the actual maps of permeable beds overlain on the structure. In spite of this, it is clear from the literature that there are significant differences between shale predicted pore pressure and in situ pressure in sand and its effects on well design. The authors interpreted observed deviations between measurements and predictions as actual differences between sand and shale pressure attributed to lateral transfer of fluids along inclined reservoirs. The elevation where shale and sand pressure are in equilibrium is called a centroid. This is an important effect as this can affect seal capacity and seal failure in dipping permeable beds. In the context of geohazards, this is also a very important effect for shallow aquifer pressured sands (SWF). Lateral transfer of fluids in these sands can easily cause pore pressure to exceed the fracture pressure resulting in a blowout. Currently, centroid

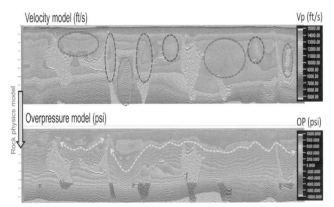

Figure 13.3 An extended large-scale velocity model converted to overpressure. The overpressure follows geology features and suggests that the slow (blue cycles) and fast (red cycles) velocity trends derived by tomography are plausible features caused by fluid migration along geologic horizons. Such models include centroid effect automatically through anisotropic velocity modeling (as done here) without having to do separate analyses. Modified after Liu et al. (2017). (A black and white version of this figure will appear in some formats. For the color version, please refer to the plate section.)

effect is handled in a *heuristic* manner – the model accounts for *only* structural elevation of the sand in relation to surrounding shales (Bowers, 1994; Stump et al., 1998; Heppard and Traugott, 1998; Liu et al., 2017). When and how that occured is irrelevant in the model. Reflection tomography/FWI based approaches using high-quality 3D seismic data such as the ones used by Liu et al. (2017) and Dutta et al. (2015) show that large-scale models do reflect centroid effect adequately without any ad hoc extra corrections. Currently, most practitioners estimate shale pressure in 1D and then apply this correction. This practice – one in which 3D effects are purportedly simulated using 1D model, is inadequate at best and definitely, *not* a best practice. The complete treatment to account for this effect should include the history of basin evolution in 3D (shale, sands, other lithologies, faults, salt and other entities) that affect concomitant buildup and dissipation of pore pressure (Hantschel and Kauerauf, 2009). We think that in the near future, a seamless workflow that includes *both* basin modeling and seismic velocity modeling and imaging for structure and pore pressure would be carried out routinely. This will require *integration* of various subdisciplines (Petmecky et al., 2009; Pradhan et al., 2017b, 2020) – a key to success in the prediction of pore pressure. Some elements of this approach are discussed in Chapter 10 and in Section 13.5. However, much remains to be done.

Integration of disciplines noted above will be hampered if it is not done by integrating teams with a common goal. It is called "buddy work" – a term used in the military to achieve victory. Same should be true in geopressure prediction. Every project should be done by a team populated by geologists, geophysicists, petrophysicists, and well designers – the latter being the beneficiaries of the project output. This is a best practice.

13.3   Physics of Pore Pressure Generation (Models)

Geopressure phenomenon is related to generation and dissipation of fluids in porous rock formations under gravitational and tectonic loading. If the rate of dissipation of fluids is not rapid enough to keep up with the production, geopressure will happen. In Chapter 3 we discussed various mechanisms that cause geopressure. It is certain that any single mechanism *cannot* explain the magnitude and distribution of geopressure in a basin and that multiple mechanisms operate in tandem. Historically, most practitioners focused on simple models such as porosity reduction with depth (compaction models). Although some investigators (e.g., Hower et al., 1976; Meissner, 1978; Bruce, 1984; Dutta, 1986, 1987b; Berg and Gangi, 1999; Hansom and Lee, 2005) addressed various aspects of fluid generation due to mineral metamorphism of argillaceous sediments and hydrocarbon generation, the subject never received the full attention it deserved. Such models are based on integration of chemical kinetics of various mineral conversion processes in geology with complete reconstruction of thermal history of the basin in 3D. In fact, many of the processes listed and discussed in Chapter 3 are thermally controlled and are coupled. This increases the complexity of the modeling of geopressure processes as we need to know the operating envelopes of these mechanisms and models to describe those causes. Hence, the modeling efforts did not extend much beyond simplistic approaches and assumptions. So a best practice scenario envisions understanding the root cause(s) of geopressure before diving into a particular pore pressure prediction project.

13.3.1   Rock Physics Compliant Models for Geopressure

In the current landscape, most of the models for pore pressure generation mechanisms and their implementation with geophysical and drilling tools are *not* compliant with the basic principles of rock physics and rock mechanics. In many cases, these models have *unphysical* end-limits and ignore lithology dependency often found in field examples. Most of the popular models are not consistently backed by the effective stress law as these ignore the loading-path dependency (thermal history dependent process) as noted by Bowers (1994) and Huffman (2002). A case in point is the widely used Eaton method (Eaton, 1975). It is subjected to criticism in all the points just raised. This method is *not* rock physics compliant: (1) the so-called normal compaction trend has unphysical porosities that is less than the "critical porosity" (Marion et al., 1992) at shallow burial depths and unphysical porosity at deeper burial depths, (2) there is no physically justified reason for the logarithm of NCT of velocity versus depth to be a straight line– it must bend at shallow and deeper depths, and (3) the model neglects lithology dependency as well as various thermal and other geological effects. There are other unphysical attributes also that we discussed in Chapters 3 and 4. The same criticisms apply to methods by Hottman and Johnson (1965), Pennebaker (1968), Reynolds (1973), and many others. *We recommend the following best practice to those who chose to use these models: make the NCT of either velocity or resistivity rock physics compliant by making sure that porosity is eliminated from the compaction model (for example, porosity as a function of effective stress) by an appropriate attribute law (for example, porosity dependence of velocity).* This was discussed in Chapter 4.

Chapter 4 also discussed various other alternatives to rock physics compliant velocity models. *We strongly recommend the use of effective stress based approaches to pore pressure prediction as discussed in* Chapter 4. *For example, Bowers's model is far more physical than the Eaton-type or Hottman-Johnson type approaches to pore pressure prediction.* Some authors (e.g., Sayers et al., 2002) have suggested the use of mudweight- and well log–based velocity based data to construct a locally calibrated empirical relationship between effective stress and velocity to predict effective stress from velocity data. There are two problems with this approach. First, one needs to be sure that the mud weights are true representation of formation pressure – at best, it is an upper limit to pore pressure. Second, we need to have physical and geological justifications for using curves such as these beyond the rage of validity of the pressure data. As the depth of investigation goes deeper, it is expected that various thermal effects will come into play and could invalidate the use of these locally calibrated curves. *Best practice of pressure prediction with a geophysical tool such as velocity or resistivity must recognize that velocity (or resistivity) of a subsurface rock is a direct function of its depositional and thermal history. A proper accounting of these require an integrated approach that involves basin modeling together with rock physics and data from wells.*

13.3.2   Pressure Prediction in Formations Other than Clastics

We discussed this is Chapter 4 with a focus toward carbonates (high-velocity rocks). We noted that the methods and models that are available in the literature for pore pressure prediction works well for clastics but not for nonclastics such as carbonates due mainly to the fact that in this type of rocks there is a loss of sensitivity of velocity dependency on effective stress. Using an approach typical for pressure prediction in shales (high dependency of velocity on effective stress), one would obtain a wrong pressure profile. The carbonates have a typically fast velocity which would appear as a low pressure interval. In reality, the correct pressure in the carbonate layer could be very high. This is because carbonate rock properties and pore pressure are significantly affected by various diagenetic processes that depend upon burial environment, thermal history, and the pore water chemistry. The porosity in carbonates is significantly affected by these processes and so the relationship between porosity and effective stress in carbonates become unrelated for most parts as opposed to that for shale. Green et al. (2016) discuss this in detail. They showed that velocity reversal in such formations are not related to geopressure but to lithology. This was noted as early as 1971 (Reynolds et al., 1971).

The above problems make pressure prediction in carbonates (and other high-velocity nonclastic rocks) very difficult with the current techniques. We discussed some promising new developments in Section 4.4. Some authors (Huffman, 2002;

Sengupta et al., 2011) have advocated the use of high-fidelity seismic velocity based on *inversion* of seismic data as a tool for pressure prediction in carbonates. This is a promising approach and needs to be pursued further for assessment whether results are consistently reliable. Since high porosity and high pore pressure are not correlated in carbonates, we are essentially dealing with the problem of predicting two variables from one input (P-wave velocity). We need another attribute to tackle the inverse problem. Could shear wave velocity or density – the two inputs that are usually not available – be the answer? It remains to be seen. Shear wave data are usually not available and density attribute is very hard to infer from inversion of seismic data. As opposed to clastics, the stiffening and softening of rock frame in carbonates are less detectable by velocity measurements. This is an added difficulty that is a serious challenge to predicting pore pressure in carbonates. Some authors (Green et al., 2016) believe that local calibration based on rock compressibility measurements on core data coupled with detailed geologic modeling and analysis to isolate the details of the carbonate formation and pressure measurements along with seismic inversion– based velocity correlations is a viable approach to pressure prediction in carbonates. In our opinion, this type of integrations are promising and certainly worth trying. We discussed this in Section 4.4.

In summary, a best practice document for geopressure analysis and prediction in carbonates and other non-clastics *cannot* be written at this time until recently proposed models are tested, new models are proposed, evaluated and case studies are published.

13.3.3   Normal, Fracture, and Overburden Pressure

Any computation of geopressure begins with establishing the normal or hydrostatic pore pressure. This baseline defines the magnitude of overpressure. For deepwater sediments, the difference between the normal and fracture pressures is very small at shallow depths. In SWF conditions, this difference must be investigated very carefully so that casing decisions can be made reliably. For this reason, normal pore pressure should be calculated as accurately as possible in the study area. We recommend that *for best practice, fluid properties from models discussed in* Chapter 1 *and* Appendix A *should be used to evaluate density. For brine, we must refrain from using a constant brine density model.*

Predrill prediction of fracture pressure is just as important as that of pore pressure. It is one of the most critical part of every well plan. Calculating these properly defines the drilling window needed for safe drilling operations as noted in Chapter 4. Uncertainties in these models arise from the input data, assumptions used in the workflow and the complexity/uncertainty of the geological conditions and rock properties. *The best practice is to use the LOT or XLOT data from analogue wells by first recognizing that the analogue well is a good match to the well to be drilled and then extend the model to the well under consideration by using a model that uses rock physics compliant Poisson's ratio as a parameter (see equation (4.47)) after correcting these to their static counterparts.* The literature indicates that the effective stress coefficient needed for Eaton-type models may not be constant (Zhang and Yin, 2017); it could be depth dependent, affected by the tectonic stress and could also be anisotropic. So variability in fracture pressure gradient calculations must be accounted for. There is no agreement on the exact nature of effective stress coefficient's dependency on depth. Fracture gradient calculations in subsalt formations are affected by salt deformations and flow. Although some authors reported that fracture gradients in subsalt wells exceeded overburden stress gradient in their data (Zhang and Yin, 2017), it is not clear whether such data (LOT, XLOT) are affected by complexity of measurements in salt that relates to anisotropic stress orientations and the presence of tectonic stress.

Both pore pressure and fracture pressure prediction require an estimation of overburden stress. Analogue wells are good source of this data. The commonly used procedure of using 1 psi/ft for overburden stress gradient estimation is not recommended. *Ideally, density log is the best data to use for overburden stress estimation. However, when this is not possible, we recommend using seismic velocity or well log–based velocity data to estimate density using some of the models discussed in* Chapter 4. In Section 4.8, we provided a guideline for estimating bulk density from various sources. A word of caution: geophysicists routinely use Gardner's relation for bulk density and velocity (Gregory, 1977). However, this relationship is based on data from measurements on well-consolidated rocks and therefore, its application to Tertiary Clastic, and especially, overpressured shale, is *not recommended*. Furthermore, we *do not* recommend using Wylie-time average equation to obtain porosity from sonic logs or seismic velocity. *Based on our experience, we recommend the use of Issler-type relation (equation (4.27)) and seismic velocity with locally calibrated coefficients as a viable approach.* Nonetheless, whenever possible, empirical models for density should be calibrated with actual density data from logs and corrections applied to overburden stress calculations as needed.

13.4   Technology for Subsurface Prediction (Tools)

Seismic velocity is a key source for pore pressure prediction before drilling. However, it is also used improperly in many cases by the practitioners of pressure prediction. While most users have a basic level of understanding of some of the velocity analysis tools that the geophysicists use, they do not invest enough in understanding the *limitations* of these tools for pore pressure prediction. This is an important issue. Even many in the seismic imaging community who deal with various velocity analysis tools routinely do not appreciate that the ensuing product (velocity) from their data processing arsenal may not be fit for pore pressure applications. Over many decades of applications, both the seismic community and the drilling community have realized that the velocity that is fit for *imaging* may not be suitable for pore pressure prediction. *There is a saying in the geophysics community that goes something like this – What looks good from far (for example, an image) may be far from good (for example, pore pressure)*!

In Chapter 4 we discussed various velocity analysis tools – conventional stacking velocity analysis, tomography, FWI, and post- and prestack inversion. Among these tools, conventional stacking velocity analysis based on, for example, Dix-type interval velocity analysis technique is the least accurate and the *least reliab*le tool. Yet these tools (interval velocities) are still routinely used for pore pressure prediction without appreciating their lack of resolution and the purpose why the stacking process was invented. *This is a bad practice. If we must use this approach, we recommend that the 20-step quality control procedure be followed as discussed in* Chapter 4. Very often, in the engineering world of pressure prediction, this may be the only option, especially in old oil fields and on-shore projects where reprocessing of seismic data may not be an option.

High-end velocity analysis tools such as tomography of various flavors, FWI and velocity from amplitude inversion with a good low-frequency velocity model are better suited for pore pressure prediction. But these tools are relatively expensive as these require additional processing, analysis and interpretation. Moreover these are complex tools and require resources for access and additional training. It is worth noting that advancement in automation of seismic data analysis significantly reduce the cost of deployment of advanced velocity analysis tools. However, still there is added cost and historically such techniques are not used as frequently for pore pressure prediction. This is not acceptable. *The additional cost of high-end and fit-for-purpose seismic velocity analysis should be judged against the cost of not doing so – less than optimal pressure images, re-drilling or side-tracking the well and in the worst case, dealing with a blowout leading to damage to life and environment.* In addition, no matter what velocity analysis tool is used, we must address a very important issue: making sure that the velocity model is geologically consistent and defendable. *Thus geologist's and interpreter's input to velocity analysis are critical for reliable pressure prediction prior to drilling. This is a best practice.*

13.4.1   Fit-for-Purpose Seismic Velocity Analysis for Pressure Prediction

The very process of extraction of velocity from seismic data involves some type of inversion (even the interval velocity analysis by the Dix-model) and therefore, are ambiguous and non-unique. In fact all velocity analysis tools are based on the ultimate criterion that the seismic gathers be flat either in the offset or the angle domain. *While this is necessary, it is not sufficient for pore pressure prediction.* We need additional constraint(s) on the velocity to make sure that the *velocity model is physical*. This means that the velocity model must resemble a *vertical* velocity profile (not easily achieved by gather flattening criterion alone), and must handle *anisotropy* properly. Furthermore, the *vertical velocity must be within the bounds of the rock physics models*. This is because seismic velocities are non-unique and even after these are made geologically consistent and highly interpretable, these usually violate the bounds required for pore pressure prediction. For example, we must ask whether *the velocities yield hydrostatic pore pressure (highest velocity that is physical) in one end and fracture pressure (lowest velocity that is physical) at the other end?* This is an additional and very important constraint that must be imposed on any velocity analysis that is used for pressure prediction. In Chapter 6 we discussed in detail a model that was developed for this purpose – using rock physics models to guide the (vertical) velocity analysis process in tomography (see Dutta et al., 2015a, Figure 2). In this approach, the model not only yields flat seismic gathers but also delivers physical velocities (bounded by pore pressure limits) and also improves the seismic image when obtained using the *same* velocity model. This is a right step toward building an Earth model and not just a velocity model. *We recommend this type of rock physics guided or assisted velocity analysis for pore pressure prediction.*

All velocity analysis must be corrected for anisotropy prior to pore pressure application. This is imperative and is a consequence, mostly due to the seismic acquisition geometry. Neglect of anisotropy causes the vertical velocity to be faster than the "true" velocity and results in lower pore pressure than the correct value. Well-constrained tomography or FWI is a viable tool; however, these do not guaranty that the interwell space is modeled properly. Geostatistical approaches are also good techniques for this, provided the models are checked against well data for reliability. Thus corrections for anisotropy require availability of well(s). An alternate approach is to use rock physics based models for anisotropy (Bachrach et al., 2011; Pradhan et al., 2017a) provided the model parameters – there are many – are calibrated as much as possible. Most commonly used models for anisotropy are with either vertical, horizontal or tilted axes of symmetry. Other anisotropy models such as the orthorhombic models are getting traction in the seismic imaging community. These models for velocity analyses should be tested for pore pressure analyses. When techniques such as these are complimented with various anisotropy measurements using borehole-centric seismic tools, we will open a new dimension in pore pressure prediction. This is not a common practice yet, but informed practices based on these concepts is along the right-track for being a best practice.

Interval velocities from seismic inversion – poststack, AVO, and GA are very good tools for estimating pore pressure in rocks provided a good *low-frequency* model is used. (Tomography and FWI are good candidates for this.) Velocities from seismic inversion provide more details for pore pressure and are eminently suited for integration with borehole-centric seismic or sonic tools.

13.4.2 Constrained-Velocity Analysis for Pressure Prediction in Geologically Complex Areas

Most of the velocity analysis procedures for pressure prediction today are suboptimal for geologically complex areas such as subsalt, areas with complex thrusting, and clastics or softer rocks overlain by much harder rocks. There are two main reasons. First, conventional seismic data may not contain useful signal for meaningful model building and imaging. Second, limited incidence angles cause large uncertainty in velocity because of insensitivity of moveout to changes in velocity. Circumventing the first problem requires innovative seismic acquisition techniques such as using ultra-long-offset cables, salt undershooting or dual coil acquisition with multiple vessels (Revolution seismic data, for example) and full-azimuth imaging. Combining ultra-long offsets with high-end processing technologies like full-waveform inversion and anisotropic reverse time migration, improve visualization in the most complex subsalt geometries, including complex salt structures, rugose salt tops, shallow salt bodies, and steeply dipping sediment areas. In Figure 13.4 an example of the new acquisition geometry and an improved image quality are shown. The full Revolution survey dataset was acquired using WesternGeco Dual Coil Shooting – full-azimuth acquisition, which employed four seismic vessels – two source and two recording boats (Figure 13.4a). The vessels were equipped with point-receiver marine seismic system. In this example, recording boats continuously shoot in a series of interlinked circles (with 6250 m radius), and data are acquired over the full azimuth, 360°, and over the full offset range. The data example (Figure 13.4b) shows a marked improvement of the subsalt images. Data such as these are very useful for velocity model building and hence, pore pressure prediction below salt.

Even with high-quality seismic data, velocity models solely based on tomography or FWI are still nonunique. Further constraints are needed as discussed in Chapter 6. We discussed a practical approach to this in that chapter as well – constraining the subsalt tomography using geology in conjunction with thermal history modeling and rock physics principles. We referred to this as *rock physics guided (or assisted) velocity modeling for migration and pore pressure prediction*. A novel feature of this technology is to use predicted pore pressure as a guide to improve the quality of the earth model in an *iterative* fashion using tools such as tomography and FWI. Thus, we produce a velocity model that not only flattens the CIP gathers, but also limits the velocity field to its physically and geologically plausible ranges. This yields both a better image and reliable pore pressure below salt. Some examples are shown in Figures 6.16–6.21 that use Revolution seismic data and Figure 14.2 that use ultra-long-offset data.

In summary, for geologically complex areas, we need fit-for-purpose seismic data with ultra-long offsets with the preferred capability of continuous azimuth coverage for proper anisotropy modeling for velocity analysis. If the data quality is not good and image is poor, the data must not be used for pore pressure analysis. We further recommend that imposing physical constraints on inversion algorithms such as tomography and FWI of seismic data and state-of-the-art migration technique such as RTM, especially for complex geologic areas, are essential to getting reliable pore pressure estimates. When this best practice is implemented, the process will also produce substantial improvement in seismic images using the same velocity model.

13.4.3   Basin Modeling as a Tool for Pore Pressure Prediction

Basin Modeling is a very useful tool for pore pressure prediction. There are numerous commercial software available in the industry. Most of these softwares use the basic concepts outlined in Chapter 10. A major advantage of basin modeling is the capability it provides to understand predictions in the context of various geological *"what-if-scenarios."* However, some software may not provide the full flexibility for defining all the constitutive relations as discussed in Chapters 3 and 10. These are compaction, chemical kinetics, diagenesis, pressure mechanisms, mass transfer models and parameters such as permeability and fluid properties and heat transfer processes such as thermal conduction,

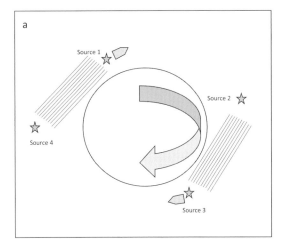

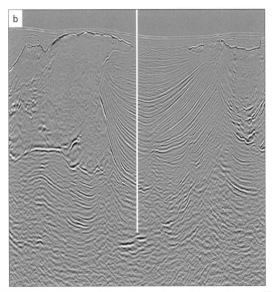

Figure 13.4 (a) Seismic data acquisition layout for multivessel long-offset circular geometry. (b) An example of the image from Revolution VIII acquisition. In this example, on the right, the right image (right of the bar) shows an enhanced stack after full-waveform inversion (FWI) plus subsalt update, while the left image (left of the bar) shows an enhanced stack from a comparable wide-azimuth dataset with FWI plus subsalt update. The improved steep dip imaging, clearer salt base, and subsalt imaging seen on Revolution VIII data give a clearer understanding of the region's geologic complexities. Images from www.multiclient.slb.com/latest-projects/north-a merica/gom_revolution.aspx.

thermal convection and phase transformation due to heating, and various heat generation processes such as basement heat flow and radioactivity model. As we showed in Chapter 10, these models are all connected and depend on each other. Therefore, pore pressure practitioners need the flexibility in defining their own models based on their experience in a given basin. A word of caution. Most algorithms have built-in system of default parameters. This is a convenient option. However, we need to question the validity of some of these options – *are these options valid in my basin? This is the beginning of a best practice.* Predicted pore pressures also need to be calibrated. This is easier said than done in basin modeling. Just obtaining some measured pressure data and posting them on the pressure versus depth curve does not constitute a full calibration. This is because the comparison is non-unique. For example, errors in permeability model may be compensated by incorrect fluid and thermal properties and still yield *fortuitously* correct pore pressure. What is the answer? We do not have the answer but offer two suggestions for beast-practice: *trust but verify and do error propagation and uncertainty analysis.* The first suggestion deals with understanding the ranges of various parameters and verifying whether these are within known valid ranges in a basin under study. Analogue wells provide good guidance for this. Once the prior range of uncertainty is defined, numerous basin model runs are simulated with these inputs, giving a range of outputs that are then manually calibrated to observed well data. This is a tedious process and requires diligence, but it is a good deterministic approach. The second and more rigorous approach requires using probability theory based analysis and Monte Carlo basin modeling simulations. Bayesian theory is a good tool for this. A Bayesian framework integrating uncertainties in basin modeling with observed well data and seismic data is described in Section 13.5 on uncertainty analysis. A probabilistic approach addresses how certain we are of our predictions from basin modeling. This does not provide a unique solution but raises the awareness of reliability of a certain prediction technique.

13.4.4   Petrophysical Well Logs for Pore Pressure Analysis

This is a mature topic, and deals with all downhole tools such as various geophysical tools, drilling tools and mudlogging devices on the rig. Most popular geophysical tools are resistivity, conductivity, density, sonic, gamma rays and neutron-gamma logs. In these methods, almost all log responses (except for radioactivity tools) are related to porosity-based indicator of pore pressure. Suffice it to say that all best practice scenarios for *processing* well logs must be followed prior to pressure prediction. (Fertl (1976) has a good discussion of the pitfalls of various geophysical (wireline) logging tools.) This includes environmental corrections such as mud-cake and mud-filtration effects as well as assessing whether key logs, for example, sonic and resistivity are conditioned for pore pressure. If using resistivity logs, we *must* account for salinity and temperature corrections. If using sonic logs, we should account for "cycle skipping" and fluid effects (e.g., effect of gas, especially, of low saturation variety). Missing or bad logs must be supplemented by synthetic logs provided such logs are checked and made rock physics compliant. Assuming all tools are used correctly and measurements are made within specified ranges of parameters such as pressure and temperature, we suggest that a *best practice is not to depend on measurements from a single tool but use indicators from multiple tools.* We recommend that pore pressure be computed from *several* petrophysical well logs – sonic, resistivity, conductivity, density, and so on. This means, for example, resistivity-based pressure indicator should be checked against those predicted using sonic tools and so on. Consistent predictions will emerge from this comparative study. This type of analysis – one in which multiple porosity indicators are used for pore pressure will yield not only better quality control of input well logs, but also of increased reliability of predicted pore pressure values from these logs.

13.4.5   Drilling Parameter Indicators of Pore Pressure

Oil and gas industry has a remarkable track record of drilling oil and gas wells safely. This resulted in various drilling parameter indicators for geopressure. We discussed some of these in Chapter 7. Penetration rate is one of the most widely used parameters in this regard. However, relating rate of penetration to pore pressure requires understanding a wide range of drilling tool related parameters and their effect on the drilling rate such as bit weight, bit diameter, rotary speed, and drill bit tooth-dullness, to name a few (Combs, 1968; Bourgoyne, 1971). We recommend a regression-type analysis for a general equation for penetration rate of formations (shales, for example). Combs (1968) proposed a parameterization of the following form:

$$R = R_0 \left(\frac{W}{3.5}\right)^a \left(\frac{N}{200}\right)^b \left(\frac{Q}{3D_h D_n}\right)^c f(P_d) f(T) \qquad (13.1)$$

Here R is penetration rate (ft/h), W is weight-on-bit (in 1000 pounds), R_0 is shale-drillability constant (ft/h) with a sharp bit at zero differential pressure, N is rotary speed (rpm), Q is flow rate (gallons per minute), D_h is bit nozzle diameter (inches), $a, b, c,$ are weight, speed and hydraulic exponents, respectively, f is a function of P_d with P_d being the differential pressure in the borehole (pounds per gallon) and T is bit-wear index. Combs (1968) and also Bourgoyne (1971) recommend using a regression analysis to determine these parameters. Bourgoyne (1971) showed that the drillability constant, R_0 is related to the bulk density of the formation by

$$\rho_b = 2.65 - 1.65 \left(\frac{S_g + \log R_0^{-6}}{S_g}\right) \qquad (13.2)$$

where ρ_b is bulk density (g/cm^3), and S_g is rock strength parameter (a value of 5.2 is recommended by the author in the Gulf of Mexico, USA). This is an empirical relation. However, this sort of analysis allows one to convert density into pseudo-porosity which in turn can be checked against density and other geophysical logs that depend on porosity. This is a bridge that links pore pressure from drilling parameters to those from geophysical well logs. *We recommend this approach – however, the predictions from drilling parameter analysis must always be checked with those from geophysical well logs. This will expand the reliability envelope of pressure prediction.*

In Chapter 12 we presented ways to construct 3D-geomechanical models (also known as the mechanical earth model or MEM). This is necessary to safely drill a well and provide a framework for update not only in real time but also during

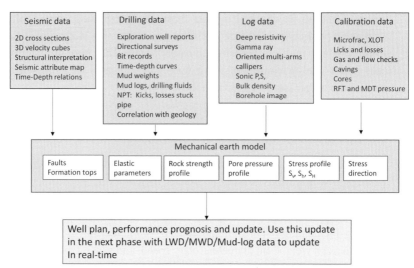

Figure 13.5 A partial list of data that contribute to building a complex geomechanical earth model that can be updated in real time. Modified after Alfred et al. (1999).

production and abandonment. Figure 13.5 provides a schematic of the process – some elements of this were also discussed in Chapter 12. *A best practice approach will be to use this type of workflow with 3D-seismic to infill the space away from a well.* This is not a common practice today as much as it should be. We think that the boundaries between various disciplines need to be blurred to produce a safe drilling outcome. This means that *all 3D-MEM be should be executed with 3D-seismic in everything (to the extent that is possible).*

13.5   Uncertainty Analysis

As mentioned multiple times, even with good quality seismic data, inversions for seismic velocities, and rock physics assisted velocities are always non-unique. Basin modeling runs can have a range of outputs depending on the range of uncertain input parameters. Data are often imperfect and insufficient; our models are approximate, and there is always natural variability and heterogeneity. All of these factors lead to uncertainty in the pore pressure predicted from seismically derived velocities, well logs, drilling parameters and basin modeling. *Hence as a best practice it is recommended to do careful uncertainty quantification in every pore pressure prediction project.* Uncertainty quantification uses the tools of probability and statistics to estimate distributions of input parameters (prior distributions) that are propagated through the models to get estimates of uncertainties in the predictions. Observations (e.g., well data or seismic kinematics) are then used to condition and constrain the prior range of outputs, giving us a possibly narrower posterior uncertainty. In the following, we describe various approaches to uncertainty quantification for pore pressure, some that do not integrate basin modeling and others that attempt to integrate the geologic history using basin models with well data, seismic data, and rock physics models for quantifying the posterior distribution of predicted pore pressure. A similar approach could be used for pressure prediction using drilling parameters.

Many empirical methods of pore pressure prediction (such as the ones described in Chapter 4) rely on the principle of compaction disequilibrium and require definition of a normal compaction trend line. Deviations from the normal trend are used as indicators of increased pore pressure. But how certain is the definition of the normal compaction trend? How does the uncertainty in the normal trend impact the pore pressure predictions? Wessling et al. (2013) addressed uncertainties in pore pressure prediction from well logs. They addressed uncertainty in the normal compaction trend by fitting multiple linear regressions to well data to estimate multiple normal compaction trends. The shallower (normally compacted) part of the resistivity log was used to define the trend. Multiple bootstrap datasets were created by windowing off different sections of the resistivity log. The windows were large enough to capture the overall linear trend. A series of linear regressions were fit to each windowed off log section to get multiple normal compaction trends. Trend line variations, and hence variations in the deviations from the trend were then propagated to the predicted pore pressure using the Eaton method (see Chapter 4). Wessling et al. (2013) present applications of their methodology to 23 data sets from different regions of the world.

Malinverno et al. (2004) present a Bayesian Monte Carlo approach to quantify and propagate uncertainties for pore pressure prediction. They use Eaton's method but instead of keeping the inputs to be deterministic, they allow the measurements and parameters to have uncertainties described by probability distributions. Monte Carlo draws from the posterior distributions of the inputs to Eaton's equation, conditioned to data, are propagated through the equation to get a distribution of predicted pore pressure. A similar workflow could also be applied using, for example, Bower's method or any of the other methods. The output of this approach is a probability distribution of pore pressure, given the available data. Malinverno et al. (2004) demonstrate examples of this methodology showing how the uncertainty in pore pressure predicted before drilling (from seismic tomographic velocities) is reduced by additional data (checkshots, mud weights, sonic logs, and pore pressure measurements) obtained while drilling. Moos et al. (2004) use a similar Monte Carlo approach to assess the impact of uncertainties in velocities and velocity-effective stress relationships on the predicted pore pressures. They use the results for quantitative risk assessment of predrill pore pressure, sealing potential, and mud window predictions.

Oughton et al. (2018) provide a coherent multivariate Bayesian framework to link all the different models and model parameters in the pore pressure prediction system. They use a sequential dynamic Bayesian network to fully model the joint conditional distribution of the data and the model parameters. At each depth level a Bayes network captures the multivariate distribution of all the variables at that depth such as density, overburden stress, vertical effective stress, porosity, lithology, sonic velocity and pore pressure. The network model at each depth is connected to the depth level above and below, thus incorporating vertical dependence. The vertical sequential model also includes a lithology transition matrix which gives the probability of each lithology (sand/shale) at each depth given the lithology at the depth above. The Bayesian network rigorously models the interactions between different quantities and can update the predictions as each type of data (e.g., porosity or sonic velocity) is observed. The outputs are the posterior distribution of the pore pressure as a function of depth, given the data and the prior distributions of input variables. The posterior distribution is obtained by Gibbs sampling which produces realizations of values from the posterior distribution. The model as presented only accounts for mechanical compaction as the pore pressure generating mechanism, (though other mechanisms could be incorporated), and only models vertical dependence. In contrast, fully 3D spatial dependence and spatial uncertainty is modeled in the workflow presented by Doyen et al. (2003) and Sayers et al. (2006). This workflow uses a 3D probabilistic mechanical earth model (P-MEM) where the variables such as porosity, clay content, seismic velocity, overburden stress and pore pressure are taken to be spatially correlated, multivariate random variables. These random variables are conditioned to available data such as the porosity and density logs as well as the smooth tomographic velocity cube. Geostatistical conditional simulations are used to generate multiple 3D cubes of P-wave velocity (V), density, porosity (ϕ) and clay content (C). The tomographic velocity is used as a smooth trend to constrain the geostatistical simulations. The overburden stress (S) is obtained by integrating the density, while velocity (V) is related to effective stress, porosity and clay content via an extended Bower's type relation in Doyen et al. (2003) given by:

$$V = a_1 - a_2\phi - a_3 C + a_4(S - P)^{a_5} \tag{13.3}$$

Here the effective stress is $(S - P)$ and $a_1, \dots, a_5$ are fitting parameters taken to be uncertain, but with spatially invariant distributions. From the simulated 3D cubes, output realizations of pore pressure are obtained using the relation

$$P = S - [(V - a_1 + a_2\phi + a_3 C)/a_4]^{1/a_5} \tag{13.4}$$

Doyen et al. (2003) compare a linearized Gaussian calculation of the posterior distribution with a full sequential simulation approach, and show that while the linearized approach is fast and analytic, the full stochastic approach is more robust and more fully captures nonlinearities in the velocity to pore pressure relation.

An important input in predrill pore pressure prediction is the tomographic seismic velocity. Addressing uncertainties in the velocity and its impact on pore pressure uncertainties should be an important part of the pore pressure prediction workflow. Caidado et al. (2012) and Osypov et al. (2013) are examples of workflows for assessing uncertainties in velocity tomography. Osypov et al. (2013) further show how this velocity uncertainty impacts uncertainty in pore pressure estimation.

As mentioned earlier, basin modeling is a useful tool for pore pressure prediction. In basin modeling, pore pressure evolution in the basin is simulated by solving differential equations for sediment compaction mechanisms and fluid flow with boundary conditions designed to simulate the physics of the problem. This enables us to model effects of common overpressure generating mechanisms such as compaction disequilibrium, aquathermal pressuring, smectite-illite transformation or kerogen maturation. However, though basin modeling presents a quantitative framework for modeling of geologic processes, a major limitation of method is the large uncertainties associated with the basin modeling input parameters including the lithologic compaction functions, lithology-specific porosity-permeability relations, present-day structural and stratigraphic model and boundary conditions such as paleo heat flow. The conventional approach to specifying these parameters is to perform a deterministic calibration of modeling outputs to well data. However, since our final goal is to quantify uncertainty in the pore pressure predictions, integrating both well and seismic data, it is necessary to rigorously quantify and propagate any basin modeling input uncertainties into the output space. De Prisco et al. (2015) identify these challenges and propagate basin modeling uncertainties into velocity models by considering certain geologic scenarios and selecting one that exhibits a best match to well data. Subsequently, several input parameter values within this scenario are perturbed to generate multiple basin modeling outputs. One crucial aspect is that, in many cases, the calibration data at the wells are also uncertain. Thus, any uncertain geologic scenarios or parameters included in the analysis should be considered in accordance with the probabilistic model for the data uncertainty.

Pradhan et al. (2020) propose an integrated Bayesian framework with geohistory and rock physics constraints, which facilitates rigorous propagation of basin modeling uncertainties into the rock velocity models. These are then constrained by the seismic kinematics and flattening of seismic gathers to get a narrower distribution of velocity posterior models, along with the corresponding posterior distribution of pore pressure conditioned to well data as well as seismic data. The workflow is shown schematically in Figure 13.6.

The Bayesian approach entails specifying prior probability distributions on uncertain basin modeling input parameters and likelihood models for well calibration data. Multiple basin modeling runs are performed using realizations of the uncertain parameters drawn through Monte Carlo sampling of the prior distributions of input uncertainties. This gives a prior distribution of basin modeling outputs (including porosity, pore pressure, and temperature). Posterior basin modeling realizations are then selected according to the likelihood computed from the well calibration data (porosity, mud weight, and temperature). These accepted posterior realizations are subsequently linked to velocity through a calibrated rock physics model. Pradhan et al. (2020) use the constant-clay model for shaly sands (Avseth et al., 2005) to compute rock velocities from basin model outputs (porosity, fractions of quartz, illite, smectite, and effective stress). These velocity models can then serve as geologically consistent and physically valid priors for the seismic velocity inversion and imaging problem. Pradhan et al. (2020) condition these velocity models by kinematic information extracted from seismic data. Prestack seismic data are imaged by performing depth migration (reverse time migration, RTM) with each prior velocity realization. Velocity models consistent with the data kinematics will generate well-focused images. Focusing quality of the

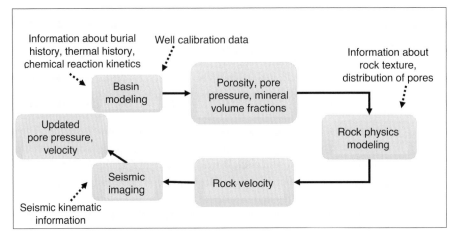

Figure 13.6 A schematic workflow for integrating geohistory and rock physics constraints with seismic imaging information to predict rock velocities for pore pressure prediction and uncertainty quantification. After Pradhan et al. (2020).

imaged reflectors are evaluated using angle domain common image gathers, and residual moveout (RMO) analysis. Velocity models that have a global summed absolute RMO error within a desired threshold are then retained as samples from the posterior distribution of velocity conditioned to seismic kinematics. These velocity realizations are already implicitly linked to the corresponding basin modeling pore pressure predictions. These corresponding pore pressure predictions represent samples from the approximate posterior distribution of pore pressure conditioned to both well calibration data as well as seismic data. These samples can be used to perform uncertainty quantification and estimate uncertainty intervals on pore pressure. Estimated pore pressure uncertainty conforms to the prior geological uncertainty on any geopressuring mechanism modeled in the basin model, and is consistent with the porosity, temperature and mudweight data collected at the well, as well as the posterior seismic velocity model.

An example from a basin in the north-central Gulf of Mexico shows an application of this integrated Bayesian methodology. The available dataset from this basin located off the coast of Louisiana consists of log data from several wells, biostratigraphic data and 3D seismic data. Pradhan et al. (2020) present their analysis on a 2D section shown in Figure 13.7. The cross section has a spatial extent of about 37 km. A well, present along the section, has density–porosity data, mudweight data and bottom-hole temperature data for basin modeling calibration. No repeat formation test (RFT) data were available, hence mudweight data was used as a guide to an upper bound for pore pressure. Gamma-ray log and biostratigraphic data, available at the well, were interpreted to identify major lithologic changes and key events on the geologic timescale. Horizons interpreted from the seismic image were used to build the structural model for the basin as shown in Figure 13.7. Ten depositional suprasalt layers to the Upper Miocene were modeled. Effects of salt movement history below the Upper Miocene

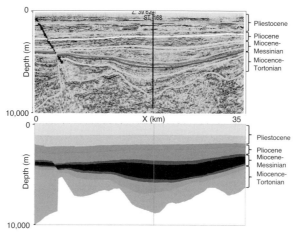

Figure 13.7 (top) The depth-migrated seismic section and interpretations of key geologic horizons. (bottom) Corresponding structural model used for basin modeling. (A black and white version of this figure will appear in some formats. For the color version, please refer to the plate section.)

was not modeled. The basin was simulated from Upper Miocene until present day using a commercial basin modeling software to numerically simulate geological processes of sedimentation of the layers, mechanical compaction, aquathermal pressuring, smectite-illite diagenesis, fluid flow and heat flow in the basin, based on differential equations as described in Chapter 10.

The first component of the workflow consists of defining the prior uncertainty on basin modeling input parameters and performing Monte Carlo basin simulations. Among the various sources of basin modeling input uncertainties Pradhan et al. (2020) consider the following: (1) uncertainty in the porosity reduction by compaction governed by porosity-depth constitutive relations for each lithology; (2) fluid flow related uncertainty modeled by permeability-porosity relations for each lithology; and (3) uncertainty in the basal heat flow for the basin. This choice was motivated by the fact that these factors significantly control the variability of present-day porosity, pore pressure and mineral volume fractions, which serve as inputs for the next step of rock physics modeling of the velocities. The porosity compaction behavior is modeled with an Athy-type law with two parameters: the initial depositional porosity and the compaction coefficient. The porosity-permeability relation modeled by a Kozeny–Carman-type relation has three uncertain parameters: the specific surface area, a scaling factor and a permeability anisotropy factor relating horizontal to vertical permeability. All of these five parameters are lithology-specific and have different prior distributions for the 10 different lithologies. Truncated Gaussian distributions were used as the prior probability distributions over the range of possible values that these parameters can assume. The depth-averaged clay volume fraction for each of the 10 lithologies derived using the gamma-ray log was used to loosely guide the specification of the prior distributions. The mean and standard deviation for each truncated Gaussian distribution were assigned based on typical parameter values available in literature (Hantschel and Kauerauf,

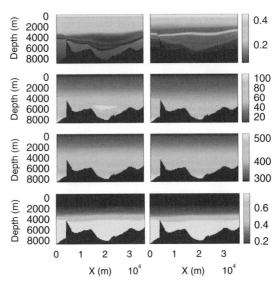

Figure 13.8 Two randomly selected prior models (out of 2500), obtained from Monte Carlo basin modeling simulations: present-day porosity (top row), pore pressure in MPa (second row), temperature in °C (third row), and illite volume fraction (bottom row). (A black and white version of this figure will appear in some formats. For the color version, please refer to the plate section.)

2009). Uncertainty in the basal heat flow was specified as a Gaussian distribution with mean and standard deviation of 47 and 3 mW/m², respectively. The uncertain prior model space for the basin thus consists of 51 parameters in total: 5 lithological parameters for each of the 10 lithologies and the basal heat flow. Monte Carlo sampling was performed to generate 2500 prior samples of these uncertain input parameters. A total of 2500 basin modeling runs were performed to simulate the porosity, pore pressure and temperature evolution from Upper Miocene to the present day. As an example, Figure 13.8 shows the present-day porosity, pore pressure, temperature, and illite volume fraction sections for two (out of the 2500) prior realizations.

The next step is to use the well calibration data (porosity, mud weight, and temperature) to determine the likelihood for the prior models and estimate the posterior distributions of the uncertain input parameters. Each data type is considered to be conditionally independent of the other, given the basin modeling input parameters. The marginal likelihood model for the porosity and temperature data samples are assumed to have a Gaussian distribution centered at the observed value, with a standard deviation reflecting the associated uncertainty. The likelihood for the mudweight data is not centered on the observed mudweight. Mudweight refers to the density of the drilling fluid and is usually kept higher than the true formation pressure which was not available for the present study. The recorded mud weights can thus be taken to be the upper limits on the pore pressure. Hydrostatic pressure represents the theoretical lower limit on the pore pressure at any given depth. Hence, the marginal likelihood distribution for mudweight data is taken to be a truncated triangular distribution with the observed mudweight and hydrostatic pressure as the upper and lower limits, respectively. Likelihoods

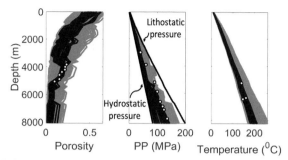

Figure 13.9 (left) Porosity, (middle) pore pressure, and (right) temperature profiles extracted at the well location from the corresponding present-day output sections of the prior basin simulations (light gray) and the posterior models (dark gray) selected based on their high likelihood. Observed calibration data are plotted as circles.

for intermediate values scale linearly between the two limits. Figure 13.9 compares the present-day porosity, pore pressure, and temperature profiles of the 2500 prior models extracted at the well location against the selected posterior models that have the highest joint data likelihoods. The porosity outputs of these posterior models are tightly constrained at depths where porosity calibration data are available and display broader range of uncertainty at shallower and deeper intervals in the absence of calibration data. Similarly, given the triangular likelihood model for mudweight, models predicting pore pressures higher than the recorded mudweight are assigned a likelihood value of zero. These posterior models are used to update the distributions of the basin modeling input parameters and are also used as inputs to rock physics modeling of velocities. The basin modeling posterior velocities serve as priors for the seismic imaging problem, which is the next step in the workflow.

To convert the posterior basin modeling outputs to velocity requires a rock physics model. Pradhan et al. (2020) use the constant-clay model (Avseth et al., 2005) for shaly sands. The model requires as inputs (in addition to end-member mineral and pore fluid properties) the porosity, quartz, smectite and illite fractions, and effective stress. These inputs come from the posterior basin models conditioned to the well calibration data. Monte Carlo simulated velocities are used for reverse time migration (RTM) of the seismic data. Residual moveout (RMO) analysis of angle domain common image gathers was used to evaluate the focusing quality of the reflectors. The basin modeling posterior velocity models were ranked according to their overall global RMO errors. Models that had errors below a tolerance threshold were selected as an approximate representation of the seismic posterior models. The seismic data was also migrated using the legacy velocity model that had been earlier estimated by ray-based tomographic inversion, without basin modeling or rock physics constraints. The RMO errors for a sizable number of the selected posterior models were lower than or comparable to the tomographic RMO error value. Figure 13.10 shows the legacy tomographic velocity model (left) and one of the selected seismic posterior velocity models (right) that had a lower global RMO error than that obtained with the legacy model. Figure 13.11 shows porosity and velocity depth profiles extracted at a particular x location of 15 km. As we saw earlier porosity calibration data are available

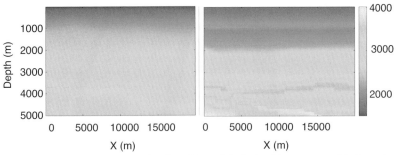

Figure 13.10 (left) Legacy velocity (in m/s) model obtained by tomographic inversion. (right) One posterior realization of basin and rock physics modeling derived velocity section. (A black and white version of this figure will appear in some formats. For the color version, please refer to the plate section.)

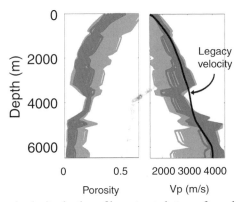

Figure 13.11 Porosity and velocity depth profiles extracted at a surface *x* location of 15 km for the basin modeling posterior models calibrated to well data likelihoods. Overlain in dark gray are the profiles extracted from the seismic posterior models derived after constraining to seismic kinematic information. Legacy velocity model trace at this location is shown in black.

only in the Miocene layers and consequently basin modeling posterior realizations (shown in gray) have significant uncertainty in the Pleistocene and Pliocene layers. Incorporation of seismic kinematic information is effective in further constraining the uncertainty in these layers as evidenced by the spread of the seismic a posteriori realizations (shown in dark grey). Note that at a depth of about 4 km, the legacy tomographic velocity seems to be outside the range of geologically plausible velocities, even though it is an adequate velocity model for flattening the seismic gathers. This workflow provides velocity models that are not only good for imaging but also are geologically and physically consistent with the basin geohistory.

Since each posterior velocity model is linked to pore pressure sections generated as outputs of basin simulations, it is also possible to update the posterior pore pressure uncertainty according to kinematic information contained in seismic data. Figures 13.12 and 13.13 depict how the prior pore pressure uncertainty was successively updated, first with basin modeling calibration using mudweight data and then with seismic data. In Figure 13.12, we compare the pore pressure depth profiles of these models extracted at a particular x location of 15 km. The gray profiles represent 2500 prior models obtained by Monte Carlo sampling of the prior without any data constraints; blue curves are basin modeling posterior models calibrated to well data only, and finally the red curves are seismic posterior models accepted after migration velocity analysis. Figure 13.13 compares the prior and posterior probability densities of pore pressure at two depths at this location. The seismic posterior distributions consistently

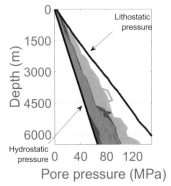

Figure 13.12 Pore pressure depth profiles extracted at surface x location of 15 km for prior models (shown in light gray), basin modeling posterior models calibrated to well data alone (shown in dark gray) and seismic posterior models (shown in medium gray), after calibration to seismic kinetic information.

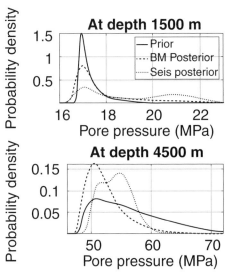

Figure 13.13 Estimates of pore pressure uncertainty at x location of 15 km and at depths listed on the top of each plot. Shown is the initial prior on the pore pressure (solid line), the updated posterior distributions after constraining to well calibration data (dashed line), and further updated posterior after constraining to seismic kinematic information (dotted line).

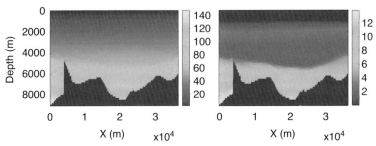

Figure 13.14 (left) Posterior mean and (right) standard deviation for pore pressure (in MPa) estimated using the posterior models accepted after migration velocity analysis (Pradhan et al., 2020). (A black and white version of this figure will appear in some formats. For the color version, please refer to the plate section.)

have more probability density for higher pore pressure values as compared to the prior and basin modeling posterior distributions. Note that at 4500 m mudweight data alone does not rule out hydrostatic pressure as a possibility. However, information from seismic data assigns negligible probability to hydrostatic pressure and shifts the probability density to higher pore pressure values. The a posteriori realizations can be used to constrain the spatial pore pressure uncertainty as shown in Figure 13.14, which shows the mean pore pressure (mean over the selected seismic posterior realizations) and the corresponding standard deviation of pore pressure.

This workflow satisfies four important criteria: (1) the model should be geologically consistent, that is, it should conform to our knowledge and beliefs about the geologic history and structure; (2) model variability should be constrained by any bounds derived from our physical understanding of the rocks; (3) the model should agree with all available data from wells and seismic measurement; and (4) modeling uncertainty should be rigorously quantified. This integrated workflow achieves a consistent modeling of earth properties and pore pressure along with the associated uncertainties. A major challenge is that the approach relies on performing multiple basin modeling, rock physics and depth migration evaluations, each of which can be computationally demanding. Considering additional sources of uncertainty such as structural uncertainty, uncertainty in interpreted horizons, or increasing the model dimensions from 2D to 3D would exacerbate the computational costs of this approach. *However, with advances in computing power this workflow should become a best practice for the future.*

14 Recent Advances in Geopressure Prediction and Detection Technology and the Road Ahead

14.1 Introduction

Before a well is drilled, seismic analysis is the most widely used tool for geopressure prediction. It is a remote sensing tool. Seismic data are acquired, processed, interpreted and inverted to obtain velocity information, which in turn is used for pressure prediction. Basin modeling is another technique for pore pressure prediction prior to drilling a well. These techniques are augmented with well logs and drilling data. We discussed these techniques in the prior chapters. Here we focus on the new advances in the prediction and detection technologies and point out the possible future directions as best as we can. However, we realize -"*It's tough to make predictions, especially about the future,*" to quote Yogi Berra (born Lawrence Peter Berra – a baseball legend who is as famous for his sports career as he is for his malapropisms!).

14.2 Seismic Technology

Seismic imaging and inversion technology is a mature technology. Velocity, acoustic and elastic attributes are key parameters out of this technology. In general, there are two sources of information in this context that have different resolution for pore pressure prediction. The first is the seismic imaging velocity, which is derived during either time or depth imaging process. Modern depth imaging velocity model building analysis uses tomography/FWI based inversion techniques to derive a velocity field. Ray based tomography is smooth and has a relatively low spatial resolution while FWI can have higher resolution at higher computational cost. The second source to derive higher-resolution velocity is from inversion based on amplitudes of seismic data (which provides higher frequency models). In earlier chapters we discussed these in the context of earth modeling. The resolution of pore pressure volumes obtained from conventional time domain interval velocity analysis is fairly coarse. Velocities obtained from tomography/FWI analysis are more detailed and geologically consistent. Hence, these are more suited for pore pressure analysis. Most practitioners of pore pressure prediction are routinely using tomography based anisotropic velocity models. This is a pleasant trend in the industry now. Although FWI method has gained traction in the industry, it has been limited mostly to imaging. FWI is rarely used for pore pressure prediction, as in many cases the cost of high-resolution FWI is higher than tomography. Furthermore, acoustic-FWI is the only option available at the industrial deployment scale; it does not directly provide density or Poisson's ratio needed for overburden stress and fracture pressure estimation, and thus the amplitudes inverted are not always reliable as they do not account for the full elastic wave propagation in the earth. Elastic-FWI is in the research stage, as the higher computational cost associated with modeling full elastic wave is a strong barrier to deployment. (This comment does not apply to 1D elastic full wave form -based inversion techniques as discussed in Chapter 5.) All FWI methods require access to large-scale computer resources, and algorithms are complex, although these are not showstoppers.

Velocities from amplitude inversion (derived from P-impedance) have higher resolution than those from either conventional velocity analysis or tomography; however, it requires a meaningful and physical "low-frequency trend" to be of use in pore pressure applications. This is because the seismic data lack low-frequency components. More and more geoscientists are using tomography based seismic amplitude inversion for obtaining high-frequency velocity models. New techniques that relax the need for detailed low-frequency modeling has been introduced in the recent years where prior compaction trends are used to invert jointly for lithology and elastic properties (Rimstad et al., 2012; Kemper and Gunning, 2014). Basin modeling attributes which can be transformed into lithology trends can be readily integrated within the AVA inversion process.

However, seismic inversion is still mostly in the geophysics domain. Engineers need to be exposed to this technology for well construction projects. This needs to be done in a multidisciplinary team of geoscientists and engineers. Although gather flattening of seismic data is and will remain as a backbone of seismic processing and imaging technology in the foreseeable future, earth modeling will require that analysts carry out the process in the geomechanics domain. As an example, in Chapter 6 we discussed how to carry out this process directly in the pore pressure domain. This is a powerful tool as geophysicists are nudged to think about such concepts as pore pressure gradient or ppg and other drilling data such as mud weights and rate of penetration logs. We are very hopeful that a required cultural shift to bridge the chasm between geophysicists and drilling-oriented engineers will be mitigated in this way.

For pore pressure prediction in higher resolution *ahead of the bit* (in real time) reverse VSP data from drill bit as a seismic source is shown to be useful (Miranda et al., 1996). However, *quiet* drill-bits are favored by drillers and hence this technology has not gained much traction outside of a few oil companies. Integration of 3D-seismic velocity models with VSP-While-Drilling data is very helpful for pressure prediction during the drilling phase (Kulkarni et al., 1999; Armstrong et al., 2004). Currently, the application is, however, done during the down-time in the drilling phase such as casing emplacement and bit change. Thus, a key challenge of this technology is to increase the rate of delivery of the data to impact drilling in real time. This is a work in progress now.

We think that in the near future much of pore pressure prediction will be done with velocities (both P- and S-waves) derived from joint analysis of kinematics and amplitudes either through a hybrid of tomography and amplitude inversion schemes or directly through FWI technology (both acoustic and elastic, when available). Furthermore, with the availability of massive compute power, and breakthroughs in sophisticated software, it is fair to say that *real-time application* of such technology during drilling phase will not be far behind. Additionally, the shift of the industry toward automation of seismic data processing and inversion practices is expected to bring a large reduction in cost, together with added robustness to imaging, tomography and inversion. With the advent of elastic-FWI, fast basin modeling, and the greater availability of multicomponent acquisition, all stress components (pore pressure, total vertical and horizontal stresses and their directions and fracture pressures) will be predicted during predrill model building and eventually, in real-time during drilling. This will make geomechanical models highly predictive and realistically applicable to a wide range of basins. The only challenge that we foresee would be the *adaptation* of such technologies. It would require a crew-change and substantial training as it would inevitably be a disruptive technology.

14.3   Models That Relate Velocity to Pore Pressure

This area is still dominated by simple empirical correlations between pore pressure and velocity such as various normal compaction trend analysis techniques. Often, this lacks the backing by fundamental physics of pressure prediction for many models – just what is a normal compaction trend when we are dealing with mixed lithologies? What about the new normal trends when rocks are altered during basin evolution and tectonism that alter the rock properties? Recent decades have seen a gradual shift from sheer empiricism to those models that seek correlation between more physics-based parameters such as effective stress and velocity (Bowers, 1994) and between porosity, effective stress and thermal history of rocks (Dutta et al., 2014; Dutta, 2016; Pradhan et al., 2017a, 2017b). Much is still ahead – applicability of models to "hard rocks" is still incomplete although high pore pressures in those rocks do occur frequently. These require availability of data from measurements on cores. Unfortunately, this has remained as sole property of individual oil companies. However, with the recent emphasis on data-driven machine learning, and with the understanding that in many cases collaborative research provides higher impact, larger datasets will be created with access available to researchers and developers of new methods. This could be a way forward. This is a pure speculation on our part at this time. However, we believe that with proper nurturing, this could be accomplished.

We noted earlier that shale-based pressure prediction techniques usually do not apply to carbonates and other hard rocks. However, we also noted that there are some encouraging *emerging* methods for geopressure prediction in these types of rocks. Some techniques are based on empirically correlating different seismic attributes with well data. As the ability to generate larger datasets is here, modern data analytics techniques that will explore complex relations between many variables jointly can improve current practices. We believe that with the advent of high-resolution, large-offset, and continuous-azimuth seismic data, together with high-end velocity models aided by ample support from computing technology, reliable pore pressure prediction in carbonates and similar high-velocity rocks would be routinely applicable in the future. This will also help in building earth models for fractured "hard" rocks including various types of basement rocks where high pressures are known to occur and pose drilling challenges.

It is now believed that the complete burial history of rocks has a significant imprint on *today's* subsurface pore pressure. In this context, we have come to recognize that various thermal effects in tectonically dormant and active basins play crucial roles in generating overpressure and in fact, in many cases, dominate the undercompaction effect. Such processes suggest not only the *burial-path* dependency of pore pressure in subsurface rocks but also require complete integration of various diagenetic processes in rocks to be integrated with various geophysical attributes. This means that the gap that currently exists between geochemistry and geophysics could be narrowed. Since tectonic effects are coupled with thermal effects due to local heat generation and contribute further to lowering of effective stress, we believe that further research is necessary to lay the foundation for this type of integration. This was also observed by Zoback (2007).

14.4   Seismic Velocity Analysis for Pore Pressure Prediction: What We Have Learned and the Road Ahead

The causes of failure to predict pore pressure reliably using seismic velocity are many. However, there are two major causes worth mentioning. The first cause deals with not accounting for the inadequacies associated with the "trend analysis methods" and extending the operating envelopes of the models outside its validity regime. We discussed this above and suggested that rock physics compliant models are better alternatives than "trend" analysis methods. The second cause deals with not recognizing the limitations of seismic velocity used as a probing tool for converting velocity to either pore pressure or effective stress. This is a knowledge-gap issue. In Chapter 5 we discussed various velocity models that can be extracted from seismic data. We addressed how these models can be used for pressure prediction including limitations and best practices. However, geophysicists have not succeeded completely in narrowing this gap. This, however, works both ways – geophysicists have not taken advantage either of the knowledge that our sister communities such as drilling and mudlogging possess. (An example: *well ballooning* as an indicator of high pore pressure in formations!)

With the advancement of high performance computing, velocity analyses techniques have focused more and more on several key factors: (1) improving resolution of velocity through data processing and imaging methods that utilize the full bandwidth, offset, and azimuth of seismic data in the prestack domain, (2) including various high-end anisotropic models for velocity analyses and imaging, (3) carrying out velocity analyses through full-waveform inversion methods. and (4) recognizing that velocity inversion techniques should be constrained by principles of rock physics, drilling data and basin history. These are giant steps forward; the pore pressure prediction methodologies are indeed backed up by good science. It is becoming increasingly possible to tackle the estimation of the full tensor-field of the elastic constants as discussed in Chapter 2 and realizing the full potential of the seismic data. This will be even more interesting with the advent of full elastic waveform inversion methods including general anisotropy. We believe that estimations of full tensor-field of elastic constants will be routine in the future.

There is a realization that integration of various rock physics based models with seismic data, well logs, and tools such as basin modeling and geomechanics is necessary for reliable geopressure analysis. This is particularly important for applications in complex geologic areas such as subsalt. A key to success in this area is to be able to link the suprasalt velocity information and extend its low-frequency trends to the subsalt domain. This sort of workflow is gaining traction in the industry. We think this would be more prevalent in the future. An example is given for subsalt pore pressure estimation using rock physics guided velocity modeling and tomography. This example is taken from Gautam et al. (2017). This study was carried out on a dual coil seismic data set that was acquired over 1800 sq mi (204 OCS blocks) in the Garden Banks region of the Gulf of Mexico. The survey area is characterized by very complex salt structures and their geometries varied geographically (Figure 14.1a). The authors used many of the concepts that we have discussed here: rock physics template, rock physics guided velocity modeling (*RPGVM*), multiple pore pressure mechanisms and gather flattening in pore pressure gradient (ppg) domain. The study suggested that (1) the legacy velocity in the deeper region was much slower than the regional Gulf of Mexico trend (Figure 14.1b), (2) the result of several migrations (Reverse Time Migration – RTM) performed with various iso-ppg trends showed improved image quality in the deeper targets with a faster trend, and (3) the subsalt low-frequency velocity trend with rock physics guided tomography led to better imaging below salt (Figure 14.2). Applications such as this with tomography and FWI for pore pressure estimation with improved imaging (and sharing the common earth model) in complex geology over large areas (over 1800 sq mi) of exploration interest would be routine.

In this book we discussed extensively the value of guiding tomography by using pore pressure as a constraint on the inversion with many examples. This approach improves the image quality as the velocity model not only conforms to geology but also yields a physical velocity model everywhere. We believe that iterative procedures similar to or better than the one described in the book will be of common use for imaging and pore pressure estimation using a *common* anisotropic velocity model. An approach like this suggests that we are inching closer toward building an *Earth model.*

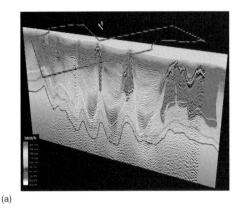
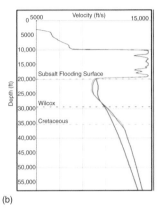

(a)   (b)

Figure 14.1 (a) Interpreted salt geometries over a large survey area (~ 1800 sq mi) in the Garden Banks of the Gulf of Mexico. (b) Velocities extracted at a pseudo-well location. Velocity before subsalt trend (red); velocity after subsalt trend (green); velocity after one pass of subsalt tomography (blue). From Gautam et al. (2017). (A black and white version of this figure will appear in some formats. For the color version, please refer to the plate section.)

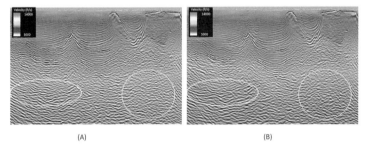

(A)   (B)

Figure 14.2 Improvements in imaging below salt. (A) Legacy image. (B) RTM migrated images based on rock physics guided velocity model of Figure 14.1b using anisotropic CIP tomography. It resulted in better images below salt in the deeper section as well as a more reliable pore pressure model using the same velocity model. Color scale for velocity: white - blue (5000 ft/s) and red (14000 ft/s). From Gautam et al. (2017). (A black and white version of this figure will appear in some formats. For the color version, please refer to the plate section.)

We believe that in the future geoscientists will focus more and more on building an *Earth model with general anisotropy.*

Full-waveform inversion is an exciting development in the field of velocity model building over the last several decades and is being used more and more for seismic imaging. The attraction of the approach is the promise of deriving high-fidelity earth models for seismic imaging from the full-waveform inversion of the acquired seismic data. The techniques uses the full bandwidth of the data. As our ability to understand and manage the complex, nonlinear inversions has developed and available computer power has grown, full-waveform inversion – even anisotropic full-waveform inversion – has become practical. Recent work by Le (2019) not only used anisotropic FWI but also a rock physics guided velocity modeling to arrive at new and exciting results. This is

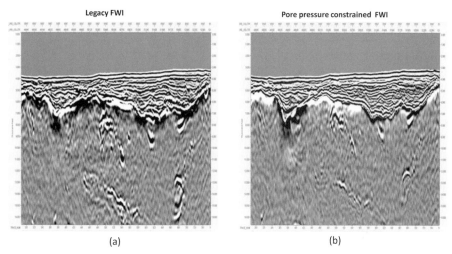

Figure 14.3 An example showing structural image enhancement. (a) Legacy FWI model. (b) FWI with pore pressure constraint.

a new frontier. This has, at least in a theoretical sense, increased the ability of the approach to outperform semblance or ray-based tomography in terms of resolving complex, small-scale structures in the earth (Brittan et al., 2013). This process was further aided by the availability of low-frequency, large-offset seismic data (Kapoor et al., 2013). The readers are referred to excellent reviews by Brossier and Virieux (2011) and Brittan et al. (2013). *We believe that in the future all seismic and pore pressure imaging will be done with FWI.* A good example is provided by Benfield et al. (2017). In Figure 14.3 we show image quality improvement when the FWI (acoustic, in this case) process is constrained by pore pressure. We believe that more innovation will happen in devising other physical constraints on the FWI process for estimating earth models. In a recent PhD dissertation at Stanford University (Le, 2019), the author concluded (we quote):

Incorporation of bound constraints derived from pore pressure and rock physics leads to more accurate velocity and anisotropy models. The final RTM image shows improvements in focusing and continuity of reflectors. Significant reductions in both vertical velocity and anisotropy in the shallow part of the model leads to an upward shift in reflectors' depth. Without constraints, model update is emphasized more in velocity than in anisotropic parameters, which results in an unrealistic and erroneous velocity model. Benefits of having the constraints are not only in terms of improving structural images but with the imposed constraints, the estimated velocity model is also more geologically consistent with the pore pressure models.

Unconventionals have changed the landscape of hydrocarbon exploration and production in the USA during the last decades. These unconventional plays are low-permeability reservoir rocks; these rocks also act as their own source and seal. Economic production from these "tight" reservoirs (organic rich shales with high TOC, coal-bed methane and tight gas sands, for example) require assistance such as stimulation by chemicals, steam injection and fracturing. Some unconventional plays such as the Haynesville shales in North Louisiana (Zhang and Wieseneck, 2011) are highly overpressured. However, unconventionals such as Barnett and Marcellus shales are normally pressured. The role of pore pressure in unconventionals is not well understood. We speculate that high pore pressure (low effective stress) could provide a favorable environment ("sweet spots") for hydrocarbon and thus any technique to provide guidance prior to drilling would be welcome. Unfortunately the techniques that we have discussed with seismic data may not provide reliable diagnostics due to the fact that velocities in these rocks are affected by multiple factors such as TOC, degree of maturation, and amount and type of free gas in the system. Whether basin modeling technology could provide guidance in locating "sweet" spots for unconventionals remain to be seen. This area remains under investigation.

14.5   Pore Pressure Prediction in Real-Time

We discussed the current status of real-time pore pressure prediction earlier (Chapter 9). The current methodology or workflow is very basic and we expect a step change here in several areas. The success here critically depends on the seamless integration of data from multiple downhole sources (VSPWD; other LWD/MWD tools; drilling data such as ROP) with a geophysical model based on seismic and petrophysical logs from offset wells. The process is currently hindered by limitations of mud-pulse telemetry; it is slow and cannot transform recorded data such as waveforms from VSPWD tool. Thus, a step change is needed here. Perhaps a wired drill pipe system is an answer that bypasses the old fashioned mud-telemetry system. These systems use fiber optics cables built into every component of the drill string, which carry electrical signals directly to the surface. These systems can transmit data at rates orders of magnitude greater than anything possible with mud-pulse or electromagnetic telemetry, both from the downhole tool to the surface and from the surface to the downhole tool (Ali et al., 2008; Olberg et al., 2008; Nygard et al., 2008). When this technology is deployed, we believe that pressure prediction in real time in every well will be a reality. We have not analyzed the cost-benefit ratio of this technology. However, we believe that the cost for not using available technology such as this to minimize drilling risk – pore pressure prediction being the most important part – in real time could be enormous. There is another area of improvement needed. Currently, the process of calibrating the static model (behind the drill bit) with real-time data from ahead of the bit is very slow. We think that this process of interrogation and prediction ahead of the bit will benefit immensely by machine learning. The challenge is to create a workflow for real-time data calibration and prediction (data labeling process in machine-learning language). *The future potential of this technology for drilling applications in real time is very high.*

The big data revolution that has swept the industry in recent years is also expected to generate a step change in our ability to predict in real-time. Much of the drilling process is instrumented and a large variety of data are being continuously recorded. The ability to use advanced data analytics techniques together with real time LWD and drill bit information is expected to generate further understanding about the relationship between the drilling process, the elastic rock properties and the state of stress and pressure in the rock.

Real-time pore pressure prediction using *basin modeling* is now a reality (Mosca et al., 2017). We discussed this in Chapter 9. We believe that such an approach has a significant value and *it will be adapted for more applications in the future.* However, the turnaround time needs to be improved considerably for this technology to impact the real-time drilling decisions (at least to within the time required to change a couple of stands of pipes; not overnight). Faster turnaround time could be a breakthrough for this technology.

14.6   Integration of Disciplines

A conceptual model of how integration on a grand scale would work is shown in Figure 14.4. The workflow starts with basin modeling to obtain outputs like pore pressure, porosity and mineral volume fractions. There are several parameter optimizations that need to be executed here and well calibration is essential. The outputs from basin modeling are subsequently linked to rock properties such as rock velocity and various anisotropic parameters (epsilon and delta for VTI models, for example) through rock physics models (Bachrach, 2011a; Dutta, 2016; Pradhan et al., 2017a). This step also includes many iterations to derive an optimum model with parameters such as rock texture, pore aspect ratio distributions, and so on. Building a rock velocity model is an important step because it can be used as a substitute for vertical velocity in lieu of the NMO velocity required in the next step. This step deals with velocity model building with anisotropic tomography/FWI workflows. This process is highly iterative and must deal with non-uniqueness of the velocity prediction. This can be partially achieved by imposing constraints from the rock physics and pore pressure modeling as discussed. The output from this step would be a velocity model that satisfies seismic kinematics as well as rock physics and basin modeling constraints. Up to this step, the process is similar to what De Prisco et al. (2015) outlined in what they termed as *geophysical basin modeling.* What is proposed here is to use the output from the current step to start the basin modeling and *re-do* the subsequent steps again to obtain a *new* earth model that is now better constrained than the one obtained in the first iteration. We envision this loop to be executed more than once. We term this loop as the *grand loop* and *we exit when not only a converged model for pore and fracture pressures are obtained but also an improved seismic image is produced.*

14.7   Data Analytics and Machine Learning

As has been mentioned earlier, in addition to integrating disciplines, recent years have seen the growing application of machine learning for integrating different types of data in earth modeling. The large quantities of different types of data (vibrations, mud pressure fluctuations, and more) that are being recorded today continuously in time, together with the large computer power available, provide opportunities to explore different types of correlations using various new set of algorithms and perhaps gain new insight to increase predictive power. Machine-learning techniques such as artificial neural networks and Bayesian classification have long been used for seismic inversion, velocity estimation, and for predicting lithofacies and porosity from seismic and well data (see, e.g., Rogers et al., 1992; Roth and Tarantola, 1994; Calderón-Macías et al.,

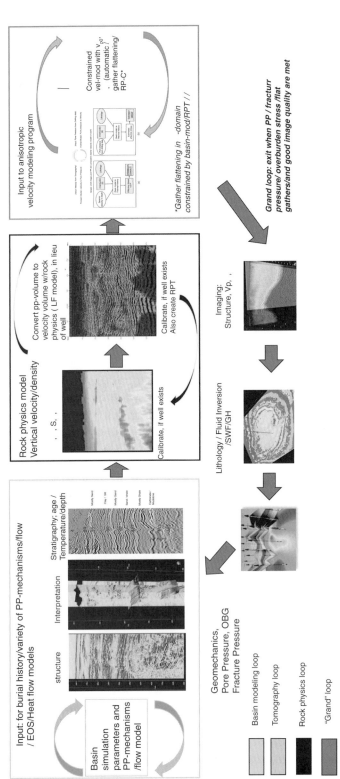

Figure 14.4 Schematic of a workflow that integrates basin modeling derived pore pressure to rock physics and seismic anisotropic velocity modeling and imaging. (A black and white version of this figure will appear in some formats. For the color version, please refer to the plate section.)

1998; Avseth and Mukerji, 2002; Poulton, 2002; Avseth et al., 2005). Bougher (2016) showed that unsupervised machine learning can be used on the well logs to get clusters that can be correlated to the lithology of the well. This adds to the increased efficiency of log interpretation. More recently, deep learning methods using neural networks with a large number of layers have evolved from the previous generation of artificial neural networks. Convolutional neural networks (CNN), one form of a deep neural network, have gained popularity after having outperformed humans in image classification tasks. Deep networks have been applied to many tasks related to seismic interpretation, such as seismic horizon and fault interpretation, seismic texture identification, seismic inversion, velocity modeling, and predicting lithofacies from seismic data. Araya-Polo et al. (2018) utilized deep CNNs for velocity tomography. Das et al. (2019) and Biswas et al. (2019) show the application of CNNs for seismic impedance inversion. Wu and McMechan (2018) used a combination of conventional full-waveform inversion with CNNs to obtain better FWI results. Zhang et al. (2018) used CNNs for seismic lithology prediction, while Das and Mukerji (2019) used CNNs to predict petrophysical properties (porosity and clay content) from prestack seismic data. Predictions of these attributes (velocities, impedances, and petrophysical properties) are useful inputs for pore pressure prediction as has been described in the preceding chapters. Yu et al. (2020) compare four different machine-learning algorithms (multilayer neural network, support vector machine, random forest, and gradient boosting machine) to predict pore pressure from log data (sonic velocity, porosity, and shale volume). The machine-learning approach builds a non-linear regression between the inputs (sonic velocity, porosity, and shale volume) and the output (effective stress), without having to specify an exact functional form for the multivariate relation. The training data consists of the petrophysical logs from wells and the computed effective stress in the normally compacted interval. Effective stress is then converted to pore pressure using the overburden stress (obtained by integrating the rock density) and the usual effective stress law. Yu et al. (2020) apply the trained models to two wells from East China Sea Shelf basin where overpressure is encountered in Paleocene/Oligocene formations, and show a good match to DST pore pressure measurements. With the availability of larger and larger data sets machine-learning algorithms can be trained to capture complex relations between various parameters to improve current understanding. Unconventional plays, which typically have a very large number of wells drilled, offer an ideal situation for applying machine-learning algorithms. In a study located in the Delaware Basin in West Texas, Naeini et al. (2019) apply deep neural networks in a supervised regression setting to predict petrophysical properties and pore pressure from various log and seismic inputs. A critical component in any supervised machine learning is selecting and quality controlling the input training data. As a part of the Delaware Basin study about 1500 wells were evaluated. The published study shows results using 13 wells. In addition to well logs, direct measurements of pore pressure taken from either dynamic fracture initiation tests, drill-stem tests (DST), or by interpretations from the drilling history were used in the training. Extensive and careful quality control of the training data included removal of anomalous data, log normalization, core calibration, as well as multiwell consistency checks. One network was trained to predict volume fractions of shale, sand, dolomite, calcite, kerogen, and porosity, given P-wave velocity, density, gamma-ray, resistivity, and neutron logs as inputs. Another network used as inputs P- and S-wave velocities, density, resistivity, and neutron log, along with volume fractions of shale, kerogen, and porosity, to predict pore pressure. Finally a third network was trained to predict pore pressure, minimum and maximum horizontal stresses, and volume fraction of kerogen, using as inputs only P- and S-wave velocity and density. This network could be applied to velocities and densities obtained from prestack seismic inversion to predict pore pressure spatially in the whole seismic date cube where the inversion results were available.

One of the challenges in using supervised learning for pore pressure prediction is that these methods are data-driven approaches requiring a large labeled training dataset. Direct pore pressure measurements from well tests are scarce due to costs and operational difficulties. Hence, the training data set has to be augmented using rock physics and geologic understanding of the causes of overpressure. The best strategies, we feel, combine the power of machine-learning algorithms with physics and geology-based understanding.

14.8   Summary

Before we leave the subject of current advances in pore pressure prediction and detection technology and the road ahead, we would like to take advantage of the pulpit of being authors and make an observation regarding the future. The petroleum industry is rather conservative. Naturally. Another Deepwater Horizon – type disaster could permanently cripple the industry. So we need to be very careful. This is, however, not an excuse for being *slow* in adapting new technology (and training to use it). The pharmaceutical industry has a good track record in this area. (It takes about seven years from concept to commercialization of a product.) For the oil and gas industry, it takes about *twice* as long! This needs to change and we think it will. The development of new energy solutions together with decarbonization processes is likely to increase the need to generate cost effective solutions to mitigate subsurface risk. We see technology playing a central role in this effort. This revolution will be aided by machine learning, GPU-assisted cloud computing, and perhaps even quantum computing! Artificial intelligence, machine learning, and big data science have made great leaps in many engineering areas and their impacts are being felt in the geosciences as well. Algorithms such as convolutional neural networks, deep learning and reinforcement learning have shown impressive performance in various tasks related to image segmentation, classification, optimization, and prediction. These machine-learning algorithms augmented by the domain knowledge of geology and geophysics will very likely change the way pore pressure is predicted from seismic, well log, and drilling data. The future is almost here, and it is bright indeed!

Appendices

Appendix A Empirical Relations for Fluid (Brine, Oil, Gas) Properties

A.1 Brine

For water with dissolved NaCl (brine), the density depends on salinity, temperature, and pressure. Based on measurements by Batzle and Wang (1992), the following expression for brine density is widely used:

$$\rho_B = \rho_w + S\{0.668 + 0.44S + 10^{-6}[300P - 2400PS + T(80 + 3T \\ -3300S - 13P + 47PS)]\} \tag{A.1}$$

where ρ_w, the density of pure water, is

$$\rho_w = 1 + 10^{-6}(-80T - 3.3T^2 + 0.00175T^3 + 489P - 2TP + 0.016T^2P \\ -1.3 \times 10^{-5}T^3P - 0.333P^2 - 0.002TP^2) \tag{A.2}$$

We note that in equations (A.1) and (A.2) pressure P is in MPa, T is in degrees Celsius, salinity S is in fractions of one (parts per million divided by 10^6), and density (ρ_B and ρ_w) is in g/cm^3. Based on Batzle and Wang (1992), the acoustic velocity of brine (in m/s) is

$$V_B = V_W + S(1170 - 9.6T + 0.055T^2 - 8.5 \times 10^{-5}T^3 + 2.6P - 0.0029TP \\ -0.0476P^2) + S^{1.5}(780 - 10P + 0.16P^2) - 1820S^2 \tag{A.3}$$

where the sound velocity in pure water V_W in m/s is (Batzle and Wang, 1992)

$$V_W = \sum_{i=0}^{4} \sum_{j=0}^{3} w_{ij} T^i P^j \tag{A.4}$$

and coefficients w_{ij} are

$$
\begin{aligned}
&w_{00} = 1402.85 &&w_{02} = 3.437 \times 10^{-3}\\
&w_{10} = 4.871 &&w_{12} = 1.739 \times 10^{-4}\\
&w_{20} = -0.04783 &&w_{22} = -2.135 \times 10^{-6}\\
&w_{30} = 1.487 \times 10^{-4} &&w_{32} = -1.455 \times 10^{-8}\\
&w_{40} = -2.197 \times 10^{-7} &&w_{42} = 5.230 \times 10^{-11}\\
&w_{01} = 1.524 &&w_{03} = -1.197 \times 10^{-5}\\
&w_{11} = -0.0111 &&w_{13} = -1.628 \times 10^{-6}\\
&w_{21} = 2.747 \times 10^{-4} &&w_{23} = 1.237 \times 10^{-8}\\
&w_{31} = -6.503 \times 10^{-7} &&w_{33} = 1.327 \times 10^{-10}\\
&w_{41} = 7.987 \times 10^{-10} &&w_{43} = -4.614 \times 10^{-13}
\end{aligned}
$$

The bulk modulus of brine K_B with no dissolved gas is related to the velocity and density in the following manner:

$$ K_B = \rho_B V_B^2 \tag{A.5} $$

Acoustic velocity of brine is, however, dependent on the amount of dissolved gas, for example, methane. Batzle and Wang (1992) showed that the following relation is valid for bulk modulus of brine with dissolved gas, K_G:

$$ \frac{K_B}{K_G} = 1 + 0.0494 R_G \tag{A.6} $$

where K_G denotes the bulk modulus of brine with dissolved gas (methane) and R_G denotes the volume of dissolved gas at standard conditions to the volume of brine and is known in the industry as the *gas–water ratio*. How much methane can be dissolved in brine? It is a function of the salinity, temperature, and pressure of brine. Batzle and Wang (1992) give the following expression for the maximum amount of methane that can be dissolved in brine:

$$ \log_{10}(R_G) = \log_{10}(0.712P|\text{T-}76.71|^{1.5} + 3676P^{0.64}) - 4 - 7.786S(T + 17.78^{-0.306}) \tag{A.7} $$

The above empirical relation is valid for $T < 250°C$. This completes the discussion on the density and velocity of brine as a function of T, P, S, and dissolved gas.

A.2   Oil

For oil the density and velocity depend on the composition of oil, temperature, pressure, and dissolved gas. Unlike brine, the amount of gas that can be dissolved is very large. Therefore any quantitative work dealing with pressure exerted by a column of oil must address whether the oil is "dead" (no dissolved gas) or "live" (with dissolved gas). Batzle and Wang (1992) also dealt with this problem and made some good recommendations that can be programmed for computation. A widely used classification of crude oil is the American Petroleum Institute's oil gravity (API gravity). It is defined as

$$API = \frac{141.5}{\rho_0} - 131.5 \tag{A.8}$$

where ρ_0 is a reference (standard) density (g/cm^3) at 15.6°C and atmospheric pressure. API gravity is a measure of how heavy or light a petroleum liquid is compared to water. Thus, a heavy oil with a specific gravity of 1.0 (i.e., with the same density as pure water at 60°F) has an API gravity of 10 API. If its API gravity is greater than 10, it is lighter and floats on water; if less than 10, it is heavier and sinks. API gravity is thus an inverse measure of a petroleum liquid's density relative to that of water (also known as specific gravity). Although API gravity is mathematically a dimensionless quantity (see equation (A.8)), it is referred to as being in "degrees" by the industry. API gravity values of most petroleum liquids fall between 10° and 70°. For very heavy oil, the API gravity is ~5°, while for light condensates, the API gravity is ~100°. In the oil industry, quantities of crude oil are often measured in metric tons. One can calculate the approximate number of barrels per metric ton for a given crude oil based on its API gravity:

$$\text{Barrel of crude oil in one metric ton} = \frac{131.5 + API}{141.5 \times 0.159} \tag{A.9}$$

For example, a metric ton of West Texas Intermediate crude oil (39.6° API) has a volume of about 7.6 barrels. Generally speaking, oil with an API gravity between 40° and 45° commands the highest prices. Above 45°, the molecular chains become shorter and less valuable to refineries. Crude oil is classified as light, medium, or heavy according to its measured API gravity. Light crude oil has an API gravity higher than 31.1° (i.e., less than 870 kg/m^3), medium crude oil has an API gravity between 22.3 and 31.1° (i.e., 870 to 920 kg/m^3), heavy crude has an API gravity below 22.3° (i.e., 920 to 1000 kg/m^3), whereas extra heavy crude oil has an API gravity below 10.0° (i.e., greater than 1000 kg/m^3).

From the measurements of Batzle and Wang (1992), for "dead oil," the effects of temperature and pressure on density are largely independent. The temperature dependence (expressed in degrees Celsius) of density at a given pressure P is

$$\rho = \rho_0 / [0.972 + 3.81 \times 10^{-4}(T + 17.78)^{1.175}] \tag{A.10}$$

The acoustic velocity V_P (in ft/s) in "dead" oil in terms of API gravity is

$$V_P = 15450(77.1 + API)^{-0.5} - 3.7T + 4.64P + 0.01150[0.36API^{0.5} - 1]TP \tag{A.11}$$

As noted earlier oil, is classified as either "dead oil" (tar, heavy oil, and bitumen) or "live oil" based on the absence or presence of gas (mainly methane CH_4) or volatile hydrocarbons dissolved in it. The industry uses the term *gas–oil ratio* (GOR), which is the volume of liberated gas to remaining oil at 15.6°C and at atmospheric pressure.

Thus GOR is expressed in liters/liter (1 L/L = 5.615 ft.^{3}/barrel). Mavko et al. (2009) documented the following for the velocity in "live" oil:

$$V_P = 2096[\rho_1/(2.6 - \rho_1)]^{0.5} - 3.7T + 4.64P$$
$$+ 0.01150[4.12(1.08/\rho_1 - 1)^{0.5} - 1]TP \tag{A.12}$$

where

$$\rho_1 = \left[\frac{\rho_0}{B_0(1 + 0.001\text{GOR})}\right] \tag{A.13}$$

$$B_0 = 0.972 + 0.00038\left[2.4\text{GOR}\left(\frac{G}{\rho_0}\right)^{0.5} + T + 17.8\right]^{1.175} \tag{A.14}$$

Here G is gas gravity. Temperature is in degrees Celsius, and pressure is in MPa. The true density of oil with dissolved gas is

$$\rho_G = \left[\frac{\rho_0 + 0.0012G \times \text{GOR}}{B_0}\right] \tag{A.15}$$

A.3   Gas

Natural gas is characterized by its gravity G, which is the ratio of gas density to air density at 15.6°C and 1 atmosphere pressure. The gravity of methane is 0.56 and may be as large as 1.8 for heavier gases. The ideal gas law states that under isothermal conditions (denoted by subscript T) the following relations hold (Batzle and Wang, 1992):

$$\overline{V} = \frac{RT_a}{P} \tag{A.16}$$

$$\rho = \frac{M}{\overline{V}} = \frac{MP}{RT_a} \tag{A.17}$$

$$V_T = \sqrt{\frac{RT_a}{M}} \tag{A.18}$$

where $\overline{V}$ is the molar volume, R is the universal gas constant, T_a is the absolute temperature $(T_a = T^0(C) + 273.15)$, and V_T is the isothermal sound velocity. In reality, we need relationships under *adiabatic* conditions and for gases that are *not* ideal. Batzle and Wang (1992) provided the following empirical relationships which are very useful. They introduced two useful quantities, pseudo-pressure P_r and pseudo-temperature P_T.

$$P_r = \frac{P}{4.892 - 0.4048G} \tag{A.19}$$

$$P_T = \frac{T_a}{94.72 + 170.75G} \tag{A.20}$$

In terms of these quantities, the density, ρ_G, and the adiabatic bulk modulus, K_G, are given as follows:

$$\rho_G \approx \frac{28.8GP}{ZRT_a} \tag{A.21}$$

$$Z = aP_r + b + E, \ E = cd \tag{A.22}$$

$$d = \exp\left\{-\left[0.45 + 8\left(0.56 - \frac{1}{T_r}\right)^2\right]\frac{P_r^{1.2}}{T_r}\right\} \tag{A.23}$$

$$c = 0.109(3.85 - T_r)^2 \tag{A.24}$$

$$b = 0.642T_r - 0.007T_r^4 - 0.52 \tag{A.25}$$

$$a = 0.03 + 0.00527(3.5 - T_r)^3 \tag{A.26}$$

$R = 8.314\,41$ J/(g mol deg) (gas constant)

$$K_G \approx \frac{P\Upsilon}{1 - P_r f / Z} \text{ with } \Upsilon = 0.85 + \frac{5.6}{P_r + 2} + \frac{27.1}{(P_r + 3.5)} - 8.7e^{-0.65(P_r+1)} \tag{A.27}$$

$$m = 1.2\left\{-\left[0.45 + 8\left(0.56 - \frac{1}{T_r}\right)^2\right]\frac{P_r^{0.2}}{T_r}\right\} \tag{A.28}$$

$f = cdm + a,$

Appendix B Basic Definitions

normal pressure: It is the pressure exerted by a static column of water of the same height as the overlying pore fluids and the same density as the pore water:

Normal pressure = (Pressure gradient of water × depth)

abnormal pressure: Pressure that is different from hydrostatic pressure (it could be higher or lower than hydrostatic pressure).

pore pressure: It is the pressure exerted by the pore fluids:

Pore pressure = Normal pressure + Over/underpressure

(units: Psi/ft. × 19.268 = ppg (pounds per gallon); ppg × 0.0519 = psi/ft.)

geopressure: Strictly speaking, it is defined as the subsurface formation fluid pressure.

overpressure: It is the excess pressure above normal pressure:

Overpressure = Pore pressure − Normal pressure

overburden stress: It is the stress exerted by the overlying pore fluid and rocks:

Overburden stress = Overburden stress gradient (psi/ft.) × Depth (ft.)

Terzaghi's relationship: It states that the total stress is jointly supported by the pore fluid and the rock matrix:

Overburden stress = Pore pressure + Effective vertical stress

effective vertical stress: It is the vertical stress applied to the rock matrix:

Effective vertical stress = Overburden stress − Pore pressure

buoyant pressure (ΔP): It is the excess pressure created in a confined reservoir by the density difference between hydrocarbons and water:

$$\Delta P = \text{(Water gradient (psi/ft.)} - \text{Hydrocarbon gradient (psi/ft.))}$$
$$\times \text{Height of hydrocarbon column (ft.)}$$

Pore pressure = Normal pressure + Overpressure (brine-filled) + Buoyancy pressure

Equivalent Terms

normal pressure = hydrostatic pressure = normal fluid pressure
pore pressure = fluid pressure = formation pressure
overpressured = abnormally high pressured
overburden stress = lithostatic stress = geostatic stress
effective vertical stress = net overburden stress = net confining stress
subnormal pressure = fluid pressure < normal pressure (depletion of formation fluid; stripped overburden)

Drilling Terminology

differential pressure: It is the difference between the hydrostatic pressure (HP) exerted by the mud column and the formation pressure (FP). (Never confuse this with the effective stress.)

balanced drilling: The hydrostatic pressure of the mud column is equal to the formation pressure (seldom happens); HP = FP.

overbalanced drilling: The hydrostatic pressure of the mud column is greater than the formation pore pressure (the usual case); HP > FP.

underbalanced drilling: The hydrostatic pressure of the mud column is less than the formation pressure (it is never planned, but "it happens," especially in thick shale!); HP < FP.

effective circulating density: It is the apparent increased mud density of a circulating mud column. When the mud is circulated, additional pressure, "backpressure," is placed against the formation due to frictional effects in the mud column (engineering terminology); ECD = MW (ppg) + Pa/(0.052 * TVD (ft.)); Pa = annular pressure (psi).

kick: It is the undesirable influx of formation fluid into the wellbore; it can cause blowouts.

lost circulation: It leads to a drop of both the fluid level and HP; if HP drops below reservoir pressure, the well "kicks."

fracture pressure: It is the pressure above which injection of fluids into the borehole will cause the formation to fracture hydraulically.

leak-off pressure: It is the pressure exerted on the rock matrix of an exposed formation causing fluid to be forced into the formation.

true vertical depth sub sea (TVDSS): This is measured from mean sea level (MSL). It does not reference KB. Kelly Bushing (KB) is the height of the Derek (drill rig) measured from the MSL.

Appendix C Dimensionless Coordinate Transformation of 1D Basin Modeling Equation

In Section 10.3.1 we presented a 1D model of shale-on-shale deposition on an impermeable base at time $t = 0$ in the moving coordinate system (Z', t) as explained in the text. One approach to solving this moving boundary problem is to perform another coordinate transformation going from the Lagrangian vertical coordinate to a fixed dimensionless coordinate, as shown in the following (see also Wangen, 2010; Lerche, 1990).

The basic compaction and flow equations describing the model are given in Chapter 10, equations (10.37) and (10.38), in the moving coordinates and are reproduced below:

$$\frac{\partial \varepsilon}{\partial t} = -\frac{\partial Q}{\partial Z'} \tag{C.1}$$

where

$$Q = \left(\frac{\kappa}{\mu}\right)\left(\frac{1}{1+\varepsilon}\right)\left(g(\rho_r - \rho_w) + \frac{\partial \sigma}{\partial Z'}\right) \tag{C.2}$$

$$\varepsilon = \frac{\phi}{1-\phi} \tag{C.3}$$

and

$$Z' = \int_0^z (1-\phi)\,dz \tag{C.4}$$

Here, z is the actual column height increasing upward, while Z' denotes the net (porosity-free) height; t is time, ϕ is the porosity with the corresponding void ratio ε, κ is the permeability of shale, μ is the brine viscosity, g is acceleration due to gravity, ρ_r is the grain density, ρ_w is the brine density, and σ is the vertical effective stress. The moving boundary conditions are

$$\phi(Z' = \Gamma t, t) = \phi_{sh} \tag{C.5}$$

$$Q(Z' = 0, t) = 0 \tag{C.6}$$

where Γ is a constant net sedimentation rate. Solution of equation (C.1) in the moving coordinate system is a complex problem, even in 1D, due to the implicit nature of the problem. However, following Wangen (2010) and Lerche (1990), A. Pradhan (personal communication, 2016) has shown that we can simplify the solution by transforming the vertical coordinate z' to *another* dimensionless fractional coordinate (a *fixed* coordinate system), (z_1, t), such that

$$z_1 = \frac{z'}{\Gamma t} \tag{C.7}$$

where Γ is the constant net sedimentation rate. We note that $0 \le z_1 \le 1$, with $z_1 = 1$ at the top sediment–water interface. In the *new* (fixed) coordinate system (z_1, t), the partial derivatives from the previous coordinate system (z', t) transform as follows:

$$\frac{\partial(\cdot)}{\partial z'} \Rightarrow \frac{1}{\Gamma t}\frac{\partial(\cdot)}{\partial z_1} ; \quad \frac{\partial(\cdot)}{\partial t} \Rightarrow \frac{\partial(\cdot)}{\partial t} + \frac{\partial(\cdot)}{\partial z_1}\frac{\partial z_1}{\partial t} = \frac{\partial(\cdot)}{\partial t} - \frac{\partial(\cdot)}{\partial z_1}\frac{z_1}{t} \tag{C.8}$$

In going from the Lagrangian coordinate to the fixed coordinate, the partial time derivate is replaced by the material derivative (Wangen, 2010).

With this transformation equations (C.1) and (C.2) become

$$t\frac{\partial \varepsilon}{\partial t} = -\frac{1}{\Gamma}\frac{\partial Q}{\partial z_1} + z_1 \frac{\partial \varepsilon}{\partial z_1} \tag{C.9}$$

and

$$Q = \left(\frac{\kappa}{\mu}\right)\left(\frac{1}{1+\varepsilon}\right)\left(g(\rho_r - \rho_w) + \frac{1}{\Gamma t}\frac{\partial \sigma}{\partial z_1}\right) \tag{C.10}$$

with the following boundary and initial conditions:

$$\varepsilon(z_1 = 1, t) = \varepsilon_{sh} \text{ (depositional void ratio of shale),} \tag{C.11}$$

$$Q(z_1 = 0, t) = 0 \text{ (no-flow boundary at the base), and} \tag{C.12}$$

$$\varepsilon(z_1, t = 0) = \varepsilon_0 \tag{C.13}$$

where ε_{sh} is the depositional void ratio of the shale at the shale–water interface and ε_0 is the void ratio of the basement, both assumed to be constant. The equations (C.9) and (C.10) can now be solved, given a constitutive relation between effective stress and void ratio. For example, using Dutta's constitutive relation (see equation (3.10)), $\sigma = \sigma_0\exp(-\beta\varepsilon)$, equation (C.10) becomes

$$Q = \left(\frac{\kappa}{\mu}\right)\left(\frac{1}{1+\varepsilon}\right)\left(g(\rho_r - \rho_w) + \frac{1}{\Gamma t}\left(-\sigma_0 e^{-\varepsilon\beta}\beta\frac{\partial \varepsilon}{\partial z_1}\right)\right) \tag{C.14}$$

The compaction equation (C.9) is augmented by two other differential equations, one for temperature (assuming a constant thermal gradient) and another to account for smectite-to-illite diagenetic transformation:

$$\frac{\partial T}{\partial z} = -\alpha \tag{C.15}$$

$$\frac{\partial \eta}{\partial T} = Ae^{-E/RT} \tag{C.16}$$

In the above equations the geothermal gradient is α, η is a kinetic parameter quantifying the extent of the smectite-to-illite reaction as a function of temperature T (that depends on time t), A is the pre-exponential factor in the Arrhenius model, E is the activation energy, and R is the gas constant (see Section 3.4.2). The kinetic parameter is related to the molar fraction of smectite $N_s(t)$ in the host rock by $\eta = -\ln(N_s(t)/N_s(0))$, where $N_s(0)$ is the initial fraction at $t = 0$. Equations (C.15) and (C.16) also need to be transformed from (z, t) to (Z', t) and finally to the (z_1, t) coordinate system following the chain rule for partial derivatives (see equations (10.30), (10.31), and (C.8)). After the coordinate transformations, equations (C.15) and (C.16) with their initial and boundary conditions in the (z_1, t) coordinate system are expressed as

$$t\frac{\partial T}{\partial t} = z_1 \frac{\partial T}{\partial z_1} + \alpha\Gamma t(1 + \varepsilon) \tag{C.17}$$

$$T(z_1, t = 0) = T_{\text{surface}}$$

$$T(z_1 = 1, t) = T_{\text{surface}}$$

where T_{surface} is the surface temperature at the depositional sediment–water interface, and

$$t\frac{\partial \eta}{\partial t} = \left[z_1 - \frac{Q(1-\varepsilon)^{-1}}{\Gamma}\right]\frac{\partial \eta}{\partial z_1} + tAe^{-E/RT} \tag{C.18}$$

$$\eta(z_1, t = 0) = 0$$

$$\eta(z_1 = 1, t) = 0$$

In this formulation, we need to solve *three* coupled differential equations (PDEs) shown in equations (C.9) for void ratio, (C.17) for temperature, and (C.18) for smectite–illite transformation, along with equation (C.10) and the appropriate boundary and initial conditions. The solutions are then transformed back to the original z-coordinate. This formulation follows parallel to what was described in Section 10.3.1. The coupled PDEs can be solved numerically by any commercial PDE solver such as the one in MATLAB. The website associated with this book provides an example MATLAB script to solve this system of equations numerically.

References

Aadnoy, B. S. and Larson, K., 1989, Method for fracture-gradient prediction for vertical and inclined boreholes, *SPE Drill Eng*, 4(2), 99–103.

Aadnoy, B. S. and Looyeh, R., 2010, *Petroleum Rock Mechanics: Drilling Operations and Well Design*, Elsevier, Amsterdam.

Adachi, J. I., Nagy, Z. R., Sayers, C. M., Smith, M. F. and Becker, D. F., 2012, Drilling adjacent to salt bodies: Definition of mud weight window and pore pressure using numerical models and fast well planning tool, SPE-159739.

Addis, M. A., 1997, The stress depletion response of reservoirs, SPE 38720.

Adisornsuapwat, K., Phuat tan, C., Anis, L., Vantala, A., Juman, R. and Boyce, B., 2013, Enhanced geomechanical modeling with advanced sonic processing to delineate and evaluate tight gas reservoirs, SPE Middle East Unconventional Gas Conference and Exhibition, Society of Petroleum Engineers, Muscat, Oman.

Ahmed, S., Khan, K., Omini, P. I., Aziz, A., Ahmed, M., Yadav, A. S. and Mohiuddin, M. A., 2014, An integrated drilling and geomechanics approach helps to successfully drill wells along the minimum horizontal stress direction in Khuff reservoirs, Abu Dhabi International Petroleum Exhibition and Conference, Abu Dhabi, UAE.

Aki, K. and Lee, W. H. K., 1976, Determination of three-dimensional velocity anomalies under a seismic array using first P arrival times from local earthquakes 1. A homogeneous initial model, *J Geophys Res*, 81. doi:10.1029/JB081i023p04381.

Aki, K. and Richards, P. G., 1980, *Quantitative Seismology: Theory and Methods*, W. H. Freeman, San Francisco.

Alam, A. H. M. S. and Pilger, R. H., 1986, Subsurface temperature investigations using seismic and borehole temperature data, Vermilion and Lafourche Parish, *Trans Gulf Coast Assoc Geol Soc*, 5(XXXVI).

Alberty, M., Hafle, M. and Minge, J., 1997, Mechanisms of shallow water flows and drilling practices for intervention, Offshore Technology Conference.

Al-Chalabi, M., 1994, Seismic velocity – a critique, *First Break*, 12, 589–596.

Al-Chalabi, M., 1995, Seismic velocities – a critique, *Int J Rock Mech Min Sci Geomech Abstr*, 32(7), 330A–33.

Al-Chalabi, M. and Rosenkranz, P. L., 2002, Velocity-depth and time-depth relationships for a decompacted uplifted unit, *Geophys Prospect*, 50, 661–664.

Aleardi, A. and Mazzotti, A., 2016, 1D full-waveform inversion and uncertainty estimation by means of a hybrid genetic algorithm – Gibbs sampler approach, *Geophys Prospect*. doi:10.1111/1365.2478.12397.

Aleotti, L., Poletto, F., Miranda, F., Corubolo, P., Abramo, F. and Craglietto, A., 1999, Seismic while-drilling technology: use and analysis of the drill-bit seismic source in a cross-hole survey. *Geophys Prospect*, 47, 25–39.

Aldred, W., Plumb, D., Bedford, I., Cook, J., Gholkar, V., Cousins, L., Minton, R., Fuller, J., Goraya, S. and Tucker, D., 1999, Managing drilling risk, *Oilfield Rev*, 11, 1–19.

Ali, A. H. A., Brown, T., Delgado, R., Lee, D., Plumb, D., Smirnov, N., Marsden, R., Prado-Velarde, E., Ramsey, L., Spooner, D., Stone, T. and Stouffer, T., 2003, Watching rocks change – Mechanical earth modeling. *Oilfield Rev*, 15, 22–39.

Ali, T. H., Sas, M., Hood, J. A., Lemke, S. R., Srinivasan, A., McKay, J., Fereday, K. S., Mondragon, C., Townsend S. C., Edwards, S. T., Hernandez, M. and Reeves, M., 2008, High speed telemetry drill pipe network optimizes drilling dynamics and wellbore placement, paper 112636, SPE/IADC Drilling Conference.

Al-Kawai, W. H., 2018, The impact of the allochtonous salt and overpressure development on the petroleum system evolution in the Thunder Horse mini-basin, Gulf of Mexico, PhD dissertation, Stanford University.

Al-Kawai, W. H. and Mukerji, T., 2016, Integrating basin modeling with seismic technology and rock physics, *Geophys Prospect*, 64, 1556–1574.

Al-Khalifa, T. and Tsvankin, I., 1995, Velocity analysis for transversely isotropic media, *Geophysics*, 60, 1550–1566.

Al-Khalifa, T., Tsavankin, I., Larner, K. and Toldi, J., 1996, Velocity analysis and imaging in transversely isotropic media: methodology and a case study, *Leading Edge*, 15, 371–378.

Al-Saeedi, M., Al-Mutairi, B., Al-Khaldy, M. and Sheeran, T., 2003, Fastest deep Marrat well in North Kuwait: case history of Raudhatain, SPE-85287.

Althaus, V. E., 1997, A new model for fracture gradient, *J Can Pet Technol*, 16(2), 99–108.

Ameen, M. S., Smart, B. G. D., Somerville, J. M. C., Hamilton, S. and Naji, N. A., 2009, Predicting rock mechanical properties of carbonates from wireline logs (a case study: Arab-D reservoir, Ghawar field, Saudi Arabia), *Mar Pet Geol*, 26(4), 430–444.

Anderson, B., Bryant, I., Luling, M., Spies, B. and Helbig, K., 1994, Oilfield anisotropy: its origins and electrical characteristics, *Oilfield Rev*. https://www.slb.com/~/media/Files/resources/oilfield_review/or s94/1094/p48_56.pdf.

Anderson, E. M., 1905, The dynamics of faulting, *Trans* Edin Geol Soc, 8, 387–402.

Anderson, J. E., Al-Khalifa, T. and Tsavankin, I., 1996, Fowler DMO and time migration for transversely isotropic media. *Geophysics*, 61, 835–844.

Anderson, R. A., Ingram, D. S. and Zanier, A. M., 1973, Determining fracture pressure gradients from well logs. *J Pet Technol*, 25(11), 1259–1268.

Anderson, T. L., 1991, *Fracture Mechanics: Fundamentals and Applications*, CRC Press, Boca Raton, FL.

Anstey, N. A. and Geyer, R. L., 1987, *Borehole Velocity Measurements and the Synthetic Seismogram*, IHRDC, Boston.

Aoyagi, K., Kazama, T., Sekiguchi, K. and Chilingarian, G. V., 1985, Experimental compaction of Na-montmorillonite clay mixed with crude oil and seawater, *Water–Rock Interact*, 49(1–3), 385–392.

Aplin, A. C. and Larter, S. R., 2005, Fluid flow, pore pressure, wettability, and leakage in mudstone cap rocks, in: Boult, P. and Kaldi, J. (eds.), *Evaluating Fault and Cap Rock Seals*, Hedberg Series 2, AAPG, Tulsa, OK, pp. 1–12.

Aplin, A. C., Yang, Y. and Hansen, S., 1995, Assessment of the compression coefficient of mudstones and its relationship with detailed lithology, *Mar Pet Geol*, 12(8), 955–963.

Aplin, A. C., Matenaar, I. F., McCarty, D. K. and van Der Pluijm, B., 2006, Influence of mechanical compaction and clay mineral diagenesis on the micro fabric and pore-scale properties of deep-water Gulf of Mexico mudstones, *Clays Clay Min*, 54(4), 500–514.

Araya-Polo, M., Jennings, J., Adler, A. and Dahlke, T., 2018, Deep-learning tomography, *Leading Edge*, 37 (1), 58–66.

Archie, G. E., 1942, The electrical resistivity log as an aid in determining some reservoir characteristics, *Pet Trans AIME*, 146, 54–62.

Armstrong, P., Durand, C., Barany, I. and Butaud, T., 2004, Seismic measurements while drilling reduce uncertainty in the deepwater Gulf of Mexico, SEG Annual Meeting.

Arrhenius, S. A., 1889, Über die Dissociationswärme und den Einfluß der Temperatur auf den Dissociationsgrad der Elektrolyte.

Asef, M. R. and Farrokhrouz, M., 2010, Governing parameters for approximation of carbonates UCS, *Electron J Geotechn Eng*, 15, 1581–1592.

Assad, F. A., LaMoreaux, J. W. and Hughes, T., 2004, *Field Methods for Field Geologists and Hydrogeologists*, Springer, New York.

Atashbari, V., 2016, Origin of overpressure and pore pressure prediction in carbonate rocks in Abadan Plain Basin, PhD dissertation, Australian School of Petroleum, University of Adelaide.

Atashbari, V. and Tingay, M., 2012, Pore pressure prediction in a carbonate reservoir, SPE 150836.

Athy, L. F., 1930a, Density, porosity, and compaction of sedimentary rocks. *AAPG Bull*, 14, 1–24.

Athy, L. F., 1930b, Compaction and oil migration, *AAPG Bull*, 14, 25–35.

Atkinson, B. K., 1987, *Fracture Mechanics of Rock*, Academic Press, New York.

Audet, D. M. and Fowler, A. C., 1992, A mathematical model for compaction in sedimentary basins, *Geophys* J Int, 110, 577–599.

Avasthi, J. M., Goodman H. E. and Janson, R. P., 2000, Acquisition, calibration and use of the in-situ stress data for oil and gas well construction and production, SPE 60320.

Avseth, P. and Mukerji, T., 2002, Seismic lithofacies classification from well logs using statistical rock physics, *Petrophysics*, 43, 70–81.

Avseth, P., Mukerji, T. and Mavko, G., 2005, *Quantitative Seismic Interpretation: Applying Rock Physics Tools to Reduce Interpretation Risk*, Cambridge University Press, Cambridge.

Ayers, A. and Theilen, F., 1999, Relationship between P- and S-wave velocities and geological properties of near surface sediments of the continental slope of the Barents Sea, *Geophys Prospect*, 47, 431–441.

Azadpour, M., Manaman, S., Kadkhodaire-Ilkhchi, A. and Sedghipour, M. R., 2015, Pore pressure gradient prediction and modeling using well logging data in one of the gas fields of South Iran, *J Pet Sci Eng*, 128, 15–23.

Bachrach, R., 2010, Applications of deterministic and stochastic rock physics modeling to anisotropic velocity model building, SEG Annual Meeting, expanded abstracts.

Bachrach, R., 2011a, Elastic and resistive anisotropy of shale: Joint effective media modeling and field observations, SEG Annual Meeting, expanded abstracts.

Bachrach,R., 2011b, Elastic and resistivity anisotropy of shale during compaction and diagenesis: Joint effective medium modeling and field observations, *Geophysics*, 76, E175–E186.

Bachrach, R., 2015, Uncertainty and non-uniqueness in linearized AVAZ for orthorhombic media, *Leading Edge*, 34(9), 1048–1056.

Bachrach, R., 2016, Mechanical compaction in heterogeneous clastic formations from plastic–poroelastic deformation principles: theory and applications, *Geophys. Prospect*, 65, 1–12.

Bachrach, R. and Paydayesh, M., 2017, Application of sequential and Kalman filters to seismic-geomechanics reservoir monitoring, SEG Annual Meeting, expanded abstracts.

Bachrach, R. and Sengupta, M., 2008, Using geomechanical modeling and wide-azimuth data to quantify stress effects and anisotropy near salt bodies in the Gulf of Mexico, SEG Annual Meeting, expanded abstracts. https://doi.org/10.1190/1.3054790.

Bachrach, R., Dvorkin, J. and Nur, A., 2000, Seismic velocities and Poisson's ratio of shallow unconsolidated sands, *Geophysics*, 65, 559–564.

Bachrach, R., Liu, Y., Woodward, M., Zradrova, O., Yang, Y. and Osypov, C., 2011, Anisotropic velocity model building using rock physics: comparison of compaction trends and check-shot-derived anisotropy in Gulf of Mexico, SEG Annual Meeting. https://doi.org/10.1190/1.3627619.

Bachrach, R., Sayers, C. M., Dasgupta, S., Silva, J. and Volterrani, S., 2013, Recent advances in the characterization of unconventional reservoirs with wide-azimuth seismic data, Unconventional Resources Technology Conference. doi:10.1190/urtec2013-030.

Bakulin, A., Zadraveva, O., Nichols, D., Woodward, M. and Osypov, K., 2009, Well-constrained anisotropic tomography, Annual Meeting of EAGE. doi:10.3997/2214-4609.201400387.

Bakulin, A., Woodward, M., Liu, Y., Zdraveva, O., Nichols, D. and Osypov, K., 2010a, Application of steering filters to localized anisotropic tomography with well data, SEG Annual Meeting, extended abstracts.

Bakulin, A., Woodward, M., Nichols, D., Osypov, K. and Zdraveva, O., 2010b, Localized anisotropic tomography with well information in VTI media, *Geophysics*, 75, D37–D45.

Balch, A. H. and Lee, M. W., eds., 1984, *Vertical Seismic Profiling – Technique, Applications, and Case Histories*, IHRDC, Boston.

Baldwin, B., 1971, Ways of deciphering compacted sediments, *J Sediment Petrol*, 41(1), 293–301.

Baldwin, B. and Butler, C. O., 1985, Compaction curves, *AAPG Bull*, 69, 622–626.

Bandyopadhyay, K., 2009, Seismic anisotropy: geological causes and its implications to reservoir geophysics, PhD thesis, Stanford University.

Banik, N. C., Banik, A., Wool, G., Schultz, G., Dutta, N. C., Marple, R., Casper, T. and Repar, N., 2002, Application of anisotropy in pore pressure prediction, EAGE Annual Conference.

Banik, N. C., Wool, G., Schultz, G., den Boer, L. and Mao, W., 2003, Regional and high resolution 3D pore-pressure prediction in deep-water offshore West Africa, SEG Annual Meeting, expanded abstracts.

Banik, N., Koesoemadinata, A., Wagner, C., Inyang, C. and Bui, H., 2013, Predrill pore pressure prediction directly from seismically derived impedance, SEG Annual Meeting, expanded abstracts.

Barker, C., 1972, Aquathermal pressuring – role of temperature in development of abnormal pressure zones, *AAPG Bull*, 56, 2068–2071.

Barker, C., 1990, Calculated volume and pressure changes during the thermal cracking of oil to gas in reservoirs, *AAPG Bull*, 74, 1254–1261.

Barker, J. W. and Meeks, W. R., 2003, Estimating fracture gradient in Gulf of Mexico deepwater, shallow, massive salt sections, SPE Annual Technical Conference and Exhibition. doi:10.2118/84552-MS.

Barker, J. W. and Wood, T. D., 1997, Estimating shallow below mudline deepwater Gulf of Mexico fracture gradients, AADE Houston Chapter Annual Technical Meeting.

Barkved, O. L., 2012, *Seismic Surveillance for Reservoir Delivery from a Practitioner's Point of View*, EAGE, Houten, The Netherlands.

Batzle, M. and Wang, Z., 1992, Seismic properties of pore fluids, *Geophysics*, 57, 1396–1408.

Bayer, U. and Wetzel, A., 1989, Compaction behavior of fine-grained sediments; examples from Deep Sea Drilling Project cores, *Geol Rundschau*, 78(3), 807–819.

Bear, J., 1972, *Dynamics of Fluids in Porous Media*, Dover, Mineola, NY.

Bell, D. W., 2002, Velocity estimation for pore pressure prediction, *AAPG Mem*, 76, 217–233.

Belloti, P. and Giacca, D., 1978, Seismic data can detect overpressures in deep drilling, *Oil Gas J*, 76, 47–52.

Benfield, N., Rambaran, V., Dowlath, J., Sinclair, T., Evans, M., Richardson, J., Ratcliffe, A. and Irving, A., 2017, Extracting geologic information directly from high-resolution full-waveform inversion velocity models – a case study from offshore Trinidad, *Leading Edge*. https://doi.org/10.1190/tle36010067.1

Ben-Menahem, A. and Singh, S. J., 1981, *Seismic Waves and Sources*, Springer, New York.

Berard, T. and Prioul, R., 2016, Mechanical Earth model, *Oil Field Rev*.

Berg, R. R., 1975, Capillary pressure in stratigraphic traps, *AAPG Bull*, 59, 939–956.

Berg, R. R. and Gangi, A. F., 1999, Primary migration by oil-generation microfracturing in low-permeability source rocks: application to the Austin chalk, Texas, *AAPG Bull*, 83, 727–756.

Berryman, J. G., 1980, Long-wavelength propagation in composite elastic media, I and II, *J Acoust Soc Am*, 68, 1809–1831.

Beyer, L. A., 1987, Porosity of unconsolidated sand, diatomite, and fractured shale reservoirs, South Belridge and West Cat Canyon oil fields, California: Section IV. Exploration methods, AAPG *Bull*, 25, 395–413.

Best, A. I., McCann, C. and Sothcott, J., 1994, The relationships between the velocities, attenuations and petrophysical properties of reservoir sedimentary rocks, *Geophys*. Prospect, 42, 151–178.

Bethke, C. M., 1989, Modeling subsurface flow in sedimentary basins, *Geol Rundschau*, 78(1), 129–154.

Bhaduri, A., Dutta, N. C., Yang, S., Dai, J., Chandrasekhar, S., Sundaram, K. M., Dotiwala, F., Samanta, B. G., Visweswara Rao, C. and Kutty, P. S. N., 2012, A novel approach to improving well positioning and drilling confidence using seismic interpretation and quantitative studies on rock physics guided depth imaging and integrated pore pressure analysis: a case study from Andaman Sea fore-arc exploration, paper P-326, 9th Biennial International Conference and Exposition on Petroleum Geophysics, Hyderabad, India.

Bigelow, E. L., 1994, Well logging methods to detect abnormal pressure, in: Fertl, W. H., Chapman, R. E. and Hotz, R. F. (eds.), *Studies in Abnormal Pressures*, Developments in Petroleum Science 38, Elsevier, New York, pp. 187–240.

Bilgeri, D. and Ademeno, E. B., 1982, Predicting abnormally pressured sedimentary rocks, *Geophys Prospect*, 30, 608–621.

Biot, M. A., 1941, General theory of three-dimensional consolidation, *J Appl Phys*, 12, 155–164.

Biot, M. A., 1956, Theory of propagation of elastic waves in a fluid-saturated porous solid. II. Higher frequency range. *J Acoust Soc Am*, 28, 179–191.

Biot, M. A., 1962, Mechanics of deformation and acoustic propagation in porous media, *J Appl Phys*, 33, 1482–1498.

Birchwood, R., Noeth, S., Tjengdrawira, M., Kisra, S., Elisabeth, F., Sayers, C., Singh, R., Hooyman, P., Plumb, R., Jones, E. and Bloys, J., 2007, Modeling the mechanical and phase change stability of wellbores drilled in gas hydrates by the Joint Industry Participation Program (JIP) Gas Hydrates Project, Phase II, SPE Annual Technical Conference.

Bishop, T. N., Bube, K. P., Cutler, R. T., Langan, R. T., Love, P. L., Resnick, J. R., Shuey, R. T., Spindler, D. A. and Wyld, H. W., 1985, Tomographic determination of velocity and depth in laterally varying media, *Geophysics*, 50, 903–923.

Biswas, R., Sen, M., Das, V. and Mukerji, T., 2019, Prestack and poststack inversion using a physics-guided convolutional neural network, *Interpretation*, 7, SE161–SE174.

Bjørkum, P. A., Oelkers, E. H., Nadeau, P. H., Walderhaug, O, and Murphy, W. M., 1998, Porosity prediction in quartzose sandstones as a function of time, temperature, depth, stylolite frequency, and hydrocarbon saturation, *AAPG Bull*, 82, 637–648.

Bjørlykke, K., 1998, Clay mineral diagenesis in sedimentary basins; a key to the prediction of rock properties; examples from the North Sea Basin, *Clay Min*, 33, 15–34.

Bjørlykke, K., 1999, Principal aspects of compaction and fluid flow in mudstones, in: Aplin, A. C., Fleet, A. J. and Macquaker, J. H. S. (eds.), *Muds and Mudstone: Physical and Fluid-Flow Properties*, Geological Society, London, pp. 73–78.

Bjørlykke, K., ed., 2005, *Petroleum Geoscience: From Sedimentary Environments to Rock Physics*, Springer, New York.

Bjørlykke, K., ed., 2015, *Petroleum Geoscience – From Sedimentary Environments to Rock Physics*, 2nd ed., Springer, New York.

Bjørlykke, K. and Høeg, K., 1997, Effects of burial diagenesis on stresses, compaction and fluid flow in sedimentary basins, *Mar Pet Geol*, 14, 267–276.

Blakely, R. J., 1995, *Potential Theory in Gravity and Magnetic Applications*, Cambridge University Press, New York.

Blangy, J. P., 1992, Integrated seismic lithologic interpretation; the petrophysical basis, PhD thesis, Stanford University.

Boatmen, W. A., 1967, Measuring and using shale density to aid in drilling wells in high-pressure areas, *J Pet Technol*, November, 1423–1429.

Boles, J. R. and Franks, S. G., 1979, Clay diagenesis in Wilcox sandstones of Southwest Texas; implications of smectite diagenesis on sandstone cementation, *J Sediment Res*, 49, 55–70.

Bortfeld, R., 1961, Approximation to the reflection and transmission coefficients of plane longitudinal and transverse waves, *Geophys Prospect*, 9, 485–502.

Boswell, R., Collett, T., McConnell, D., Frye, M., Shedd, W., Godfriaux, P., Dufrene, R., Mrozewski, S., Guerin, G., Cook, A., Shelander, D., Dai, J. and Jones, E., 2009, Initial results of Gulf of Mexico Gas Hydrate Joint Industry Project Leg II Logging-While-Drilling Operations, Annual GCSSEPM Perkins Conference.

Boult, P. J., 1993, Membrane seal and tertiary migration pathways in the Bodalla South oilfield, Eronmanga Basin, Australia, *Mar Pet Geol*, 10, 3–13.

Bourbie, T., Coussy, o. and Zinszner, B., 1987, *Acoustics of Porous Media*, Gulf, Houston, TX.

Bourgoyne, A. T., 1971, A graphic approach to overpressure detection while drilling, *Pet Eng*, 43(9), 76–78.

Bourgoyne, A. T., Jr., Millheim, K. K., Chenevert, M. E. and Young, F. S., Jr., 1991, *Applied Drilling Engineering*, SPE Textbook Series 2, Society of Petroleum Engineers, Houston, TX.

Bougher, B. B., 2016, Machine learning applications to geophysical data analysis, doctoral dissertation, University of British Columbia.

Bowers, G. L., 1994, Pore pressure estimation from velocity data: accounting for overpressure mechanisms besides undercompaction, SPE 27488.

Bowers, G. L., 1995, Pore pressure estimation from velocity data – accounting for overpressure mechanisms besides undercompaction, SPE Drilling and Completion.

Bowers, G. L., 2001, Determining an appropriate pore-pressure estimation strategy, paper 13042-MS, Offshore Technology Conference. http://dx.doi.org/10.4043/13042-MS.

Bowers, G. L., 2002, Detecting high overpressure, *Leading Edge*, 21, 174–177.

Brace, W. F., Walsh, J. B. and Frangos, W. T., 1968, Permeability of granite under high pressure, *J Geophys Res*, 73, 2225–2236.

Bradford, I. D. R., Fuller, J., Thompson, P. J. and Walsgrove, T. R., 1988, Benefits of assessing the solids production risk in a North Sea Reservoir using elastoplastic modeling, SPE 47360.

Bradley, J. S., 1975, Abnormal formation pressure, *Bull AAPG*, 59, 957–973.

Bradley, J. S. and Powley, D. E., 1994, Pressure compartments in sedimentary basins: a review, *AAPG Mem*, 61, 3–26.

Brandt, H., 1955, A study of the speed of sound in porous granular media, *J Appl Mech*, 22, 479–486.

Bratton, T. and Bornemann, T., 1999, Logging-while-drilling images for geomechanical, geological and petrophysical images, SPWLA Annual Logging Symposium.

Breckels, I. M. and van Eekelen, H. A. M., 1982, Relationship between horizontal stress and depth in sedimentary basins, *J Pet Technol*, 34(9), 2191–2199.

Bredehoeft, J. D. and Hanshaw, B. B., 1968, On the maintenance of anomalous fluid pressures: I. Thick sedimentary sequences, *Geol Soc Am Bull*, 79, 1097–1106.

Bredehoeft, J. D., Wesley, J. B. and Fouche, T. D., 1994, Simulations of the *Origin* of *Fluid Pressure, Fracture Genera*tion, and the *Movement* of *Fluids* in the *Uinta Basin*, Utah: AAPG.

Brennan, B. J. and Stacey, F. D., 1978, Frequency dependence of elasticity of rock – test of velocity dispersion, *Nature*, 268, 220–222.

Brevik, I., Callejon, A., Kahn, P., Janak, P. and Ebrom, D., 2011, Rock physicists step out of the well location, meet geophysicist and geologists to add value in exploration analysis, *Leading Edge*, 30, 1382–1391.

Brigaud, F., Chapman, D. S. and Douaran, S. L., 1990, Estimating thermal conductivity in sedimentary basins using lithologic data and geologic well logs, AAPG *Bull*, 74, 1459–1477.

Brittan, J., Bai, J., Delome, H., Wang, C. and Yingst, D., 2013, Full waveform inversion – the state-of-the-art, *First Break*, 31, 75–81.

Brossier, R. and Virieux, J., 2011, Lecture notes on full waveform inversion, SEISCOPE Consortium, http://seiscope.oca.eu.

Brown, J. P., Fox, A., Sliz, K., Hanbal, I., Planchart, C., Davies, G. and McFarlane, M., 2015, Pre-drill and real-time pore pressure prediction: lessons from a sub-salt, deep water wildcat well, Red Sea, Society of Petroleum Engineers, https://doi.org/10.2118/172743-MS.

Brown, R. and Korringa, J., 1975, On the dependence of the elastic properties of a porous rock on the compressibility of pore fluid, *Geophysics*, 40, 608–616.

Bruce, B. and Bowers, G. L., 2002, Pore pressure terminology, *Leading Edge*, 21, 170–173.

Bruce, C. H., 1984, Smectite dehydration – its relation to structural development and hydrocarbon accumulation in Northern Gulf of Mexico basin, *Am Assoc Pet Geol Bull*, 68, 673–683.

Burland, J. B., 1990, On the compressibility and shear strength of natural clays, *Géotechnique*, 40, 329–378.

Burnham, A. K., 2017, *Global Chemical Kinetics of Fossil Fuels: How to Model Maturation and Pyrolysis*, Springer, New York.

Burst, J. F., 1969, Diagenesis of Gulf Coast clayey sediments and its possible relation to petroleum migration, *Bull AAPG*, 53, 73–93.

Byrd, T. M., Schneider, J. M., Reynolds, D. J., Alberty, M. W. and Hafle, M. E., 1996, Identification of "flowing water sand" drilling hazards in the deepwater Gulf of Mexico, Offshore Technology Conference.

Caidado, C. C., Hobbs, R. W. and Goldstein, M., 2012, Bayesian strategies to assess uncertainty in velocity models, *Bayesian Anal*, 7, 211–234.

Calderón-Macías, C., Sen, M. and Stoffa, P., 1998, Automatic NMO correction and velocity estimation by a feedforward neural network, *Geophysics*, 63, 1696–1707.

Campbell, J., 1997, Fast-track development: the evolving role of 3-D seismic data in deepwater hazards assessment and site investigation, Offshore Technology Conference.

Cappa, F. and Rutqvist, J., 2012, Modeling of coupled deformation and permeability evolution during fault reactivation induced by deep underground injection of CO_2, *Int J Green Gas Conserv*, 5, 336–346.

Carcione, J. M. and Helle, H. B., 2002, Rock physics of geopressure and prediction of abnormal pore fluid pressures using seismic data, *Recorder Can Soc Explor Geophys*, 27(7), 9–30.

Carman, P. C., 1937, Fluid flow through granular beds, *Trans Inst Chem Eng London*, 15, 150–166.

Carmichael, R. C., 1988, *Practical Handbook of Physical Properties of Rocks & Minerals*, 1st ed., CRC Press, Boca Raton, FL.

Carver, R. E., 1968, Differential compaction as a cause of regional contemporaneous faults, *Bull Am Assoc Pet Geol*, 52, 414–419.

Castagna, J. P., Batzle, M. L. and Eastwood, R. L., 1985, Relationship between compressional shear wave velocities in clastic silicate rocks, *Geophysics*, 50, 570–581.

Castagna, J. P., Batzle, M. L. and Kan, T. K., 1993, Rock physics – the link between rock properties and AVO response, in: Castagna, J. P. and Backus, M. M. (eds.), *Offset Dependent Reflectivity – Theory and Practice of AVO Analysis*, Investigations in Geophysics 8, Society of Exploration Geophysicists, Tulsa, OK, pp. 135–171.

Chang, C., Zoback, M. D. and Khaksar, A., 2006, Empirical relations between rock strength and physical properties in sedimentary rocks, *J Pet Sci Eng*, 51, 223–237.

Chapman, R. E., 1994a, Abnormal pore pressures: Essential theory, possible causes, and sliding, in: Fertl, W. H., Chapman, R. E. and Hotz, R. F. (eds.), *Studies in Abnormal Pressures*, vol. 38, Elsevier, The Netherlands, pp. 51–91.

Chapman, R. E., 1994b, The geology of abnormal pore pressures, in: Fertl, W. H., Chapman, R. E. and Hotz, R. F. (eds.), *Studies in Abnormal Pressures*, vol. 38, Elsevier, The Netherlands, pp. 19–49.

Cheng, A. H.-D., 2016, *Poroelasticity*, Springer, Berlin.

Cheng, C. H., Paillet, F. L. and Pennington, W. D., 1992, Acoustic waveform logging – Advances in theory and application, *Log Anal* 33, 239.

Chilingar, G. V. and Knight, L., 1960, Relationship between pressure and moisture content of kaolinite, illite, and montmorillonite clays, *Am Assoc Pet Geol Bull*, 44, 101–106.

Chilingar, G. V., Serebryakov, V. A. and Robertson, J. O., 2002, *Origin and Prediction of Abnormal Formation Pressures*, Developments in Petroleum Science 50, Elsevier Science, Berlin.

Chiu, S. K. L. and Stewart, R. R., 1987, Tomographic determination of three-dimensional seismic velocity structure using well logs, vertical seismic profiles, and surface seismic data, *Geophysics*, 52, 1085–1098.

Chopra, S. and Huffman, A., 2006, Velocity determination for pore pressure prediction, *Recorder Can Soc Explor Geophys*, 31(4).

Chopra, S. and Kuhn, O., 2001, Seismic inversion, *Recorder Can Soc Explor Geophys*, 26(1), 10–14.

Chuhan, F. A., Kjeldstad, A., Bjørlykke, K. and Hoeg, K., 2002, Porosity loss in sand by grain crushing; experimental evidence and relevance to reservoir quality, *Mar Pet Geol*, 19(1), 39–53.

Chuhan, F. A., Kjeldstad, A., Bjørlykke, K. and Hoeg, K., 2003, Experimental compression of loose sands; relevance to porosity reduction during burial in sedimentary basins, *Can Geotech J*, 40, 995–1011.

Chukwu, G. A., 1991, The Niger Delta complex basin: Stratigraphy, structure and hydrocarbon potential, *J Pet Geol*, 14, 211–220.

Clark, V. A., 1992, The effect of oil under in-situ conditions on the seismic properties of rocks, *Geophysics*, 57, 894–901.

Clark, S. P., Jr., 1966, Thermal conductivity, *AAPG Mem*, 97, 459–582.

Clayton, C. and Hay, S. J., 1994, Gas migration mechanisms from accumulation to surface, *Bull. Geol. Soc. Denmark*, 41, 12–23.

Colmenares, L. B. and Zoback, M. D., 2002, A statistical evaluation of intact rock failure criteria constrained by polyaxial test data for five different rocks, *Int J Rock Mech Min Sci*, 39, 695–729.

Colombo, D., Rovetta, D., Turkoglu, E., MacNeice, G. and Sandoval-Curriel, E., 2017, Multiscale hierarchical seismic – CSEM joint inversion for subsalt depth imaging in the Red Sea, Annual SEG Meeting, https://doi.org/10.1190/segam2017-17558705.1.

Combs, G. F., 1968, Prediction of pore pressure from penetration rate, SPE 2162.

Connolly, P., 1999, *Elastic impedance, Leading Edge*, 18, 438–452.

Constant, W. D. and Bourgoyne, A. T., 1988, Fracture-gradient prediction for offshore wells, *SPE Drill Eng*, 3, 136–140.

Coussy, O, 2004, *Poromechanics*, 2nd ed., Wiley, Hoboken, NJ.

Cowland, A. P., 1996, Drill site geohazard identification facilitated by rework of suitable existing 3-D seismic data volumes, Offshore Technology Conference.

Crampin, S., 1981, A review of wave motion in anisotropic and cracked elastic-media, *Wave Motion*, 3, 343–391.

Cumberland, D. J. and Crawford, R. J., 1987, *The Packing of Particles*, Elsevier, The Netherlands.

Curtis, J., 2015, Drilling uncertainty prediction technical section (DUPTS), 45th General Meeting of ISCWSA.

Dahl, B. and Yukler, A., 1991, The role of petroleum geochemistry in basin modeling of the Oseberg area, North Sea, in: Merrill, R. K. (ed.), *Source and Migration Processes and Evaluation Techniques*, AAPG, Tulsa, OK, pp. 237–251.

Dai, J., Xu, H., Snyder, F. and Dutta, N., 2004, Detection and estimation of gas hydrates using rock physics and seismic inversion: Examples from the northern deepwater Gulf of Mexico, *Leading Edge*, 23, 60–66.

Dai, J., Snyder, F., Gillespie, D., Koesoemadinata, A. and Dutta, N., 2008a, Exploration for gas hydrates in the deepwater Northern Gulf of Mexico: Part I. A seismic approach based on geologic model, *Mar Pet Geol*, 25, 830–844.

Dai, J., Banik, N., Gillespie, D. and Dutta, N., 2008b, Exploration for gas hydrates in the deepwater northern Gulf of Mexico: Part II. Model validation by drilling, *Mar Pet Geol*, 25, 845–859.

Daines, S. R., 1982a, Aquathermal pressuring and geopressure evaluation, *AAPG Bull*, 66, 931–939.

Daines, S. R., 1982b, The prediction of fracture pressures for wildcat wells, *J Pet Technol*, 34, 863–872.

Darcy, H., 1856, *Les Fontaines Publiques de la Ville de Dijon*, Dalmont, Paris.

Das, V. and Mukerji, T., 2019, Petrophysical properties prediction from pre-stack seismic data using convolutional neural networks, SEG Annual Meeting, expanded abstracts.

Das, V., Pollack, A., Wollner, U. and Mukerji, T., 2019, Convolutional neural network for seismic impedance inversion, *Geophysics*, 84, R869–R880.

De Gennaro, V., 2012, Solutions for 3D coupled geomechanical and HF modelling in UG reservoirs, 4th EAGE Conference and Exhibition, Workshops.

DeMartini, D. C., Beard, D. C., Danburg, J. S. and Robinson, J. H., 1976, Variations of seismic velocities in sandstones and limestones with lithology and pore fluid at simulated in situ conditions, EGPC Exploration Seminar.

Deming, D., Sass, J. H. and Lachenbruch, A. H., 1996, Heat flow and subsurface temperature, North Slope of Alaska, *US Geol Surv Bull*, 2142, 21–44.

Deo, B., Ramani, K., Dutta, N., Dai, J. and Peng, C., 2014, Rock physics guided velocity modeling and reverse-time migration for subsalt pore pressure prediction: Study in Green Canyon, deep water Gulf of Mexico, International Geophysical Conference and Exposition.

Deo, B., Jenning, C., Afi, M., Sedeek, M., Shelander, D., Ramani, K., Hussein, H., Tarek, N., Dutta, N. C., Srivastava, R., Singh, T., Pradhan, B. K. and Kumar, R., 2016, Integrated workflow for shallow hazard detection using imaging, inversion, pore-pressure prediction, and interpretation on 3d seismic data, SEG Annual Meeting, extended abstracts.

De Prisco, G., Thanoon, D., Bachrach, R., Brevik, I., Clark, S. A., Corver, M. P., Pepper, R. E., Hantschel, T., Helgesen, H. K., Osypov, K. and Leirfall, O. K., 2015, Geophysical basin modeling: Effective stress, temperature, and pore pressure uncertainty, *Interpretation*, 3, SZ27–SZ39.

de Kok, R., Dutta, N., Khan, M. and Mallick, S., 2001, Deepwater geohazard analysis using prestack inversion, SEG International Annual Meeting, extended abstracts.

De Stefano, M., Andreas, F. G., Re, S., Virgillo, M. and Snyder, F., 2011, Multiple domain simultaneous joint inversion of geophysical data with application to subsalt imaging, *Geophysics*, 76, R69–R80.

Dickey, P. A., Shriram, C. R. and Paine, W. R., 1968, Abnormal pressures in deep wells of southwestern Louisiana, *Science*, 160, 608–615.

Dickinson, G., 1951, Geological aspects of abnormal pressures in the Gulf Coast Region of Louisiana, USA, World Petroleum Congress.

Dickinson, G., 1953, Geological aspects of abnormal reservoir pressures in Gulf Coast region of Louisiana, USA, *Bull AAPG*, 37, 410–432.

Digby, P. J., 1981, The effective elastic moduli of porous granular rocks, *ASME J Appl Mech*, 48, 803–808.

Dix, C. H., 1955, Seismic velocities from surface measurements, *Geophysics*, 20, 68–66.

Domenico, S. N., 1977, Elastic properties of unconsolidated porous sand reservoirs, *Geophysics*, 42, 1339–1368.

Dow, W. G., 1984, Oil source beds and oil prospect definition in the Upper Tertiary of the Gulf Coast, *Gulf Coast Assoc Geol Soc Trans*, 34, 329–339.

Dowdle, W. L. and Cobb, W. M., 1975, Static formation temperature from well logs – an empirical method, *J Pet Technol*, 27, 1326–1330.

Doyen, P. M., Malinverno, A., Prioul, R., den Boer, L. D., Psaila, D., Sayers, C. M., Noeth, S., Hooyman, P., Smit, T. J. H., van Eden, C. and Wervelman, R., 2003, Seismic pore-pressure prediction with uncertainty using a probabilistic mechanical Earth model, SEG Annual Meeting, expanded abstracts.

Drucker, D. C. and Prager, W., 1958, Soil mechanics and plastic analysis for limit design, *Q Appl Math*, 10, 157–165.

Duffaut, K. K., Hokstad, R. K. and Wiik, T., 2018, A simple relationship between thermal conductivity and seismic interval velocity, *Leading Edge*, 37, 381–385.

Duijndam, A. J. W., 1988, Bayesian estimation in seismic inversion, Part I: Principles, *Geophys Prospect*, 36, 878–898.

Durmishyan, A. G., 1974, Compaction of argillaceous rocks, *Int Geol Rev*, 16, 650–653.

Dutta, N. C., 1984, Shale compaction and abnormal pore-pressures; a model of geopressures in the Gulf Coast Basin, *Geophysics*, 49, 660.

Dutta, N .C., 1986, Shale compaction, burial diagenesis, and geopressures: a dynamic model, solution, and some results in thermal modeling in sedimentary basins, in: J. Burrus (ed.), *Thermal Modeling in Sedimentary Basins*, Technip, Paris, pp. 149–172.

Dutta, N. C., 1987a, *Geopressure*, SEG Geophysics Reprint Series 7, SEG, Tulsa, OK.

Dutta, N. C., 1987b, Fluid flow in low permeable media, in: *Migration of Hydrocarbons in Sedimentary Basins*, Technip, Paris, pp. 567–595.

Dutta, N. C., 1997a, A brief review of recent technologies used for geopressure evaluation and an attempt to develop a unified model, BP Exploration Report HO97.0004, Western Hemisphere.

Dutta, N. C., 1997b, Pressure prediction from seismic data: Implications for seal distribution and hydrocarbon exploration and exploitation in the deepwater Gulf of Mexico, in: Moller-Pedersen, P. and Koestler, A. G. (eds.), *Hydrocarbon Seals: Importance for Exploration and Production*, Norwegian Petroleum Society Special Publication 7, Elsevier, Singapore, pp. 187–199.

Dutta, N. C., 2002a, Deepwater geohazard prediction using prestack inversion of large offset P-wave data and rock model, *Leading Edge*, 21(2), 193–198.

Dutta, N. C., 2002b, Geopressure prediction using seismic data: Current status and the road ahead, *Geophysics*, 67, 2012–2041.

Dutta, N. C., 2015, Velocity modeling and pore pressure imaging using tomography, rock physics and burial diagenesis of shale, Society of Petroleum Geophysicists Annual Meeting. Jaipur, India, June,

Dutta, N. C., 2016, *Effect of chemical diagenesis on pore pressure in argillaceous sediment*, special issue, Leading Edge, 35, 523–527.

Dutta, N. and Dai, J., 2009, Exploration for gas hydrates in a marine environment using seismic inversion and rock physics principles, *Leading Edge*, 28, 792–802.

Dutta, N. C. and Khazanehdari, J., 2006, Estimation of formation fluid pressure using high-resolution velocity from inversion of seismic data and a rock physics model based on compaction and burial diagenesis of shales, *Leading Edge*, 25, 1528–1539.

Dutta, N. and Mallick, S., 2010, Method for shallow water flow detection, US Patent 7,672,824.

Dutta, N. C. and Ode, H., 1979, Attenuation and dispersion of compressional waves in fluid-filled porous rocks with partial gas saturation (White model) – Part I; Biot theory; Part II: Results, *Geophysics*, 44, 1777–1805.

Dutta, N. C., Kahn, M. and Gelinsky, S. G., 2001, Seismic predrill pore pressure imaging using a deepwater rock model, EAGE Annual Meeting, extended abstracts.

Dutta, N. C., Zimmer, M. and Prasad, M., 2002a, Rock physics based overpressure detection, technical abstract, AAPG Annual Meeting.

Dutta, N. C., Boreland, W. H., Leaney, W. S., Meehan, R. and Nutt, W. L., 2002b, Pore pressure prediction ahead of the bit: An integrated approach, *AAPG Mem*, 76, 165–169.

Dutta, N., Mukerji, T., Prasad, M. and Dvorkin, J., 2002c, Seismic detection and estimation of overpressure Part I: The rock physics basis, *Recorder Can Soc Explor Geophys*, 27(7), 29–57.

Dutta, N., Mukerji, T., Prasad, M. and Dvorkin, J., 2002d, Seismic detection and estimation of overpressure, Part II: Field applications, *Recorder Can Soc Explor Geophys*, 27(7), 58–73.

Dutta, N., Utech, R., Nafie, T. and Bedingfield, J., 2002e, Geohazard detection in deepwater clastics basin: Seismic technique with application to deepwater Mediterranean, AAPG Annual Meeting, Cairo.

Dutta, N. C., Utech, R. W. and Shelander, D., 2010, Role of 3D seismic for shallow hazards assessment in deepwater sediments, *Leading Edge*, 29, 930–942.

Dutta, N. C., Dai, J., Yang, S., Duan, L., Ramirez, A., Bachrach, R. and Mukherjee, A., 2011, Methods and devices for transformation of collected data for improved visualization capability, US Patent 13,179,461.

Dutta, N. C., Yang, S., Dai, J., Bhaduri, A., Chandrasekhar, D., Firoze, D. and Rao, C. V., 2012, Seismic guided drilling: A new technology for look-around and look-ahead of the bit prediction during drilling, paper IPA12-G-018, Indonesian Petroleum Association Annual Convention and Exhibition.

Dutta, N. C., Yang, S., Dai, J., Chandrasekhar, S., Dotiwala, F. and Rao, C. V., 2014, Earth model building using rock physics and geology for depth imaging, *Leading Edge*, 33, 1136–1152.

Dutta, N., Deo, B., Liu, Y. K., Krishna, R., Kapoor, J. and Vigh, D., 2015a, Pore-pressure-constrained, rock-physics-guided velocity model building method: Alternate solution to mitigate subsalt geohazard, *Interpretation*, 3, SE1–SE11.

Dutta, N. C., Yang, S., Liu, Y., Lawrence, C. and Cue, J., 2015b, Rock physics guided velocity modeling and reverse-time migration for pore pressure prediction and depth imaging in complex areas, paper IPA15-G-008, Indonesian Petroleum Association Annual Convention and Exhibition.

Dvorkin, J. and Mavko, G., 2006, Modeling attenuation in reservoir and non-reservoir rocks, *Leading Edge*, 25, 194–197.

Dvorkin, J. and Nur, A., 1996, Elasticity of high-porosity sandstones: Theory for two North Sea datasets, *Geophysics*, 61, 1363–1370.

Dvorkin, J., Mavko, G. and Nur, A., 1991, The effect of cementation on the elastic properties of granular material, *Mech* Mater, 12, 207–217.

Dvorkin, J., Nur, A. and Yin, H., 1994, Effective properties of cemented granular material, *Mech* Mater, 18, 351–366.

Dvorkin, J., Mavko, G. and Nur, A., 1999a, Overpressure detection from compressional- and shear-wave data, *Geophys Res Lett*, 26, 3417–3420.

Dvorkin, J., Moos, D., Packwood, J. and Nur, A., 1999b, Identifying patchy saturation from well logs, *Geophysics*, 64, 1756–1759.

Dzevanshir, R. D., Buryakovsky, L. A. and Chilingarian, G. V., 1986, Simple quantitative evaluation of porosity of argillaceous sediments at various depths of burial, *Sediment Geol*, 46(3–4), 169–175.

Eaton, B. A., 1969, Fracture gradient prediction and its application in oilfield operations, *J Pet Technol*, 21 (10), 25–32.

Eaton, B. A., 1972, The effect of overburden stress on geopressures prediction from well logs, *J Pet Technol*, 747, 929–934.

Eaton, B. A., 1975, The equation for geopressure prediction from well logs, SPE paper 5544.

Eaton, B. A., 1976, Graphical method predicts geopressure worldwide, *World Oil*, 182, 51–56.

Eaton, B. A., 1970, How to drill offshore with maximum control, *World Oil*, 171, 73–77.

Eaton, B. A. and Eaton, T. L., 1997, Fracture gradient prediction for the new generation, *World Oil*, 218, 93–100.

Eaton, L. F., 1999, Drilling through deepwater shallow-water flow zones at Ursa, 1999, paper SPE/IADC 52780, SPE/IADC Drilling Conference.

Eberhart-Phillips, D., Han, D.-H. and Zoback, M. D., 1989, Empirical relationships among seismic velocity, effective pressure, porosity, and clay content in sandstone, *Geophysics*, 54(1), 82–89.

Eberl, D. and Hower, J., 1976, Kinetics of illite formation, *Bull Geol Soc Am*, 87, 1326–1330.

Ebrom, D., Heppard, P., Mueller, M. and Thomsen, L., 2003, Pore-pressure prediction from S-wave, C-wave, and P-wave velocities, SEG Annual Meeting.

Eissa, A. and Kazi, A., 1988, Relation between static and dynamic Young's moduli of rocks, *Int J Rock Mech Min Sci Geomech Abstr*, 25, 479–482.

Elliot, S. E. and Wiley, B. F., 1975, Compressional velocities of partially saturated, unconsolidated sands, *Geophysics*, 40, 949–954.

Ellis, D. V. and Singer, J. M., 2008, *Well Logging for Earth Scientists*, 2nd ed., Springer, Berlin.

Engelhardt, W. V. and Gaida, K. H., 1963, Concentration changes of pore solutions during the compaction of clay sediments, *J Sediment Petrol*, 33, 919–930.

Esmersoy, C., Ramirez, A., Teebeny, S., Yangjun, L., Shih, S., Sayers, C., Hawthorn, A. and Nessim, M., 2013, A new, fully integrated method for seismic geohazard prediction ahead of the bit while drilling, *Leading Edge*, 32, 1222–1233.

Fan, Z. Q., Jin, Z.-H. and Johnson, S. E., 2010, Subcritical propagation of an oil-filled penny-shaped crack during kerogen–oil conversion, *Geophys J Int*, 182, 1141–1147.

Fan, Z. Q., Jin, Z.-H. and Johnson, S. E., 2012, Modelling petroleum migration through microcrack propagation in transversely isotropic source rocks, *Geophys J Int*, 190, 179–187.

Fatti, J. L., Smith, G. C., Vail, P. J., Strauss, P. J. and Levitt, P. R., 1994, Detection of gas in sandstone reservoirs using AVO analysis: A 3D seismic case history using the Geostack technique, *Geophysics*, 59, 1362–1376.

Fertl, D. H. and Timko, D. J., 1971, Parameters for identification of overpressured Formations, paper SPE 3223, Society of Petroleum Engineers 5th Conference on Drilling and Rock Mechanics.

Fertl, D. H., Chilingar, G. V. and Robertson, J. O., 2002, Drilling parameters, in: Chilingar, G. V., Serebryakov, V. A. and Robertson, J. O., (eds.), *Origin and Prediction of Abnormal Formation Pressures*, Developments in Petroleum Science 50, Elsevier, New York, pp. 151–167.

Fertl, W. H., 1976, *Abnormal Formation Pressures*, Developments in Petroleum Science 2, Elsevier, Amsterdam.

Fertl, W. H., Chapman, R. E. and Hotz, R. F., 1994, *Studies in Abnormal Pressure*, Elsevier, Forbes.

Fjaer, E., Holt, R. M., Horsrud, P., Raaen, A. M. and Risnes, R., 2008, Stresses around boreholes: Borehole failure criteria, in: *Petroleum Related Rock Mechanics*, 2nd ed., Elsevier, Amsterdam, pp. 146–148.

Fomel, S. and Grechka, V., 2001, Weak Anisotropy Approximation for VTI media, report CWP-372, Center for Wave Phenomena, Colorado School of Mines.

Foster, J. R. and Whalen, H. E., 1966, Estimation of formation pressure from electrical surveys – offshore, Louisiana, *J Pet Technol*, 18, 165–171.

Foster, W. R., 1981, The smectite–illite transformation: Its role in generating and maintaining geopressure, Annual Meeting of the Geological Society of America.

Foster, W. R. and Custard, H. C., 1980, Smectite–illite transformation role in generating and maintaining geopressure, abstract, *AAPG Bull*, 64, 708.

Fossum, A. F. and Fredrich, J. T., 2002, Salt mechanics primer for near-salt and sub-salt deepwater Gulf of Mexico field developments, Sandia National Laboratory Report SAND2002-2063.

Fowler, W. A., 1970, Pressure, hydrocarbon accumulation and salinities – Chocolate Bayou field, Brazoria County, Texas, *J Pet Technol*, 22, 411–432.

Fredrich, J. T., Engler, B. P., Smith, J. A., Onyia, E. C. and Tolman, D. N., 2007, Predrill estimation of subsalt fracture gradient: Analysis of the Spa prospect to validate nonlinear finite element stress analyses, SPE/IADC Drilling Conference.

Freed, R. L., 1979, Shale mineralogy of the No. 1 Pleasant Bayou Geothermal Test Well: A progress report, Fourth Geopressure and Geothermal Energy Conference.

Freed, R. L., 1981, Shale mineralogy and burial diagenesis of Frio and Vicksburg Formations in two geopressured wells, McAllen Ranch Area, Hidalgo County, Texas, *Trans Gulf Coast Assoc Geol Soc*, 31, 289–293.

Fuck, R. F., Bakulin, A. and Tsvankin, I., 2009, Theory of traveltime shifts around compacting reservoirs: 3D solutions for heterogeneous anisotropic media, *Geophysics*, 74, D25–D36.

Furlow, W., 1998, Shallow water flows: How they develop; what to do about them, Offshore, September, 70.

Gabitto, J. and Tsouris, C, 2010, Physical properties of gas hydrates: A review, *J Thermodyn*, Article 271291.

Gabrysch, R. K., 1967, Development of ground water in Houston District, Texas, 1961–65, Texas Water Development Board Report 63.

Garcia, A. and MacBeth, C., 2013, An estimation method for effective stress changes in a reservoir from 4D seismic data, *Geophys Prospect*, 61, 803–816.

Gardner, G. H F., Gardner, L. W. and Gregory, A. R., 1974, Formation velocity and density – The diagnostic basics for stratigraphic traps, Geophysics, 39, 770–780.

Gassmann, F., 1951, Elasticity of porous media: Uber die elastiziat poroser medien, *Vierteljahrsschr Naturforsch Gesellschaft*, 96, 1–23.

Gautam, S., Dai, J., Rosa-Perez, N., De L. and Jalbert, A., 2017, Rock physics based velocity modeling for reducing subsalt velocity uncertainty, SEG International Exposition.

Gjøystdal, H. and Ursin, B., 1981, Inversion of reflection times in three dimensions, *Geophysics*, 46, 972–983.

Glezen, W. H. and Lerche, I., 1985, A model of regional fluid flow: Sand concentration factors and effective lateral and vertical permeabilities, *Math* Geol, 17, 297–315.

Golubev, A. A. and Robinovich, G. Y., 1976, Resultaty primeneia apparatury akusticeskogo karotasa dlja predeleina proconstych svoistv gormych porod na mestorosdeniaach tverdych isjopaemych, *Prikl Geofiz Moskova*, 73, 109–116.

Goodman, R. E., 1989, *Introduction to Rock Mechanics*, Wiley, New York.

Goulty, N. R., 1998, Relationship between porosity and effective stress in shales, *First Break*, 16, 413–419.

Goulty, N. R., 2004, Mechanical compaction behavior of natural clays and implications for pore pressure estimation, *Pet Geosci*, 10, 73–79.

Goulty, N. R, Sargent, C., Andras, P. and Aplin, A. C., 2016, Compaction of diagenetically altered mudstone – Part 1: Mechanical and chemical contributions, *Mar Pet Geol*, 77, 703–713.

Grauls, D., Dunand, J. P. and Beaufort, D., 1995, Predicting abnormal pressure from 2-D seismic velocity modeling, OTC Conference.

Green, S., O'Connor, S. A. and Edwards, A. P., 2016, Predicting pore pressure in carbonates: A review, search and discovery, paper 41830 (2016), Middle East Geosciences Conference and Exhibition.

Greenberg, J., 2008, Seismic while drilling keeps bit turning to right while acquiring key real-time data, *Drilling Contractor*, March, 44–45.

Gregory, A. R., 1977, Aspects of rock mechanics from laboratory and log data that are important to seismic interpretation, *AAPG Mem*, 26, 15–46.

Gretener, P. E., 1969, Pore pressure: Fundamentals, general ramifications and implications for structural geology, Continuing Education Course Note Series 4, AAPG.

Gretener, P. E., 1981, *Pore Pressure: Fundamentals, General Ramifications and Implications for Structural Geology*, Suppl. 1979, Course Note, AAPG, Tulsa, OK.

Griffith, A. A., 1920, The phenomena of rupture and flow in solids, *Philos Trans R Soc London, Ser A*, 221, 163–198.

Gueguen, Y. and Palciauskas, V., 1994, *Introduction to the Physics of Porous Media*, Princeton University Press, Princeton, NJ.

Haidary, S. A., Shehri, H. A., Abdulraheem, A., Ahmed, M. and Alqam, M. H., 2015, Wellbore stability analysis for trouble free drilling, SPE Kuwait Oil and Gas Show and Conference.

Haimson, B. C. and Fairhurst, C., 1967, Initiation and extension of hydraulic fractures in rocks, *SPE J*, 7(3), 310–318.

Hale, D., 2011, Structure-oriented smoothing and semblance, CWP 635, Center for Wave Phenomena, Colorado School of Mines.

Hall, P. L., 1993, Mechanisms of overpressuring – an overview, in: Manning, D. A. C., Hall,P. I. and Hughes, C. R. (eds.), *Geochemistry of Clay-Pore Fluid Interactions*, Chapman and Hall, London, pp. 265–315.

Ham, H. H., 1966, New charts help estimate formation pressures, *Oil Gas J*, 65(51), 58–63.

Hamilton, E. L., 1971a, Elastic properties of marine sediments, *J Geophys Res*, 76, 579–604.

Hamilton, E. L., 1971b, Prediction of in situ acoustic and elastic properties of marine sediments, *Geophysics*, 36, 266–284.

Hamilton, E. L., 1976a, Shear-wave velocity versus depth in marine sediments: A review, *Geophysics*, 41, 985–996.

Hamilton, E. L., 1976b, Variations of density and porosity with depth in deep-sea sediments, *J Sediment Petrol*, 46, 280–300.

Hampshire, K. and MacGregor, A., 2017, Pore pressure prediction while drilling: 3D Earth model in the Gulf of Mexico, *AAPG Bull*, online first, doi:10.1306/0605171619617050.

Han, D.-H., 1986, Effects of porosity and clay content on acoustic properties of sandstones and unconsolidated sediments, PhD thesis, Stanford University.

Han, D.-H., Nur, A. and Morqan, D., 1986, Effects of porosity and clay content on wave velocities in sandstone, *Geophysics*, 51, 2093–2107.

Hansen, S., 1996, A compaction trend for Cretaceous and tertiary shales on the Norwegian shelf based on sonic transit times, *Pet Geosci*, 2, 159–166.

Hansom, J. and Lee, M., 2005, Effects of hydrocarbon generation, basal heat flow and sediment compaction on overpressure development: A numerical study, *Pet Geosci*, 11, 353–360.

Hantschel, T. A. and Kauerauf, A., 2009, *Fundamentals of Basin Modeling*, Springer, Berlin.

Hardage, B. A., 1985, *Vertical Seismic Profiling, Part A: Principles*, 2nd ed., Pergamon, Oxford.

Hardage, B. A., 1992, *Crosswell Seismology and Reverse VSP*, Geophysical Press, London.

Hardage, B. A., 1994, Seismic prediction of overpressure conditions ahead of the bit in real drill time, in: Fertl, W. H., Chapman, R. E. and Hotz, R. F. (eds.), *Studies in Abnormal Pressures*, Elsevier, Amsterdam, pp. 241–250.

Hardage, B., 2009, Seismic-while-drilling: Techniques using the drill bit as the seismic source, *Search and Discovery*, Article 40411.

Harkins, K. L. and Baugher, J. W., 1969, Geological significance of abnormal formation pressures, *J Pet Technol*, 21, 961–966.

Harper, D., 1969, New findings from overpressured detection curves in tectonically stressed beds, Society of Petroleum Engineers Meeting.

Hart, B. S., Flemings, P. B. and Deshpande, A., 1995, Porosity and pressure: Role of compaction disequilibrium in the development of geopressures in a Gulf Coast Pleistocene basin, *Geology*, 23, 45–48.

Hatchell, P. J., van den Beukel, A., Molenaar, M. M., Maron, K. P., Kenter, C. J., Stammeijer, J. G. F., van den Velde, J. J. and Sayers, C. M., 2003, Whole earth 4D: Monitoring geomechanics: 73rd Annual International Meeting, SEG, expanded abstracts.

Hearst, J. R., Nelson, P. H. and Paillet, F. L., 2000, *Well Logging for Physical Properties: A Handbook for Geophysicists, Geologists, and Engineers*, 2nd ed., Wiley, Hoboken, NJ.

Hedberg, H. D., 1936, Gravitational compaction of clays and shales, *Am J Sci*, 31, 241–287.

Heidbach, O., Rajabi, M., Reiter, K., Ziegler, M. and WSM Team, 2016, World Stress Map Database Release 2016. V. 1.1. GFZ Data Services. http://doi.org/10.5880/WSM.2016.001.

Heppard, P. D. and Traugott, M., 1998, Use of seal, structural and centroid information in pore pressure prediction, abstract, American Association of Drilling Engineers Forum on Pore Pressure Regimes in Sedimentary Basins and Their Predictions.

Hermanrud, C., Cao, S. and Lerche, I., 1990, Estimates of virgin rock temperature derived from BHT (bottom-hole temperature) measurements – bias and error, *Geophysics*, 55, 924–931.

Herwanger, J. and Koutsabeloulis, N., 2011, *Seismic Geomechanics: How to Build and Calibrate Geomechanical Models Using 3D and 4D Seismic Data*, EAGE, Stavanger, Norway.

Hettema, M. H. H., Schutjens, P. M. T. M., Verboom, B. J. M. and Gussinklo, H. J., 2000, Production-induced compaction of sandstone reservoirs: The strong influence of field stress, S*PE Reservoir Evaluation and Engineering*, 3, 342–347.

Hickman, S., Sibson, R. and Bruhn, R., 1995, Introduction to special section: Mechanical involvement of fluids in faulting, *J Geophys Res*, 100, 12831–12840.

Hill, A. W., 1996, The use of exploration 3-D data in geohazard assessment: Where does the future lie? Offshore Technology Conference.

Hill, D., Sonika, S., Lowden, D., Paydayesh, M., Barling, T. and Branston, M., 2017, Full-field 4D image modeling to determine a reservoir monitoring strategy, *First Break*, 35, 77–88.

Hinch, H. H., 1973, The physical properties of shale, shale hydration, and the nature of the shale-water system, Amoco Prod. Co. Report F73-G-17.

Hinze, W. J., von Frese, R. R. B., Saad, A. H., 2013, *Gravity and Magnetic Exploration: Principles, Practices, and Applications*, Cambridge University Press, Cambridge.

Holbrook, P., 2004, The primary controls over sediment compaction, *AAPG Mem*, 21–32.

Holbrook, P. W., Maggiori, D. A. and Hensley, R., 1994, Real-time pore pressure and fracture pressure determination in all sedimentary lithologies, SPE paper 26791.

Horne, S. and Leaney, S., 2001, Polarization and slowness component inversion for TI anisotropy, *Geophys Prospect*, 48, 779–788.

Hornby, B. E., Schwartz, L. M. and Hudson, J. A., 1994, Anisotropic effective-medium modeling of the elastic properties of shales, *Geophysics*, 59, 1570–1583.

Horner, D. R., 1951, Pressure build-up in wells, Third World Petroleum Congress.

Horsrud, P., 2001, Estimating mechanical properties of shale from empirical correlations, *SPE Drill Completion*, 16(2), 68–73.

Hottman, C. E. and Johnson, R. K., 1965, Estimation of formation pressures from log-derived shale properties, *J Pet Technol*, 17, 717–723.

Houbolt, J. J. H. C. and Wells, P. R. A., 1980, Estimation of heat flow in oil wells based on a relation between heat conductivity and sound velocity, *Geol Mijnbouw*, 59, 215–224.

Howard, J. J., 1981, Lithium and potassium saturation of illite/smectite clays from interlaminated shales and sandstones, *Clays Clay Min*, 29, 136–142.

Hower, J., Eslinger, E. V., Hower, M. E. and Perry, E. A., 1976, Mechanism of burial metamorphism of argillaceous sediment. I. Mineralogical evidence, *Geol Soc Am Bull*, 87, 725–737.

Huang, Y., Bai, B., Quan, H., Huang, T., Xu, S. and Zhang, Y., 2011, Application of RTM 3D angle gathers to wide-azimuth data subsalt imaging, *Geophysics*, 76, 1–8.

Hubbert, M. K. and Rubey, W. W., 1959, Role of fluid pressure in mechanics of overthrust faulting, I. Mechanics of fluid filled porous solids and its application to overthrust faulting, *Bull Geol Soc Am*, 70, 115–166.

Hubbert, M. K. and Willis, D. G., 1957, Mechanics of hydraulic fracturing, *Pet Trans AIME*, 210, 153–168.

Hubral, V. and Krey, T., 1980, Interval velocities from seismic reflection time measurements, Society of Exploration Geophysicists.

Hudson, J. A. and Harrison, J. P., 1997, *Engineering Rock Mechanics:* An Introduction to the Principles, 1st ed., Pergamon Press, London.

Huffman, A. R., 1998, The future of pressure prediction using geophysical methods, in: Pressure *Regimes* in *Sedimentary Basi*ns an*d Their Predic*tion, SEG Conference *Proceedings*, SEG, Houston, pp. 217–233.

Huffman, A. R., 2002, The future of pressure prediction using geophysical methods, *AAPG Mem*, 76, 217–233.

Huffman, A. R. and Bowers, G. (eds.), 2002, *Pressure Regimes in Sedimentary Basins and Their Prediction*, AAPG Memoir 76, AAPG, Tulsa, OK.

Huffman, A. R. and Castagna, J. P., 1999, Rock physics and mechanics considerations for shallow water flow characterization, Shallow Water Flow Conference.

Huffman, A. and Castagna, J., 2001, Petrophysical basis for shallow-water flow prediction using multi-component seismic data, *Leading Edge*, 20, 1030–1052.

Hunt, J. M. (ed.), 1996, *Petroleum Geochemistry and Geology*, 2nd ed., W. H. Freeman, New York.

Hunt, J. M., Whelan, J. K., Eglinton, L. B. and Cathles, L. M., III, 1994, Gas generation – a major cause of deep Gulf Coast overpressures, *Oil Gas J*, 92(29), 59–63.

Hussein, S. A. and Alnajm, F. M., 2019, Estimation of minimum and maximum horizontal stresses from well log: A case study in Rumaila Oil Field, Iraq, *Am J Geophys Geochem Geosyst*, 5(3), 78–90.

Huyen, B., Smith, M., Graham, J., Snyder, F. and Singh, S. K., 2012, Workflow enables first successful GOM subsalt post-stack inversion, Hart Energy E&P, August.

Inglis, C. E., 1913, Stresses in plates due to the presence of cracks and sharp corners, *Trans Inst Naval Architects*, 55, 219–241.

Irwin, G. R., 1958. Fracture, in: *Encyclopedia of Physics*, vol. VI, Springer, New York, pp. 551–590.

Issler, D. R., 1992, A new approach to shale compaction and stratigraphic restoration, Beaufort-Mackenzie Basin and Mackenzie Corridor, northern Canada, *Am Assoc Pet Geol Bull*, 76, 1170–1189.

Jackson, J. D., 1999, *Classical Electrodynamics*, Wiley, New York.

Jaeger, J. C., Cook, N. G. W. and Zimmerman, R. W., 2007, *Fundamentals of Rock Mechanics*, 4th ed., Blackwell, Malden, MA.

Jageler, A. H., 1976, Improved hydrocarbon reservoir evaluation through use of borehole gravimeter data, *J Pet Technol*, 28, 709–718.

Japsen, P., Dysthe, D. K., Hartz, E. H., Stipp, S. L. S., Yarushina, V. M. and Jamtveit, B., 2011, A compaction front in North Sea chalk, *J Geophys Res*, 116, B11208.

Jin, Z. H., Johnson, S. E. and Fan, Z. Q., 2010, Subcritical propagation and coalescence of oil-filled cracks: Getting the oil out of low-permeability source rocks, *Geophys Res Lett*, 37(1), L01305.

Jizba, D. L., 1991, Mechanical and acoustical properties of sandstones and shales, PhD thesis, Stanford University.

Jones, O. T., 1944, The compaction of muddy sediments, *Q J Geol Soc London*, 100, 137–160.

Jones, S. M., 1995, Velocities and quality factors of sedimentary rocks at low and high effective pressures, *Geophys J Int*, 123, 774–780.

Jones, S., 2010, Tutorial: Velocity estimation via ray-traced based tomography, *First Break*, 28, 45–52.

Jordan, J. R. and Shirley, O. J., 1966, Application of drilling performance data to overpressure detection, *J Pet Technol*, 18, 1387–1394.

Jorgensen, G. J. and Kisabeth, J. L., 2000, Joint 3D inversion of gravity, magnetic and tensor-gravity fields for imaging salt formations in the deepwater Gulf of Mexico, abstract, SEG Annual Meeting.

Jowett, E. C., Cathles, L. and Davis, B. W., 1992, Predicting depths of gypsum dehydration in evaporitic sedimentary, *AAPG Bull*, 77, 402–413.

Kan, T. K. and Sicking, C. J., 1994, Pre-drill geophysical methods for geopressure detection and evaluation, in: Fertl, W. H., Chapman, R. E. and Hotz, R. F. (eds.), *Studies in Abnormal Pressures*, Developments in Petroleum Science 38, Elsevier, New York, pp. 155–186.

Kan, T. K. and Swan, H. W., 2001, Geopressure prediction from automatically derived seismic velocities, *Geophysics*, 66, 1937–1946.

Kankanamge, T., 2013, Pore pressure and fracture pressure modelling with offset well data and its application to surface casing design of a development well; deep panuke gas pool offshore nova scotia, MSc thesis, Dalhousie University.

Kao, J., Tatham, R. H. and Murray, P. E., 2010, Estimating pore pressure using compressional and shear wave data from multicomponent seismic nodes in Atlantis Field, deepwater Gulf of Mexico, SEG Annual Internal Meeting, expanded abstracts.

Kapoor, S., Vigh, D. and Wiarda, E., 2013, Full waveform inversion around the world, paper WE-11-03, EAGE Exhibition and Conference.

Karl, R., 1965, Gesteinsphysikalische Parameter (Schallgeschwindigkeit und Wärmeleitfähigkeit), *Freiberger Forschungs*, C197, 7–76.

Karig, D. E. and Hou, G., 1992, High-stress consolidation experiments and their geologic implications, *J Geophys Res*, 97(1), 289–300.

Katahara, K., 2006, Overpressure and shale properties: Stress unloading or smectite-illite transformation? SEG Annual Meeting, extended abstracts.

Katahara, K., 2009, Lateral earth stress and strain, *SEG Tech Prog Exp Abstr*, 28, 2165–2169.

Kaufman, A. A., Alekseev, D. and Oristaglio, M., 2014, *Principles of Electromagnetic Methods in Surface Geophysics*, Methods in Geochemistry and Geophysics 45, Elsevier, New York.

Keaney, G., Li, G. and Williams, K., 2010, Improved fracture gradient methodology understanding the minimum stress in Gulf of Mexico, paper ARMA-10-177, US Rock Mechanics Symposium.

Kemper, M. and Gunning, J., 2014, Joint impedance and facies inversion–seismic inversion redefined, *First Break*, 32(9), 89–95.

Kennedy, C. and Holster, W. T., 1966, Pressure-volume-temperature and phase relations of water and carbon dioxide, in: *Handbook of Physical Constants*, Geol Soc Am Mem 97, pp. 225–241.

Kennett, J. P., Cannariato, K. G., Hendy, I. L. and Behl, R. J., 2003, *Methane Hydrates in Quaternary Climate Change: The Clathrate Gun Hypothesis*, American Geophysical Union, Washington, DC.

Khan, K., Al-Awadi, A., Dashti, Q., Kabir, M. R. and Aziz, M. R., 2009, Understanding overpressure trends helps optimize well planning and field development in tectonically active areas in Kuwait, SPW 122631.

Khatchikian, A., 1996, *Deriving Reservoir Pore-Volume Compressibility from Well Logs*, SPE Advanced Technology Series 5, SPE, Philadelphia, PA.

Khazanehdari, J. and Dutta, N., 2006, High-resolution pore pressure prediction using seismic inversion and velocity analysis, Society of Exploration Geophysicists Annual Meeting, expanded abstracts.

Kılıç, A. and Teymen, A., 2008, Determination of mechanical properties of rocks using simple methods, *Bull Eng Geol Environ*, 67, 237–244.

King, M. S., 1983, Static and dynamic elastic properties of rock from the Canadian Shield, *Int J Rock Mech Min Sci Geomech Abstr*, 20, 237–241.

Kirsch, G., 1898, Die theorie der elastizitat und die bedurfnisse der festigkeitslehre, *Z Ver Dtsch Ing*, 42, 797–807.

Klimentos, T. and McCann, C., 1990, Relationships among compressional wave attenuation, porosity, clay content, and permeability in sandstones, *Geophysics*, 55, 998–1014.

Knoll, L., 2016, The process of building a mechanical earth model using well data, MSc thesis, Department of Petroleum Geology, Montan Universitat.

Koefoed, O., 1955, On the effect of Poisson's ratios of rock strata on the reflection coefficients of plane waves, *Geophys Prospect*, 3, 381–387.

Koefoed, O., 1962, Reflection and transmission coefficients for plane longitudinal incident waves, *Geophys Prospect*, 10, 304–351.

Komac, B. and Zorn, M., 2013, Geohazards. In: Bobrowsky, P. T. (ed.), *Encyclopedia of Natural Hazards*, Springer, Dordrecht, Netherlands, pp. 289–295.

Korosi, A. and Fabuss, B. M., 1968, Viscosity of liquid water from 25 to 150 degree. measurements in pressurized glass capillary viscometer, *Anal Chem*, 40, 157–162.

Koutsabeloulis, N. and Zhang, X., 2009, 3D reservoir geomechanical modeling in oil/gas field production, SPE-126095-MS.

Kozeny, J., 1927, Ueber kapillare Leitung des Wassers im Boden, *Sitzungsber Akad Wiss Wien*, 136(2a), 271–306.

Kuchuk, F., Ayan, C. and Radeka, A., 1996, A revolution in reservoir characterization, *Middle East Well Eval Rev*, 6, 41–55.

Kugler, H. G., 1933, Contribution to the sedimentary volcanism in Trinidad, *J Inst Pet Technol*, 19, 760–772.

Kulkarni, R., Meyer, J. H. and Sixta, D., 1999, Are pore pressure related drilling problems predictable? The value of using seismic before and while drilling, Society of Exploration Geophysicists International Exposition and Annual Meeting.

Kumar, R., Al-Saeed, M., Al-Kandiri, J. M., Verma, N. K. and Al-Saqran, F., 2010, Seismic based pore pressure prediction in a West Kuwait field, Society of Petroleum Engineers, expanded abstracts.

Kunze, K. R. and Steiger, R. P., 1991, Extended leak-off tests to measure in situ stress during drilling, in: Roegiers, J.-C. (ed.), *Rock Mechanics as a Multidisciplinary Science*, Balkema, Rotterdam, pp. 35–44.

Lacy, L. L., 1997, Dynamic rock mechanics testing for optimized fracture designs, SPE paper 38716-MS.

LaFehr, T. R. and Nabighian, M. N., 2012, *Fundamentals of Gravity Exploration*, SEG, Tulsa, OK.

LaFehr, T. R., 1983, Rock density from borehole gravity surveys, *Geophysics*, 48, 341–356.

Lahann, R., 2000, Impact of diagenesis on compaction modeling and compaction equilibrium, Drilling and Exploiting Overpressured Reservoirs: A Research Workshop for the Millennium.

Lahann, R. W., 2002, Impact of smectite diagenesis on compaction modeling and compaction equilibrium, *AAPG Mem*, 76, 61–72.

Lailly, P., 1983, The seismic inverse problem as a sequence of before stack migrations, Conference on Inverse Scattering, Theory and Application, Society for Industrial and Applied Mathematics, extended abstracts.

Lake, L. W. and Fanchi, J. R., 2007, *Petroleum Engineering Handbook*, Society of Petroleum Engineers, Richardson, TX.

Lal, M., 1999, Shale stability: Drilling fluid interaction and shale strength, SPE Latin American and Caribbean Petroleum Engineering Conference.

Lander, R. H. and Walderhaug, o., 1999, Predicting porosity through simulating sandstone compaction and quartz cementation, *AAPG Bull*, 83, 433–449.

Landrø, M. and Stammeijer, J., 2004, Quantitative estimation of compaction and velocity changes using 4D impedance and traveltime changes, *Geophysics*, 69, 949–957.

Larsen, G. and Chilingar, G. V., 1983, *Diagenesis in Sediments and Sedimentary Rocks: 2, Introduction, Developments in Sedimentology 25B*, Elsevier, New York.

Lashkaripour, G. R., 2002, Predicting mechanical properties of mudrock from index parameters, *Bull Eng Geol Environ*, 61, 73–77.

Laughton, A. S., 1957, Sound propagation in compacted ocean sediments, *Geophysics*, 22, 233–260.

Law, B. E., 1984, Relationships of source-rock, thermal maturity, and overpressuring to gas generation and occurrence in low permeability Upper Cretaceous and Lower Tertiary rocks, greater Green River Basin, Wyoming, Colorado, and Utah, in: Woodward, J., Meissner, F. F. and Clayton, J. L. (eds.), *Hydrocarbon Source Rocks of the Greater Rocky Mountain Region*, Rocky Mountain Assoc. Geol., Denver, pp. 469–490.

Law, B. E. and Spenser, C. W., 1998, Abnormal pressures in hydrocarbon environments, *AAPG Mem*, 70, 1–11.

Law, B. E., Spencer, C. W., Charpentier, R. A., Crovelit, R. A., Mast, R. F., Dolton, D. L. and Wandrey, C. J., 1989, Estimates of gas resources in overpressured low permeability Cretaceous and Tertiary sandstone reservoirs, Greater Green River Basin, Wyoming, Colorado and Utah, in: *Wyoming Geological Association Fortieth Field Conference Guidebook*, Wyoming Geological Association, Denver, CO, pp. 39–61.

Lawn, B. R. and Wilshow, T. R., 1975, *Fracture of Brittle Solids*, Cambridge University Press, New York.

Le, H., Pradhan, A., Dutta, N., Biondi, B., Mukerji, T. and Levin, S. A., 2017, Building pore pressure and rock physics guides to constrain anisotropic waveform inversion, Stanford Exploration Project Report.

Le, H., Pradhan, A., Dutta, N. C., Biondi, B., Mukerji, T. and Levin, S. A., 2018, Rock physics guided velocity model building, International Annual Meeting of SEG.

Le, H., 2019, Anisotropic full-waveform inversion with pore pressure constraints, PhD thesis, Geophysics, Stanford University.

Le, K. and Rasouli, V., 2012, Determination of safe mud weight windows for drilling deviated wellbores: A case study in the North Perth Basin, *WIT Trans Eng Sci*, 81, 83–95.

Leach, W. G., 1994, Distribution of hydrocarbons in abnormal pressure in South Louisiana, USA, in: Fertl, W.H., Chapman, R. E. and Hotz, R. F. (eds.), *Studies in Abnormal Pressures*, Elsevier, New York, pp. 391–428.

Leach, W. G. and Fertl, W. H., 1990, The relationship of formation pressure and temperature to lithology and hydrocarbon distribution in Tertiary sandstones, International Well Logging Symposium.

Lee, M.-K. and Williams, D. D., 2000, Paleo-hydrology of the Delaware basin, western Texas: Overpressure development, hydrocarbon migration, and ore genesis, *AAPG Bull*, 84, 961–974.

Lerche, I., 1990, *Basin Analysis: Quantitative Methods*, 2 vols., Academic Press, New York.

Levin, F. K., 1971, Apparent velocity from dipping interface reflections, *Geophysics*, 36, 310–316.

Lewis, G. N., 1908, The osmotic pressure of concentrated solutions and the laws of the perfect solution, *J Am Chem Soc*, 30, 668–683.

Li, H., 2014, Compositional upscaling for individual models and ensembles of realizations, PhD thesis, Stanford University.

Li, S. and Purdy, C. C., 2010, Maximum horizontal stress and wellbore stability while drilling: Modeling and case study. SPE Latin American and Caribbean Petroleum Engineering Conference.

Li, Y., Nichols, O. D. and Bachrach, R., 2011, Anisotropic tomography using rock physics constraints, EAGE Conference and Exhibition.

Li, Y., Biondi, B., Mavko, G. and Nichols, D., 2015, Integrated VTI model building with seismic, geological and rock physics data, 77th EAGE Conference and Exhibition.

Li, Y., Biondi, B., Clapp, R. and Nichols, D., 2016, Integrated VTI model building with seismic data, geological information, and rock-physics modeling. Part 1: Theory and synthetic test, *Geophysics*, 81, C177–C191.

Li, S. and Burdy, C. C., 2010, Maximum horizontal stress and wellbore stability while drilling: Modeling and case study, SPE Latin American and Caribbean Petroleum Engineering Conference.

Lin, W., Yamamoto, K., Ito, H., Masago, H. and Kawamura, Y., 2008, Estimation of minimum principal stress from an extended leak-off test onboard the Chikyu drilling vessel and suggestions for future test procedures, *Sci Drill*, 6, 43–47.

Lindseth, R. O., 1979, Synthetic sonic logs – A process for stratigraphic interpretation, *Geophysics*, 44, 3–26.

Liu, H., 2017, *Principles and Applications of Well Logging*, 2nd ed., Springer, New York.

Liu, Y., Dutta, N. C., Vigh, D., Kapoor, J., Hunter, C., Saragoussi, E., Jones, L., Yang, S. and Eissa, M. A., 2016, Basin-scale integrated earth-model building using rock-physics constraints, *Leading Edge*, 35, 141–145.

Liu, Y., O'Briain, M., Hunter, C., Jones, L. and Saragoussi, E., 2018, Improving hydrocarbon exploration with pore pressure assisted earth model building, *Interpret J*, 6, 1–24.

Lizarralde, D. and Swift, S., 1999, Smooth inversion of VSP traveltime data, *Geophysics*, 64, 659–661.

Long, M. D. and Silver, P. G., 2009, Shear wave splitting and mantle anisotropy: Measurements, interpretations, and new directions, *Surv Geophys*, 30, 407–461.

López, J. L., Rappold, P. M., Ugueto, G. A., Wieseneck, J. B. and Vu, C. K., 2004, Integrated shared earth model: 3D pore-pressure prediction and uncertainty analysis, *Leading Edge*, 23, 52–59.

Luo, X. R. and Vasseur, G., 1996, Geopressuring mechanism of organic matter cracking: Numerical modeling, *AAPG Bull*, 80, 856–874.

Luo, X. and Vasseur, G., 1992, Contributions of compaction and aquathermal pressuring to geopressure and the influence of environmental conditions, *AAPG Bull*, 76, 1550–1559.

Ma, X. Q., 2001, Global joint inversion for the estimation of acoustic and shear impedances from AVO derived P- and S-wave reflectivity data, *First Break*, 19, 557–566.

MacBeth, C., Li, X. Y., Crampin, S. and Mueller, M. C., 1992, Detecting lateral variability in fracture parameters from surface data, Annual International Meeting, expanded abstracts.

Mackenzie, A. S. and Quigley, T. M., 1988, Principles of geochemical prospect appraisal, *AAPG Bull*, 72, 399–415.

Mackenzie, A. S., Leythaeuser, D., Muller, P., Radke, M. and Schaefer, R. G., 1986, Generation and migration of petroleum in the Brae area, Central North Sea, 3rd Conference on Petroleum Geology of NW Europe.

MacQueen, J. D., 2007, High-resolution density from borehole gravity data, Society of Exploration Geophysicists Technical Program, expanded abstracts.

Magara, K., 1968, Compaction and migration of fluids in Miocene mudstone, Nagaoka plain, Japan, *AAPG Bull*, 52, 2466–2501.

Magara, K., 1975, Reevaluation of montmorillonite dehydration as a cause of abnormal pressure and hydrocarbon migration, *AAPG Bull*, 59, 292–302.

Magara, K., 1980, Comparison of porosity–depth relationships of shale and sandstone, *J Pet Geol*, 3, 175–185.

Makhous, M. and Galushkin, Y., 2004, *Basin Analysis and Modeling of the Burial, Thermal and Maturation Histories in Sedimentary Basins*, Technip, Paris.

Malinverno, A., Sayers, C. M., Woodward, M. J. and Bartman, R. C., 2004, Integrating diverse measurements to predict pore pressures with uncertainties while drilling, SPE 90001.

Mallick, S., 1995, Model-based inversion of amplitude-variations-with-offset data using a genetic algorithm, *Geophysics*, 60, 939–954.

Mallick, S., 1999, Some practical aspects of prestack waveform inversion using a genetic algorithm: An example from east Texas Woodbine gas sand, *Geophysics*, 64, 326–336.

Mallick, S., 2001, AVO and elastic impedance, *Leading Edge*, 20, 1094–1104.

Mallick, S. and Dutta, N. C., 2002, Shallow water flow prediction using prestack waveform inversion of conventional 3D seismic data and rock modeling, *Leading Edge*, 21, 675–680.

Mallick, S., Huang, X., Lauve, J. and Ahmad, R., 2000, Hybrid seismic inversion: a reconnaissance tool for deepwater exploration, *Leading Edge*, 9(11), 1230–1237.

Malvern, L. E., 1969, *Introduction to the Mechanics of Continuous Medium*, Prentice Hall, Englewood Cliffs, NJ.

Mann, D. M. and Mackenzie, A. S., 1990, Prediction of pore fluid pressures in sedimentary basins, *Mar Pet Geol*, 7, 55–65.

Mao, W., Fletcher, R. and Deregowski, S., 2000, Automated interval velocity inversion, SEG International Annual Meeting, expanded abstracts.

Marion, D. A., Nur, A., Yin, H. and Han, D., 1992, Compressional velocity and porosity in sand-clay mixtures, *Geophysics*, 57, 554–562.

Marion, D. P., 1990, Acoustic, mechanical, and transport properties of sediments and granular materials, PhD thesis, Stanford University.

Marsden, J. R., 1999, *Geomechanics*, Imperial College, London.

Marshak, S. and Woodward, N. B., 1988, Introduction to cross-section balancing, in: Marshak, S. and Mitra, G. (eds.), *Basic Methods of Structural Geology*, Prentice Hall, Englewood Cliffs, NJ, pp. 303–332.

Masters, J. A., 1979, Deep basin gas trap, Western Canada, *AAPG Bull*, 63, 152–181.

Masters, J. A., ed., 1984, Elmworth – case study of a deep basin gas field, *AAPG Mem*, 38, 1–33.

Matava, T., Keys, R., Foster, D. and Ashabranner, D., 2016, Isotropic and anisotropic velocity-model building for subsalt seismic imaging, *Leading Edge*, 35, 240–245.

Matthews, W. R. and Kelly, J., 1967, How to predict formation pressure and fracture gradient, *Oil Gas J*, 65, 92–106.

Mavko, G. and Mukerji, T., 1995, Seismic pore space compressibility and Gassmann's relation, *Geophysics*, 60, 1743–1749.

Mavko, G. and Saxena, N., 2016, Rock-physics models for heterogeneous creeping rocks and viscous fluids, *Geophysics*, 81, D427–D440.

Mavko, G., Mukerji, T. and Dvorkin, J., 1998, *The Rock Physics Handbook*, 1st ed., Cambridge University Press, Cambridge.

Mavko, G., Mukerji, T. and Dvorkin, J., 2009, *The Rock Physics Handbook*, 2nd ed., Cambridge University Press, Cambridge.

Mavko, G., Mukerji, T. and Dvorkin, J., 2020, *The Rock Physics Handbook*, 3rd ed., Cambridge University Press, Cambridge.

Mayuga, M. N., 1970, Geology and development of California's giant – Wilmington Oil Field, *AAPG Mem*, 14, 158–184.

McCann, C. and McCann, D. M., 1969, The attenuation of compressional waves in marine sediments, *Geophysics*, 34, 882–892.

McClan, D. M. and Entwisle, D. C., 1992, Determination of rock mass from geophysical well logs, *Geol Soc Spec Publ*, 65, 315–325.

McCubbin, D. G. and Patton, J. W., 1981, Burial diagenesis of illite/smectite, a kinetic model, AAPG Annual Meeting.

McCulloh, T. H., Kandle, G. R. and Schoellhamer, J. E., 1968, Application of gravity measurements in wells to problems of reservoir evaluation, Society of Professional Well Log Analysts Annual Logging Symposium.

McMechan, G. A., 1983, Seismic tomography in boreholes, *Geophys J Int*, 74, 601–612.

McPeek, L. A., 1981, Eastern Green River Basin: A developing giant gas supply from deep overpressured Upper Cretaceous sandstones, *AAPG Bull*, 65, 1078–1098.

Meade, R. H., 1966, Factors influencing the early stages of the compaction of clays and sands-review, *J Sediment Petrol*, 36, 1085–1101.

Meehan, R., Nutt, W. L. and Dutta, N. C., 1998, Drill-bit seismic: A drilling optimization tool, SPE 39312.

Meissner, F. F., 1978, *Petroleum Geology of the Bakken Formation, Williston Basin, North Dakota and Montana*, Montana Geological Society, Billing, MT.

Melas, F. and Friedman, G. M., 1992, Petrophysical characteristics of the Jurassic Smackover Formation, Jay field, Conecuh Embayment, Alabama and Florida, *AAPG Bull*, 76, 81–100.

Mese, A., Dvorking, J. and Shinglaw, J., 2000, An investigation for disposal drill cuttings into unconsolidated sandstones and clayey sands, DOE Report 15172-1 (OSTI ID: 761985).

Militzer, H. and Stoll, R., 1973, Einige Beiträige der Geophysik zur primädatenerfassung im Bergbau, *Neue Bergbautechnik*, 3, 21–25.

Miller, T. W., Luk, C. H. and Olgaard, D. L., 2002, The interrelationships between overpressure mechanisms and in-situ stresses, *AAPG Mem*, 76, 13–20.

Minshull, T. A. and White, R., 1989, Sediment compaction and fluid migration in the Makran accretionary prism. Journal of Geophysical Research, *B, Solid Earth and Planets*, 94(6), 7387–7402.

Miranda, F., Aleotti, L., Abramo, F., Poletto, F., Craglietto, A., Persoglia, S. and Rocca, F., 1996, Impact of the Seismic While Drilling technique on exploration wells. *First Break*, 14, 55–68.

Miranda, F., Abramo, F., Poletto, F. and Comelli, P., 2000, Processes for improving the bit seismic signal using drilling parameters, European Patent Application 00201332.4–2213.

Mitchell, W. K. and Nelson, R. J., 1988, *A Practical Approach to Statistical Log Analysis*, SPWLA, San Antonio, TX.

Momper, J. A., 1980, Oil expulsion: A consequence of oil generation, *AAPG Stud Geol*, 35, 181–191.

Mondol, N. H., 2015, Well logging: Principles and uncertainties, in: Bjørlykke, K. (ed.), *Petroleum Geoscience*, Springer, Berlin, pp. 385–426.

Mondol, N. H., Bjørlykke, K., Jahren, J. and Hoeg, K., 2007, Experimental mechanical compaction of clay mineral aggregates – changes in physical properties of mudstones during burial, *Mar Pet Geol*, 24, 289–311.

Mondol, N. H., Fawad, M., Jahren, J. and t Bjørlykke, K., 2008, Synthetic mudstone compaction trends and their use in pore pressure prediction. *First Break*, 26, 43–51.

Moorkamp, M., Lelièvre, P. G., Linde, N. and Khan, A., eds., 2016, *Integrated Imaging of the Earth: Theory and Application*, American Geophysical Union, Washington, DC.

Moos, D. and Zwart, G., 1998, Predicting pore pressure from porosity and velocity, SEG Conference.

Moos, D., Zoback, M. D. and Bailey, L., 2001, Feasibility study of the stability of openhole multilaterals, Cook Inlet, Alaska, *SPE Drill Completions*, 16, 140–145.

Moos, D., Peska, P., Ward, C. and Brehm, A., 2004, Quantitative risk assessment applied to pre-drill pore pressure, sealing potential, and mud window predictions, Gulf Rocks 2004, 6th North America Rock Mechanics Symposium, Houston, TX.

Morton, J. P., 1985, Rb-Sr dating of diagenesis and source age of clays in Upper Devonian black shales of Texas, *GSA Bull*, 96, 1043–1049.

Mosca, F., Djordjevic, O., Hantschel, T., McCarthy, J., Krueger, A., Phelps, D., Akintokunbo, T., Joppen, T., Koster, K., Schupbach, M., Hampshire, K. and MacGregor, A., 2017, Pore pressure prediction while drilling: 3D Earth model in the Gulf of Mexico, *AAPG Bull*, 102, 691–708.

Mouchet, J. P. and Mitchell, A., 1989, *Abnormal Pressures While Drilling: Origins, Prediction, Detection and Evaluation*, Technip, Paris.

Muskat, M., 1937, *The Flow of Homogeneous Fluids through Porous Media*, McGraw-Hill, New York.

Naeini, E. Z., Green, S., Russell-Hughes, I. and Rauch-Davies, M., 2019, An integrated deep learning solution for petrophysics, pore pressure, and geomechanics property prediction, *Leading Edge*, 38, 53–59.

Najibi, A. R., Ghafoori, M., Lashkaripour, G. R. and Asef, M. R., 2015, Empirical relations between strength and static and dynamic elastic properties of Asmari and Sarvak limestones, two main oil reservoirs in Iran, *J Pet Sci Eng*, 126, 78–82.

Neuzil, C. E., 2000, Osmotic generation of "anomalous" fluid pressures in geological environments, *Nature*, 403, 182–184.

Nguyen, T., 2013, Well Design Course PE-413, New Mexico Technical University, Department of Petroleum Engineering.

Nind, C. J. M., MacQueen, J. D. and Wasylechko, R., 2013, *The Borehole Gravity Meter: Development and Results; 10th Biennial International Conference & Exposition*, Society of Petroleum Geophysicists, Kochi, India.

Nur, A. M., 1969, Effect of stress and fluid inclusions on wave propagation in rock, PhD thesis, MIT.

Nur, A., 1971, Effects of stress on velocity anisotropy in rocks with cracks, *J Geophys Res*, 76, 2022–2034.

Nur, A., 1991, Critical porosity, elastic bounds, and seismic velocities in rocks, *Tech Prog Abstr EAEG*, 53, 248–249.

Nur, A. and Byerlee, J. D., 1971, An exact effective stress law for deformation of rocks with fluids, *J Geophys Res*, 76, 6414–6419.

Nur, A. M. and Wang, A., 1989, *Seismic and Acoustic Velocities in Reservoir Rocks*, vol. 1, Experimental Studies, Geophysics Reprint Series 10, Society of Exploration Geophysics, Tulsa, OK.

Nur, A. M., Marion, D. and Yin, H., 1991, Wave velocities in sediments, in: Hovem, J. M., Richardson, M. D. and Stoll, R. D. (eds.), *Shear Waves in Marine Sediments*, Kluwer, New York, pp. 131–140.

Nur, A., Mavko, G., Dvorkin, J. and Gal, D., 1995, Critical porosity: The key to relating physical properties to porosity in rocks, SEG Annual Meeting, expanded abstracts.

Nur, A., Mavko, G., Dvorkin, J. and Galmudi, G., 1998, Critical porosity: A key to relating physical properties to porosity in rocks, *Leading Edge*, 17, 357–362.

Nuttall, H. E., Guo, T.-M., Schrader, S. and Thakur, D. S., 1983, Pyrolysis kinetics of several key world oil shales, in: *Geochemistry and Chemistry of Oil Shales*, American Chemical Society, Washington, DC, pp. 269–300.

Nye, J. F., 1985, *Physical Properties of Crystals*, Oxford University Press, New York.

Nygard, R., Gutierrez, M., Gautam, R. and Hoeg, K., 2004, Compaction behavior of argillaceous sediments as function of diagenesis, *Mar Pet Geol*, 21, 349–362.

Nygard, V., Jahangir, M., Gravem, T., Nathan, E., Evans, J. G., Reeves, M., Wolter H. and Hovda, S., 2008, A step change in total system approach through wired-drillpipe technology, SPE 112742.

O'Connor, S. A., Swarbrick, R. E., Clegg, P. and Scott, D. T., 2008, Pore pressure profiles in deep water environments: Case studies from around the world, AAPG Annual Meeting.

O'Connor, S., Swarbrick, R. E. and Lahann, R., 2011, Geologically-driven pore fluid pressure models and their implications for petroleum exploration: Introduction to thematic set, *Geofluids*, 11, 343–400.

Ohen, H. A., 2003, Calibrated wireline mechanical rock properties method for predicting and preventing wellbore collapse and sanding, SPE 82236.

Olberg, T. S., Laastad H., Lesso, B. and Newton, A., 2008, The utilization of the massive amount of real-time data acquired in wired-drillpipe operations, SPE 112702.

Olofsson, B., Probert, T., Kommedal, J. H. and Barkved, O. I., 2003, Azimuthal anisotropy from the Valhall 4C 3D survey, *Leading Edge*, 22(12), 1228–1235.

Oriji, A. and Ogbonna, J. A., 2012, New fracture gradient prediction technique that shows good results in Gulf of Guinea wells, Abu Dhabi International Petroleum Conference and Exhibition.

Ortoleva, P. J., 1994, Basin compartments and seals, *AAPG Mem*, 61, 333–348.

Osborne, M. J. and Swarbrick, R. E., 1997, Mechanisms which generate overpressure in sedimentary basins: A reevaluation, *AAPG Bull*, 81, 1023–1041.

Ostermeier, R. M., Pelletier, J. H., Winker, C. D., Nicholson, J. W., Rambow, F. H. and Cowan, K. M., 2002, Dealing with shallow-water flow in the deepwater Gulf of Mexico, *Leading Edge*, 21, 660–668.

Ostrander, W. J., 1984, Plane wave reflection coefficients for gas sands at non-normal angles of incidence, *Geophysics*, 49, 1637–1648.

Osypov, K., Yang, Y., Fournier, A., Ivanova, N., Bachrach, R., Yarman, C. E., Yu, Y, Nicholes, D. and Woodward, M., 2013, Model uncertainty quantification in seismic tomography: Method and applications, *Geophys Prospect*, 61, 1114–1134.

Oughton, R. H., Wooff, D. A., Hobbs, R. W., Swarbrick, R. E. and O'Connor, S. A., 2018, A sequential dynamic Bayesian network for pore-pressure estimation with uncertainty quantification, *Geophysics*, 83, D27–D39.

Padina, S., Churchill, D. and Bording, R. P., 2006, Travel time inversion in seismic tomography, www .cs.mun.ca/~dchurchill/pdf/HPCSPaper.pdf.

Paillet, F. and Cheng, C.-H., 1991, *Acoustic Waves in Boreholes*, CRC Press, Boca Raton, FL.

Palciauskas, V. V. and Domenico, P. A., 1989, Fluid pressures in deforming rocks, *Water Resour Res*, 25, 203–213.

Palaz, I. and Marfurt, K. (eds.), 1997, *Carbonate Seismology*, Geophysical Developments Series 6, SEG, Tulsa, OK, pp. 29–52.

Pan, W., Innanen, K. A. and Geng, Y., 2018, Elastic full-waveform inversion and parametrization analysis applied to walk-away vertical seismic profile data for unconventional (heavy oil) reservoir characterization, *Geophys J Int*, 213, 1934–1968.

Parker, R. L., 1994, *Geophysical Inverse Theory*, Princeton University Press, Princeton, NJ.

Peacock, S., Westbrook, G. K. and Bais, G., 2009, *S*-wave velocities and anisotropy in sediments entering the Nankai subduction zone, offshore Japan, *Geophys J Int*, 180, 743–758.

Pennebaker, E. S., 1968, Seismic data indicate depth, magnitude of abnormal pressure, *World Oil*, 166(7), 73–78.

Perez, M. A., Clyde, R., D'Ambrosio, P., Israel, P., Leavitt, T., Johnson, C. and Williamson, D., 2008, Meeting the subsalt challenge, *SLB Oil Field Rev*, 20(3), 32–46.

Perry, E. A., 1969, Burial diagenesis of Gulf Coast pelitic sediments, PhD dissertation, Case Western Reserve University.

Perry, E. A. and Hower, J., 1970, Burial diagenesis in Gulf Coast pelitic sediments, *Clays Clay Min*, 18, 165–177.

Perry, E. A. and Hower, J., 1972, Late-stage dehydration in deeply buried pelitic sediments, *AAPG*, 56, 2013–2021.

Peters, K. E., 2009, Getting started in basin and petroleum system modeling, American Association of Petroleum Geologists CD-ROM #16, AAPG Data pages.

Peters, K. E. and Nelson, P. H., 2009, Criteria to determine borehole formation temperatures for calibration of basin and petroleum system models, *Search and Discovery*, Article 40463.

Petmecky, R. S., Albertin, M. L. and Burke, N., 2009, Improving subsalt imaging using 3D-basin modeling derived velocities, *J Mar Pet Geol*, 26, 457–463.

Phelps, D., Akintokunbo, J. T. T., Koster, K., Schupbach, M. and De Gennaro, V., 2012, Solutions for 3D coupled geomechanical and HF modelling in UG reservoirs, EAGE Conference and Exhibition.

Pickett, G. R., 1963, Acoustic character logs and their applications in formation evaluation, *J Pet Technol*, 15, 659–667.

Pigott, J. D. and Tadepalli, S. V., 1996, Direct determination of clastic reservoir porosity and pressure from AVO inversion, SEG Annual International Meeting, extended abstracts.

Pigott, J. D., Shrestha, R. K. and Warwick, R. A., 1990, Direct determination of carbonate reservoir porosity and pressure from AVO inversion, SEG Annual International Meeting, extended abstracts.

Pilkington, P. E., Fracture gradient estimates in Tertiary basins, *Pet Eng Int*, 8, 138–148.

Pittman, E. D. and Larese, R. E., 1991, Compaction of lithic sands: Experimental results and applications, *Bull AAPG*, 75, 1279–1299.

Pixler, B. O., 1945, Some recent developments in mud analysis logging, SPE 2026.

Plessix, R. E. and Perkins, C., 2010, Full waveform inversion of a deep water ocean bottom seismometer data set, *First Break*, 28, 71–78.

Plumb, R., Edwards, S., Pidcock, G., Lee, D. and Stacey, B., 2000, The mechanical earth model concept and its application to high-risk well construction projects, IADC/SPE Drilling Conference.

Plumb, R., Edwards, S., Pidcock, G., Lee, D. and Stacey, B., 2013, The mechanical earth model concept and its application to high-risk well construction projects, IADC/SPE Drilling Conference, New Orleans, LA.

Polleto, F. B. and Miranda, F., 2004, Fundamentals of drill-bit seismic, in: *Seismic while Drilling*, vol. 35, Elsevier, New York, pp. 422–429.

Potter, P. E., Maynard, J. B. and Pryor, W. A., 2012, *Sedimentology of Shale: Study Guide and Reference Source*, Springer, New York.

Poulton, M. M., 2002, Neural networks as an intelligence amplification tool: A review of applications, *Geophysics*, 67, 979–993.

Pouya, A., Irini, D. M., Violaine, L. V. and Daniel, G., 1998, Mechanical behavior of fine grained sediments: Experimental compaction and three-dimensional constitutive model, *Mar Pet Geol*, 15, 129–143.

Powers, M. C., 1967, Fluid release mechanisms in compacting marine mud rocks and their implications in oil exploration, *Bull AAPG*, 51, 1240–1254.

Powley, D. E., 1993, Shale compaction and its relation to fluid seals: Sec. III, quarterly report, GRI contract 5092–2443.

Pradhan, A., Mukerji, T. and Dutta, N. C., 2017a, Effect of compaction and smectite-illite transition on velocity anisotropy, SEG International Exposition, Technical Program expanded abstracts.

Pradhan, A., Scheirer, A. H., Dutta, N. C., AlKawai, W. H., Le, H. Q. and Mukerj, T., 2017b, Integrated constraints for basin modeling and seismic imaging: application to the northern Gulf of Mexico basin, BPSM Affiliates' Meeting.

Pradhan, A., Dutta, N. C. and Mukerji, T., 2018, Basin modeling and rock physics constraints for seismic imaging, Stanford Rock Physics Report F5.

Pradhan, A., Le, H. Q., Dutta, N. C., Biondi, B. and Mukerji, T., 2020, A Bayesian framework for seismic velocity and earth modeling with basin modeling, rock physics and imaging constraints, *Geophysics*, 85: ID19-ID34.

Prasad, M., 1988, Experimental and theoretical considerations of velocity and attenuation interactions with physical parameters in sands, PhD thesis, Kiel University.

Prasad, M., 2002, Acoustic measurements in sands at low effective pressure: Overpressure detection in sands, *Geophysics*, 67, 405–412.

Prasad, M. and Manghnani, M. H., 1997, Effect of pore pressure and differential pressure of compressional wave velocity and quality factor in Berea and Michigan sandstones, *Geophysics*, 62, 1163–1176.

Prasad, M. and Meissner, R., 1992, Attenuation mechanisms in sands: Laboratory versus theoretical (Biot) data, *Geophysics*, 57, 710–719.

Prioul, R., Bakulin, A. and Bakulin, V., 2004, Nonlinear rock physics model for estimation of 3D subsurface stress in anisotropic formations: Theory and laboratory verification, *Geophysics*, 69, 415–425.

Pritchard, D. M. and Lacy, K. D., 2011, Deepwater well complexity – The new domain, Deepwater Horizon Study Group Working Paper, *Prospecting*, 32, 998–1015.

Quijada, M. F. and Stewart, R. R., 2007, Density estimations using density-velocity relations and seismic inversion, CREWES Research Report 19.

Rabia, H., 1985, *Oilwell Drilling Engineering: Principles and Practice*, Graham and Trotman, Gaithersburg, MD.

Raiga-Clemenceau, J., Martin, J. P. and Nicoletis, S., 1988, The concept of acoustic formation factor for more accurate porosity determination from sonic transit time data, SPWLA Annual Logging Symposium.

Ramm, M., 1992, Porosity-depth trends in reservoir sandstones: Theoretical models related to Jurassic sandstones in offshore Norway, *Mar Pet Geol*, 9, 553–567.

Rao, C. V. and Chandrashekar, S., 2013, Syn-dill seismic imaging through seismic guided drilling: A case history from East coast India deep water example, paper P431, International Conference and Exposition.

Rao, M. V., 2002, Method of drilling in response to other publications looking ahead of drill bit, US Patent 6,480,118 B1.

Raymer, L. L., Hunt, E. R. and Gardner, J. S., 1980, An improved sonic transit time-to-porosity transform, SPWLA Annual Logging Symposium.

Reagan, M. and Moridis, G., 2010, Numerical studies for the characterization of recoverable resources from methane hydrate deposits, Annual Gas Hydrates Meeting.

Rector, J. W., 1990, Utilization of drill-bit vibrations as a downhole seismic source, PhD thesis, Stanford University.

Rehm, W. A. and McClendon, R., 1971, Measurement of formation pressure from drilling data, SPE 3601.

Reuss, A., 1929, Berechnung der fliessgrense von mischkristallen auf grund der plastizitatsbedinggung fur einkristalle, *Z Angew Math Mech*, 9, 49–58.

Reynolds, E. B., 1970, Predicting overpressured zones with seismic data, *World Oil*, 171(5), 78–82.

Reynolds, E. B., 1973, The application of seismic techniques to drilling techniques, SPE preprint 4643.

Reynolds, E. B., May, J. E. and Klaveness, A., 1971, Geophysical aspects of abnormal fluid pressures, in: *Abnormal Subsurface Pressure, a Study Group Report 1969–1971*, Houston Geological Society, Houston, TX, pp. 31–47.

Reynolds, J. M., 1997, *An Introduction to Applied and Environmental Geophysics*, Wiley, New York.

Rimstad, K., Avseth, P. and Omre, H., 2012, Hierarchical Bayesian lithology/fluid prediction: A North Sea case study, *Geophysics*, 77, B69–B85.

Rochon, R. W., 1968, *The Effect of Mud Weight in Mud Logging Gas Anomalies*, Monarch Logging Co., San Antonio, TX.

Rogers, S. J., Fang, J. H., Karr, C. L. and Stanley, D. A., 1992, Determination of lithology from well-logs using a neural network, *Am Assoc Pet Geol Bull*, 76, 792–822.

Ronen, S., Rokkan, A., Bouraly, R., Valsvik, G., Larson, L., Ostenvig, E., Paillet, J., Dynia, A., Matlosg, A., Brown, S., Drummie, S., Holden, J., Koster., J., Monk, D. and Swanson, D., 2012, Imaging shallow gas drilling hazards under three Forties field platforms using ocean bottom nodes, *Leading Edge*, 31, 465–469.

Roscoe, K. H. and Burland, J. B., 1968, On the generalized stress-strain behavior of wet clays, in: Heyman, J. and Leckie, F. A. (eds.), *Engineering Plasticity*, Cambridge University Press, Cambridge, pp. 535–609.

Roth, G. and Tarantola, A., 1994, Neural networks and inversion of seismic data, *J Geophys Res*, 99, 6753–6768.

Rowan, M. G., 1993, A systematic technique for the sequential restoration of salt structures, *Tectonophysics*, 228, 331–348.

Rubey, W. M. and Hubbert, M. K., 1959, Role of fluid pressure in mechanics of overthrust faulting. II. Overthrust belt in geosynclinals area of Western Wyoming in light of fluid-pressure hypothesis, *Geol Soc Am Bull*, 70, 167–206.

Rutherford, S. R. and Williams, R. H., 1989, Amplitude-versus-offset variations in gas sands, *Geophysics*, 54, 680–688.

Sainson, S., 2017, *Electromagnetic Seabed Logging: A New Tool for Geoscientists*, Springer, New York.

Salehi, S. and Mannon, T., 2013, Application of seismic frequency based pore pressure prediction in well design: Review of an integrated well design approach in deepwater Gulf of Mexico, *J Geol Geosci*, 2, 125.

Sarker, R. and Batzle, M., 2008, Effective stress coefficient in shales and its applicability to Eaton's equation, *Leading Edge*, 27, 798–804.

Sasaki, A., 1987, A reliability of Horner-plot method for estimating static formation temperature from well log data, *J Jpn Assoc Pet Technol*, 52(3), 23–32.

Satti, I. A., Ghosh, D. and Yusoff, W. I. W., 2015, 3D predrill pore pressure prediction using basin modeling approach in a field in Malay Basin, *Asian J Sci*, 8, 24–31.

Savich, A. I., 1984, Generalized relations between static and dynamic indices of rock deformability, *Power Technol. Eng.*, 18, 394–400.

Sayers, C. M., 2006, An introduction to velocity-based pore-pressure prediction, *Leading Edge*, 25, 1496–1500.

Sayers, C. M., 2010, *Rocks under Stress*, Society of Exploration Geophysicists/European Association of Geoscientists and Engineers, Tulsa, OK.

Sayers, C., Woodward, M. J. and Bartman, R. C., 2001, Predrill pore-pressure prediction using 4-C seismic data, *Leading Edge*, 21, 1056–1059.

Sayers, C. M., Johnson, G. M. and Denyer, G., 2002, Predrill pore-pressure prediction using seismic data, *Geophysics*, 67, 1286–1292.

Sayers, C. M., den Boer, L. D., Nagy, Z. R. and Hooyman, P. J., 2006, Well-constrained seismic estimation of pore pressure with uncertainty, SEG Annual Meeting, expanded abstracts.

Schatz, J. F. and Simmons, G., 1972, Thermal conductivity of Earth materials at high temperatures, *J Geophys Res*, 77, 6966–6983.

Schlumberger, 2016, Real-time drilling geomechanics reduces NPT, www.slb.com/~/media/Files/dcs/product_sheets/geomechanics/geomechanics_rt_ps.pdf.

Schneider, F., Coussy, O. and Dormieux, L., 1998, A mechanical modelling of the primary migration, *Rev Inst Fran Pétrole EDP Sci*, 53, 151–161.

Schoenberg, M., Muir, F. and Sayers, C., 1996, Introducing ANNIE: A simple three-parameter anisotropic velocity model for shales, *J Seismic Explor*, 5, 35–49.

Schowalter, T. T., 1979, Mechanics of secondary hydrocarbon migration and entrapment, *AAPG Bull*, 63, 723–760.

Schreiber, B. C., 1968, Sound velocity in deep sea sediments, *J Geophys Res*, 73, 1259–1268.

Schuster, D. C., 1995, Deformation of allochthonous salt and evolution of related salt–structural systems, eastern Louisiana Gulf Coast, *AAPG Mem*, 65, 177–198.

Sclater, J. G. and Christie, P. A. F., 1980, Continental stretching; an explanation of the post-Mid-Cretaceous subsidence of the central North Sea basin, *J Geophys Res*, 85, 3711–3739.

Scott, D. and Thomsen, L. A., 1993, A global algorithm for pore pressure prediction, SPE 25674.

Scott, T. E. and Abousleiman, Y., 2004, Acoustical imaging and mechanical properties of soft rock and marine sediments, Final Technical Report 15302, DOE Award DE-FC26-01BC15302.

Sen, M. K. and Stoffa, P. L., 1991, Nonlinear one-dimensional seismic waveform inversion using simulated annealing, *Geophysics*, 56, 1624–1638.

Sen, M. K. and Stoffa, P. L., 1992, Rapid sampling of model space using genetic algorithms: Examples from seismic waveform inversion, *Geophys J Int*, 108, 281–292.

Sengupta, M., Dai, J., Volterrani, S., Dutta, N., Rao, N. S., Al-Qadeeri, B. and Kidambi, K. K., 2011, Building a seismic-driven 3D geomechanical model in a deep carbonate reservoir, SEG Annual Meeting.

Serra, O., 2008, *The Well Logging Handbook*, Technip, Paris.

Shamsipour, P., Marcotte, D., Chouteau, M. and Keating, P., 2010, 3D stochastic inversion of gravity data using cokriging and cosimulation, *Geophysics*, 75, I1–I10.

Shea, W. T., Schwalbach, J. R. and Allard, D. M., 1993, Integrated rock-log evaluation of fluvio-lacustrine seals, in: Ebanks, J., Kaldi, J., Vavra, C. (eds.), *Seals and Traps: A Multidisciplinary Approach*, AAPG Hedberg Research Conference, unpublished abstract.

Shelander, D., Dai, J., Bunge, G., McConnell, D. and Banik, N., 2010a, Predicting saturation of gas hydrates using pre-stack seismic data, Gulf of Mexico, *Mar Geophys Res*, 31, 39–57.

Shelander, D., Dai, J., Bunge, G., Collett, T., Boswell, R. and Jones, E., 2010b, Predictions of Gas Hydrates using pre-stack seismic data, Deepwater GOM, AAPG Annual Convention.

Sheriff, R. E. and Geldart, L. P., 1995, *Exploration Seismology*, 2nd ed., Cambridge University Press, Cambridge.

Shuey, R. T., 1985, A simplification of the Zoeppritz equations, *Geophysics*, 50, 609–614.

Siggins, A. F., Dewhurst, D. N. and Tingate, P. R., 2001, Stress path, pore pressure and micro structural influences on Q in Camarvon basin sandstones, Offshore Technology Conference.

Singha, D. K. and Chatterjee, R., 2014, Detection of overpressure zones and a statistical model for pore pressure estimation from well logs in the Krishna-Godavari Basin, India, *Geochem Geophys Geosyst*, 15, 1009–1020.

Sirgue, L., Barkved, O. I., Van Gestel, J. P., Askim, O. J. and Kommedal, J. H., 2009, 3D waveform inversion on Valhall wide-azimuth OBC, EAGE Conference and Exhibition, expanded abstracts U038.

Skempton, A. W., 1944, Notes on the compressibility of clays, *Q J Geol Soc London*, 100(Parts 1 & 2), 119–135.

Skempton, A. W., 1970, The consolidation of clays by gravitational compaction, *Q Geol Soc London*, 125 (Part 3), 373–411.

Smith, D. A., 1966, Theoretical considerations of sealing and non-sealing faults, *AAPG Bull*, 50, 363–374.

Smith, J. E., 1970, The dynamics of shale compaction and evolution of pore fluid pressures, *Math Geol*, 3, 239–263.

Smith, N. J., 1950, The case for gravity data from boreholes, *Geophysics*, 15, 605–636.

Smorodinov, M. I., Motovilove, E. A. and Volkov, V. A., 1970, Determination of correlation relationships between strength and some physical characteristics of rocks, *Proc Int Soc Rock Mech*, 1, 1–19.

Sone, H. and Zoback, M. D., 2014, Time-dependent deformation of shale gas reservoir rocks and its long-term effect on the in situ state of stress, *Int J Rock Mech Min Sci*, 69, 120–132.

Song, L., 2012, Measurement of minimum horizontal stress from logging and drilling data in unconventional oil and gas, MSc thesis, University of Calgary.

Spencer, C. W., 1987, Hydrocarbon generation as a mechanism for overpressuring in Rocky Mountain region, *Bull AAPG*, 71, 368–388.

Stewart, R. R., 1984, VSP interval velocities from traveltime inversion, *Geophys Prospect*, 32, 608–628.

Stoffa, P. L. and Sen, M. K., 1991, Nonlinear multi-parameter optimization using genetic algorithms: Inversion of plane-wave seismograms, *Geophysics*, 56, 1794–1810.

Stone, D. G., 1983, Predicting pore pressure and porosity from VSP data, SEG Annual International Meeting, Technical Program abstract.

Stork, C., 1992, Reflection tomography in the post migrated domain, *Geophysics*, 57, 680–692.

Storvoll, V., Bjørlykke, K. and Mondol, N. H., 2005, Velocity-depth trends in Mesozoic and Cenozoic sediments from the Norwegian shelf, *AAPG Bull*, 89, 359–381.

Strandenes, S., 1991, Rock physics analysis of the Brent Group Reservoir in the Oseberg Field, Stanford Rock Physics and Borehole Geophysics Project.

Streeter, V. L., 1966, *Fluid Mechanics*, McGraw-Hill. New York.

Stump, B. B., Flemings, P., Finkbeiner, T. and Zoback, M. D., 1998, Pressure differences between over-pressured sands and bounding shales of the Eugene Island 330 field (off-shore Louisiana, USA) with fluid flow induced by sediment loading, *AAPG Mem*, 22, 83–92.

Stunes, S., 2012, Methods of pore pressure detection from real-time drilling data, MSc thesis, Norwegian Institute of Science and Technology.

Suman, A., 2013, Joint inversion of production and time-lapse: Seismic application to Norne field, PhD dissertation, Stanford University.

Suman, S., 2009, Toward understanding and modeling compressibility effects on velocity gradients in turbulence, PhD thesis, Texas A & M University, College Station, TX.

Sun, Q., 2012, *Fracture Mechanics*, Elsevier, New York.

Surdam, R. S., Dunn, T. L., McGowan, D. B. and Heasler, H. P., 1989, Conceptual models for the prediction of porosity evolution with an example from the Frontier Sandstone, Bighorn Basin, Wyoming, in: *Petro Genesis and Petrophysics of Selected Sandstone Reservoirs of the Rocky Mountain Region*, Rocky Mountain Association of Geologists, Denver, CO.

Swarbrick, R. E., 2002, Challenges of porosity-based pore pressure prediction, *CSEG Recorder*, 27(7), 74–77.

Swarbrick, R. E. and Osborne, M. J., 1998, Mechanisms that generate abnormal pressures: An overview, *AAPG Mem*, 70, 13–34.

Swarbrick, R. E., Osborne, M. J. and Yardley, G. S., 2002, Comparison of overpressure magnitude resulting from the main generating mechanisms, *AAPG Mem*, 76, 1–12.

Swarbrick, R. E., Osborne, M. J. and Yardley, G. S., 2014, Identifying the presence of secondary over-pressure generating mechanisms in Jurassic shales, UK-Sector, Central North Sea, EAGE Conference and Exhibition.

Sweeney, J. J., Burnham, A. K. and Braun, R. L., 1987, A model of hydrocarbon generation from type I kerogen: Application to Uinta Basin, Utah, *Bull AAPG*, 71, 967–985.

Szydlik, T., Helgesen, H. K., Brevik, I., De Prisco, G., Anthony, C., Kvamme, O. L., Duffaut, K., Stadtler, C. and Cogan, M., 2015, Geophysical basin modeling: Methodology and application in deepwater Gulf of Mexico, *Interpretation*, 3, SZ49–SZ58.

Tamimi, N., Tsvankin, I. and Davis, T. L., 2015, Estimation of VTI parameters using slowness-polarization inversion of P and SV waves, *J Seismic Explor*, 24, 455–474.

Taner, M. T. and Koehler, F., 1969, Velocity spectra – Digital computer derivation and applications of velocity functions, *Geophysics*, 34, 859–881.

Tang, X.-M. and Cheng, A., 2004, *Quantitative Borehole Acoustic Methods, Handbook of Geophysical Exploration: Seismic Exploration*, vol. 24, Pergamon, New York.

Tarantola, A., 1984, Linearized inversion of seismic reflection data, *Geophys Prospect*, 32, 998–1015.

Tarantola, A., 1987, *Inverse Problem Theory: Methods for Data Fitting and Model Parameter Estimation*, Elsevier, New York.

Tarantola, A., 2005, *Inverse Problem Theory and Methods for Model Parameter Estimation*, SIAM, Philadelphia.

Terzaghi, K., 1923, Die Berechnung der Durchlass igkeitsziffer des Tones aws dem Verlanf der Hydrodynamischen Spannungsercheinungen, *Sb Akad Wiss Wien*, 132, 125–138.

Terzaghi, K., 1925, Principles of soil mechanics. IV. Settlement and consolidation of clay, *Eng News-Rec.*, 95, 874–878.

Terzaghi, H., 1943, *Theoretical Soil Mechanics*, John Wiley, New York.

Terzaghi, K. and Peck, R. P., 1968, *Soil Mechanics in Engineering Practice*, Wiley, New York.

Thomas, E. C. and Stieber, S. J., 1975, The distribution of shale in sandstones and its effect upon porosity, Annual SPWLA Logging Symposium.

Thomas, L. K., Katz, D. L. and Ted, M. R., 1968, Threshold pressure phenomena in porous media, *Trans SPE*, 243, 174–184.

Thomsen, L., 1986, Weak elastic anisotropy, *Geophysics*, 51, 1954–1966.

Thomsen, L., 2002, Understanding seismic anisotropy in exploration and exploitation, SEG-EAGE Distinguished Instructor Series 5.

Thurston, R. N. and Brugger, K., 1964, Third-order elastic constants and the velocity of small amplitude elastic waves in homogeneously stressed media, *Phys Rev*, 133, A1604–A1610.

Tillner, E., Shi, J.-Q., Bacci, G., Nielsen, C. M., Frykman, P., Dalhoff, F. and Kempka, T., 2014, Coupled dynamic flow and geomechanical simulations for an integrated assessment of CO_2 storage impacts in a saline aquifer, *Energy Procedia*, 63, 2879–2893.

Timko, D. J. and Fertl, W. H., 1972, How downhole temperatures, pressures affect drilling: Predicting hydrocarbon environments with wireline data, *World Oil*, 175(5).

Timur, A., 1987, Acoustic logging, in: Bradley, H. B. (ed.), *Petroleum Engineering Handbook*, Society of Petroleum Engineers, Richardson, TX, pp. 51–112.

Tissot, B. P. and Welte, D. H., 1978, *Petroleum Formation and Occurrence: A New Approach to Oil and Gas Exploration*, Springer, New York.

Toksoz, M. N., Cheng, C. H. and Timur, A., 1976, Velocities of seismic waves in porous rocks, *Geophysics*, 41, 621–645.

Toksoz, M. N., Johnston, D. H. and Timur, A., 1978, Attenuation of seismic waves in dry and saturated rocks: I. Laboratory measurements, *Geophysics*, 44, 681–690.

Toldi, J., 1985, Velocity analysis without picking, SEG Technical Program, expanded abstracts.

Tosaya, C., 1982, Acoustic properties of clay bearing rocks, PhD dissertation, Stanford University.

Traugott, M., 1997, Pore pressure and fracture pressure determinations in deepwater, Deepwater Technology Supplement to *World Oil*.

Traugott, M. O. and Heppard, P. D., 1994, Prediction of pore pressure before and after drilling – taking the risk out of drilling overpressured prospects, *AAPG Hedberg Res Conf*, 70, 215–246.

Tsvankin, I., 1995, Normal moveout from dipping reflectors in anisotropic media, *Geophysics*, 60, 268–284.

Tsvankin, I., 1996, P-wave signatures and notation for transversely isotropic media: An overview, *Geophysics*, 61, 467–483.

Tsvankin, I., 1997, Anisotropic parameters and P-wave velocity for orthorhombic media; geophysics, *Geophysics*, 62, 1292–1309.

Tsvankin, I. and Thomsen, L., 1994, Non-hyperbolic reflection moveout in anisotropic media, *Geophysics*, 59, 1290–1304.

Tsavankin, I., Gaiser, J., Grechka, V., van der Baan, M. and Thomsen, L., 2003, Seismic anisotropy in exploration and reservoir characterization: An overview, www.cwp.mines.edu/Meetings/Project10/cwp-642P.pdf.

Tsuji, T., Dvorkin, J., Mavko, G., Nakata, N., Matsuoka, T., Nakanishi, A., Kodaira, S. and Nishizawa, O., 2011, VP/VS ratio and shear-wave splitting in the Nankai Trough seismogenic zone: Insights into effective stress, pore pressure, and sediment consolidation, *Geophysics*, 76, WA71–WA82.

Tura, A. and Lumley, D. E., 1999, Estimating pressure and saturation changes from time-lapse AVO data, SEG Annual Meeting, expanded abstracts.

Ungerer, P. L. and Mudford, B. S., 1992, A two-dimensional model of overpressure development and gas accumulation in Venture field, Eastern Canada, *AAPG Bull*, 76, 318–338.

Valle, R. D., Kerdan, T., Leon, A., Renteria, J. and Diaz, M., 2017, Seismic attenuation workflow for lithology and fluid interpretation, AAPG/SEG International Conference & Exhibition Search and Discovery, 42035.

Van Heerden, W. L., 1978, General relations between static and dynamic moduli of rocks, *Int J Rock Mech Min Sci Geomech Abstr*, 24(6), 381–385.

Vasseur, G., Djeran, M. I., Grunberger, D., Rousset, G., Tessier, D. and Velde, B., 1995, Evolution of structural and physical parameters of clays during experimental compaction, *Mar Pet Geol*, 12, 941–954.

Vavra, C. L., Kaldi, J. G. and Sneider, R. M., 1992, Geological applications of capillary pressure: A review, *AAPG Bull*, 76, 840–850.

Velde, B., 1996, Compaction trends of clay-rich deep sea sediments, *Mar Geol*, 133(3–4), 193–201.

Vieira, F., Liconga, N., Santos, C., Bonfim, O., Navarro, N. and Jones, R. L., 2014, The Dynamic kill case study, SPE-170292-MS.

Vigh, D., Starr, E. W. and Kapoor, J., 2009, Developing earth model with full waveform inversion, *Leading Edge*, 28, 432–435.

Vigh, D., Moldoveanu, N., Jiao, K., Huang, W. and Kapoor, J., 2013, Ultra long-offset data acquisition can complement full-waveform inversion and lead to improved sub-salt imaging, *Leading Edge*, 32, 1116–1122.

Vigh, D., Lewis, W., Parekh, C., Jiao, K. and Kapoor, J., 2015, Introducing well constraints in full waveform inversion and its applications in time-lapse seismic measurements, SEG Annual Meeting.

Virieux, J. and Operto S., 2009, An overview of full-waveform inversion in exploration geophysics, *Geophysics*, 74, 127–152.

Vutukuri, V. S., 1978, *Handbook on Mechanical Properties of Rocks*, vol. II, Trans Tech Publications, Clausthal, Germany.

Voigt, W., 1910, *Lehrbuch der Kristallphysik*, Leipzig University, Berlin.

Walker, C. W., 1976, Origin of Gulf Coast salt-dome cap rock, *AAPG Bull*, 60, 2162–2166.

Walton, K., 1987, The effective elastic moduli of a random packing of spheres, *J Mech Phys Solids*, 35, 213–226.

Wang, B., Pann, K. and Meek, R. A., 1995, Macro velocity model estimation through model-based globally-optimized residual-curvature analysis, SEG Technical Program expanded abstracts.

Wang, H. F., 2000, *Theory of Linear Poroelasticity with Applications to Geomechanics and Hydrogeology*, Princeton University Press, Princeton, NJ.

Wang, Z., 1997, Seismic properties of carbonate rocks, in: Palaz, I. and Marfurt, K. J. (eds.), *Carbonate Seismology*, Society of Exploration Geophysicists, Tulsa, OK.

Wang, Z. and Nur, A. (eds.), 2000, *Seismic and Acoustic Velocities in Reservoir Rock*, vol. 3, Geophysics Reprint Series 19, Society of Exploration Geophysicists, Tulsa, OK.

Wang, Z., Wang, H. and Cates, M. E., 2001, Effective elastic properties of solid clays, *Geophysics*, 66, 428–440.

Wangen, M., 1992, Pressure and temperature evolution in sedimentary basins, *Geophys J Int*, 110, 601–603.

Wangen, M., 1993, A finite element formulation in Lagrangian coordinates for heat and fluid flow in compacting sedimentary basins, *Int J Numer Anal Methods Geomech*, 15, 705–733.

Wangen, M., 2010, *Physical Principles of Sedimentary Basin Analysis*, Cambridge University Press, Cambridge.

Ward, C. D., Coghill, K. and Broussard, M. D., 1995, Brief: Pore- and fracture-pressure determinations: Effective stress approach, SPE 30141.

Waters, K. H., 1978, *Reflection Seismology*, Wiley, New York.

Watts, N. L., 1987, Theoretical aspects of cap-rock and fault seals for single and two-phase hydrocarbon columns, *Mar Pet Geol*, 4, 274–307.

Weakley, R. R., 1989, Recalibration techniques for accurate determinations of formation pore pressures from shale resistivity, SPE 19563.

Weaver, C. E., 1979, Geothermal alteration of clay minerals and shales: Diagenesis, Technical Report ONWI-21, Georgia Institute of Technology.

Weingarten, J. S. and Perkins, T. K., 1995, Prediction of sand production in gas wells: Methods and Gulf of Mexico case studies, *J Pet Technol*, 47, 596–600.

Weiren, L., Yamamoto, K., Ito, H., Masago, H. and Kawamura, Y., 2008, Estimation of minimum principal stress from an extended leak-off test onboard the Chikyu drilling vessel and suggestions for future test procedures, *Sci Drill*, 6, 43–47.

Weller, J. M., 1959, Compaction of sediments, *AAPG Bull*, 43, 273–310.

Wells, J. D. and Amafuele, J. O., 1985, Capillary pressure and permeability relationships in tight gas sands, SPE 13879.

Welte, D. H., Horsfield, B. and Baker, D. R. (eds.), 1997, *Petroleum and Basin Evolution*, Springer, New York.

Wessling, S., Pei, J., Dahl, T., Wendt, B., Marti, S. and Stevens, J., 2009, Calibrating fracture gradients – an example demonstrating possibilities and limitations, International Petroleum Technology Conference.

Wessling, S., Bartetzko, A. and Tesch, P., 2013, Quantification of uncertainty in a multistage /multi-parameter modeling workflow: Pore pressure from geophysical well logs, *Geophysics*, 78, WB101–WB112.

White, A. J., Traugott, M. O. and Swarbrick, R. E., 2002, The use of leak-off tests as means of predicting minimum in situ stress, *Pet Geosci*, 8, 189–193.

White, B. G., Larson, M. and Iverson, S. R., 2004, Origin of mining-induced fractures through macroscale distortion, Gulf Rocks 2004, North American Rock Mechanics Symposium.

White, J. E., 1975, Computed seismic speeds and attenuation in rocks with partial gas saturation, *Geophysics*, 40, 224–232.

Wilhelm, R., 1998, Seismic pressure-prediction method solves problem common in deepwater Gulf of Mexico, *Oil Gas J*, 41, 15–20.

Winker, C. D. and Booth, J. R., 2000, *Sedimentary Dynamics of the Salt-Dominated Continental Slope, Gulf of Mexico: Integration of Observations from the Seafloor, Near-Surface, and Deep Subsurface, Deep-Water Reservoirs of the World* by Paul Weimer, *SEPM Soc Sediment Geol*, 20, 1059–1086.

Winker, C. D. and Stancliffe, R. J., 2007, Geology of shallow-water flow at Ursa: 1. Setting and causes, Offshore Technology Conference.

Wojtanowicz, A. K., Bourgoyne, A. T., Zhou, D. and Bender, K., 2000, Strength and fracture gradients for shallow marine sediments, final report, US MMS, Herndon, VA.

Wood, A. B., 1941, *A Text Book of Sound*, Macmillan, Cambridge.

Woodward, M., Nichols, D., Zadraveva, O., Whitfield, P. and Johns, T., 2008, A decade of tomography, *Geophysics*, 73, VE5–VE11.

Wu, Y. and McMechan, G. A., 2018, Feature-capturing full-waveform inversion using a convolutional neural network, SEG Technical Program expanded abstracts.

Wyllie, M. R. J., 1983, *Fundamentals of Well Log Interpretation*, Academic Press, New York.

Wyllie, M. R. J. and Gregory, A. R., 1953, Formation factors of unconsolidated porous media: Influence of particle shape and effect of cementation, *Trans Am Inst Mech Eng*, 198, 103–110.

Wyllie, M. R. J., Gregory, A. R. and Gardner, L. W., 1956, Elastic wave velocities in heterogeneous and porous media, *Geophysics*, 21, 41–70.

Wyllie, M. R. J., Gregory, A. R. and Gardner, G. H. F., 1958, An experimental investigation of factors affecting elastic wave velocities in porous media, *Geophysics*, 23, 459–493.

Wyllie, M. R. J., Gardner, G. H. F. and Gregory, A. R., 1963, Studies of elastic wave attenuation in porous media, *Geophysics*, 27, 569–589.

Wyllie, P. J. (ed.), 1967, *Ultramafic and Related Rocks*, Wiley, New York.

Yamamoto, K., 2003, Implementation of the extended leak-off test in deep wells in Japan, in: Sugawara, K., et al. (ed.), *Proceedings of the Third International Symposium on Rock Stress, Kumamoto '03, Rotterdam*, Balkema, Netherlands, pp. 225–229.

Yang, X.-S., 2006, *Theoretical Basin Modeling*, Diggory Press, London.

Yang, Y. and Aplin, A. C., 2004, Definition and practical application of mudstone porosity – effective stress relationships, *Pet Geosci*, 10, 153–162.

Yang, Y. and Mavko, G., 2018, Mathematical modeling of microcrack growth in source rock during kerogen thermal maturation, *AAPG Bull*, 102, 2519–2535.

Yilmaz, O., 1987, Seismic data processing, in: *Investigations in Geophysics*, no. 2, Society of Exploration Geophysicists, Tulsa, OK.

Yilmaz, O., 2001, Seismic data analysis: Processing, inversion, and interpretation of seismic data, in: *Investigations in Geophysics*, no. 10, Society of Exploration Geophysicists, Tulsa, OK.

Yin, H., 1992, Acoustic velocity and attenuation in rocks: Isotropy, intrinsic anisotropy, and stress-induced anisotropy, PhD dissertation, Stanford University.

Yin, H., Han, D. H. and Nur, A., 1988, Study of velocity and compaction on sand-clay mixtures, Stanford Rock and Borehole Project, vol. 33.

York, P. L., Prichard, D. M., Dodson, J. K., Dodson, T., Rosenberg, S. M., Gala, D. and Utama, B., 2009, Eliminating non-productive time associated with drilling through trouble zones, Offshore Technology Conference.

Young, R. A., Pankratov, A. B. and Greve, J. F., 2004, Method of seismic signal processing, US Patent 6681185.

Yu, F., Jin, Y., Chen, K. P. and Chen, M., 2014, Pore-pressure prediction in carbonate rock using wavelet transformation; *Geophysics*, 79, D243–D252.

Yu, H., Chen, G. and Gu, H., 2020, A new multivariate pore-pressure prediction method based on machine learning, *Comput Geosci*, in press.

Zamora, M., 1989, New method predicts fracture gradient, *Petroleum Engineer International*, September, pp. 38–47.

Zamora, M., 1972, Slide rule correlation aids "d" exponent use, *Oil Gas J, December 18.*

Zhang, G., Wang, Z. and Chen, Y., 2018, Deep learning for seismic lithology prediction, *Geophys J Int*, 215, 1368–1387.

Zhang, J., Standifird, W. B. and Lenamond, C., 2008, Casing ultradeep, ultralong salt sections in deep water: a case study for failure diagnosis and risk mitigation in record-depth well, SPE Annual Technical Conference and Exhibition.

Zhang, J., 2011, Pore pressure prediction from well logs: Methods, modifications, and new approaches, *Earth Sci Rev*, 108, 50–63.

Zhang, J. and Wieseneck, J., 2011, Challenges and surprises of abnormal pore pressure in shale gas formations, SPE 145964.

Zhang, J. and Yin, S., 2017, Fracture gradient prediction: An overview and an improved method, *Pet Sci*, 14, 720–730.

Zhang, Y. and Zhang, J., 2017, Lithology-dependent minimum horizontal stress and in situ stress estimate, *Tectonophysics*, 703–704, 1–8.

Zhdanov, M. S., 2017, *Foundations of Geophysical Electromagnetic Theory and Methods*, 2nd ed., Elsevier, New York.

Zhou, Z.-Z., Howard, M. and Mifflin, C., 2011, Use of RTM full 3D subsurface angle gathers for subsalt velocity update and image optimization: Case study at Shenzi field, *Geophysics*, 76, WB27–WB39.

Zimmer, M., Prasad, M. and Mavko, G., 2002, Pressure and porosity influences on VP-VS ratios in unconsolidated sands, *Leading Edge*, 21, 178, 183.

Zimmerman, R. W., 1991, *Compressibility of Sandstones*, Development in Petroleum Sciences 29, Elsevier Science, New York.

Zoback, M. D., 2007, *Reservoir Geomechanics*, Cambridge University Press, Cambridge.

Zong, Z., Yin, X. and Wu, G., 2012, Elastic impedance variation with angle inversion for elastic parameters, *J Geophys Eng*, 9, 247–260.

Index